African Landscapes

Interdisciplinary Approaches

STUDIES IN HUMAN ECOLOGY AND ADAPTATION

Series Editors: Daniel G. Bates,
Hunter College – City University of New York, New York, New York

Ludomir Lozny,
Hunter College – City University of New York, New York, New York

AFRICAN LANDSCAPES
Edited by Michael Bollig and Olaf Bubenzer

AS PASTORALISTS SETTLE:
Social, Health, and Economic Consequences of the Pastoral Sedentarization in Marsabit District, Kenya
Elliot Fratkin and Eric Abella Roth

RISK MANAGEMENT IN A HAZARDOUS ENVIRONMENT:
A Comparative Study of Two Pastoral Societies
Michael Bollig

SEEKING A RICHER HARVEST:
The Archaeology of Substenence Intensification, Innovation and Change
Edited by Tina L. Thurston and Christopher T. Fisher

STAYING MASAAI?:
Livelihoods, Conversation, and Development in the East African Rangelands

For more information about this series, including the most recent titles, please visit the series homepage at www.springer.com/series/6877

A Continuation Order Plan is available for this series. A continuation order will bring delivery of each new volume immediately upon publication. Volumes are billed only upon actual shipment. For further information please contact the publisher.

African Landscapes

Interdisciplinary Approaches

Edited by

Michael Bollig

Institute of Cultural and Social Anthropology
University of Cologne
Cologne, Germany

and

Olaf Bubenzer

Department of Geography
University of Heidelberg
Heidelberg, Germany

 Springer

Michael Bollig
University of Cologne
Germany
michael.bollig@uni-koeln.de

Olaf Bubenzer
University of Heidelberg
Germany
olaf.bubenzer@geog-uni-heidelberg.de

ISBN: 978-0-387-78681-0   e-ISBN: 978-0-387-78682-7
DOI: 10.1007/978-0-387-78682-7

Library of Congress Control Number: 2008939989

Cover illustration: (*Top*) Figure 11.3 in book. Panorama of two Kaoko hunters posed against landscape. Courtesy National Archives of Namibia, A450 Hahn Collection, (*Bottom*) Figure 7.9 in book. Decoration over the gate of the inner façade of the temple of Esna, showing the nocturnal sun personified by the god Khnum (Photograph by Dagma Budde, Mainz).

Printed on acid-free paper

springer.com

Preface

Landscape has been a crucial concept to produce, store, and to present knowledge on human–environment interactions in various academic disciplines and in works of art. It has been bemoaned that the concept is ambiguous and inadequate for scientific discourses because of its vagueness, its equivocalness, and its proneness to ideological cooptation. The sheer number of attempts at a more precise definition bespeaks the uneasiness scientists feel when dealing with the term. Despite its analytical shortcomings the use of landscape as a key concept to analyse and interpret human–environment interaction is rather increasing than decreasing.

Not only have the cultural studies discovered the term with its unrivalled appeal to stress boundedness, integration, and heterogeneity at the same time, but also anthropology, cultural studies, and history have undergone a spatial turn during the last two decades integrating the landscape concept into their disciplinary lexicon. By refocussing on the landscape concept historians and anthropologists emphasise that environments and historical and cultural processes are linked by a great number of interrelated feedback loops. Landscapes are not merely scenery and stage but are intimately interwoven with history and culture. At the same time the concept is reevaluated in the geosciences where it had been discarded since the 1970s in favour of more problem-centred and less ambiguous concepts. Landscape also has a continued appeal to artists expressing their thoughts and feelings about man's placement in and interaction with nature (Schama, 1995).

The widespread use of the landscape concept corresponds with an era in which global environmental change has indeed changed most natural landscapes into heavily used environments. Various land-use activities have transformed large parts of the globe's surface and human activities have appropriated one third to one half of global ecosystem production (Foley et al., 2005, p. 570). Croplands and pastures constitute major parts of the planet's surface. The clearing of tropical forests may lead to drier and warmer regional climates in the near future, whereas the clearing of boreal forests may result in cooler climates in the North (Nemani et al., 2003).

Escobar's claim (1999, p. 1) that we have entered an epoch which is defined by the sense of being "after nature" is as true as his tenet that geoecological processes are increasingly reshaped by human activities and constituted by discursive practices. Escobar's claim resounds with the wording of Noble Prize laureate Paul Crutzen, who has named the recent geological phase "anthropocene" (Crutzen & Steffen, 2003). Crutzen and Escobar emphasise the increasing human dependence on these very processes and resources and the growing understanding that major environmental processes are beyond the control of humans: even if we succeed in reducing CO_2 output the effects of global warming will transform landscapes profoundly over the next decades to come (IPCC, 2007). Glaciers will vanish and coastal areas will become inundated; some deserts may expand and others shrink. Large dams, water carriers, and the expansion of megacities transform landscapes as much as the artificial exclusion of humans from specific sites and entire biomes designated as parks and wilderness areas. However, our potential to control and correct geoecological processes is very limited. Escobar's emphasis that environments resist being fully coopted by humans, is borne out by numerous contemporary reports on major catastrophes and the increasing vulnerability of ecosystems.

As human–environment feedback loops define most of the basic stressors that constitute "human life" in relation to both biology and geoecological processes, there is a need for an interdisciplinary approach to landscape research. It is odd that most of the current literature is usually linked more or less clearly to one scientific field, either being affiliated with the natural sciences or the humanities. True interdisciplinary approaches to landscapes bridging this gap are exceedingly rare. It is here that this volume wants to make an impact: over the past ten years the contributors to this volume have cooperated in an interdisciplinary programme – the Collaborative Research Centre ACACIA (Arid Climate, Adaptation and Cultural Innovation in Africa) – dealing with the interrelation between cultural processes and geoecological dynamics in Africa's arid areas. The concept 'landscape' has been crucial in all projects, be they Egyptological, Africanist, anthropological, geographical, botanical, historical, or archaeological.

The attempt to work along a unified definition of the landscape concept was given up early on. Rather it was deemed to be more rewarding to have each discipline explore its own access to the topic and from there explore bridges between different disciplinary approaches. The belief in a diversity of landscape approaches made it necessary to explicate the epistemological fundamentals of one's own conceptual base. However, there has been a basic understanding that 'for constructivists, the challenge lies in learning to incorporate into their analyses the biophysical basis of reality; for realists it is examining their frameworks from the perspective of their historical constitution' (Escobar, 1999, p. 3).

Michael Bollig and Olaf Bubenzer

Literature

Crutzen, P.J. & Steffen, W. (2003) How long have we been in the Anthropocence era? *Climatic Change*, 61, 251–257.

Escobar, A. (1999). After nature. Steps to an antiessentialist political ecology. *Current Anthropology*, 40, 1–30.

Foley, J . et al. (2003). Global consequences of land use. *Science*, 309, 570–574.

Nemani R. et al. (2003) Climate-driven increases in global terrestrial net primary production from 1982 to 1999. *Science*, 300, 1560–153.

Contents

Contributors

Laura E. Bleckmann, University of Leuven, Leuven, Belgium

Michael Bollig, Institute of Cultural and Social Anthropology, University of Cologne, Cologne, Germany

Andreas Bolton, Department of Geography, University of Cologne, Cologne, Germany

Inge Brinkman, African Studies Center, Leiden University, Leiden, Netherlands

Olaf Bubenzer, Department of Geography, University of Heidelberg, Heidelberg, Germany

Frank Darius, Department of Geography, University of Cologne, Cologne, Germany

Phillipe Derchain, Seminar of Egyptology, University of Cologne, Cologne, Germany

Ute Dieckmann, Institute of Cultural and Social Anthropology, University of Cologne, Cologne, Germany

Jan-Bart Gewald, African Studies Center, Leiden University, Leiden, Netherlands

Nina Gruntkowski, Department of Geography, University of Cologne, Cologne, Germany

Patricia Hayes, History Department, University of the Western Cape, Bellville, South Africa

Michael Herb, Department of Geography, University of Bonn, Bonn, Germany

Birte Kathage, Africa Consulting, Cologne, Germany

Jekura U. Kavari, University of Namibia, Windhoek, Namibia

Karin Kindermann, Department of Geography, University of Cologne, Cologne, Germany

Tilman Lenssen-Erz, African Research Center, University of Cologne, Germany

Anja Lindstädler, Botanical Institute, University of Cologne, Cologne, Germany

Jörg Linstädter, African Research Center, University of Cologne, Germany

Gunter Menz, Department of Geography, University of Bonn, Bonn, Germany

Wilhelm J.G. Mölig, African Studies Center, Leiden University, Leiden, Netherlands

Jan Richters, Department of Ecology, Technische Universität Berlin, Berlin, Germany

Heiko Riemer, African Research Center, University of Cologne, Germany

Martin Rössler, Institute of Cultural and Social Anthropology, University of Cologne, Cologne, Germany

Thomas Widlock, Max Planck Institute for Psycholinguistics, Nijmegen, Netherlands

List of Tables

List of Figures

Introduction

Visions of Landscapes: An Introduction

Michael Bollig

In the following chapter I trace the history of the landscape concept. I especially focus on the tradition of the landscape concept in German thought, mainly because I think that the changing interpretations of the concept and the multiple layers of meaning are rewarding to an archaeology of landscape concepts. The landscape concept in German thought resounds with ambiguities and puts the risks of ideological cooptation into focus. In contrast to this the meaning of landscape seems to have been more stable in the English and French tradition. On the other hand the German landscape concept has profoundly influenced the thinking about landscapes in diverse sciences. Franz Boas for example took on the idea of the landscape from Herder and the early nineteenth century geographer Carl Ritter (Boas, 1887, pp. 137ff.), who both opted for a holistic approach to human–environment relations and the historical dimension of these relations.

The idea that a peculiar landscape, a people, and its history were intimately linked and unique became a key idea in early twentieth century U.S. anthropology. The founding father of academic German geography, Friedrich Ratzel, had a profound effect on Frederick Jackson Turner's ideas on the relation amongst geology, biology, and history (N. Finzsch, personal communication, 2007). Ratzel (1882/1891) had argued in his two volumes on anthropogeography that human societies can only be successful as long as they adapt to geophysical givens of specific landscapes. Turner applied these ideas to his historical model of the American frontier (Turner, 1920) which was largely influential in American historiography. Phenomenological approaches to landscapes such as advocated by one of the founding fathers of U.S. American geography, Carl O. Sauer (see Sauer,

1

M. Bollig, O. Bubenzer (eds.), *African Landscapes*,
doi: 10.1007/978-0-387-78682-7, © Springer Science+Business Media, LLC 2009

1925), are linked to similar approaches in German geography (see, e.g., Hettner, 1923, but also earlier writers such as Carl Ritter).

Carl Troll, a highly influential geographer in Germany between the 1930s and 1960s coined the term landscape ecology (Lauer, 1976) in an attempt to refocus the concept, which had been tarnished with fascist and racist ideas during the early decades of the twentieth century. Influenced by aerial photographs that, for the first time, were available with a wide coverage, he underlined a natural sciences perspective to landscapes and saw the analysis of flows and feedback loops between different elements of the system as the key to the analysis of landscapes (Troll, 1970).

1. ETHYMOLOGY

Today we are confronted with two rather dissimilar definitions of the term landscape: whereas the *Oxford English Dictionary* explains landscape as 'view', 'prospect', and 'representation' of 'natural inland scenery', a standard German dictionary (*Duden Wörterbuch*) defines the German word *Landschaft* as 'part, section of land surface . . . shaped in a particular way with regard to its external appearance.' (quoted after Luig & von Oppen, 1997, p. 9). Much of the recent debate on this term gives evidence of a cleft between naturalistic and mentalistic approaches and I spend some time explaining from where the different meanings come. During the past 1000 years the concept was enriched with various layers of meaning linking political, ecological, and cognitive aspects of human–environment relations and through the course of history the landscape concept has absorbed meanings in different European historical contexts.

The term *Landschaft* (landscape) is derived from the Germanic verbal abstract *skapi* which can be translated with composition or nature, *skapi* itself being derived from *skapjan*, to create (Müller, 1977, p. 4). Words to which the verbal abstract was attached were either abstract nouns, collective nouns, or expressions of spatiality such as the old English *burhscipe* (urban area) or *nidscipe* (badness) or the old Saxon term *heriskepi*, as part of a territory. In old German the term *landscaf* alluded to the quality of a larger settlement area (Müller, 1977, p. 6). It then referred to the social institutions of an area and in an extended meaning to the land in which such norms were adhered. There is little evidence that the old German *landscaf* referred to natural features of a settlement area. In several medieval texts *landscaf* is used as a translation for the Latin term *provincia*. Notker (ad 950–1022), a monk working from St. Gallen, defines *landscaf* as '*prouincia is diu lántscaf. Regio díû gibíûrda. Mánige regiones mugen sîn in êinero prouincia.*'(transl.: '*Provincia* is the landscape. *Regio* is the inhabited land/landscape. Several regions can constitute a province', quoted after Müller, 1977, p. 6). In contrast to this the Latin term *terra* referring to the earth's surface, is rarely translated with *lantscaf*. Already in medieval German the older meaning of norms and standards of a people shaping (creating) a particular *lant*, was gradually replaced by a new use of the term. Then *lantschaft* referred to the knights and/or the political elite of a given area. In Gottfried of Straßburg's *Tristan* (1980, p. 394) the narrator, for example, says '*Do kom al die lantschaft/ und volkes ein so michel kraft*' (transl. 'There came all the knights of the region and so many people') thereby clearly juxtaposing the *lantschaft* (the knights of the region) from the *volk* (the people). From here it was just a small step to the late medieval use of the term as 'council of estates' or 'council of political representatives of an area'. In the German language this meaning of *landschaft* was retained well into the nineteenth century and a dictionary from the 1830s (*Allgemeine deutsche Realencyklopädie*, 1830, p. 427) still gives this meaning.

During the Renaissance the content of the term changes once again. During the early Renaissance in Northern Italy urban elites attached strong aesthetic connotations to adjoining rural areas and a special way of painting landscapes emerged (Luig & von Oppen, 1997, p. 10; Daniels & Cosgrove, 1988, p. 5), which subjects were rural scenes, nature as viewed by urban elites. Whereas in the Middle Ages landscapes had remained in the background of paintings and had been reduced to allegoric representations, in the early sixteenth century landscapes became the focus of paintings. It is first in the southwest of Germany that landscape (*landschafft*) takes this alternative aesthetic meaning. In 1525 Albrecht Altdorfer painted *Donaulandschaft mit Schloss bei Wörth* (Danube Landscape with Castle near Wörth) which is reckoned to be the first painting which has a landscape as its sole focus (see Figure 1). It is perceivable that during this period the concept landscape gained a new meaning, implicating the view of a detached but interested observer in a rural landscape.

New technical aids such as the *camera obscura* and the *velum* were used to create a three-dimensional space and to depict landscapes in perspective. Furthermore the development of oil painting in the Netherlands allowed for a more intense portrayal of the play of light and shadow in a landscape. From the middle of the sixteenth century Dutch landscape painting flourished connected to names such as Pieter Brueghel and Hieronymus Bosch in an older school of landscape painters and van Goyen and van Ruisdael in a later school of landscape artists. The Dutch use of landscape (*landschap*) in connection with artistic but naturalistic depictions of landscapes is then transported into English, where we find this use of the term during the latter part of the sixteenth century (Hirsch, 1995, p. 2; Forman & Godron, 1986, pp. 5ff.). Whereas in German alternative meanings of landscape coexisted, in English landscape took on the meaning of a visual representation of rural scenery.

Whereas the aesthetic use of the term dominated in English, in the German language the term also became applied to segments of the earth and closely tied to what was conceptualised as nature. Luig and von Oppen (1997, p. 12) point out that at the same time the concept 'nature', originally describing the quality of something, became an abstract singular denoting prediscursive aspects of life and was conceptualised as an antidote to culture (see also Casimir, 2007). This wedge between nature and culture then became an essential part of the landscape

Figure 0.1. Albrecht Altdorfer, Danube Landscape with Castle near Wörth (Courtesy Bildarchiv Preußischer Kulturbesitz) (*See also Color Plates*)

concept over the next centuries. Only when nature was differentiated from culture on epistemological grounds could the landscape concept take on connotations of natural structures and processes. Josuua Maaler's Latin dictionary edited in 1561 in Zürich takes *landschafft* as land (in contrast to water) and translates the term with the Latin *tractus* which is a smaller well-defined spatial unit (Müller, 1977, p. 9) (and not with *provinicia* as with Notker some 500 years earlier). Parallel to landscape painting, cartography became another, scientific way to approach and represent the structures and details of landscapes in the sixteenth and seventeenth centuries (on the development of cartography as an early branch of geography see Bagrow & Skelton, 1964; Harley & Woodward, 1987; Burnett, 1999; see also Figure 2).

Bollig and Heinemann (2002) trace, for the southwestern part of the African continent, how maps containing fictitious and narrated elements characteristic for the seventeenth and eighteenth centuries were replaced by more naturalistic maps during the second part of the nineteenth century.

An aesthetic interpretation of the landscape concept prevailed and during the Romantic period, landscape representations were not only popular in writing, painting, and music (Kuzniar, 1988) but also became imbued with a specific morality. Landscape was good to think about and – in contemporary terms – allusions to one's rootedness in the landscape were politically correct. Especially in Germany landscape poetry became emblematic for the political movement propagating a German national identity. Whereas earlier Renaissance paintings had mainly portrayed encultured landscapes with buildings and other signs of human settlement, now wilderness as much as settled landscapes became topics for popular presentations.

Figure 0.2. Pieter Brueghel; The Fall of the Icarus (Courtesy Bildarchiv Preußischer Kulturbesitz) (*See also Color Plates*)

Landscape painters of the Romantic period sought the sublime in landscapes; rarely were their paintings meant to be objective representations of landscapes (Kuzniar, 1988). They were meant to symbolise transcendent relations between humans, God, and nature and were media for introspection and religious experience and on the other hand, Romantic landscape paintings contained implicit political messages: German nationalism, for example, finds its expression in the choice of special landscapes or landscape elements (e.g., oak, fir tree). Tieck's description of the Alps or Caspar David Friedrich's paintings of rural scenes (Körner, 1995, see also Rössler, this volume) described awe-inspiring nature and pointed out correlations between human emotions, national identity, and particular features of a landscape. Landscapes were imbued with a political meaning as ideas on the agreement between the character of a nation and the features of the landscape it inhabited became prominent (see, e.g., Heinrich von Kleist, Clemens Brentano in Apel, 2000).

The link between landscape and identity became a salient topic of the arts but also of discourses in incipient academic disciplines such as history and geography. Von Humboldt used the concept in the 1840s as an umbrella term to describe the 'totality of a part of the earth' (Humboldt, 1845) and pleaded for a naturalistic approach to landscapes. His landscape depictions contained many scribblings specifying elevation, topography, and vegetation (see Figure 3).

In the 1880s Grimm and Grindel (1885, p. 131) defined *Landschaft* as (1) region, land complex in relation to its position and natural characteristics and (2) the artistic representation of such a region (Grimm & Grindel, 1885, p. 131). Luig and von Oppen (1997, p. 11) argue that at the turn of the twentieth century land-scape identities in Europe became popularised and politicised, when at the same time the landscape concept was elevated to a key concept of scientific geoecologi-cal enquiry. Whereas in the German language the natural sciences approach to the landscape concept came to coexist with a politicised (landscape as identity) and with an artistic and at times metaphysical meaning (landscapes as reflection of transcended or psychological realities), in the English language the artistic mean-ings of the term predominated clearly (as it did in French).

The German geographer Hard (1969) made an effort to account for the rise and the demise of the landscape concept in various disciplines in Western Europe since the late nineteenth century. He sampled titles of books and articles which had incorporated the landscape concept and then counted the different types of meanings that occurred. An initial finding showed that since the 1880s the use of the landscape concept had increased rapidly. Apparently the concept was enthusiastically embraced by intellectuals with a background in the humanities. Only since the 1920s did geography as a discipline take on the cover term as a leading concept. Taking off from the perspective Alexander von Humboldt had cherished, Passarge (1921) and Hettner (1923) developed *Landschaftskunde* – although with different approaches – to a dominating branch of geography. They looked for clearly discernible types of landscape and tried to understand their peculiar characteristics (see also definition of *Landschaft* in Meyers Lexikon, 1939, p. 210). Passarge (1921, p. 217) emphasises the natural components of the land-scape and maintains that a landscape is established by features of the 'solid earth

Figure 0.3. Humboldt, 'Geographie der Pflanzen in den Tropenländern' (geography of plants in the tropic countries; Courtesy Bildarchiv Preußischer Kulturbesitz) (*See also Color Plates*)

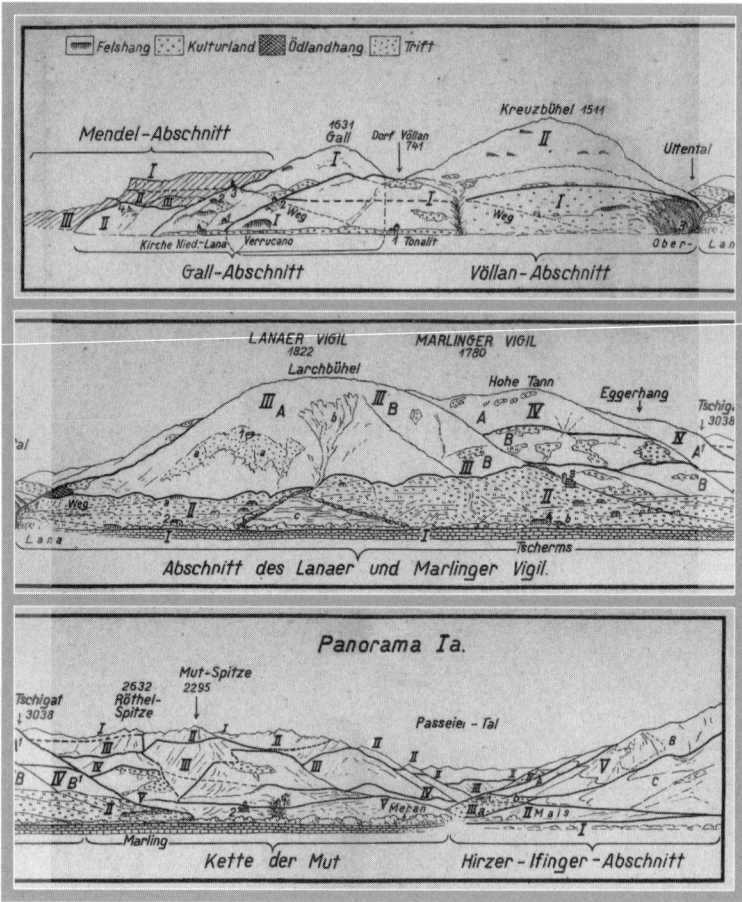

Figure 0.4. Panorama illustrating clearly discernible types and peculiar characteristics
of landscape (Passarge, 1921)

surface' (*feste Erdrinde*; topography, rock stratum, soils, and climate). A landscape
in Passarge's methodology is set up by various discernible parts (see Figure 4).

Hettner (1923) criticises Passarge for only focussing on inorganic nature and
vegetation when discussing landscapes and proposes that human interaction with
geoecological processes must be part of the landscape analysis. The differentiation
between cultural and natural landscapes became a major classificatory achievement of this period (see especially Sauer, 1925). Hard shows that in Germany
between 1910 and 1940 landscape became a dominant concept in the humanities,
in geography, and in artistic presentations. Progressively the landscape concept
was (mis)used to present the vitality and forces connecting a physical landscape

and a population and its identity. The holistic and integrative perspective of the concept and the harmony between humans and environment it allegedly conveyed became attractive in a rapidly modernising and politically torn setting.

The emphasis on the landscape concept may be interpreted as an antidote to modernity as landscape suggested the rootedness of a population and the *longue durée* of its national identity (Luig & von Oppen, 1997, p. 12). Populations and physical surroundings were presented as if they were inextricably linked and constituted each other. The fate of a people and a landscape were tightly interwoven entering upon a blood and soil mystique. Exemplary titles from this period are 'Reichtum deutscher Landschaft' (The Wealth of the German Landscape), 'Landschaft und nordische Seele' (Landscape and the Nordic Soul), and 'Die Seele deutscher Landschaft' (The Soul of the German Landscape). The *Volk* was rooted in the soil and in the landscape. To protect the landscape, which in an almost metaphysical sense became the base of the nation, a National Nature Conservation Law (*Reichsnaturschutzgesetzgebung*) not only protected but also developed landscapes according to fascist visions. Especially in landscape gardening Nordic landscapes were to be presented and foreign non-Germanic traces of landscapes and vegetation were to be excluded.

The NSDAP regime, however, did not only advertise the nationalisation of gardens, it was also adamant in reshaping entire landscapes. Especially after the *Reichsnaturschutzgesetz* had been stipulated in 1935 the NSDAP regime sought to build a new academic discipline *Landschaftsgestaltung* (landscape design) from garden architecture and horticultural technology (Troll, 1947, p. 8). Wiepking-Jürgensmann, since 1934 holder of the chair at the Institute for Garden Design at the Agricultural University of Berlin, saw the dawn of the 'golden age of the German landscape'. According to him the German race had always been close to nature and he stated that: 'Over and over again the love for plants and the landscape bursts forth from our blood, and the harder we try and the more seriously we search for reason, the more we must come to realise that the feeling we have towards a harmonic landscape and that the feeling of being related to the plants belongs to biological laws innate in our being' (both quotes after Gröning & Wolschke-Bulmahn, 1987, p. 136). He dreamt of the heroic tasks ahead of landscape designers.

Landscape planning according to fixed nationalist ideals became a major task. Mäding (1943, quoted by and translated by Gröning & Wolschke-Bulmahn, 1987, p. 137) postulated: 'Germans will be the first occidental people to form their own spiritual environment in the landscape and, thereby, for the first time in the history of mankind they will develop a life style in which a people consciously determines the local conditions for its physical and mental life.' These ideas were closely connected to the genocidal cleansing of landscapes, when, for example, a national park was established in Poland. Landscapes were to be Germanised which not only meant the enforced removal of nonethnic Germans but also the change of physical attributes of landscapes. Gröning and Wolschke-Bulmahn (1987, p. 144) show German cultural landscapes (Figure 5) which were designed

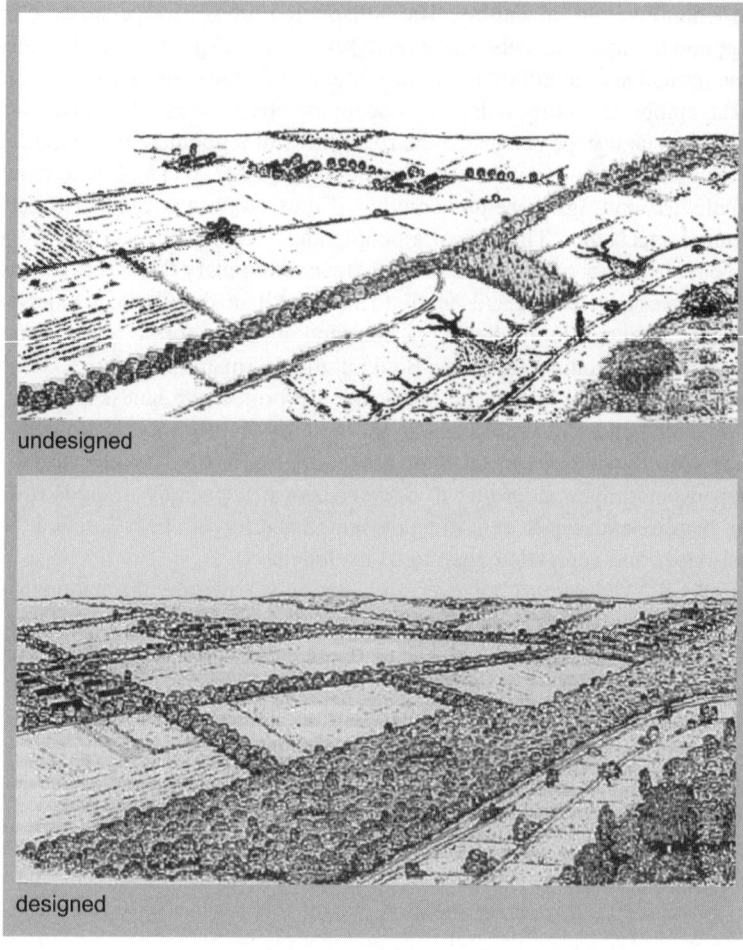

Figure 0.5. Designed and undesigned German culture landscape. (Gröning &
Wolschke-Bulmahn, 1987; Courtesy Lit.-Verlag)

to be implemented in the occupied areas of Eastern Europe. It was especially the
hedge-fenced landscape of northern Germany which was thought to embrace the
essentials of a Germanic landscape.

There has been little critical research on how fascist sciences made use of
the landscape concept (but see Schama, 1995) but there is little doubt that the
concept as read in the early twentieth century easily lent itself to thorough poli-
tisation and ideological cooptation.

Immediately after the Second World War leading geographers, notably Carl
Troll (1947), differentiated between a legitimate scientific approach in geography
and a misguided politicised version of geography. The experience of World War

II and the usurpation of the landscape idea by the fascist regime in Germany led to the demise of holistic readings of the landscape concept especially in geography and the denial of landscape as a key concept in the humanities. In contrast to this, an approach concentrating on the physical and topographical properties of landscapes became the main version of landscape geography in Germany.

Definitions of landscape from within the geosciences in the 1950s and 1960s still stressed the holistic perspective (Troll, 1950, p. 165; Schmidthüsen & Bobek, 1949, p. 119) but actual research was prioritising geoecological, hydrological, and morphological research. After World War II landscape ecology developed rapidly as a scientific and analytic reading of the concept. The landscape concept was then mainly used by physical geographers who took the term as an umbrella concept addressing different geoecological processes (Luig & von Oppen, 1997, p. 14). Landscape ecology originated in geography and biology. Pioneer landscape ecologists such as Troll (1947, pp. 24ff.; 1950; 1968) saw landscape ecology as the study of the physicobiological relationships that govern the different spatial units of a region.

In the 1960s the landscape concept came under major critique from within the geosciences. Definitions of landscape could not convincingly solve the question of borders (see Menz & Richters, this volume) and increasing methodological sophistication in both physical geography and in social geography made it more and more difficult to integrate information at a higher level. Furthermore, it was asserted that landscapes do not have an objective reality and that research in human–environment interactions needed a firmer epistemological stand than the landscape concept offered: Luig & von Oppen (1997, p. 14) report that 'the rise of social and economic geography, in particular, triggered in Germany a radical critique of "landscape geography", attacking its spatial holism as essentialist, almost metaphysical, its political implications as conservative, and its substantive understanding of landscape as premodern'. It was especially social geography which completely rejected the landscape concept and opted for problem-oriented research.

In contrast to this physical geography retained the concept and developed a sophisticated methodology to study the various components of the ecosystem (e.g., Dikau et al., 1999; see also Bubenzer this volume). In ecology the landscape concept became a standard tool to define the spatial circumferences of a research area. Ecologists nowadays use two ways to represent landscapes: landcover types are depicted in polygons or as raster lattices representing the landscape as a grid. A highly interesting third way is explored by Urban and Keit (2001) who present a landscape as a set of nodes (e.g., habitat patches) connected to some degree by linkages that join nodes functionally (see Figure 6).

The connectivity of such landscape networks is then analysed by methods based on graph theory. The journal, *Landscape Ecology*, bundles a lot of the theoretical discussion taking place in this direction (e.g., Gardner et al., 1987). The problem of defining boundaries of landscapes is still a crucial theoretical

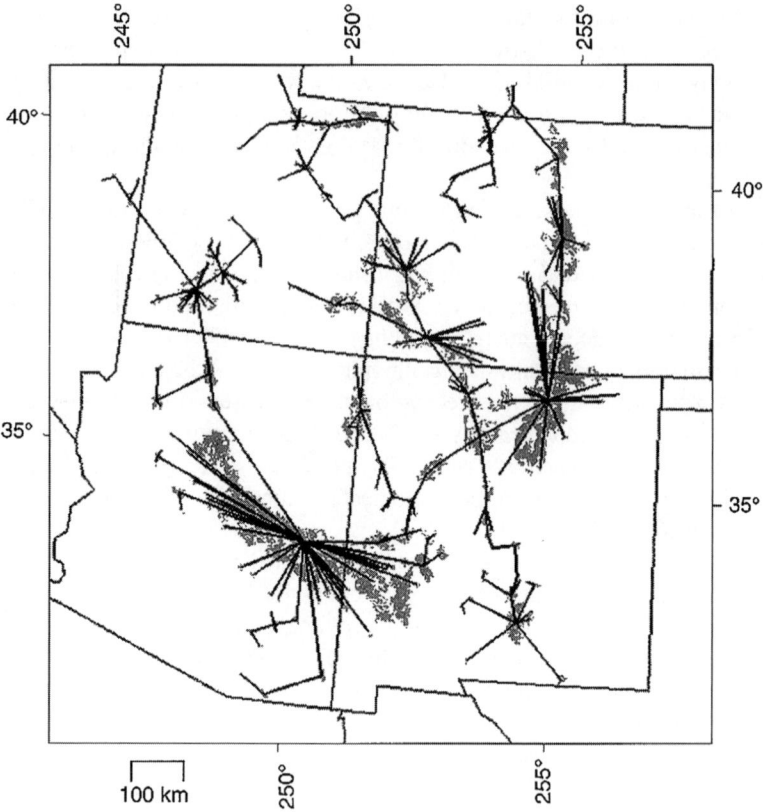

Figure 0.6. Representation of a landscape of potential habitat (green) of the Mexican
spotted owl (Strix occidentalis lucida) in the Southwest (United States) as an island
model and a graph (Urban & Keitt, 2001; Courtesy Ecological Society of America, ESA)

issue in this line of thought (Hansen & di Castri, 1992). Menz and Richter (this
volume) explore the potential of remote-sensing-based methodology to define
landscape boundaries. Bolten et al. (this volume) follow a statistical approach
combining archaeological findings, topographical structures, and ecological
features.

The use of the landscape concept in France and Great Britain differed a
lot. In France the concept *paysage* remained confined to literature critique,
aesthetics, and belletristic literature. French geography did not make much use of
the concept. However, in the historiography of the Annales School landscapes
became a key element of historical analysis: Braudel assumed that certain lifestyles in the Mediterranean had a *longue durée* because the geographical features
rendering the arena for these lifestyles had stayed similar for a long period of
time. Not much unlike Hettner he attributed formative power to landscapes which

were constitutive for the *milieu* of a society. When portraying the history of the Mediterranean in the period between 1550 and 1600 Braudel (1990/1966) spends many pages describing the geographical characteristics of his unit of historical analysis such as topography, climate, and vegetation. Geography, cultures, and politics were inextricably linked in the historiography of Braudel et al. (1990).

However, Braudel and also his compatriots (see also Aymard, 1990, original 1985) rarely made use of the landscape concept; although talking occasionally about the Mediterranean landscape, they often prefer the concept milieu instead of landscape as does Braudel in the preface to the first edition to *La Méditerranée*: 'La première (partie) met en cause une histoire quasi immobile, celle de l'homme dans ses rapports avec le milieu qui l'entouré . . . '. In England the use of the term is again different: about 80% of the titles Hard found (between 1923 and the end of the 1960s) belonged to either landscape gardening or textbooks for painters and artists and aesthetics (Hard, 1969, p. 260; see also Steinhardt, 2000). In England as in France, the landscape concept was rarely used by geographers and also in the humanities there was little trace of the spiritual bonding of populations and environments which was so typical for the use of the concept in Germany between 1900 and 1950.

The landscape concept has experienced a true revival in the humanities since the late 1980s. Anthropology, history, but also archaeology have taken up the concept vigorously. Some proponents of the concept even speak of a 'spatial turn' in the historical sciences since the early 1990s. Although the landscape concept was abolished in social geography, it became popular in historical sub-disciplines such as historical ecology and environmental history. Whereas the latter discipline, rapidly developing since the 1980s (see work by Bunn, 1996; Carruthers, 2005; Ranger, 1999; Harries, 1997; Griffiths & Robin, 1997), dealt with the discursive appropriations and factual changes in management practices when imperial powers invaded African, Australian, and American landscapes, historical ecology rather looked at long time sequences and followed up human–environment interactions through the centuries (e.g., Schmidt, 1993; McGovern, 1994; Crumley, 1994b).

In archaeology the analysis of noneconomic human–land relations brought about an interpretative challenge. Tilley (1994) looked for the reconstruction of prehistoric landscape experiences. Constructed landscapes, conceptualised landscapes, and ideational landscapes were differentiated: constructed landscapes are characterised by the human creation of markers such as monuments but also subtle constructions; conceptualised landscapes 'are mediated through and to some extent constitutive of social processes' and ideational landscapes are both imaginative and emotional and the concept is used to elicit the emic view: 'Ideational landscapes may provide moral messages, recount mythic histories, and record genealogies. . . . ' (Knapp & Ashmore, 1999, pp. 10–12). Notably the gap between landscape archaeologists and postprocessual archaeologists widened the more the latter group expanded its research into the ideational and conceptual realm and the more the first group refined its natural sciences methodology to settlement and explored new analytical ways to explore human–environment interactions of the prehistoric past. The interaction between different spatial scales and demographic dynamics within specified spatial boundaries, as much as the transformation of the landscape itself became new targets for innovative research (Zimmermann et al., 2004; see also Lenssen-Erz & Linstädter; Riemer and Bolten et al. this volume; Meurers-Balke et al., 1999).

In the humanities the landscape concept became applied to very different issues. Appadurai (1992, pp. 296–297) proposed analysing global cultural flows with the concepts ethnoscapes, mediascapes, technoscapes, finances-capes, and ideoscapes. He made use of the suffix *scape* 'to indicate first of all that there are not objectively given relations which look the same from every angle of vision, but rather that they are deeply perspectival constructs, inflected very much by the historical, linguistic, and political situatedness of different sorts of actors, for these landscapes are eventually navigated by agents who both experience and constitute larger formations in part by their own sense of what these landscapes offer.

These landscapes thus, are the building blocks of what, extending Benedict Anderson, I would like to call 'imagined worlds' that is, the multiple worlds which are constituted by the historically situated imaginations of persons and groups spread around the globe. We have become accustomed to terms such as media landscape, cinematic landscape, feminist landscape, hazard scapes, and business scape (there are many more versions). Besides a conceptual stretching of the landscape concept, new virtual technologies bring about new appropriations of landscapes. The accessibility of remote sensing data and three-dimensional virtual landscapes via Google have revolutionised our relation to landscapes in the past decade. In 2005 digital maps of the earth went online and all landscapes of the globe were accessible for any Internet user. The maps offered on the net can be combined with various datasets. Locative media may give an insight into the position of dentists in an urban landscape but also inform about atrocities in the Sudan (see Figure 7). Far from being a neutral presentation of the Earth's surface also virtual landscapes are imbued with power relations (see also Thrift, 2006).

Google-Earth in cooperation with the Holocaust Memorial Museum in Washington produced maps of landscapes of violence in the Sudanese Darfur region. Specific symbols stand for different types of violence. Zooming on certain spots it is possible to click on icons which then reveal photographs of destroyed villages or original commentaries of victims. Virtually the visitor may move through a landscape of destruction not reflecting the fact that hyperrealistic depictions are also constructions (but not necessarily fabrications).

The recent appropriation of the landscape concept by highly diverse discourses has almost deflated it to something devoid of any biogeophysical content. It has become a habit to suffix '-scape' to nouns whenever a cover term for a complex, situated, and (if vaguely) bounded network of items is discussed. Terms such as mindscape, soundscape, and political landscape are very much

Figure 0.7. Google-Earth presentation of the Darfur conflict (Courtesy Google-Earth in cooperation with the Holocaust Memorial Museum in Washington) (*See also Color Plates*)

in vogue. The increasingly loose use of the landscape concept is mirrored in the UNESCO's definition of landscapes for heritage protection. 'Clearly defined' landscapes which were 'designed and created intentionally' (such as parklands, monumental structures, etc.), 'organically evolved' landscapes which 'began as a particular socio-economic, administrative or religious initiative which evolved subsequently in association with and response to the natural environment' are differentiated. There are 'relict organically evolved landscapes' and 'continuing landscapes' according to UNESCO's definition. Finally 'associative cultural landscapes are identified by such features as sacred promontories, or religious settlements in outstanding landscapes' (Knapp & Ashmore, 1999, p. 9). The increasingly loose use of the concept makes it necessary to reassess its value for specific scientific debates. I do not propose a new definition of the concept here but rather outline scientific perspectives of interdisciplinary research.

2. PERSPECTIVES ON LANDSCAPES

The preceding passages have shown continuities and discontinuities in the use of the landscape concept in various disciplines. Lines of continuity run along the following trajectories. (1) Landscape always conveys a holistic perspective on a section of the environment, (2) the unit being analysed and/or represented is usually characterised by heterogeneity and yet it is homogeneous enough to differ from other similar units, and (3) it focusses attention on the organic and inorganic materialisation of human–environment interactions and human presentations/ perceptions of these interactions. Crumley identifies landscape together with concepts such as scale, region, boundary, diversity, and organisational structure as cornerstones of an emerging language that bridges the social and the natural sciences and prepares for the analysis of complex human–environment relations (Crumley, 1994a, p. 13). It is mainly in relation to other broad concepts that the landscape concept develops its scientific attraction. Landscape and time is one such trajectory, landscape and power another one. Landscape and nature, or in the wording of Arturo Escobar landscape 'after nature' (Escobar, 1999), is a third important field for interdisciplinary contact. These three fields of scientific inquiry – landscape and temporality, landscape and power, landscape and nature – overlap to varying degrees. In the following text I delineate three trajectories which seem to offer the most promising aspects for future disciplinary and inter-disciplinary landscape research.

2.1. Landscape, Time, and Memory

Ingold sees a focus on the temporality of landscapes as a way to move the focus of research beyond the 'sterile opposition between the naturalistic view of the landscape as a neutral, external backdrop to human activities, and the culturalistic view that every landscape is a particular cognitive or symbolic ordering of space' (Ingold, 1993, p. 152). Landscapes are process, both in a geoecological sense and in a cultural sense! Ingold goes expressly against definitions of landscape which stress an inherently static view of landscape as a perspective on the environment. He points out that 'what appear to us fixed forms of the landscape, passive and unchanging unless acted upon from outside, are themselves in motion, albeit on a scale immeasurably slower and more majestic than that on which our own activities are conducted' (Ingold, 1993, p. 164). Bender (2002) made a similar point when stressing that landscapes are materialisations of very different temporalities: (1) there is landscape as solid geology which addresses evolutionary time, (2) there is the topography of landscapes with its great time depth materialising geoecological as well as human interventions, (3) there are the seasonally changing vegetational patterns of landscapes, and (4) there is the landscape which directly mirrors human land-use (Bender, 2002, p. 103). Zooming in on these points Bender describes landscapes as 'time materialising' (Bender, 2002, p. 103). Such a broad approach engages geologists as much as philosophers in a meaningful debate. Ingold's and Bender's broad vision of the temporality of a landscape are reminiscent of James Lyell's discovery of the temporality of the earth's surface. Lyell who influenced eminent evolutionists such as Darwin and Spencer was the first to acknowledge the archival qualities of sediments and other layers of the earth. In his *Principles of Geology* (1830) Lyell argued that

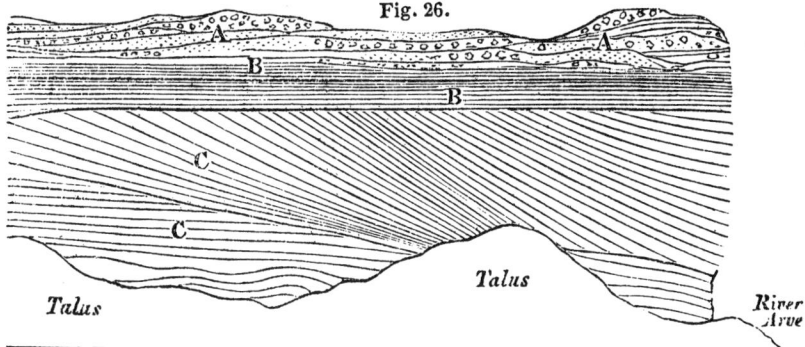

Fig. 26.

Section of a sand-bank in the bed of the Arve at its confluence with the Rhone, showing the stratification of deposits where currents meet.

Figure 0.8. James Lyell's discovery of the archival character of sediments (Lyell, 1830)

the layers observable in the present suffice to account for the geomorphological past and give evidence of the long time span over which the earth has developed (see Figure 8).

Lyell's book was of considerable importance for the development of evolutionism and both Darwin and Spencer took to his ideas of evolutionary processes. In fact Darwin saw human bodies as archives of evolutionary trends and Spencer and other evolutionists such as Tylor and Morgan regarded societies as archives of the past. For geologists and archaeologically working geographers Lyell's ideas have become part and parcel of their daily work. Sediments have stored abundant information: when in 2001 a geographical subprogramme of the ACACIA programme was able to recover a four metre long drilling core from Lac Ounianga in Northern Chad the varves were fine enough to give a record of the climatological events of more or less every year over a period of roughly 3000 years (Kröpelin, 2007; Kröpelin & Kuper, 2008). Meurers-Balke et al. (1999), using pollen remains as an archive of landscape changes in the Rhinelands, show how each epoch since the area was first settled by farmers some 7000 years ago had a typical landscape and that this landscape left its specific fingerprint in the pollen archive.

These processes are often neither purely natural nor are they purely manmade. Abel and Blaikie (1989) and Watts (1988), for example, have shown how the dynamics of a political economy come to interact with specific edaphic and hydrological features of the environment and how increasing poverty and vulnerability lead to environmental degradation and landscape level changes. In fact, much of the political ecology debate of the last decade (Kottak, 1999; for an early summary see Bryant, 1992) runs in this direction. Historical ecology addresses landscape temporality directly as it 'traces the ongoing dialectical relations between human acts and acts of nature, made manifest in the landscape. Practices are maintained or modified, decisions are made and ideas are given shape, a landscape retains the physical evidence of these mental activities' (Crumley, 1994a, p. 9). It is in this intersection that landscape as an archive of history and landscape as an expression of power relations converges.

Landscapes are not only archives in themselves but they are also used as stores of personal and collective memories on various levels. Many societies hold that the landscape somebody is born in has a direct bearing on his or her character; that is, that human beings embody characteristics of the landscape in which they dwell. The character of a landscape resonates in the body and character of individuals and societies. This is an idea widely shared by many cultures. Ibn Khaldun (1958), for example, saw the superiority of desert-dwelling communities being conditioned by the austerity of their environment. Böck and Rao (2000) describe that concepts of nurturance, the consumption of food, and with this the consumption of the qualities of the soil, are of crucial importance to understanding links between descent ideology and location in India. Rao (2000, p. 115) explains that the Kashmir Bakkrwal regard a person's blood as being influenced by the food eaten and that through food, properties of soil, water, air, and altitude are absorbed into human bodies. She concludes that 'places are thought to effect *mijaj* (temperament, innate temper) through the kinds of water that flows through them, the properties of their crops. . . . All these properties . . . are stored in the body and transmitted over generations through blood.'

Many authors have argued that deterritorialisation is a key element of globalisation: in such a setup recourse to the close relationship amongst locality, character, and culture may become a key element in identity strategies focussing on rootedness. Rössler (2003) shows how the Makassar of Eastern Indonesia use landscape metaphors to express political stands; they too emphasise the link between landscape and character and in a poem cited by Rössler young men claim to stand as brave and unshaken as the central volcano of the region. It is especially indigenous communities which have recently resorted to the rhetorics of 'rootedness in the landscape' to prove their stand, for example, in land-right cases (for examples from Southern Africa see Bollig & Berzborn, 2004; Bollig & Heinemann, 2002).

Landscapes are also intensely linked with subjective recollections of the past and with collective memories. Ingold draws upon this relation when proposing a *dwelling perspective* according to which 'the landscape is constituted as an enduring record of – and testimony to – the lives and works of past generations who have dwelt within it, and in so doing, have left there something of themselves' (Ingold, 1993, p. 152; see also Dieckmann, this volume). Küchler (1993, p. 85) proceeds in the same direction when she says that 'landscape becomes the most important *aide-mémoire* of a culture's knowledge and understanding of its past and future'. She argues that, for example, painted landscapes of the western tradition are landscapes of memory whereas in many non-Western contexts landscapes are implicated as templates in the process of memory work: landscapes are treated as memory (Küchler, 1993, pp. 85–86, see also Knapp & Ashmore, 1999, p. 13).

Memory is spatialised and memory is often attached to spatial categories. Aboriginal Australians see landscapes as materialised presentations of a distant mythical past in which their totemic ancestors settled the land (see Strehlow, 1970 who speaks of 'totemic landscapes'; see also Layton, 1995 and Taçon, 1999). Landscape features are ordered along song-lines, giving evidence of the lives and journeys of ancestors. Horowitz shows that the history of the Kanak is inscribed in landscape features, which are used to memorise the mythical adventures and passages of human and spiritual ancestors (Horowitz, 2001, p. 237). Ranger (1997) and Colson (1997) show, in two southern African contexts, that rituals tie together collective memory and landscape. Bollig (1997, see also this volume) and Widlok (1998) show that the placement of graves within landscapes in two Namibian contexts gives expression to collective memories. Assmann (1999, pp. 38ff.) argues that collective memory always strives for spatialisation. He claims that cognitive maps as networks of places form a grid to which collective memories are attached. These maps are mnemotopes, media for collective memory (Assmann, 1999, p. 60; see also Hayes et al., this volume). In order to perform such functions it is necessary that landscapes are categorised in a detailed way and in a specific language. A differentiated lexical domain on specific landscape features seems to be the conceptual basis for a mnemotope (see also Levinson, 1996, pp. 358ff.). Mark and Turk (2003) show the detailed way in which the Australian Yindjibarndi categorise the landscape in which they live. They show that the 'basic level categories in a language must be tuned to the variations in the particular environment in which a speech community lives, and to the ways in which that environment affords various activities essential to life, if it is to provide the common terms needed in every-day speech' (Mark & Turk, 2003, p. 12). The contributions by Möhlig and Kathage in this volume argue along similar lines. Both contributions show how the elaboration of specific domains – in the one example mountains and in the other one riparian environments – is linguistically structured, whereas Kavari and Bleckmann show how a culture proceeds from categorisation towards using a landscape as memory.

2.2. Landscapes, Power, and Identity

The link between collective memories and landscapes is of course not only mnemonic but also intensely political. The reference to the traces of past generations of settlers left in the landscape creates an aura of legitimacy and authenticity. The placement of monuments and other marks (such as graves and shrines) on the landscape emphasises this quest. Rights of exclusive use, rights of administration, and rights of transmission are directly implicated by specific readings of a mnemotope. Landscape features, natural or manmade, can then be taken as direct evidence for the legitimacy of claims. Land rights in West Africa are connected to migrational histories and the traces they leave in the landscape (Lentz, 2003; Dafinger, 2005). The visiting of ancestral graves among the Namibian Herero has direct implications for their claims to a restoration of ancestral lands which fell to German settlers after the 1904 genocide (Förster, 2004). Simon Schama's *Landscape and Memory* (1995) makes the great effort to disentangle the intertwining of political history and landscape concepts in Europe and North America. Drawing on the art historian Aby Warburg, Schama is especially interested in the *longue durée* of specific characteristics of collective memory attached to the landscape and their political uses and misuses.

A lot of literature coming from the field of environmental history deals with the multifarious ways in which colonial conquest restructured visions of the landscape. Colonial mapping and photography are interpreted as meaningful efforts of colonial *in-scription,* described aptly by Johannes Fabian (1983, p. 24):

> Colonial expeditions were not just a form of invasion; nor were their purpose inspection. They were determined efforts at *in-scription.* By putting regions on a map and native words on a list, explorers laid the first and deepest foundations for colonial power. By giving proof of the 'scientific' nature of their enterprise they exercised power in a pure and subtle form – as the power to name, to describe, to classify.

Especially colonial photography has been interpreted in this direction in earlier publications. In a fine-grained interpretation of landscape photography, Patricia Hayes shows the interpenetration of cognition and colonial ideology: wide open spaces and a subtle play between foreground and background are emblematic icons of colonial discourses and indirect rule (Hayes, 2000).

Of course power relations do not only materialise in landscapes where colonial powers usurp hegemony. Although the focus on imperial encapsulation by environmental historians has furthered our understanding of colonial human–environment relations, it has neglected and perhaps even downplayed the effects that internal power dynamics have had on landscapes (but see Hayes' contribution to this volume). The archaeology of rock art has interpreted the placement of paintings within wider landscapes as indicative for the meaning of drawings (Lewis-Williams, 1987; see also Lenssen-Erz & Linstädter, this volume). Forager communities inscribed their visions of power onto the landscape: they sought control over natural processes and fauna to lower environmental uncertainty.

The fixation of holy places, places of power, and shrines which connect the living to supernatural beings and ancestors are another way of inscribing power relations onto a landscape. Colson's treatment of places of power (which are natural) and shrines (which are created by people) in southern Africa (Colson, 1997) exemplifies the dealings with landscape in an agrarian, subsistence-oriented system in which lineage-based social organisation is the focal point of social structure. Colson alleges that places of power are constructed as being outside of history, presenting continuity and permanence in the landscape. Their often imposing physical shape is added up by intense sensual qualities, such as whispering winds, echoes, or an island character within the larger landscape. They are permanent features of landscape, regarded as inherently sacred and loci of spir-

The pastoral Pokot of northern Kenya, with whom I completed fieldwork in the late 1980s and early 1990s, knew of such places of power: one of them was an extinct volcano, Mt. Paka. In the crater's interior the whispering winds foretold the rain. People living there and elders would occasionally gather in the crater for rainmaking rituals. Everything to do with the height of the imposing volcano and its crater was *kablawach* (secret) and was not talked about often. People climbing into the crater should be spiritually clean (*tilil*) and respectful (*tekotön*). Clearly the landscape these pastoralists dwelled in carried a moral code. The Pokot knew of other such places of power: a green island in the midst of blackish lava streams was such a place where rainmaking rituals took place. The place, Nakurkur, is a small green spot in a landscape of solidified black lava at the foot of the extinguished volcano, Mt. Paka. The sharp dark silhouettes of the rock stand out in magnificent contrast to the green grass and the high trees of Nakurkur itself. Such places are icons of cultural memory and enduring gerontocratic power. During a rainmaking ritual I had the chance to participate in, ritual elders stressed that the ground and the rocks of Nakurkur had often been soaked with the blood of sacrificial oxen. Society and nature had interpenetrated each other in ritual to the benefit of the people living on the land. Todokin's blessing spoken on the 23rd of September 1993 in Nakurkur addressed essential landscape features directly and stressed the agency of mountains and the interpenetration of soil and sacrifice.

> The sacred mountain the Njemps pray to,
> that sacred mountain shall help the cattle
> I order that mountain, be clean, be clean and come together with what?
> with the clouds and what else? . . .
> arise mountain and open the armpits, the armpits, the armpits
> open them now as . . .
> all those who ate that oxen, you are still in this land
> we were given the sacred stomach contents
> those stomach contents which are still in the waters of Tilam,
> the land shall be clean, the land shall be clean, the land shall be soft....
> Let the blood flow on the land down to where? to Suguta
> become soft [i.e., have more grass] mountain Silali,
> have more grass mountain Paka . . .
> I say 'land become soft'.
> Be blessed! Be blessed! Be blessed!

itual power. Their cults usually override narrow kinship affiliations, underwriting the idea of a moral world order reflected in the landscape. If humans go to these places the travelling takes a sacred character, and becomes a form of pilgrimage. Schlee (1989) describes how the pastoral Rendille travel from their homes in northern Kenya to specific sacred mountains in southern Ethiopia to conduct important rituals which are closely tied to these specific landscapes

The majority of studies interested in Africa's environmental history look at colonial projections and colonial attempts at inscription onto the landscape. Ranger (1997, p. 38) is certainly right when he pleads for more research on the vitality of local discourses on landscapes. In his intriguing history of the Matopos landscape in southern Zimbabwe Ranger (1997) shows how the colonial conquest coopted this landscape. The fighting between rebels and British troops taking place in the hills as a consequence of the Ndebele rebellion was critical to the European imaginative appropriation of the Matopos hills. A colonial observer alleged that here British soldiers had 'written with their blood on these imperishable rocks a glorious and authentic page of English history'. The burying of Cecil Rhodes in the hills made the 'Matopos . . . the iconic landscape par excellence of white Rhodesia' (Ranger, 1997, p. 64). Ranger's account is outstanding as it does not only portray the Western account but also tries to find evidence of dynamic African appropriations of the same landscape. Ranger assumes that rain ceremonies took place in the Matopos hills and that these rituals were later incorporated into the Mwali cult which constructed a radically different idea of the landscape than that of white painters and writers. Ranger also treats African painters connected to mission stations who placed the great events of the Bible into the landscape of the Matopos in their paintings.

However, African landscapes did not easily lend themselves to colonial hegemonic inscriptions. Whereas in Fabian's vision African landscapes are fairly easily inscribed, Harries (1997) and also Ranger (1997) show that Europeans found it difficult to imbue meaning into African landscapes. Ranger wonders why there are no nineteenth century paintings of the Matopos Hills although Frank Oates and Thomas Baines, both famous painters, traversed the hills but did not paint them. For the painters the landscape needed historical and cultural associations which the Matopos hills obviously only gained for European eyes after the suppression of the Ndebele rebellion in the hills. Harries (1997) observes something similar with early missionary approaches to southern African landscapes. African landscapes could not be structured according to European principles. The 'parched and colourless African landscape' (Harries, 1997, p. 178) was inaccessible for the perception of European travellers. Early missionaries often portrayed the African landscape as 'uniform, monotonous, without end . . . ' (Harries, 1997, p. 180) and apparently lacked the visual conventions to structure African landscapes. One is reminded of Tacitus' description of the wide forested areas of Germania as *informen terris,* shapeless or dismal lands, the antidote to the well-structured Mediterranean landscape (Schama, 1995, p. 81). Brinkman (this volume) reports that Portuguese administrators described the remote southeastern part of Angola with similar terms.

In their volume, *The Colonizing Camera*, Hartmann et al. (1998) show how societies and landscapes were framed visually according to colonial expectations. Landau (1998) in the same volume shows how sexual aspirations and phantasies were visually inscribed onto the landscape. Visual presentations – painted, drawn, and later photographed – were a powerful tool in the colonisation of consciousness. Visual presentations of people and landscape of northwestern Namibia, habitually labelled as Kaokoveld or Kaokoland, are a key to the understanding of colonial hegemonic visions and the practice of colonial rule. Miescher and Rizzo (2000) describe in detail how in German times landscape photography dominated and was then gradually replaced by visual presentations of an allegedly stable 'tribal society' during South African colonial times, privileging the photography of women who were rendered as embodied icons of tribal harmony. The landscape topos was carried on in colonial photography (Hayes, 2000) and the specific landscape characteristics such as wideness, ruggedness, and wilderness were cherished. The colonial landscape/aboriginality discourse has entered contemporary discussions and depictions of the region. In recent photography the Namibian Himba pastoral community is framed as an indigenous community and fused with the environment as their close physical relation to the landscape is stressed both in writing and photographs (Bollig & Heinemann, 2002).

Of course, attempts at the inscription of African landscapes have not ended with colonial times. Inscription is perhaps more implicit and relies more on media and the rhizomatic extension of international organisations (governmental and nongovernmental) than on direct administrative power nowadays. The state and a host of national and international nongovernmental organisations has spun its net over societies and landscapes. There is hardly a village, even in remote areas of the Kalahari, which is not organised as a community-based organisation, not linked to some nongovernmental organisation, and not receiving donor funds from one or the other side. Model landscapes are conceptualised in which modernised regions characterised by metropolitan areas and lands of highly mechanised agriculture are juxtaposed to wilderness areas. Transfrontier parks are nowadays added to already existing national parks to extend wilderness areas. The recently opened Limpopo Transboundary Park including the Kruger Park and adjacent protection zones in Mozambique and Zimbabwe comprises some 100,000 km^2, an area almost double the size of Switzerland (Whande & Suich 2008).

Most of the literature has dealt with rural landscapes. Lately, however, the landscape concept has also been extended to urban settings. The apartheid city in South Africa was one subject of a landscape approach to urban settings. In order to gain control of rapidly growing urban settings the Natives Urban Area Act, which stipulated a resettlement of the black population in separate locations, was issued in 1923. Mixed urban quarters were destroyed. This policy was radicalised after 1948 through the Group Areas Act. The townships of the black population were relocated some distance away from the city centre. Within these townships quarters were arranged according to ethnic affiliation. Buffer strips divided black, coloured, and white living quarters. According to an ideology of rationality these newly planned urban landscapes usually addressed as townships were uniform: only one type of house – NE 51/6 – was built in Soweto 65,564 times in a short period of time (Reddy, 2000, pp. 142–147). Gewald (this volume) shows how the

South African apartheid state not only demolished living quarters of Windhoek's Old Location when forcefully resettling black and coloured inhabitants to the outskirts of the town, but built new roadworks in a way that the old mnemotope was efficiently destroyed. However, in general, studies on power and the shaping of townscapes are still infrequent and the few studies which exist focus on larger metropolitan areas. Pinther (2005), for example, discusses the emergence and transformation of the urban landscapes of Accra and Kumasi in Ghana; visions of urbanity and cultural identity informed the layout of quarters, the network of roads, and specific architecture.

2.3. Landscape After Nature

The holistic understanding of landscape as a link between nature and culture so prominent in the first part of the twentieth century (see, e.g., Hettner, 1923; Sauer, 1925) has been discarded in recent decades. Landscape ecology and physical geography have developed as highly specialised disciplines with a natural sciences methodology (see Bubenzer, this volume); on the other hand approaches within anthropology, archaeology, and history have used the landscape concept without taking geoecological factors into account. Simon Schama's book, *Landscape and Memory*, is perhaps a good example as Schama states that: 'Landscapes are culture before they are nature; constructs of the imagination projected onto wood and water and rock' (Schama, 1995, p. 61). The book then mainly deals with cultural imaginations and political history and has only little to say on natural processes.

   Escobar (1999, p. 2) criticises that postmodernists and poststructuralists 'have to hastily come to think that since there is no nature outside of history, there is nothing natural about nature'. He holds that it is necessary to acknowledge 'both the constructedness of nature in human contexts . . . and nature in the realist sense, that is, the existence of an independent order of nature' (ibid. 3). There is a growing awareness in a post-postmodern science-scape that the mechanisms and dynamics of nature display autonomy, resistance, and resilience. Gottdiener (1995, p. 30) pointed in the same direction when saying that 'culture is not simply understood as a system of signification, but as a sign system articulating with exo-semiotic processes' and Soule and Lease (1995) speak of a prediscursive nature which is only partially captured by human rationality, thought, and emotion. In the following I pinpoint approaches in historical and political ecology and outline their use of the landscape concept.

   Historical ecology is defined as the 'study of the structure, function, and change of a heterogeneous land area composed of interacting ecosystems . . . landscape history is the study of past ecosystems by charting the change in landscape over time' (Crumley, 1994a, p. 6, see also Balée, 1998; for a review of pertinent literature). Winterhalder (1994, p. 19) argues in the same direction when he claims that 'a complete explanation of ecological structure and function must involve reference to the actual sequence and the timing of the causal events that produced them.' Ecosystems are characterised by (1) their systemic qualities (there are complex interactions or connections among parts), (2) their history, (3) their spatiality, and (4) their nonlinear character (key interactions may be characterised by lags and thresholds).

Stability and resilience are key concepts to analyse such systems. Stability is defined as the propensity of a system to attain or retain an equilibrium condition of steady-state or stable oscillation. Resilience refers to the ability of a system to maintain its structure and patterns of behaviour in the face of disturbance (Holling, 1973).

The patchiness of a landscape is another key variable to explain the dynamics of ecosystems (see also the contribution by Linstädter, this volume). Patches are defined as 'localised discontinuities in the landscape which affect behaviour' or as 'ecologically distinct locality in the landscape which is problem- and organism-defined, relative to the behaviour, size, mobility, habits, and perceptive capabilities of the population being studied' (Winterhalder, 1994, p. 33). Although patchiness summarises the state of environmental heterogeneity in space, concepts such as stability and resilience help to characterise the range and regularity of variation of environmental states in time. Linstädter (this volume) and Reid et al. (2000) explore methodologies of how to describe the history of disturbance regimes and land-cover changes along timelines showing the consequences of land-use/land-cover changes in a Namibian and an Ethiopian habitat.

There are a good number of studies on how humans have shaped their respective environments over long spans of time (Balée, 1998). It is especially archaeologists and anthropologists who have dealt with this topic. Widgren and Sutton (2004) offer an interesting approach of how to capture human investment or noninvestment in landscapes when discussing 'islands of intensive agriculture' in East Africa. They argue that over generations intensified farming created permanent gardenlike landscapes, by such devices as construction of irrigation furrows, levelling of fields, and stone terracing (Widgren & Sutton, 2004, p. 1). These landscapes are 'the accumulated results of labour over a long time period'. Such landesque capital is the result of generations of sculpting the landscape (Widgren & Sutton, 2004, pp. 8–9).

The case of ancient Egypt is a good example of how landesque capital is accumulated over many generations to the benefit of each future generation. Hassan (1994), for example, gives a good description of how the cultural landscape of the Egyptian Nile Valley was almost totally transformed over a period of 3000 years during the Old, Middle, and New Kingdoms. Irrigation channels and channels for the drainage of swamps structured the landscape. Increasing agricultural specialisation in the Old Kingdom brought about changes of the riverine landscape (Hassan, 1994, p. 161). The development of public and private dykes and a fully fledged irrigation system began in the First Intermediate Period and was enforced during the Second Kingdom. Temples and human habitations framed this setup (see Herb & Derchain, this volume).

Östberg (2004), Börjeson (2004), and Watson & Schlee (2004) neatly describe how complex irrigation operations and terracing transformed the landscapes the Kenyan Marakwet, the Tanzanian Iraqw, and Ethiopian Konso (all

three decentralised, clan-based societies) dwelled in, showing that the accumulation of landesque capital is not the prerogative of states. Tiffen et al. (1994)
describe how a combination of population pressure, rampant soil erosion, and the
accessibility of capital through migrant labour led to the establishment of terracing in the Kenyan Machakos area. In a period of two decades (1970s and 1980s)
the Machakos landscape changed from an area characterised by environmental
degradation to a neatly managed landscape. Although pastoral populations affect
landscapes in a more ephemeral way, they leave structures (such as graves,
Bollig, this volume) and permanently modify the landscape by specific modes of
grazing (Linstädter, this volume).

If we introduce landesque capital to understand the relation amongst labour,
technology, and landscape we must also introduce landesque mortgage. Just as
there are outstanding examples of longstanding investment into landscapes, there
are case studies on how landscape properties are used in unsustainable ways.
The archaeologist Schmidt (1993, p. 100), for example, outlines the history of
processes of landscape formation in an area once under gallery and woodland
forest in northwestern Tanzania's Buhaya District. He finds that already early
Bantu settlers destroyed large parts of the primeval forest. Sutton (2004) shows
the development of a major irrigation complex in central Tanzania (Engaruka)
from at least the fifteenth till the seventeenth century. It was possibly the substantial
deforestation of the escarpment which affected the flows of the crucial rivers
negatively, which led to the demise of Engaruka (Sutton, 2004, p. 124). Diamond
(2004) presents the ancient Maya, the Anasazi, and the Greenland Viking societies as showcases to demonstrate the detrimental effects of elites pushing the
exploitation of a landscape to its limit, forcing the environment they live in into
a collapse. Linstädter (this volume), Oba and Kaitira (2006), Oba and Cotile
(2001), and Lykke (2000) analyse the linkages between environmental degradation and the perception of landscapes in different African pastoral settings.

The colonisation of non-European areas between 1750 and 1900 brought
grave structural changes to the landscapes inhabited by indigenous societies
in the Americas, Siberia, Africa, and Australia (see, e.g., Denevan, 1992 for
the Americas). The inclusion into globally operating empires implicated grave
changes in environmental management. In a modernist vision plantations and
settler farms and a rescheduling of indigenous modes of production towards com-
moditised agricultural production became essential. In this vision zones of intensive use were juxtaposed to zones of complete protection. Large schemes such
as the Gezirah scheme for cotton production in the Sudan were implemented,
however, wildernesses were inscribed onto the landscape. Carruthers (1995)
shows how the original farming population of the South African Kruger Park was
expulsed to create the impression of wilderness (see also Dieckmann, this volume
on a similar case in northern Namibia).

A new and very productive field of research is how European perceptions
have had an impact upon environmental policies and how such perceptions
were continued into the present and influenced resource/landscape management

in the contemporary world. Misconceptions – the lies of the land – have had a direct impact via colonial administrations on natural resource management and eventually on the landscape itself. Pyne (1997) analyses the origins and effects of the colonial prohibitions to use fire when managing tropical savanna and forest landscapes. Indeed the use of fire has been attested to have contributed to the shaping of savanna landscapes: hunter and gatherers as much as herders used fires to destroy unwanted vegetation and to fertilise soils before the rains. The colonial paranoia of fires stems from the condemnation of fires in nineteenth century European agricultural sciences. European farmers still using fire to clear fields (such as in the German Black Forest) were seen as remnants of a premodern past and the use of fires by autochthonous communities was regarded as a further proof of their wasteful engagement with natural resources.

Neumann (1997) describes how German forestry departments put large tracts of forest under protection in Tanganyika laying the base for the later natural parks programme. Fairhead and Leach (1997, p. 194) argue that 'forestry and social science analyses have served to frame and support each other in such a way as to give each greater weight, as well as to underwrite approaches to conservation which, in appropriating forest landscapes, have divested landholders and users of resource control.' They trace the emergence and institutionalisation of a colonial forestry science in tropical West Africa pointing to the close cooperation amongst English, French, and German forestry experts. The common view was that human habitation had replaced the primary forest and that farming and firesetting had been essential in making a savanna landscape. When patches of forest were found in the savanna, it was thought that these were the remainders of vast primeval forests of the past.

The deforestation hypothesis linked to assumptions on the close relationship between deforestation, desertification, and famine became a leading motive in colonial forest policy (Grove, 1995). Reconstructing the history of the forest Fairhead and Leach refute the arguments that earlier extensions of the West African tropical forest, for example, in Sierra Leone, were vast. In stark contrast to colonial assumptions early travellers report on a landscape bare of forests and describe how local farmers intentionally encouraged the growth of forest around their villages, partly as a fortification, partly as a measure connected to fallowing (Fairhead & Leach, 1997, p. 204).

In a similar manner colonial environmental scientists strongly believed in the malevolent consequences of overstocking for savanna landscapes. It was held that African pastoral communities accumulate large herds of cattle and small stock beyond any rational level in order to amass prestige within their communities. The consequences were perceived as grave and irreversible for the environment: deserts encroached upon pastures and acacia shrub replaced more useful grassland vegetation. Already in the 1930s and 1940s a veritable discourse on expanding deserts developed (Troll, 1947). Anderson (2000) showed that in Kenya visions of British administrators on the African savanna in the 1930s were much influenced by their exposure to the American Dust Bowl during training seminars. Although situations were not strictly comparable they saw the Dust Bowl phenomenon everywhere in northern Kenya and prescribed enforced destocking and restrictive grazing schemes as a measure. The myth that pastoralists overexploit pastures through high stocking rates has pervaded the thinking of many development projects in pastoral areas in the second half of the twentieth century. Scientists have lately pointed out that dryland savannas with a rate of climatic variability of higher than 30 per cent (or less than 300 mm annual average) do not primarily react to grazing pressure but are rather driven by stochastic rainfall events (Behnke & Scoones, 1993).

In the present all local communities develop their interaction with the environment within a global framework, although with very different outcomes. In some areas of the globe the state and a host of national and international nongovernmental organisations run capital-intensive development efforts and through nongovernmental organisations or community-based organisations engage with the landscape. Governments seeking unexploited resources target rural areas for major mechanised agricultural schemes that transform entire landscapes (cf. Manger, 1988, p. 169). In Tanzania the pastoral landscapes of the Maasai and Barabaig have partially been transformed into wheat farms during the last two decades (Ndagala, 1994; Lane, 1994). The Sudanese Hadendowa lost important grazing in the fertile Gash valley to food and cotton growing agricultural schemes (Salih, 1994).

Large dams for the construction of hydroelectric energy have transformed landscapes all over the continent and led to major resettlement programs for local populations: for the High Dam at Aswan more than 100,000 were resettled between 1963 and 1969, for Cabora Bassa dam in Mozambique 25,000 in 1974, for the Kainji dam in Nigeria 44,000 in 1967–1968, for the Kariba dam in Zambia and Zimbabwe 57,000 in 1958, for the Koussou dam in Ivory Coast 75,000 in 1970, and for the Akasombo dam in Ghana 80,000 in the years following 1967 (de Wet, 2000, p. 6; Diaw & Schmidt-Kallert, 1990; McCully, 1996, pp. 322–332; Scudder, 2006; Scudder & Colson, 1972). Other megaprojects such as national parks have also required large numbers of people to move (e.g., Carruthers, 2005; Neumann, 1997). In the course of huge government programs such as the compulsory villagisation campaigns in Ethiopia or Tanzania millions have been involuntarily resettled: 5–12 million peasants were to be resettled in Ethiopia from 1985 to 1988 (Rahmato, 1984) and at least 5 million Tanzanians were relocated from 1973 to 1976 in the course of the *ujamaa* programme (Hyden, 1980, p. 130; Scott, 1998, p. 223). De Wet estimates that at least 25 million have been villagised from 1900 to 1990 in Africa. In South Africa millions have been relocated due to apartheid policies and the so-called betterment schemes (de Wet, 1995, pp. 26–28). All these resettlement schemes left landscape level changes behind in the regions from where resettlers came and in the regions to which resettlers were brought.

2.4. The Contributions to This Volume

The volume falls into four sections: contributions to Part I adopt a natural sciences perspective, whereas papers in Part II take a historical vantage point. Part III has a social-anthropological outlook and contributions to Part IV deal with the relations amongst language, cognition, and landscape.

Bubenzer starts the first section with an introduction into geosciences approaches to landscapes. He portrays the landscape concept as central and at the same time as contested in the history of geography. He argues that the revolution of geographical data management through GIS, which allows for a complex modelling of interdependent landscape factors on the basis of digital elevation models, holds the potential for the revival of a holistic approach to landscapes. Linstädter, taking off from a landscape ecology perspective, discusses the biophysical and social causes and consequences of landscape heterogeneity in northern Namibia. She hypothesises that a sustainable utilisation of a savanna landscape implies a disturbance regime that resembles the geobiophysical dynamics occurring in such a landscape under natural conditions. Lenssen-Erz and Linstädter give a short overview of Anglophone landscape archaeology and the equivalent in German-speaking archaeology. Whereas the German tradition favours an empirical approach close to landscape ecology, the more recent British tradition focuses on the archaeology of mindscapes in order to offer a fuller understanding of prehistoric life worlds.

Lenssen-Erz and Linstädter propose an outline to a methodology which takes into account both geobiophysical givens and intellectual appropriations. In two archaeological case studies – located at Namibia's Brandberg and at Egypt's Jebel Auweinat – they demonstrate the potential of their hybrid approach. Riemer applies the landscape concept to a specific archaeological case study: the Great Sand Sea/Egypt which is characterised nowadays by the great uniformity of giant dunes rising up to 100 m and running up to 500 km from north to south. Prehistoric sites of the Sand Sea are about 150–250 km away from permanent water sources, to which also prehistoric forager populations had to withdraw during dry seasons. Exotic goods point to the embeddedness of a local forager population in wider exchange networks. The contribution of Bolten, Bubenzer, Darius, and Kindermann links up directly with Riemer's contribution and focuses on the same region. In a joint effort of archaeology, geomorphology, and ecology the prehistoric land use potential of two specific locations of the Western Desert of Egypt is reconstructed. Finally, a model of interacting key variables affecting general landscape development of this part of the Sahara during the Early and Mid-Holocene is presented.

Menz and Richters propose a remote-sensing-based methodology in order to reach an objective definition and delineation (boundary demarcation) of specific landscapes thereby addressing one of the salient problems of geographical landscape definitions in the 1950s and 1960s. Based on a digital elevation model subsets of varying sizes from two study areas in northern Namibia are used to produce a multilayered dataset. It is shown that within the spatially differentiated continuum of the landscape there are continuity areas which are separated by borders with discontinuous and rapidly changing value distributions.

The second section of the book deals with historiographic approaches to landscape research. Four contributions draw a direct line between landscape and memory mainly working with oral traditions and written testimonies. Whereas Herb and Derchain concentrate their analysis on inscriptions and pictorials in tombs and temples of Ancient Egypt, Hayes analyses the appropriation of the northern Namibian floodplain by the local elite and the first European travellers in the mid-nineteenth century. Brinkmann deals with recollections of Angolan refugees in Northern Namibia and shows how nostalgic recollections are intensely inter-woven with the creation of a landscape remembered as home. Gewald's contri-bution treats the history of Windhoek's urban landscape which emerged under conditions of colonial segregation and colonial obsessions with sanitation and control. All four chapters deal with the conceptualisation of landscape and the relation amongst landscapes, power, and memory.

The most popular ancient Egyptian term characterising the riverine landscape was *kmt*, 'the black', a term coming into use at the end of the Old Kingdom, that is, around 2200 bc. The term referred to the large alluvial areas of the Nile Valley, as after the annual inundations the alluvials were covered with fertile layers of dark mud. 'The black' was juxtaposed to *dsrt* 'the red' which was used as a designation for the adjoining deserts from around 2400 bc onwards. Although the topographic features of landscapes were rarely (if ever) made a topic in these pictorial programmes, the activities of people within the landscape were the leading topic. Hayes explores the interrelationship between historical accounts of royal succession and power and the local floodplain ecologies on the one hand and appropriations of the same landscape by the mid-nineteenth century European explorers, Galton and Anderson. Whereas Ovambo marked trees and graves as powerful sites of hypomnesia, Galton and Anderson erased autoch-thonous readings of landscape through a discourse of blankness.

Brinkmann works with a set of oral recollections that are communicated among refugees from southeastern Angola nowadays living near the town Rundu in northern Namibia. In stark contrast to the Western scientific presentations of southeastern Angola as a dull and barren country, in their recollections the refugees remember the agricultural affluence of their former homes and stress the fertile soils and the abundance of perennial water courses. This landscape changes into a landscape of suffering when recollections of the war are repre-sented. The war dissolves links between parts of the landscape as town and bush come to stand in opposition: the Portuguese administration concentrated on the control of towns and on settling people forcefully in larger heavily controlled settlements, whereas the guerrilla called the bush their home. South Africa's urban landscape bears witness to a history of segregation and the attempt of the state to coopt collective memories. The enforced removal of major parts of Windhoek's black population from the so-called Old Location in the 1960s and the destruction of this part of the town is an example of the political appropriation of an urban landscape by a state. The newly built township Katutura displayed an orderly layout and a well controllable and according to apartheid's classifications systematically shaped town-scape. Gewald finishes his historical treatise of Windhoek's townscape by critically analysing the public architecture of newly independent Namibia focussing on the intentions and meanings connected to the gigantic Heroes' Acre monument at the outskirts of the city.

Starting the third section Rössler offers an overview of anthropological studies of landscape. He emphasises that in anthropology the systematic and comparative study of landscapes is still young. Over the past two decades, however, the attention towards the complex relation between culture and landscape has dramatically increased. Landscapes are researched as symbolic systems and the close relation between landscape and memory has been constantly emphasised. Only within the fields of historical and cultural ecology have anthropologists attempted to explore the relation between humans and geobiophysical processes.

The other three contributions of this section deal with the constitution of landscape in ritual and social practice. Bollig shows how collective memory is linked to landscape in and through ancestral ritual among the Himba of northern Namibia and Grunkowski explores the close links between landscape perception and social identity. She studies the Namibian Topnaar (a Nama-speaking community) living along the lower reaches of the Khuiseb river within a desert environment and commercial farmers of European descent living at the upper reaches of the Khuiseb. Dieckmann compares landscape visions of the Etosha National Park of northern Namibia and travels through the landscape both with tourists and former dwellers: Hai//om foragers who have been expelled from the park to create a pristine natural landscape inhabited by wildlife only.

Bollig's contribution shows how kinship and landscape are linked in ancestral commemoration rituals. By commemorating their ancestors the Himba inscribe genealogical networks onto the landscape but also make profound political statements on solidarity within the descent group and about landrights.

In her comparative study of landscape perception of farmers and indigenous herders Gruntkowski found that many farmers constantly stressed the aesthetic value of the land and related their decision to live there as conscious and related to their special relation to nature. Farmers were often deeply emotionally affected by the wilderness of the landscape. The Topnaar do not share the commercial farmers' concepts of wilderness nor their ideas on landscape aesthetics. For them the river, its banks, and immediate surroundings are imbued with historical and cultural meaning. Dieckmann deals with two sets of traditions: the view of the (usually) Western tourist of the Etosha pan and the ideas of Hai//om of their former homeland. Western aesthetical perspectives of an 'empty' landscape and a general idea of an ideal African scenery are evoked, for example, in Internet chats on visits to the region. Photography is the epitome of Western landscape perception. The view of the Hai//om is totally different: for them the landscape is constituted by a multitude of features (topographical, vegetational, faunistic, and historical in the sense of including humans) organised around a network of paths both of human and of wildlife origin. Whereas for tourists temporality and history are excluded from their perceptions of the landscape, for the Hai//om the landscape is intensely imbued with different layers of temporality.

Widlok offers a highly interesting methodological approach to settlement and landscape which is of use for anthropologists and archaeologists alike. He argues that the division between settlement and landscape is deeply entrenched in European thought and also in the worldview of many agrarian societies. The comparative use of permeability maps presents a promising route towards the cross-fertilisation of both strands. Permeability is defined as the way in which space allows or prevents humans from passing through places. The permeability concept is relevant to make sense of the fuzzy zone where landscape and settlement merge. The method presented concentrates on describing varying degrees of permeability. Widlok distinguishes the number of access routes and their connectedness, differentiates unipermeable (single-access) places from multipermeable (multiple-access) places. Widlok demonstrates the application of this approach in case studies from northern Namibia, juxtaposing the permeability of a forager settlement (Hai//om) and of a centralised Ovambo polity.

Three linguistic and sociolinguistic contributions deal with ways of conceptualising the landscape. Conceptualisation is understood as 'a cognitive process which transforms human experience into mental representations of knowledge, that is, into concepts' (Kathage, this volume). Lexicalisation is seen as a linguistic process of encoding conceptual information. Concepts of landscape are organised in semantic fields which are imaged as multidimensional and historically dynamic networks of lexical items that are hierarchically and horizontally organised. Based on case studies from Namibia (Herero, Mbukusuhu, Rumanyo) the contributions seek relations of inclusion and exclusion in the organisation of landscape vocabulary. Möhlig finds that whereas in Otjiherero the vocabulary for mountain formations is highly differentiated, the river-dwelling Rumanyo-speakers, living at the Kavango River in Namibia's north, have a large vocabulary addressing the riverine landscape.

In her study of the Namibian language Mbukushu Kathage shows conceptual transfers to be of great importance: concrete entities are employed in order to understand and explain less concrete entities. As a source entity parts of the body are frequently used to label parts of the landscape: in Mbukushu, for example, the term for navel is extended to the meaning of pan, a typical landscape feature of the region. Metaphor and metonymy are further strategies to generate landscape terminology. In metaphors the choice of a source entity is due to features it has in common with the target entity; for example, the threshing floor (*thindanda*) and the riverbed (*rundanda*) both have the firm ground as a common quality. In metonymy a source to label particular features is extended to describe an entire landscape.

Both contributions take care to analyse specific strategies to generate landscape vocabulary. Möhlig, for example, shows how in the two Bantu languages shifts from one noun class to another can be used to generate landscape vocabulary.

Kavari and Bleckmann's analysis of praise songs of a pastoral-nomadic community shows that not only statelike political structures and political elites are able to canonise the links between memory and landscape: the texts they decipher are highly standardised, packed with metaphor, and formulate a condensed view of a place's history. Kavari and Bleckmann's herders attach aesthetic and historic importance to those landscape features which were significant for livestock husbandry. In praise poems of places (there are also poems for people and cattle) landscapes are used as mnemonic devices. The praise songs frequently combine metaphorical allusions to a place's specific topography and hydrology and combine these with the evocation of people.

Acknowledgements To write this introduction was a long and arduous task. Of course, I profited greatly from the many talks we had within the ACACIA programme. Michael Schnegg, Hanjo Berressem, and Ursula Peters read the manuscript or parts of it and supplied important comments. Margret Szoellesi-Janze gave helpful comments on Braudel's use (or nonuse) of the landscape concept. Norbert Finzsch pointed out the epistemological differences of the historical ecology and the environmental history approaches and pointed out the influence of German geographers on American historiography. Franziska Bedorf took great care to search for literature in hidden corners and assisted in giving the chapter its final shape. Lutz Meyer-Ohlendorf did a lot to organise the final layout of all contributions.

REFERENCES

Abel, N.O.J. & Blaikie, P. (1989). Land degradation, stocking rates and conservation. Policies for the communal rangelands of Botswana and Zimbabwe. *Land Degradation and Rehabilitation*, 1, 101–123.

Allgemeine deutsche Realencyklopädie für gebildete Stände (Conversations-Lexikon) in 12 Volumes. L.-M. (1830) Leipzig: Brockhaus.

Anderson, D. (2000). Depression, dust bowl, demography and drought. The colonial state and soil conservation in East Africa during the 1930s. *African Affairs*, 83, 321–343.

Apel, F. (2000). Deutscher Geist und deutsche Landschaft. Eine Topographie. München: Knaus.

Appadurai, A. (1992). Disjuncture and difference in the global cultural economy. In M. Featherstone (Ed.), *Global Culture. Nationalism, Globalisation and Modernity* (pp. 295–310). London: Sage.

Assmann, J. (1999). *Das kulturelle Gedächtnis. Schrift, Erinnerung und politische Identität in frühen Hochkulturen*. München: Beck.

Aymard, M. (1990). Lebensräume. In F. Braudel, G. Duby, & M. Aymard (Eds.), *Die Welt des Mittelmeeres. Zur Geschichte und Geographie kultureller Lebensformen*. Frankfurt: Fischer.

Bagrow, L. & Skelton, R.A. (1964). *History of Cartography*. Cambridge: Watts.

Balée, W. (Ed.) (1998). *Advances in Historical Ecology*. New York: Columbia University Press.

Behnke, R. & Scoones, I. (1993). Rethinking range ecology. Implications for rangeland management in Africa. In R. Behnke, I. Scoones & C. Kerven (Eds.), *Range Ecology at Disequilibrium. New Models of Natural Variability and Pastoral Adaptation in African Savannas* (pp. 1–30). London: Overseas Development Institute.

Bender, B. (2002). Time and landscape. *Current Anthropology*, 43, 103–112.

Boas, F. (1887). The study of geography. *Science*, IX (Suppl), 137–142.

Böck, M. & Rao, A. (2000). Introduction: Indigenous models and kinship theories: An introduction to a South Asian perspective. In M. Böck & A. Rao (Eds.), *Culture, Creation, and Procreation. Concepts of Kinship in South Asian Practise* (pp. 1–52). Oxford: Berghahn.

Bollig, M. (1997). *When War Came the Cattle Slept. Himba Oral Traditions*. Köln: Köppe.

Bollig, M. & Berzborn, S. (2004). The making of local traditions in a global setting: Indigenous peoples'organisations and their effects at the local level in southern Africa. In P. Probst & G. Spittler (Eds.), *Between Resistance and Expansion. Explorations of Local Vitality in Africa* (pp. 297–330). Münster: LIT.

Bollig, M. & Heinemann, H. (2002). Nomadic savages, ochre people and heroic herders – Visual presentations of the Himba of Namibia's Kaokoland. *Visual Anthropology*, 15, 267–312.

Börjeson, L. (2004). The history of Iraqw intensive agriculture, Tansania. In M. Widgren & J. Sutton (Eds.), *Islands of Intensive Agriculture in Eastern Africa* (pp. 68–104). Oxford: James Currey.

Braudel F. (1966). *La Méditerranée et le monde méditerranéen a l'époque de Philippe II*. Bd. 1–2. Paris: Livre de Poche.

Braudel, F. (1990). *Das Mittelmeer und die mediterrane Welt in der Epoche Philipps II*. Frankfurt: Fischer.

Braudel, F., Duby, G. & Aymard, M. (Eds.) (1990). Die Welt des Mittelmeeres. Zur Geschichte und Geographie kultureller Lebensformen. Frankfurt: Fischer.

Bryant, R. (1992). Political ecology. An emerging research agenda in third-world studies. *Political Geography*, 11, 12–36.

Bunn, D. (1996). Comparative barbarism: Game reserves, sugar plantations and the modernisation of the South African landscape. In K. Darian-Smith, L. Gunner & S. Nuttall (Eds.), *Text, Theory and Space. Land, Literature and History in South Africa and Australia* (pp. 37–49). London: Routledge.

Burnett, D.G. (1999). The history of cartography and the history of science. *Isis*, 90(4), 775–780.

Carruthers, J. (1995). *The Kruger National Park: A Social and Political History*. Pietermaritzburg: University of Natal Press.

Carruthers, J. (2005). Africa's environmental history. *IHDP Update, Focus Environmental History*, 2, 8–9.

Casimir, M. (2007). *Culture and the Changing Environment. Uncertainty, Cognition, and Risk Management in Cross-Cultural Perspective*. Oxford: Berghahn.

Colson, E. (1997). Places of power and shrines of the land. *Paideuma*, 43, 47–57.

Crumley, C. (1994a). Historical ecology: A multidimensional ecological orientation. In C. Crumley (Ed.), *Historical Ecology. Cultural Knowledge and Changing Landscapes* (pp. 1–16). Santa Fe, NM: SAR Press.

Crumley, C. (1994b). The ecology of conquest. Contrasting agropastoral and agricultural societies. adaptation to climatic change. In C. Crumley (Ed.), *Historical Ecology. Cultural Knowledge and Changing Landscapes* (pp. 183–201). Santa Fe, NM: SAR Press.

Dafinger, A. (2005). *Anthropologie des Raumes*. Köln: Köppe.

Daniels, S. & Cosgrove, D. (1988). Introduction: Iconography and landscape. In D. Cosgrove & S. Daniels (Eds.), *The Iconography of Landscape* (pp. 1–10). Cambridge: University Press.

Denevan, W. (1992). The pristine myth: The landscape of the Americas in 1492. *Annals of the Association of American Geographers*, 82, 369–385.

De Wet, C. (1995). *Moving Together, Drifting Apart. Betterment Planning and Villagisation in a South African Homeland*. Johannesburg: Witwatersrand University Press.

De Wet, C. (2000). The experience with dams and resettlement in Africa. Contributing paper to the *World Commission on Dams, Cape Town*.

Diamond, J. (2004). *Collapse: How Societies Choose to Fail or Succeed*. New York: Viking.

Diaw, K. & Schmidt-Kallert, E. (1990). *Effects of Volta Lake Resettlement in Ghana – A Reappraisal After 25 Years*. Hamburg: Institut für Afrikakunde.

Dikau, R., Friedrich, K. & Leser, H. (1999). Die Aufnahme und Erfassung Landschaftsökologischer Daten. In H. Zepp & M.J. Müller (Eds.), *Landschaftsökologische Erfassungsstandards*. Forschungen zur Deutschen Landeskunde, Band 244, Flensburg: Deutsche Akademie für Landeskunde.

Escobar, A. (1999). After nature. Steps to an antiessentialist political ecology. *Current Anthropology*, 40, 1–30.

Fabian, J. (1983). *Time and the Other: How Anthropology Makes Its Object*. New York: Columbia University Press.

Fairhead, J. & Leach, M. (1997). Deforestation in question: Dialogue and dissonance in ecological, social and historical knowledge of West Africa. *Paideuma*, 43, 193–225.

Forman, R. & Godron, M. (1986). *Landscape Ecology*. New York: John Wiley and Sons.

Förster, L. (2004). Zwischen Waterberg und Okakarara: Namibische Erinnerungslandschaften. In L. Förster, D. Henrichsen & M. Bollig (Eds.), *Namibia – Deutschland: Eine geteilte Geschichte. Widerstand, Gewalt, Erinnerung* (pp. 164–179). (Ethnologica Neue Folge, Vol. 24). Köln: Rautenstrauch Joest Museum.

Gardner, R., Milne, B.T., Turner, M. & O'Neill, R. (1987). Neutral models for the analysis of broad-scale landscape pattern. *Landscape Ecology*, 1, 19–28.

Gottdiener, M. (1995). *Postmodern Semiotics: Material Culture and the Forms of Postmodern Life*. Cambridge: University Press.

Griffiths, T. & Robin, L. (1997). *Ecology and Empire. Environmental History of Settler Societies*. Edinburgh: Keele University Press.

Grimm, J. & Grindel, W. (1885). *Deutsches Wörterbuch L-M*. Leipzig: Hirzel.

Gröning, G. & Wolschke-Bulmahn, J. (1987). Politics, planning and the protection of nature: political abuse of early ecological ideas in Germany, 1933–1945. *Planning Perspectives*, 2, 127–148.

Grove, R.H. (1995). *Green Imperialism: Colonial Expansion, Tropical Island Edens and the Origins of Environmentalism, 1600–1860*. Cambridge: University Press.

Hansen, A. & Di Castri, F. (Eds.) (1992). *Landscape Boundaries: Consequences for Biotic Diversity and Ecological Flows*. New York: Springer.

Hard, G. (1969). Die Diffusion der Idee der Landschaft. Präliminarien zu einer Geschichte der Landschaftsgeographie. *Erdkunde*, XXII, 249–264.

Harley, J.B. & Woodward, D. (1987). *The History of Cartography*. Chicago: University of Chicago Press.

Harries, P. (1997). Under alpine eyes. Constructing landscape and society in late pre-colonial South-East Africa [in thematic section 'the making of African landscape']. *Paideuma*, 43, 171–191.

Hartmann, J.S. et al. (Eds.) (1998). *The Colonizing Camera: Photographs in the Making of Namibian History*. Athens: Ohio University Press.

Hassan, F. (1994). Population ecology and civilization in ancient Egypt. In C. Crumley (Ed.), *Historical Ecology. Cultural Knowledge and Changing Landscapes* (pp. 155–182). Santa Fe, NM: School of American Research Press.

Hayes, P. (2000). Camera Africa. Indirect rule and landscape photographs of Kaoko, 1943. In G. Miescher & D. Henrichsen (Eds.), *New Notes on Kaoko*. Basel: Basler Afrika Bibliographien.

Hettner, A. (1923). Methodische Zeit- und Streitfragen. *Geographische Zeitschrift*, 29, 37–59.

Hirsch, E. (1995). Introduction. Landscape: Between place and space. In E. Hirsch & M. O'Hanlon (Eds.), *The Anthropology of Landscape. Perspectives on Place and Space* (pp. 1–30). Oxford: University Press.

Holling, C.S. (1973). Resilience and stability of ecological systems. *Annual Review of Ecology and Systematics*, 4, 1–23.

Horowitz, L.S. (2001). Perceptions of nature and responses to environmental degradation in New Caledonia. *Ethnology*, 40, 237–250.

von Humboldt, A. (1845). Kosmos. Entwurf einer physischen Weltbeschreibung, Bd. 1–5, Stuttgart, Tübingen: Cotta.

Hyden, G. (1980). *Beyond Ujamaa in Tanzania. Underdevelopment and an Uncaptured Peasantry*. Berkeley, CA: University Press.

Ingold, T. (1993). The temporarlity of the landscape. *World Archaeology*, 25, 152–174.

Khaldun, I.M. (1958). *An Introduction to History*. Translated from the Arabic by Franz Rosenthal. 3 Vols. New York: Pantheon Books.

Knapp, B. & Ashmore, W. (1999). Archaeological landscapes: Constructed, conceptualised, ideational. In W. Ashmore & B. Knapp (Ed.), *Archaeologies of Landscape* (pp. 1–32). London: Blackwells.

Körner, J.L. (1995). *Caspar David Friedrich and the Subject of Landscape*. New Haven, CT: Yale University Press.

Kottak, C.P. (1999). The new ecological anthropology. *American Anthropologist*, 101, 23–35.
Kröpelin, S. (2007). High resolution climate archives in the Sahara (Ounianga, Chad). In O. Bubenzer, A. Bolten & F. Darius (Eds.), *Atlas of Cultural and Environmental Change in Arid Africa*. Cologne: Heinrich-Barth-Institut.
Küchler, S. (1993). Landscape as memory: The mapping of process and its representation in a melanesian society. In B. Bender (Ed.), *Landscape. Politics and Perspectives* (pp. 85–106). Oxford: Berg.
Kuzniar, A. (1988). The vanishing canvas: Notes on German romantic landscape aesthetics. *German Studies Review*, XI, 359–376.
Landau, P. (1998). Hunting with gun and camera: A commentary. In W. Hartmann, J. Silvester & P. Hayes (Eds.), *The Colonising Camera. Photographs in the Making of Namibian History* (pp. 151–155). Cape Town: University Press.
Lane, C. (1994). Pastures lost: Alienation of Barabaig land in the context of land policy and legislation in Tanzania. *Nomadic Peoples*, 34/35, 81–94.
Lauer, W. (1976). Carl Troll – Naturforscher und Geograph. *Erdkunde*, 30, 1–7.
Layton, R. & Titchen, S. (1995). Uluru: An outstanding Australian aboriginal cultural landscape. In B. von Droste, H. Plachter, & M. Rössler (Eds.), *Cultural Landscapes of Universal Value*. (pp. 174–181). New York: Gustav Fischer Verlag Jena.
Lentz, C. (2003). This is Ghanaian territory: Land conflicts on a West African border. *American Ethnologist*, 30(2), 273–89.
Levinson, S. (1996). Language and space. *Annual Reviews in Anthropology*, 25, 353–382.
Lewis-Williams J.D. (1987). Paintings of power: Ethnography and rock art in southern Africa. In M. Biesele, R. Gordon & R. Lee (Eds.), *The Past and Future of !Kung Ethnography. Critical Reflections and Symbolic Perspectives* (pp. 55–87). Essays in Honour of Lorna Marshall. Quellen zur Khoisan-Forschung 4. Hamburg.
Luig, U. & von Oppen, A. (1997). Landscape in Africa. Process and vision: An introductory essay. *Paideuma*, 43, 7–46.
Lyell, C. (1830). *Principles of Geology*. London: Hilliard, Gray & Co.
Lykke, A.M. (2000). Local perceptions of vegetation change and priorities for conservation of woody-savanna vegetation in Senegal. *Journal of Environmental Management*, 59, 107–120.
Manger, L. (1988). Traders, farmers and pastoralists: Economic adaptations and environmental problems in the Southern Nuba Mountains. In D. Johnson & D. M. Anderson (Eds.), *The Ecology of Survival. Case Studies from Northeast African History* (pp. 155–171). Boulder, CO: Westview Press.
Mark, D.M. & Turk, A.G. (2003). Landscape categories in Yindjibarndi: Ontology, environment, and language. In W. Kuhn, M. Waboys & S. Timpf (Eds.), *Spatial Information Theory. Foundations of Geographic Information Science* (pp. 31–49). Berlin: Springer-Verlag.
McCully, P. (1996). *Silenced Rivers. The Ecology and Politics of Large Dams*. London: Zed.
McGovern, T. (1994). Management for extinction in Norse Greenland. In C. Crumley (Ed.), *Historical Ecology. Cultural Knowledge and Changing Landscapes* (pp. 127–154). Santa Fe, NM: School of American Research Press.
Meurers-Balke, J. et al. (1999). Landschafts- und Siedlungsgeschichte des Rheinlandes. In K. H. Knörzer et al. (Eds.), *Pflanzenspuren. Archäobotanik im Rheinland: Agrarlandschaft und Nutzpflanzen im Wandel der Zeiten* (pp. 11–66). Köln: Rheinland-Verlag.
Meyers Lexikon (1939). 8. Edition Leipzig: Bibliographisches Institute AG.
Miescher G. & Rizzo, L. (2000). Popular pictorial constructions of Kaoko in the twentieth century: Continuities and discontinuities. In G. Miescher & D. Henrichsen (Eds.), *New Notes on Kaoko* (pp. 10–47). Basel: Basler Afrika Bibliographien.
Müller, G. (1977). Zur Geschichte des Wortes Landschaft. In A. Wallthor & H. Quirin (Eds.), *Landschaft als interdisziplinäres Forschungsproblem* (pp. 4–12). Münster: Aschendorff.
Ndagala, D. (1994). Pastoral territory and policy debates in Tanzania. *Nomadic Peoples*, 34/35, 23–36.
Neumann, R.P. (1997). Forest rights, privileges and prohibitions. Contextualising state forestry policy in colonial Tanganyika. *Environment and History*, 3, 45–68.
Oba, G. & Cotile, D. (2001). Assessment of landscape level degradation in southern Ethiopia: Pastoralists versus ecologists. *Land Degradation & Development*, 12, 461–475.

Oba, G. & Kaitira, L.M. (2006). Herder knowledge of landscape assessment in arid rangelands in northern Tanzania. *Journal of Arid Environments*, 66, 168–186.

Östberg, W. (1994). The expansion of Marakwet Hill-Furrow irrigation in the Kerio valley of Kenya. In M. Widgren & J. Sutton (Eds.), *Islands of Intensive Agriculture in Eastern Africa* (pp. 19–48). Oxford: James Currey.

Oxford English Dictionary, (1989). *Vol VIII. Interval-Lecie*. Oxford: University Press.

Passarge, S. (1921). *Die Landschaft*. Leipzig: Verlag von Quelle und Meyer.

Pinther, K. (2005). Imaginäre Stadtlandschaften in Ghana. PhD Thesis, University of Cologne.

Pyne, S. (1997). Frontiers of fire. In K.L. Darian-Smith, L. Gunner & S. Nuttall (Eds.), *Text, Theory and Space. Land, Literature and History in South Africa and Australia* (pp. 19–34). London: Routledge.

Rahmato, D. (1984). *Agrarian Reform in Ethiopia*. Uppsala: Scandinavian Institute of African Studies.

Ranger, T. (1997). Making Zimbabwean landscapes. Painters, projectors and priests. In U. Luig & A. von Oppen (Eds.), *The Making of African Landscapes. Paideuma*, 43, 59–74.

Ranger, T. (1999). *Voices from the Rocks. Nature, Culture and History in the Matopos Hills of Zimbabwe*. Oxford: James Currey.

Rao, A. (2000). Blood, milk and mountains: Marriage practice and concepts of predictability among the Bakkarwal of Jammu and Kashmir. In M. Böck. & A. Rao (Eds.), *Culture, Creation, and Procreation. Concepts of Kinship in South Asian Practise* (pp. 101–134). Oxford: Berghahn.

Ratzel, F. (1891). *Anthropogeographie – Die geographische Verbreitung des Menschen*. Stuttgart: Engelhorn.

Reddy, T. (2000). *Hegemony and Resistance. Contesting Identities in South Africa*. Aldershot: Ashgate.

Reid, R. (2000). Land-use and land-cover dynamics in response to changes in climatic biological and socio-political forces: The case of southwestern Ethiopia. *Landscape Ecology*, 15, 339–355.

Rössler, M. (2003). Landkonflikt und politische Räumlichkeit: Die Lokalisierung von Identität und Widerstand in der nationalen Krise Indonesiens. In B. Hauser-Schäublin & M. Dickhardt (Eds), *Kulturelle Räume – räumliche Kultur* (pp. 171–220). Münster: LIT.

Salih, H.M. (1994). Struggle for the delta: Hadendowa conflict over land rights in the Sudan. *Nomadic Peoples*, 34/35, 147–158.

Sauer, C.O. (1925). The morphology of landscapes. *University of California Publications in Geography*, 2, 19–54.

Schama, S. (1995). *Landscape and Memory*. New York: Alfred A. Knopf.

Schlee, G. (1989). *Identities on the Move: Clanship and Pastoralism in Northern Kenya*. Manchester: University Press.

Schmidt, P. (1993). Historical ecology and landscape transformation in Eastern Equatorial Africa. In C. Crumley (Ed.), *Historical Ecology. Cultural Knowledge and Changing Landscapes* (pp. 99–122). Santa Fe, NM: School of American Research Press.

Schmidthüsen, J. & Bobek, H. (1949). Die Landschaft im logischen System der Geographie. *Erdkunde*, 3, 112–140.

Scott, J. (1998). *Seeing Like a State. How Certain Schemes to Improve the Human Condition Have Failed*. New Haven, CT: Yale University Press.

Scudder, T. (2006). *The Future of Large Dams. Dealing with Social, Environmental, Institutional and Political Costs*. London: Earthscan.

Scudder, T. & Colson, E. (1972). The Kariba dam project. Resettlement and local initiative. In R. Bernard & P. Pelto (Eds.), *Technology and Social Change* (pp. 40–69). New York, Macmillan.

Soule, M. & Lease, G. (Eds.) (1995). *Reinventing Nature? Responses to Postmodern Deconstruction*. Washington, DC: Island Press.

Steinhardt, U. (2000). *Mensch und Natur – Gedanken zum Landschaftsbegriff und zum Umgang mit Landschaft*. Thema 4(2), www.tu-cottbus.de/BTU/Fak2/TheoArch/Wolke/deu/Themen/992/ Steinhardt 2000, 1–14.

Straßburg, G. von(1980). *Tristan*. Nach dem Text von Friedrich Ranke neu herausgegeben und ins neuhochdeutsche übersetzt, mit einem Stellenkommentar und einem Nachwort von Rüdiger Krohn. Stuttgart: Reclam.

Strehlow, T.G.H. (1970). Geography and the totemic landscape in central Australia: A functional study. In R.M. Berndt (Ed.), *Australian Aboriginal Anthropology* (pp. 172–189). Perth: University of Western Australia Press.

Sutton, J. (2004). Engaruka. In M. Widgren & J. Sutton (Eds.), *Islands of Intensive Agriculture in Eastern Africa* (pp. 114–132). Oxford: James Currey.

Taçon, P. (1999). Identifying ancient sacred landscapes in Australia: From physical to social. In W. Ashmore & B. Knapp (Eds.), *Archaeologies of Landscape* (pp. 33–57). London: Blackwells.

Thrift, N. (2006). Space. *Theory, Culture & Society*, 23, 139–155.

Tiffen, M., Mortimore, M. & Gichuki, F. (1994). *More People, Less Erosion: Environmental Recovery in Kenya*. London: Wiley.

Tilley, C. (1994). *A Phenomenology of Landscape. Places, Paths and Monuments*. Oxford: Berg.

Troll, C. (1947). Die Geographische Wissenschaft in Deutschland in den Jahren 1933 bis 1945. *Erdkunde*, 1, 3–48.

Troll, C. (1950). Die geographische Landschaft und ihre Erforschung. *Studium Generale*, 3, 163–181.

Troll, C. (1970). Landschaftsökologie (Geoecology) und Biogeoeconologie. Eine terminologische Studie. *Revue Roumaine de Géologie. Série de Géographie*, 14, 9–18.

Turner, F.J. (1920). *The Frontier in American History*. New York: Henry Holt .

Urban, D. & Keitt, T. (2001). Landscape connectivity: A graph-theoretic perspective. *Ecology*, 82, 1205–1218.

Watson, E.E. & Schlee, G. (Eds.) (2004). *Changing Identifications and Alliances in North-East Africa*. London: James Currey.

Watts, M.J. (1988). Coping with the market. Uncertainty and food security among Hausa peasants. In I. de Garine & G. Harrison (Eds.), *Coping with Uncertainty in Food Supply* (pp. 260–289). Oxford: University Press.

Whande, W & Suich, H. (forthcoming). Transfrontier conservation initiatives in southern Africa– Observations from the Great Limpopo Transfrontier Conservation Area. In Suich, H & Child, B with Anna Spenceley (Eds.), *Evolution and Innovation in Wildlife Conservation: Parks and Game Ranches to Transfrontier Conservation Areas*. London: Earthscan.

Widgren, M. & Sutton, J. (2004). *Islands of Intensive Agriculture in Eastern Africa*. Oxford: James Currey.

Widlok, T. (1998). Unearthing culture. Khoisan funerals and social change. *Anthropos*, 93, 115–126.

Winterhalder, B. (1994). Concepts in historical ecology. The view from evolutionary ecology. In C. Crumley (Ed.), *Historical Ecology* (pp. 17–41). Santa Fe, NM: School of American Research Press.

Zimmermann, A. (2004). *Landschaftsarchäologie II. Überlegungen zu Prinzipien einer Landschaftsarchäologie*. Mainz: Römisch Germanische Kommission.

Part I

Arid Landscapes: Detection and Reconstruction – Perspectives from Earth Sciences and Archaeology

Chapter 1

Landscape in Geography and Landscape Ecology, Landscape Specification, and Classification in Geomorphology

OLAF BUBENZER

At first, mainly with respect to the German literature but with comparisons to some recent Anglophone papers, an excursive overview of the development and the meaning of the term 'landscape' in geography and landscape ecology is given. This is followed by a description of how the surface of a given landscape can be specified and classified within geomorphology as a basis for further investigations in the natural sciences and humanities.

1.1. INTRODUCTION

The term 'landscape' has been used for more than 150 years within the multidisciplinary geoscientific and bioscientific terminology. Alexander von Humboldt (1845/47) dealt with it and defined landscape as 'the total character (or impression) of a region.' Since this period until the present scientists have been engaged in questions such as: is it possible to describe and classify a landscape objectively? Which factors determine a landscape?

M. Bollig, O. Bubenzer (eds.), *African Landscapes*,
doi: 10.1007/978-0-387-78682-7_1, © Springer Science + Business Media, LLC 2009

Geography with its integrative-holistic approach allows an interdisciplinary analysis. The earth's surface with its relief is fundamental for many processes in the geosystem[1] which form and change the landscape. Therefore, this contribution presents the terrain analysis and the terrain classification in an exemplary fashion.

1.2. LANDSCAPE IN GEOGRAPHY

'Landscape' in general and regional geography[2] is a central but not undisputed term (Klink, 2002). Etymologically the term dates from the Old High German language [*lantscaft*, eighth century] and means 'part of a country' or 'region' [from Latin *regio*] (see Bollig, this volume). Therefore it indicates a geographical coherent area with a definite character (Steinhardt, 2000).

Geographers who study landscape history determine the 'landscape' as their real field of research (see below), whereas other scientists disapprove of the holistic term because of its inaccuracy. In Germany, Hard (1970) criticises the transfer of the term from colloquial language into scientific language. Against the background of new integral observational and analytical methods such as remote sensing and geographical information systems (GIS) Widacki (1994) demands a renunciation of the 'landscape and geocomplex paradigma.'

Within applied geography, the landscape represents the highest integration level of the geographical space and includes all abiotic, biotic, and anthropogenic (technical) components. In this sense landscape can be defined as a segment of the geosphere of any size but which is determined by a homogeneous structure and an interaction system of its components (interaction of substances and processes) (Klink, 2002).

1.2.1. Evolution

Early scientific definitions of 'landscape' came from Middle and East Europe and were developed both in geography and biology (Bastian, 2001a). If we follow the reviews of Naveh (1995), Haase (1996), Leser (1997), Steinhardt (2000), Barsch et al. (2000), and Bastian (2001a, b) it is possible to outline the following evolution.

At the beginning the previously mentioned complex and holistic view of 'landscape' prevailed (e.g., 'landscape as the total character of a region,' Alexander von Humboldt, 1845/47). After this, a period of subdivision and analysis of the

---

[1] Geosystem as the spatial–temporal interrelation of its factors 'climate, rock, groundwater and soil water, soil, relief, animal, vegetation and human'.

[2] The general geography aims in toward the deduction of rules (nomothetical). The regional geography (*"Länderkunde"*) describes and explains unique phenomena which are penetrably penetrable only ideographically.

geofactors followed according to the scale of investigation (Passarge, 1919–1921, 1933; Troll, 1939; Bobek & Schmithüsen, 1949; Lautensach, 1953; Neef, 1963, 1967). For example, Neef (1967) understands landscape to be a segment of the earth's surface with a homogeneous structure and an interrelation of its compartments. Therefore, this development provided a basis for applied questions with respect to landscape evaluation and landscape diagnosis (Haase, 1978; Mannsfeld, 1983; Haase & Richter, 1980; Marks et al., 1989).

This period of increasing quantification of central landscape elements entered a second phase at the end of the 1980 with the revolutionary development in computer hardware and software that allowed the input, management, analysis, and presentation of ever-increasing amounts of data with spatial reference within geographical information systems (Duttmann, 1999). A secondary effect was a certain renaissance of the 'Länderkundliches Schema' (Hettner, 1932; Wardenga, 2001) which made possible the quantification and networking of separate factors such as rock, terrain, climate, hydrography, flora, fauna, soil, human, and economics, as well as politics in the form of singular layers in a geographical information system (GIS).

During the last 10 to15 years, especially in connection with the Global Change Program, humans are no longer viewed only in their physical-material existence (as part of the biosphere), but more as thinking and acting beings in the landscape. This has considerably broadened the previous opinion that landscape is a product of the environment and land use (Tress & Tress, 2000). Therefore, landscape is defined as a holistic term, as a spatial and mental entity that is a result of the interaction of the subsystem's geosphere (lithosphere and hydrosphere), biosphere, and anthroposphere/noosphere (Steinhardt, 2000; Bastian, 2001a). Such a definition combines a multitude of scientific, social, aesthetic, and psychological aspects in a 'theory of the landscape in a broader sense' and abolishes the contradiction between the natural sciences and the humanities in the philosophical sense (Bastian, 2001a, p. 46). One effect of this development is the derivation and application of terms such as 'resilience', 'sensitivity', 'vulnerability', and 'sustainability'(Stocking & Munaghan, 2001).

A fundamental aid to solving the transformation problems between the environment and the society (and vice versa) is the definition of landscape functions and landscape potentials (see also the contribution of Bolten et al., this volume). Landscape functions means not only fluxes of energy, nutrients, or species between the landscape elements or patch–matrix interactions (Forman, 1981), but capacities of the landscape for human societies (compare Bastian & Röder, 1996). To fulfil such demands, real interdisciplinary studies are required, which combine different academic disciplines with contrasting paradigms, so that disciplinary boundaries must be crossed in order to obtain new knowledge (Tress et al., 2003). A further goal is to overcome the differences between qualitative and quantitative approaches in order to bring together social and natural sciences. With regard to the ACACIA project this means that interdisciplinary research was done during all working steps, from the discussion of adequate methods, during the combined field research, the data analysis, and the publications.

1.3. 'LANDSCAPE' IN LANDSCAPE ECOLOGY

The development of the discipline landscape ecology from its roots to the present is reviewed by Klink et al. (2002). Schmithüsen (1976) determines landscape as 'synergy'. Thus one landscape can be differentiated from another according to its structure and its acting processes. A landscape area is defined as an isomorphic segment of the geosphere because of its whole content (total character in the sense of Humboldt) (Schmithüsen, 1976). In the landscape ecology the term 'landscape ecosystem' replaces the holistic term 'landscape' (Leser, 1997). Following Bastian (2001b), outstanding features of landscape ecology are structures, processes, and changes; spatial and hierarchical aspects; and the complexity of different factors in a landscape. Regarding the landscape as a complex structure of functions needs the inclusion of abiotic patterns and processes (Löffler & Steinhardt, 2007). Similarly to 'areas of unspoiled nature (*Naturraum*)'[3] landscape can also be defined as a limited spatial interaction system that can be represented in different dimensions (Klink, 2002).

The main aspects of the geosynergetic–ecological interaction system within the landscape ecology are structure, function, and dynamic (Haase, 1996). The structure is the spatial mosaic of the landscape that is defined by its components (composition) as well as by its shape and order (configuration). The function characterizes the active processes within the landscape, that is, in this context it is not the utilizability of the landscape, but rather the recent fluxes of energy, matter, and organisms. Temporal changes in the landscape can be ascertained as a result of this dynamic (Syrbe, 2002).

The characterisation of the landscape with the terms 'structure', 'function', and 'change' is also used in the Anglo-American literature (Forman & Godron, 1986; Turner & Gardner, 1991; see also the contribution of Bolten et al., this volume). In North America and Europe the discipline of landscape ecology has been mainly developed more from ecosystem research. The use of computers, remote sensing, GIS, and simulation models plays an important role. Thus, effective quantitative methods allow the investigation of pattern–process relations because these techniques allow broad-scale studies and a much easier measurement of landscape structure variables (Fahrig, 2005).

The structure of a given landscape is the result of the former and the recent formative processes. On the other hand, the spatial distribution and the strength of the processes were, and are, driven by the landscape structure. For actual investigations it is only possible to observe the recent functionality (Syrbe, 2002). Former geoecological relations are solely recognisable in relicts of the landscape structure (Haase, 1976). Therefore, a detailed structural analysis enables the derivation of concrete results of actual processes as well as retrospective views and prognosis for the future. Structural and processural approaches are

---

[3] An area of unspoiled nature (*'Naturraum'*") is defined as a horizontal or vertical space–time structure of the natural geofactors and their interrelation.

complementary because without real process observations, derivations from the landscape structure are hypothetical and only with the help of structural knowledge is it possible to make statements and a future prognosis of wider complex landscape units (Syrbe, 2002). Within the last few years landscape ecology operates more and more in a holistic manner with respect to abiotic, biotic, and anthropogenic structures and processes to solve problems of the human–nature complex. Based on the fact that landscape exists along a continuum of scales (e.g., Turner et al., 2001), the functionality of landscape systems can only be successfully understood if both small-scaled processes and large-scaled processes are considered (Löffler & Steinhardt, 2007 after Müller & Fath, 1998). To exemplify this for one abiotic landscape factor the following chapter describes the specification and classification of the earth's surface within geomorphology.

1.4. SPECIFICATION AND CLASSIFICATION OF LANDSCAPES IN GEOMORPHOLOGY

1.4.1. Definition and Approaches

Geomorphology [from Old Greek *gé*, earth; *morphé*, shape; and *logos*, word, knowledge] is the science of the shape of the solid earth surface, its developing factors and processes. Therefore, it is the science of the georelief of the earth's crust (Brunotte, 2002). The georelief builds an important geofactor as a boundary layer between the lithosphere and the pedosphere as well as between the atmosphere and hydrosphere and as a 'basis' for the biosphere. It can be defined as the surface of the subsoil body (Kugler, 1974) and thus as a spatial continuum whose appearance is the product of the georelief development, that is, the effects of the geomorphic processes of the past and the present (Dikau & Schmidt, 1999).

According to Kugler (1964) and Leser (1977, 1993) the geomorphic inspection of the terrain is possible with respect to its characteristics (spatially and habitually), to its building material (substantially), and/or to its genesis (historic-genetically, dynamic-prognostically).

This subdivision indicates the following main approaches of geomorphology.

- Description of forms (see Chapter 4.2) with qualitative (geomorphography) and quantitative methods (geomorphometry)
- Explanation of the general genesis of the forms (geomorphogenesis)
- Resulting geomorphic processes (geomorphodynamic) which are generated by the media involved (ice, snow, water, air, or biotic activities)
- Dating of the forms
- Their stratigraphical classification (geomorphochronology)

According to Dikau (1996) two approaches exist for a fundamental understanding of form–process relations:

1.   Landforms are deduced according to rules and assumptions which are based on physical laws. The background is the functional effect of formative processes which act in special circumstances. Several single studies with respect to biogenic, aeolian, fluvial marin, limnic, glacial, nival, gravitative, cryogenic, cosmogenic, and technogenic/anthropogenic processes exist (literature survey, e.g., in Summerfield, 1991; Ahnert, 1996; Press & Siever, 2003). Additionally, it is possible to subdivide thresholds as well as positive and negative feedback (e.g., Brunsden, 1996; Thomas & Simpson, 2001; Bubenzer, 2001).

2.   The surface as the result of its genesis, that is, the effect of the formative processes which have been effective over a longer period (e.g., Tricart, 1965; Büdel, 1981; Choreley et al., 1984; Bremer, 1989). Thereby the formative process is deduced from the landform itself and is abducted due to actual analogous conclusions and recent geomorphometrical, lithostratigraphical, and pedological observations (Dikau, 1996, p. 17).

In summary it can be said that landforms are the result of one or several formative phases (single phase or multiphase) as well as one or several formation processes (monogenetic or polygenetic). Considering the worldwide occurrence of definite landforms with their mean sizes and lifetimes, some dependencies are obvious (Figure 1.1). Larger landforms exist in general over longer durations than smaller ones. Small landforms are more frequent, are superimposed on older and larger ones, and are functionally connected with them (Dikau, 1989).

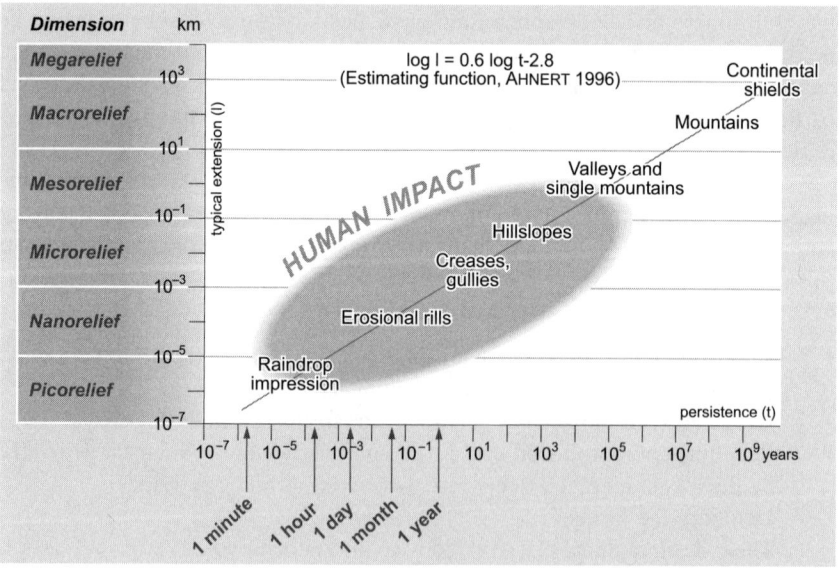

Figure 1.1. Relation between size and persistence of landforms as well as the approximate human impact (Changed after Dikau, 1988 and Ahnert, 1996)

The importance of georelief within the environment was already recognized by von Richthofen (1877). For him the earth's surface is a central research object and he defines geography as the science of the earth's surface and its casually related materials and appearances. Penck (1894) is regarded as the founder of terrain analysis. Milne (1935, catena-principle) and Jenny (1941, toposequence)[4] stress the close relationship between the geofactors soil and relief. Leser (e.g., 1997) emphasizes the control function of georelief for the dynamic of processes in the interrelation of biotic and abiotic factors. According to Rohdenburg (1989) georelief influences the solar net radiation (mainly through aspect and slope) and the redistribution of water and nutrients, and therefore it directly and indirectly dominates the landscape structure.

Regional geomorphology takes an inventory of the whole terrain forms of a given part of the earth's surface, which is derived from the natural division, the drainage system, or – with respect to applied and regional geography – from the political territories, and reconstructs the landscape history. In contrast to this, general geomorphology pursues the laws of relief development under different aspects according to geomorphogenetic factors, for example, rock and tectonic, climate, gravitation, and general and relative height ranges. Therefore, several subdisciplines have arisen, such as structural geomorphology, climatic geomorphology, and applied geomorphology. With regard to the active processes, the geographical position, and the altitudinal belt or the climatic zone they can be further subdivided into glacial and periglacial geomorphology, coastal and high mountain geomorphology, arid and tropic geomorphology, and so on. In Europe geomorphology is institutionalised in geography; in northern America it is anchored in geology (Brunotte, 2002).

1.4.2. Georelief Specification and Classification

Firstly, geomorphology sees georelief as an association of terrain units with different geometry, topology, structure, and size, which are founded on the subsoil. A detailed ubiquitary applicable description and classification of the terrain is a fundamental precondition for the understanding of any form–process relation; this is provided by the geomorphography and the georeliefclassification (or terrain analysis).

1.4.2.1. Geomorphography

Geomorphography is the science of the qualitative and quantitative specification and analysis of the genetic–topological attributes of georelief (Dikau & Schmidt, 1999). According to this definition it also contains geomorphometry. With the aid of this method landforms can be clearly classified, former and recent processes can be estimated, and future processes forecast.

---

[4]Catenae and toposequences are relief sequences with characteristic soil profiles which are considerably affected by the geomorphodynamic processes (erosion, denudation, transport, grading, and aggradation).

For the spatial terrain description, position (geographic coordinates) and height (above sea level) are primary information (Bork & Dalchow, 2000). Often more important than these are the relative (topological) neighbouring connections of individual relief elements to each other or within a superordinate terrain unit, for example, the definition of a central, marginal, boundary, culmination, ridge, anticlinal, flat, deep, upper slope, middle slope, lower slope, or a pediment position. If a digital elevation model is available, it is possible to calculate the height (Z-value) for each given surface coordinate (X-, Y-values). By means of descriptive statistical methods, further important values such as relief energy (maximum height range in a regular grid), height ranges according to watersheds or drainage lines, as well as catchment areas and flow accumulations are calculable (see Bolten et al., this volume).

The basis dimensions for a quantitative terrain specification are slope, aspect, and curvature (Figure 1.2). The slope angle (measured in degrees or percent) is very important for both the analysis and the prognosis of geomorphic and pedological processes. For example, the slope influences decisive gravitative,

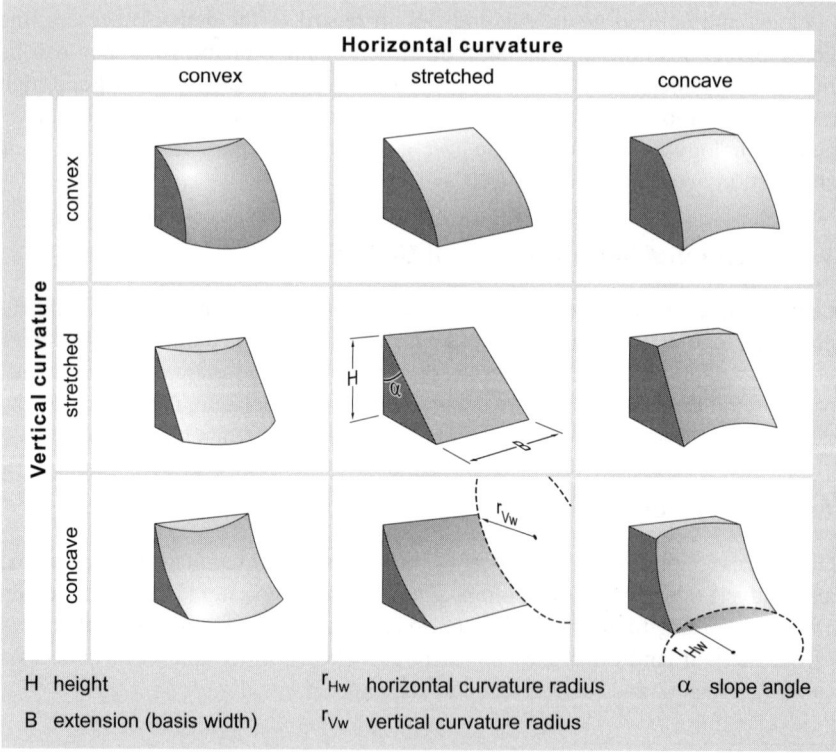

| | H | height | ^rHw | horizontal curvature radius | α | slope angle |
| | B | extension (basis width) | ^rVw | vertical curvature radius | | |

Figure 1.2. Direction, type, strength of curvature, slope, extension, and height of landform elements (Adapted from Dikau, 1988)

fluvial, and gravitative–cryogenic processes such as landslides, solifluction, and soil erosion. Moreover, the slope plays an important rule for climatological questions, for example, the calculation of flows of cold air or the land use (boundary gradients for the employment of machines, etc.).

The aspect is determined by the slope direction and is measured in degrees with respect to the compass rose. It is important for climatic circumstances (radiation input, sunny or shady side, upwind or downwind position) which again influence geomorphologic processes.

The curvature of the terrain is divided into two main directions, the vertical curvature (in the direction of the slope perpendicular to the contour lines) and the horizontal curvature (parallel to the contour lines). Its dimension is described with theoretical circles (in Figure 1.2 shown as ellipses due to the perspective distortion), which are lined up either vertically or horizontally against the surface of the measured terrain. The vertical curvature shows the change of the slope, and the horizontal curvature the change of the aspect. The possible types of curvature are convex (positive radius values), concave (negative radius values), or stretched (infinite radius values).

1.4.2.2. Georelief Classification

The classification of the terrain is a geomorphographical method which aims to specify the spatial homogeneity due to a regular arrangement of the terrain attributes (Dikau & Schmidt, 1999). The basic assumption is the empirically gained and absolutely comprehensible finding that structured components prevail on all spatial levels. This means that formative and changing processes rather lead to structured forms of the boundary surface layer than to chaotic patterns (Dikau, 1996). Examples are the worldwide occurrence of distinct landforms such as barchans, v-shaped valleys, escarpments, beach terraces, and the like.

A review of geomorphometric classification models is given by Dikau & Schmidt (1999). One polyhierarchical model which was developed by Kugler (1964, 1974) on the basis of approaches by Penck (1894), Passarge (1933), and Spiridonov (1973) and which was further developed by Barsch (1969, 1976) and Dikau (1988) is presented here. The advantage of this structural taxonomical model is that it allows the division of all landforms into simple landform elements and facets with the result of a complete geometrically correct illustration of reality (Figure 1.3). It allows the integration of homogeneous geomorphic elements within higher hierarchical levels. Thus the partial units have both different complexity and different dimensions. With the aid of an edged ridge, Figure 1.3 explains the landform units of this model in a hierarchical order.

Landform facets (relief facets). Landform facets are the simplest and smallest units which are homogeneous in slope, aspect, and curvature (maximum variations of 1°). From the ecological viewpoint they can be treated as homogeneous sites. In low mountain ranges, for example, homogeneous inclined and exposed slope segments are landform facets.

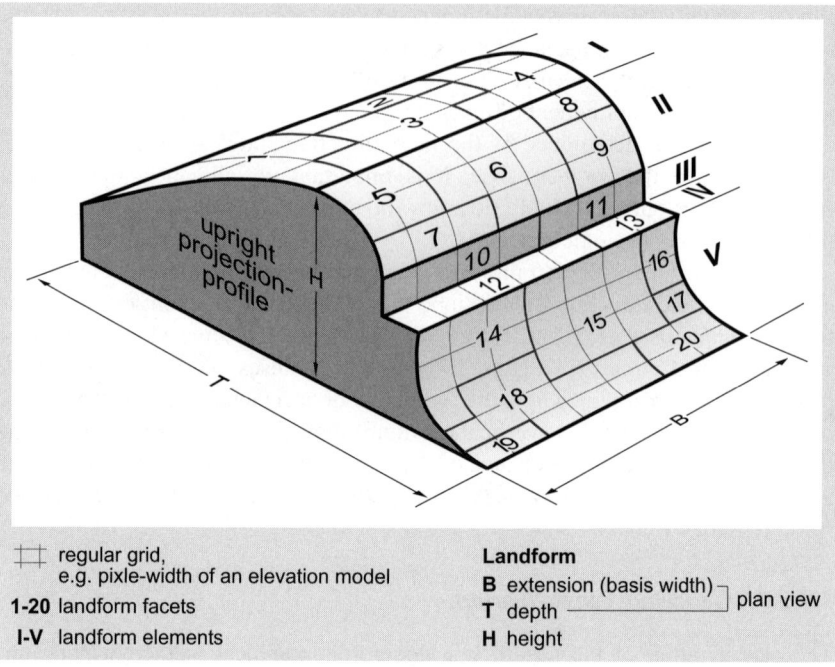

Figure 1.3. Box model (schematic) of structural taxonomical elements as basis for the georelief classification

  Landform elements (relief elements). Landform elements have uniform curvatures and are composed solely of landform facets. The dynamic of the runoff and the soil water, the soil substrate, the pedogenesis, and the vegetation as well as the local climate vary particularly on the level of the landform elements (Bork & Dalchow, 2000). They are described especially by the curvature and are restricted by so-called curvature change lines. Slopes and edges (see also Figure 1.2) are typical landform elements.
  Landforms (relief forms). Landforms are homogeneous with regard to their figure type (upright projection and plan view). Individual positive forms (mountains, hills, ridges, etc.) or negative forms (valleys, basins, calderas, etc.) are typical examples.
  Landform associations (relief associations). Landform associations consist of landforms, landform elements, and landform facets. For example, a low mountain range with its mountains, valleys, slopes, and plains is a complex landform association. Landscape units such as the Alps in Europe or the Appalachian mountains or Death Valley in the United States of America are examples of this. Characteristic landform associations can also be seen in the working areas of ACACIA, for example, the Kaokoveld in Namibia or the Great Sand Sea in Egypt (see other contributions in this volume).

1.4.2.3. Model Application

It is possible to specify, measure, and classify the georelief directly in the field as well as indirectly with the usage of digital elevation models which can be processed from stereoscopic aerial and satellite image data (see contribution of Bolten et al., this volume). The chosen methodology is defined by the scale of investigation and the complexity of the landscape (e.g., lowland, low mountain range, high mountain). Within the scope of the priority program 'Detailed geomorphic mapping' of the German Research Foundation (Leser & Stäblein, 1978; Brunotte et al., 1980; Hagedorn & Lehmeier, 1983; Spönemann & Lehmeier, 1989) different slope classes were developed for different scales and landscapes (see also Dikau et al., 1999). With regard to the georelief classification it can be deduced that, in general, within larger scales (up to 1:25,000) landform elements and smaller landforms are represented whereas smaller scales require the mapping of higher taxonomical units.

Landform facets and landform elements can be specified and differentiated automatically with powerful calculation algorithms in computer programs, already partly integrated as modules in geographical information systems. Also the calculation of derived values such as relief energy, slope gradient classes, and so on are possible. In contrast to this, considerable research and development is required for an objective and reproducible automatic differentiation of complex landforms and landform associations (Dikau et al., 1999). At present the best results are obtained from a terrain analysis which combines field mapping and the evaluation of digital elevation models which have been produced from remote sensing data (stereoscopic aerial, satellite, or laserscan data).

If the relief of a landscape is specified and classified in the described manner an objective and comparable foundation is given for both further natural and social studies.

1.5. CONCLUSION AND OUTLOOK

Geography offers a methodological spectrum that is suitable for a synopsis of different data with a landscape reference. Especially geomorphology, with its approaches, has techniques which make an objective comparison of terrain and the understanding of form–process relations as a basis for multidisciplinary investigations possible. All given landforms can be taken apart into the smallest units and therefore they can be specified and classified.

Future areas within geomorphology are the exact definition and quantification of interrelations between geomorphometric parameters on a physical basis as well as their automation with the use of computers. Further considerable progress can be expected from the areawide terrain analysis with digital elevation models on the basis of remote sensing data in geographical information systems or special process models.

However, the method of the geomorphographic–geomorphometric georelief classification has not yet sufficient standards (Dikau & Schmidt, 1999, p. 235), especially for complex landforms and landform associations. Multiscale dynamic process modelling of space–time systems (4D-modelling) is still in its infancy.

In the year 1993, Lousma remarked that an integral compilation and representation of information – the 'Geographic Information Science' – is a crucial challenge for the 21st century in order to understand our complex living space and the environment. After more than ten years this statement can be confirmed. Geography (including geomorphology and landscape history) as well as landscape ecology meet this challenge and are able to contribute to the decoding of the complex human–environment interrelation.

REFERENCES

Ahnert, F. (1996). *Einführung in die Geomorphologie.* Stuttgart: Eugen Ulmer.

Barsch, D. (1969). Studien zur Geomorphogenese des Zentralen Berner Jura. *Basler Beiträge zur Geographie*, 9.

Barsch, D. (1976). Das GMK-Schwerpunktprogramm der DFG: Geomorphologische Detailkartierung in der Bundesrepublik Deutschland. *Zeitschrift für Geomorphologie, N.F.*, 20, 488–498.

Barsch, H., Billwitz, L., & Bork, H.-R. (Eds.) (2000). *Arbeitsmethoden in Physiogeographie und Geoökologie.* Gotha, Stuttgart: Klett.

Bastian, O. (2001a). Landschaftsökologie – auf dem Wege zu einer einheitlichen Wissenschaftsdisziplin? *Naturschutz und Landschaftsplanung*, 33(2/3), 41–51.

Bastian, O. (2001b). Landscape ecology – Towards a unified discipline? *Landscape Ecology*, 16, 757–766.

Bastian, O. & Röder, M. (1996). Beurteilung von Landschaftsveränderungen anhand von Landschaftsfunktionen. *Naturschutz und Landschaftsplanung*, 28, 302–312.

Bobek, H. & Schmithüsen, J. (1949). Die Landschaft im logischen System der Geographie. *Erdkunde*, 3, 112–120.

Bork, H.-R. & Dalchow, C. (2000). Reliefaufnahme. In H. Barsch, L. Billwitz, & H.-R. Bork (Eds.), *Arbeitsweisen in Physiogeographie und Geoökologie* (pp. 143–172). Gotha, Stuttgart: Klett.

Bremer, H. (1989). *Allgemeine Geomorphologie.* Berlin: Borntraeger.

Brunotte, E. (2002). Geomorphologie. In E. Brunotte, H. Gebhardt, M. Meurer, P. Meusburger, & J. Nipper (Eds.), *Lexikon der Geographie, II* (pp. 27–28). Heidelberg: Spektrum Akademischer Verlag.

Brunotte, E., Garleff, K., & Wahle, H. (1980). Neue morphographische und geomorphologische Karten aus dem südniedersächsischen Bergland. *Neues Archiv für Niedersachsen*, 29, 85–86.

Brunsden, D. (1996). Geomorphological events and landform change. *Zeitschrift für Geomorphologie. N. F.*, 40, 289–303.

Bubenzer, O. (2001). Fluviale Systeme: Steuerungsfaktoren, Reaktionszeiten, Schwellenwerte, Rückkopplungen. In E. Brunotte, H. Gebhardt, M. Meurer, P. Meusburger, & J. Nipper (Eds.), *Lexikon der Geographie, I* (pp. 395–398). Heidelberg: Spektrum Akademischer Verlag.

Büdel, J. (1981). *Klima-Geomorphologie.* Berlin: Borntraeger.

Choreley, R.J., Schumm, S.A., & Sudgen, D.E. (1984). *Geomorphology.* London: Routledge.

Dikau, R. (1988). Entwurf einer geomorphographisch-analytischen Systematik von Reliefeinheiten. *Heidelberger Geographische Bausteine*, 5.

Dikau, R. (1989). Application of a digital relief model to landform analysis in geomorphology. In J. Raper (Ed.), *Three Dimensional Application in Geographic Information Systems* (pp. 51–77). London: Taylor & Francis.

Dikau, R. (1996). Geomorphologische Reliefklassifikation und -analyse. *Heidelberger Geographische Arbeiten*, 104, 15–23.

Dikau, R. & Schmidt, J. (1999). Georeliefklassifikation. In R. Schneider-Sliwa, D. Schaub, & D. Gerold (Eds.), *Angewandte Landschaftsökologie. Grundlagen und Methoden* (pp. 217–244). Berlin: Springer.

Dikau, R., Friedrich, K., & Leser, H. (1999). Die Aufnahme und Erfassung landschaftsökologischer Daten. In H. Zepp & M.J. Müller (Eds.), *Landschaftsökologische Erfassungsstandards*. Forschungen zur deutschen Landeskunde, 244 (pp. 29–167). Flensburg.

Duttmann, R. (1999). Geographische Informationssysteme (GIS) und raumbezogene Prozessmodellierung in der Angewandten Landschaftsökologie. In R. Schneider-Sliwa, D. Schaub, & D. Gerold (Eds.), *Angewandte Landschaftsökologie. Grundlagen und Methoden* (pp. 181–199). Berlin: Springer.

Fahrig, L. (2005). When is landscape perspective important? In A. Wiens & R. Moss (Eds.), *Issues and Perspectives in Landscape Ecology* (pp. 3–10). Cambridge: Cambridge University Press.

Forman, R.T.T. (1981). Interaction among landscape elements: A core of landscape ecology. *Proc. Int. Congr. Neth. Soc. Landscape Ecol* (pp. 35–48). Veldhoven, Pudoc: Wageningen.

Forman, R.T.T. & Godron, M. (1986). *Landscape Ecology*. Wiley: New York.

Haase, G. (1976). Die Arealstruktur chorischer Naturräume. *Petermanns geographische Mitteilungen*, 120(2), 130–135.

Haase, G. (1978). Zur Ableitung und Kennzeichnung von Naturpotentialen. *Petermanns Geographische Mitteilungen*, 122(2), 113–125.

Haase, G. (1996). Geotopologie und Geochorologie – Die Leipzig-Dresdener Schule der Landschaftsükologie. In G. Haase & E. Eichler (Eds.), *Wege und Fortschritte der Wissenschaft; Sächsische Akademie der Wissenschaften zu Leipzig* (pp. 201–229). Berlin: Akademie Verlag GmbH.

Haase, G. & Richter, H. (1980). *Geographische Landschaftsforschung als Beitrag zur Lüsung von Landeskultur- und Umweltproblemen.* Sitz.ber. Akad. d. Wiss. DDR, Math.-Nat.- Technik, 5 N.

Hagedorn, J. & Lehmeier, F. (1983). Zur Konzeption der Geomorphologischen Karte 1:25.000 (GMK 25) aufgrund von Kartierungserfahrungen im Niedersächsischen Bergland. *Forschung zur deutschen Landeskunde*, 220, 63–81.

Hard, G. (1970). Die 'Landschaft' der Sprache und die 'Landschaft' der Geographen. *Colloquium Geographicum*, 11, Bonn.

Hettner, A. (1932). Das länderkundliche Schema. *Geographischer Anzeiger*, 33, 1–6.

Humboldt, A.v. (1845/47). *Kosmos, Entwurf einer physischen Weltbeschreibung.* J. Cotta. Stuttgart 1845/47. Reprint 2006, Eichborn.

Jenny, H. (1941). *Factors of Soil Formation*. Reprint 1994. New York: Dover.

Klink, H.-J. (2002). Landschaft. In E. Brunotte, H. Gebhardt, M. Meurer, P. Meusburger, & J. Nipper (Eds.), *Lexikon der Geographie, II* (pp. 304–305). Heidelberg: Spektrum Akademischer Verlag.

Klink, H.-J., Potschkin, M., Tress, B., Tress, G., Volk, M., & Steinhardt, U. (2002). Landscape and landscape ecology. In O. Bastian & U. Steinhardt (Eds.), *Development and Perspectives of Landscape Ecology* (pp. 1–47). Dordrecht/Boston /London: Springer Netherland.

Kugler, H. (1964). Die geomorphologische Reliefanalyse als Grundlage großmaßstäbiger geomorphologischer Kartierung. *Wiss. Veröff. d. dt. Inst. f. Länderkunde, N.F.*, 21/22, 541–655.

Kugler, H. (1974). *Das Georelief und seine kartographische Modellierung.* Diss. B, Martin-Luther-Universität. Halle/Wittenberg.

Lautensach, H. (1953). Der geographische Formenwandel. Studien zur Landschaftssystematik. *Colloquium Geographicum*, 3.

Leser, H. (1977). *Feld- und Labormethoden der Geomorphologie*. Berlin: Ulmer.

Leser, H. (1993). *Geomorphologie*. Braunschweig: Westermann.

Leser, H. (1997). *Landschaftsökologie. Ansatz, Modelle, Methodik, Anwendung*. 4. Aufl. Stuttgart: Ulmer.

Leser, H. & Stäblein, G. (1978). Legende der Geomorphologischen Karte 1:25.000 (GMK 25). 3. Fassung im DFG-Schwerpunktprogramm. *Berliner Geographische Abhandlungen*, 30, 79–90.

Löffler, J. & Steinhardt, U. (2007). Fundamentals, paradigms and visions in landscape ecology. In J. Löffler & U. Steinhardt (Eds.), *Landscape Ecology. Colloquium Geographicum 28*. pp. 1–10 Bonn.

Lousma, J.R. (1993). Rising to the challenge: The role of information systems. *Photogrammetric Engineering & Remote Sensing*, 59(6), 957–959.

Mannsfeld, K. (1983). Landschaftsanalyse und Ableitung von Naturraumpotentialen. *Abhandlungen der Sächsischen Akadademie der Wissenschaften zu Leipzig, Math.-Nat. Klasse*, 55(3).

Marks, R., Müller, M.J, Leser, H., & Klink, H.-J. (1989). Anleitung zur Bewertung des Leistungsvermögens des Landschaftshaushaltes (BA LVL). *Forschung zur deutschen Landeskunde*, 229.

Milne, G. (1935). Some suggested units of classification and mapping particularly for East African Soils. *Soil Research*, 4(3), 183–198.

Müller, F. & Fath, B. (1998). The physical basis of ecological goal functions – An integrative discussion. In F. Müller & M. Leupelt (Eds.), *Eco Targets, Goal Functions, and Orientors* (pp. 269–285). New York/Berlin: Springer.

Naveh, Z. (1995). Interactions of landscapes and cultures. *Landscape and Urban Planning*, 32, 43–54.

Neef, E. (1963). Dimensionen geographischer Betrachtungen. *Forschungen und Fortschritte*, 37, 361–363.

Neef, E. (1967). *Die theoretischen Grundlagen der Landschaftslehre*. Gotha, Leipzig: Haack VEB.

Passarge, S. (1919–1921). *Die Grundlagen der Landschaftskunde. Ein Lehrbuch und eine Anleitung zu landschaftskundlicher Forschung und Darstellung*. 3 Bde. Hamburg: Haack VEB.

Passarge, S. (1933). *Einführung in die Landschaftskunde*. Leipzig: Haack VEB.

Penck, A. (1894). *Morphologie der Erdoberfläche*. Stuttgart: Engelhorn.

Press, F. & Siever, R. (2003). *Allgemeine Geologie*. Heidelberg/Berlin/Oxford: Spektrum Akademischer Verlag.

Richthofen, F.v. (1877). Die heutigen Aufgaben der wissenschaftlichen Geographie. In F.v Richthofen, *China*, 1, 729–733.

Rohdenburg, H. (1989). *Landschaftsökologie – Geomorphologie*. Cremlingen-Destedt: Catena.

Schmithüsen, J. (1976). *Allgemeine Geosynergetik. Grundlagen der Landschaftskunde*. Berlin/New York: Gruyter.

Spiridonov, A.I. (1973). Physiognomic landscape features as indicators of origin and development of the landscape. In A.G. Chikishev (Ed.), *Landscape Indicators*. New York/London: Plenum.

Sponemann, J. & Lehmeier, F. (1989). Geomorphologische Kartographie in der Bundesrepublik Deutschland: Normierung und Weiterentwicklung. *Erdkunde*, 43(2), 77–85.

Steinhardt, U. (2000). Mensch und Natur – Gedanken zum Landschaftsbegriff und zum Umgang mit Landschaft. *Wolkenkuckucksheim. Internationale Zeitschrift Theorie und Wissenschaft der Architektur* Jg.5, H.1 (http://www.theo.tu-cottbus.de/Wolke/deu/Themen/001/Steinhardt/steinhardt.html).

Stocking, M. & Munaghan, N. (2001). *Handbook for the Field Assessment of Land Degradation*. London: Earth Publications.

Summerfield, M.A. (1991). *Global Geomorphology*. New York: Prentice-Hall.

Syrbe, R.-U. (2002). Entwicklung von Struktur und Funktionsweise in der Landschaft. In G. Haase & K. Mannsfeld (Eds.), *Naturraumeinheiten, Landschaftsfunktionen und Leitbilder am Beispiel von Sachsen (= Forsch. zur dt. Landeskunde 250: 19–26)*. Flensburg.

Thomas, F.T. & Simpson, I.A. (Eds.) (2001). Landscape sensitivity: Principles and applications in northern cool temperate environments. *Catena (Special Issue)* 42(2–4), 80–383.

Tress, B. & Tress, G. (2000). Eine Theorie der Landschaft. Plenarvortrag IALE-Deutschland, Nürtingen, 20.-22.7.2000, Tagungsband: 14–15. Nürtingen.

Tress, B., Tress, G., & Fry, G. (2003). Potential and limitations of interdisciplinary and transdisciplinary landscape studies. In B. Tress, G. Tress, A. van der Valk, & G. Fry (Eds.), *Interdisciplinary and Transdisciplinary Landscape Studies: Potential and Limitations* (pp. 182–192). Wageningen: Delta Series 2.

Tricart, J. (1965). *Principe et methods de la géomorphologie*. Paris: Masson.

Troll, C. (1939). Luftbildplan und ökologische Bodenforschung (Aerial photography and ecological studies of the earth). *Zeitschrift der Gesellschaft für Erdkunde*, 241–298.

Turner, M.G. & Gardner, R.H. (Eds.) (1991). *Quantitative Methods in Landscape Ecology (= Ecological Studies 82)*. New York: Springer.

Turner, M.G., Gardner, R.H., & O'Neill, R.V. (Eds.) (2001). *Landscape Ecology – Pattern and Processes*. New York: Springer.

Wardenga, U. (2001). Zur Entwicklung des länderkundlichen Ansatzes. *Beiträge zur Regionalen Geographie*. 53, 7–35, Leipzig: Institut für Länderkunde.

Widacki, W. (1994). The end of the geocomplex paradigm in physical geography? In A. Richling, E. Malinowska, & J. Lechnio (Eds.), *Landscape Research and Its Applications in Environmental Management* (pp. 109–113) Warsaw: Warsaw University.

Chapter 2

Towards a Reconstruction of Land Use Potential: Case Studies from the Western Desert of Egypt

ANDREAS BOLTEN, OLAF BUBENZER, FRANK DARIUS, AND KARIN KINDERMANN

This chapter is situated in the field among archaeology, geomorphology, and ecology. Two case studies from different east-Saharan landscape units classify and analyse archaeological, geoscientific, and remote-sensing data of Early and Mid-Holocene archaeological sites. The section combines the approaches of landscape ecology and landscape archaeology. The aim is a parameterisation of the research areas with respect to structural and ecological features. The data were used within a Geographical Information System (GIS), a hydromodelling, and statistical software. The analysis allows an indication of the observed landscape parameters that are essential for the location of the sites within each time slice. Therefore, the study broadens the understanding of the man–environment relationships.

With the help of this integral and autochthonous landscape inspection it is possible to reconstruct the past potential of the utilisation of such arid landscapes. Such an approach also helps in locating new archaeological sites within landscape units. At the end a first suggestion for a model of interacting key variables and the general landscape development of the Western Desert during the Early and Mid-Holocene is presented.

M. Bollig, O. Bubenzer (eds.), *African Landscapes*,
doi: 10.1007/978-0-387-78682-7_2, © Springer Science+Business Media, LLC 2009

2.1. INTRODUCTION

This chapter represents an attempt to generalise and calculate the available data for selected research areas of ACACIA[1] in the Western Desert of Egypt, which were collected on archaeological surveys between 1995 and 2002.

The research concentrated on the Early and Mid-Holocene of the Western Desert of Egypt when more humid climatic conditions allowed people to live in what is presently a hyperarid landscape. Outside the Nile Valley and the oases, the archaeological, botanical, and geoscientifical results of the research programs 'Besiedlungsgeschichte der Ostsahara' (B.O.S.) and ACACIA (both University of Cologne projects) show a complex picture regarding both the history of Holocene land use (Kindermann & Bubenzer, 2007; Kuper, 1989; Gehlen et al., 2002; Kuper, 2002, 2006; Kuper & Kröpelin, 2006; Riemer, this volume) and palaeoecological conditions (Besler, 2002; Bubenzer & Riemer, 2007; Bubenzer et al., 2007b; Kindermann et al., 2006; Riemer, 2006) in the timeframe approximately between 9000 and 5000 cal bc.

Within the Sahara, especially in Egypt, only macro- or mesoscalic palaeo-ecological studies relating to the recent or former land use potential exist (e.g., Adams & Faure, 1997; Alaily, 1993; Anhuf, 1997; Anhuf & Frankenberg, 2000; Bornkamm & Darius, 1999; de Noblet-Ducoudré et al., 2000; El Kady et al., 1995; Kehl & Bornkamm, 1993; Prentice et al., 2000; Schulz et al., 2001; Pachur & Altmann, 2006; Wendorf et al., 2001). Examples of archaeological reconstructions of ancient land use by use of GIS are given, for instance, by Spikins (2000, Northern England) and Farshad (2001, Iran). In contrast to the existent studies which often have the aim of reconstructing former landscapes and their use potential by an actualistic comparison with similar recent landscapes (e.g., Neumann, 1989), the attempt presented here works exclusively with the autoch-thonous data. The methods of data derivation and analysis of this new approach are described in detail in the appendix.

The archaeological and geomorphological field observations, generated during some 20 years of field research in the Western Desert of Egypt, were the starting point of this study. The spatial distribution of archaeological sites shows a distinct pattern over a wide range of spatial scale (Figure 2.1). It seems to be that Early and Mid-Holocene sites are mainly found in characteristic geomorphological positions, namely in association with drainage lines and depressions which supply a surplus of water (Bubenzer & Riemer, 2007). To prove this hypothesis an archaeological dataset from the two contrasting regions Djara (Egyptian Limestone Plateau) and Regenfeld (Great Sand Sea) were analysed in combination with environmental

---

[1] The results presented in this article chapter are based on the collaborating research of the projects A1 'Climatic Change and Human Settlement Between the Nile Valley and the Central Sahara' and E1 'GIS-Based Atlas of Holocene Land Use Potential for Selected Areas.' The archaeological database derives from the fieldwork of project A1, whereas the project E1 integrates the results of geoscientific disciplines with those of archaeology and social sciences (Bubenzer & Bolten, 2003; Bubenzer et al., 2007 a).

Figure 2.1. Archaeological sites within the Western Desert of Egypt recorded during archaeological surveys by B.O.S. and ACACIA (1980–2002). Displayed are three different spatial scale factors according to the scale discussion. 1: Landsat 5 image (30 m resolution); 2: ASTER image (15 m resolution); 3: Quickbird image (0.61 m resolution) (*See also Color Plates*)

information derived from topography, hydrography, and geology. A special focus was emphasised on the interrelationship between spatial configuration and the physical properties of the landscape. The scale of investigation (1:50,000) was predefined by the density of the archaeological evidence and the resolution of remote sensing data. An important advantage of this chorological scale[2] – which ranges approximately between 1:25,000 and 1:200,000 – is that it already generalises the complex reality and therefore represents a model of natural structures (Bastian, 1999).

The classification of the research areas according to structural and ecological features was an important step. For this aim, areawide detailed relief data, elevation models, and the (palaeo-) hydrography are fundamental (see contribution of Bubenzer, this volume). Due to the lack of topographic maps in these areas and point-specific field information, remote-sensing data had to be used in addition to field measurements. Particularly the distribution of archaeological artefacts and the geomorphological analysis of landforms were of crucial importance to establish an idea of the palaeolandscape.

This chapter proceeds along the following points of argumentation.

1.   Definition of land use potential in terms of the project.
2.   Methodology to derive parameters, which describe the land use potential; this includes the extraction of a digital elevation model from ASTER satellite data as a base for several steps of the analysis (see box).
3.   Compilation and analysis of the derived dataset with statistical methods.
4.   Brief conclusion with an outlook for further investigations.

2.2. FORMER LAND USE POTENTIAL

In general 'land use potential' can be defined as the capacity of land to sustain specific types of utilisation for a longer period of time without significantly degrading resources. In feasibility studies 'potential' means the capacity or the maximum productivity of a factor or a system of factors with regard to quantity and quality of output extraction in a defined acquisition period. According to the FAO Framework for Land Evaluation the present-day use potential of land depends on both biophysical and socioeconomic conditions (e.g., Kutter et al., 1997). Key elements are determined in the first place by climate, soil, and landforms, because, for example, the range and potential yield of crops are functions of these conditions. Beyond this, the degree to which the natural potential of a landscape can be tapped by the land users depends on technology, knowledge, and labor, as well as on people's aspirations (comp., e.g., Tress et al., 2003).

---

[2] Within the investigation of whole landscapes the chorological scale is able to delimit or to order natural units. In the landscape ecology four aspects are distinguishable: the topological, chorological, regional, and geospherical dimensions (comp. Bubenzer, this volume).

The research object of complex physical geography or landscape ecology (Troll, 1939, 1966; Bastian & Steinhardt, 2002) is the Earth's surface (v. Richthofen, 1877; Penck, 1894), understood as a three-dimensional layer including the lithosphere, atmosphere, and hydrosphere as well as the biosphere, which develops in relation to the former three (Gardner, 1977; Leser, 1997). A particular position is occupied by the anthroposphere or noosphere (from Greek *noos*, mind) because human beings live not only in the three-dimensional Euclidian physical-geographical space of land and water systems, but also in a conceptual sphere of the human mind and consciousness (Naveh, 1995). When reconstructing the land use potential of palaeolandscapes it is virtually impossible to access the latter. Therefore, the present study is mainly restricted to physical parameters. The first step, however, has to be an examination of recent landscape structures with their typical elements including evolution and thus their former function and change over time (Turner & Gardner, 1990; Haase, 1996). Synonyms for landscape elements are patch, ecotope, biotope, cell facies, habitat, or site. The landscape structure as the spatial mosaic of these elements is determined by their composition and configuration (Turner & Gardner, 1990; Syrbe, 2002).

During the Early and Mid-Holocene the more favorable climate of the northeastern Sahara was the most important factor with the growth of plants and that allowed people to live at least occasionally within an arid to periodically or episodically semiarid landscape. On the basis of precipitation amounts of 50–100 mm per annum (Neumann, 1989; Haynes, 2001; Schild & Wendorf, 2001), outside of the Nile Valley and of the oases, sufficient (fresh-)water occurred only in geomorphological positions with a surplus of run-off or seepage water (Bubenzer & Riemer, 2007). Therefore, the overall analysis of the georelief is a crucial point. The georelief is a product of landscape evolution. It conditions the redistribution of water and nutrients, and therefore dominates landscape structure directly and indirectly (e.g., Rohdenburg, 1989).

The soil is another important factor with its characteristic texture, thickness, salinity, and availability of nutrients. It becomes clear that the terrain characteristic (e.g., slope or exposition) also strongly influences the capability of land to support various types of vegetation (Barbour et al., 1999). In the region considered, most of these characteristics are strongly influenced by the geological setting, the georelief, and by the exogenous geomorphological processes, for example, the work of wind and water. Finally, the bedrock builds the basis for the landscape, is more or less resistant against weathering, and therefore influences evolution of landforms and soil generation and the availability of raw material for stone tools is a function of the geological setting (Kinderman et al., 2006).

A reconstruction of biomass production, food web structure, or biodiversity within former ecosystems is possible only by assessing the key factors, physical constraints as well as biotic requirements, which were active during different phases of landscape development.

2.3. STUDY AREAS

In particular, the present xeric (from Greek *xeri*, dry) landscape of the Western Desert of Egypt demonstrates an excellent test arrangement for research into the relation between man and environment. Both the onset of the Holocene humid phase and its termination are plumbable without larger disturbances by modern human activities. Taking into account that the archaeological sites are mostly located on the surface, an areawide detection is possible. Finally the sites are within restricted areas that allow an investigation in the chorological scale. For the purpose of this chapter two archaeological study areas with different geomorphological settings were chosen.

2.3.1. Djara

The Djara region (approximately 200 m a.s.l.; Figure 2.2) is part of the hyperarid Western Desert and lies in the center of the Egyptian Limestone Plateau (also named the Abu Muhariq Plateau). This plateau consists predominantly of Eocene marine carbonate rocks with minor shale intercalations. Its strata and surface dip gently to NNE.

The relief documents a karstic landscape with rounded hilltops, flat depressions, and drainage channels resulting from former wetter climate phases. The Pleistocene as well as modern hyperaridity led and leads to partly strong wind abrasion as well as dune formation (e.g., the famous 'Abu Muhariq' dune belt), serir, and hamada surfaces. The depressions and wadi channels are currently covered with sparse vegetation such as shrubs and a few tufts of grass and small herbs, which are independent of groundwater and depend on rare precipitation events that cause run-off. In the Early to Mid-Holocene humid phase (approximately from 9000 to 5000 cal bc) precipitation amounted to 50–100 mm per annum. This led to an accumulation of water and playa sediments in the depressions (Bubenzer & Hilgers, 2003). Until the present it was not clear whether the rains fell in winter or summer months but archaeobotanical findings support the idea of precipitation in both seasons (Kinderman et al., 2006).

The prehistoric settlement area of Djara embraces more than 150 archaeological sites, situated in a well-defined area of 10 by 5 km (Figure 2.2). Most of these sites were found and surveyed next to shallow depressions, often with living vegetation on playa sediments. The radiocarbon dates, taken from anthropogenic features, give – except for an Early Holocene unit (7700–6700 cal bc) – evidence for a main settlement duration between 6400 and 5300 cal bc. Stone tools establish the largest group of artefacts in all inventories, whereas ceramic are remarkably rare (Kindermann, 2004). On the base of diagnostic tool types and additional ^{14}C-dates two Mid-Holocene units, labelled as Djara A (6400–5900 cal bc) and Djara B (5800–5300 cal bc), are distinguishable. Neither vegetation nor fauna were abundant during the Mid-Holocene. Presumably these conditions supported highly mobile populations, with a predominant hunter-gatherer subsistence. After Djara B a distinct drop-off in the number of ^{14}C-dates marks the onset of modern hyperaridity (Gehlen et al., 2002; Kindermann, 2003, 2006; Kindermann et al., 2006).

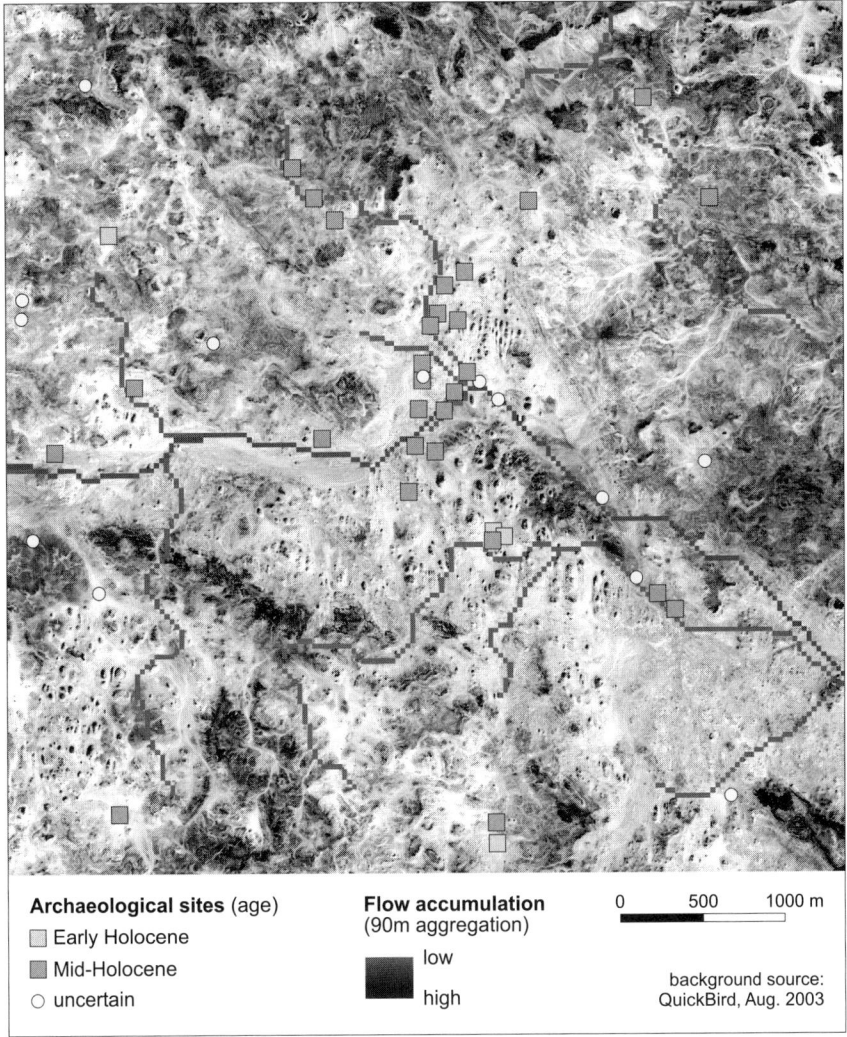

Figure 2.2. Archaeological sites and flow-accumulation in the Djara area. The flow-accumulation is derived from the digital elevation model. Clearly visible is the accumulation of sites along the main drainage line in the centre of the figure (*See also Color Plates*)

2.3.2. Regenfeld

The Regenfeld area (approximately 400 m a.s.l.; Figure 2.3) is located in the southern part of the Great Sand Sea, Western Desert of Egypt. The bedrock consists of Cretaceous shale, silt, and sandstones (Nubian Sandstone) dipping gently to the north. Besides the deflated Holocene playa remnants in the corridors, Pleistocene megadunes (draa with heights up to 70 m above the corridors) with riding modern longitudinal silk dune (height: approximately 15 m) are

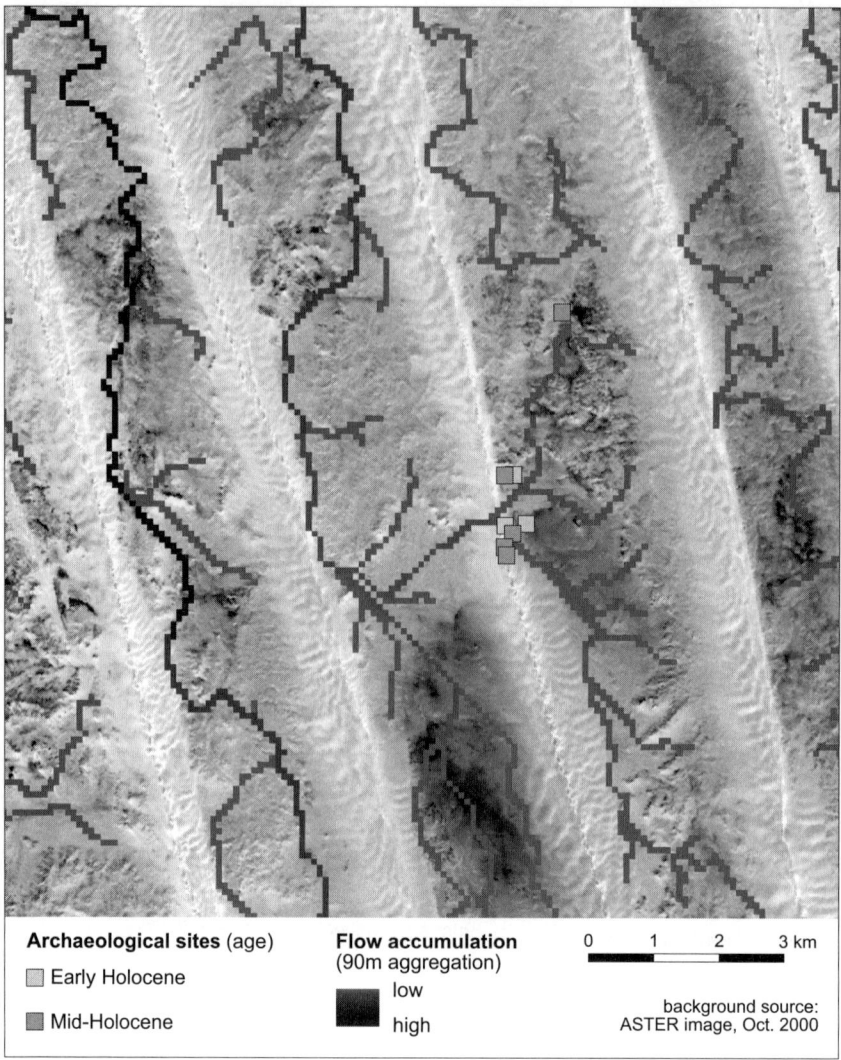

Figure 2.3. Archaeological sites and flow-accumulation in the Regenfeld area.
The flow-accumulation is derived from the digital elevation model. For better understanding the flow-accumulation raster is aggregated from 30 to 90 m pixel resolution. Clearly
visible is the incidence of archaeological sites at the calculated dune penetration point
(*See also Color Plates*)

formative (Riemer, 2000; Besler, 2002). Actual vegetation (mostly *Stipagrostis*
spp.) is restricted to the lower dune flanks. From the geomorphological point
of view the favorable situation of the Regenfeld area in the Holocene humid
phase is based on a surplus of water from the megadunes and from a small sand
sheet which reaches from the south into this area (Bubenzer & Riemer, 2007).

These sand accumulations are able to store large quantities of water over a long time after a precipitation event. Because the megadunes are built of single grains a deep penetration of the water is possible. In addition the building of capillaries is prevented and therefore the evaporation is diminished. A reddening of the playa sediments (generation of hematite minerals) corroborates the assumption that a northward expansion of monsoonal summer rains occurred during the Holocene optimum. The annual amount of precipitation was approximately 100 mm (Bubenzer & Besler, 2005). The accumulation of playa sediments took place in the endorheic pans by sedimentation in seasonal or ephemeral lakes, which developed after surface run-off. For the Great Sand Sea groundwater was of no relevance for humans, thus the psammitic silicate deposits of the playas are the result of precipitation, run-off, and seeping water from the draa bodies during the Holocene humid phase.

After the mostly hyperarid Pleistocene, human occupation started in the Western Desert of Egypt around 8800–8700 cal bc with the beginning of the Holocene humid phase (Kröpelin, 1993; Kuper & Kröpelin, 2006). The Regenfeld area can be subdivided into four archaeological phases, three Early Holocene units (Regenfeld A–C) and a Mid-Holocene one, labelled Regenfeld D (Gehlen et al., 2002). Whereas units Regenfeld A–C are distinguished by different lithic-tool kits, one of the most important characteristics of unit D (approximately 6500–5400 cal bc) is the introduction of pottery and the abundance of grinding stones. All bones identified on the archaeological sites belong to wild animals and verify a hunter-gatherer subsistence.

2.4. DATA PREPARATION AND EXPLORATIVE STATISTICS

Although several basic structures within the spatial patterning of the archaeological sites can be discovered simply by visual inspection of overlays, an ordination analysis of site attributes and landscape features leads to a deeper understanding of the landscape and vegetation patterns as a driving factor of spatial occupation patterns.

Environmental variables, which have not been measured in the field, were prepared from remote-sensing datasets for the assessment of their strength as predictors for site distributions (see Appendix for details on the applied methods and resources).

The explanatory power of the derived environmental variables was assessed using Canonical Correspondence Analysis (CCA; ter Braak, 1986, 1994). This multivariate ordination technique is attractive here as it takes two related data matrices as input: the first (main) one typically consisting of incidences or abundances of attributes within a set of sites whereas the second matrix contains environmental variables measured in the same sample units. The ordination within attribute space is constrained by multiple regression on variables in the secondary matrix. CCA is implicitly based on the chi-squared distance and expects a unimodal relation of site attributes to the set of explanatory variables.

The result of the CCA is shown in a bi- or triplot. The resulting ordination is a product of the variability of both the environmental and the archaeological attribute data. Site scores and attribute scores are plotted on the same graph using different scales. The angle and length of the arrows show the direction and strength of the relationship between ordination scores and environmental variables. Its aim is to sort the variables and display their relations to identify dependencies between environmental and archaeological variables.

Our main matrix was set up to hold the information on artefacts from the field surveys, compiled and classified to binary variables by Karin Kindermann and Heiko Riemer (Table 2.1). In both study areas, Djara and Regenfeld, a multilevel strategy was applied. The investigations combined large-scale reconnaissance surveys, local systematic surveys within the areas of research, and detailed archaeological analysis of the sites. Field work comprised excavations and complete collections of artefacts spread out on the surface (Riemer, this volume). To systemise the survey data, a standard form was used for the registration of archaeological features. Excavations were established to separate relevant chronostratigraphic units and to reconstruct the site's spatial configuration and centers of activity using high-resolution excavation grids. The stone artefacts, included in the following statistical analysis, are material from surfaces as well as from excavations. Age categories were derived from typochronological classification of diagnostic artefacts and from ^{14}C-dates.

The second matrix contained the environmental information for the archaeological sites, represented by the remote-sensing and GIS-derived elements. Variables

Table 2.1. Archaeological attributes, which have been used for the main ordination matrix; abbreviations, categories, and number (*n*) of respective cases by region

Archaeological site descriptors			Djara (n=158)	Regenfeld (n=43)
Variable	*Explanation*	*Category*		
AGE	Classified age	Early Hol.	11	5
		Mid-Hol.	77	38
		Uncertain	70	0
SIZE	Area of sites	Small	108	29
		Middle	45	11
		Large	5	3
DENS	Density of artefacts	Isolated	90	16
		Low	41	20
		High	27	7
BLANK	Blank production	Present	96	28
TOOL	Tools	Present	85	23
ARROW	Arrow heads	Present	16	10
ADZE	Adzes	Present	26	0
GRIND	Grinding stones	Present	44	30
HEARTH	Hearths	Present	117	10
OES	Ostrich egg shell artef.	Present	11	5

Table 2.2. Environmental variables, which have been used for the secondary ordination matrix; abbreviations, categories, and number (*n*) of all classified locations for the background

Environmental site descriptors:	Djara (n = 5,524), Regenfeld(r = 27,174)
Variable	*Explanation*
LAT	UTM Latitude position
LONG	UTM Longitude position
ALT	Altitude from DEM (m)
SLOPE	Inclination (deg)
EXP_180	Aspect (north–south: 1...0...-1)
EXP_090	Aspect east–west 1...0...-1)
TOPO index	Topographic position within 150 m neighborhood: TOPO = (H-min(HĐ))* $n(x < H)/24$
FLOAC	Flow accumulation through site pixel (30 m)
FLOACX	Flow accumulation through next channel pixel (30 m)
HYDRO index	Maximum flow accumulation through site neighborhood (210 m)
DIST	Low cost distance to next channel (m)
GEO$	Geological unit (from digitized map)

were log-transformed where necessary, and standardised to zero mean and unit variance (Table 2.2).

Frequency distributions of the environmental variables were calculated, which allowed a comparison of site attribute spectra with the respective regional background (neutral model). Differences would normally then be confirmed statistically by bootstrap techniques.

The spatial resolution of all data presented is 30 m. Although CCA is reported to be rather insensitive to autocorrelation, all analyses were repeated with a coarse-grained subset (resolution 210 m) in order to detect possible bias induced by the clumped distribution of the archaeological sites. Neither the frequency distributions nor the CCA results showed a marked difference to the original dataset.

All analyses were done using the programs SYSTAT 10 (Systat software Inc.), CANOCO 4 (ter Braak & Smilauer, 1998), and PC-ORD 4 (McCune & Mefford, 1999).

2.5. RESULTS AND DISCUSSION

Archaeological remains within the study areas are generally, more often than expected from the statistical background, located at lower positions (depression area or wadi) within the topological sequence, close to hydrological features receiving considerable run-off from surrounding terrain (Figure 2.4). Frequency distributions of sites dated as Mid-Holocene, which constitute the majority of the investigated places, show a tendency to bimodality in comparison to the regional background (e.g., arrowheads, ostrich egg shell artefacts (OES), adzes, and grinding stones).

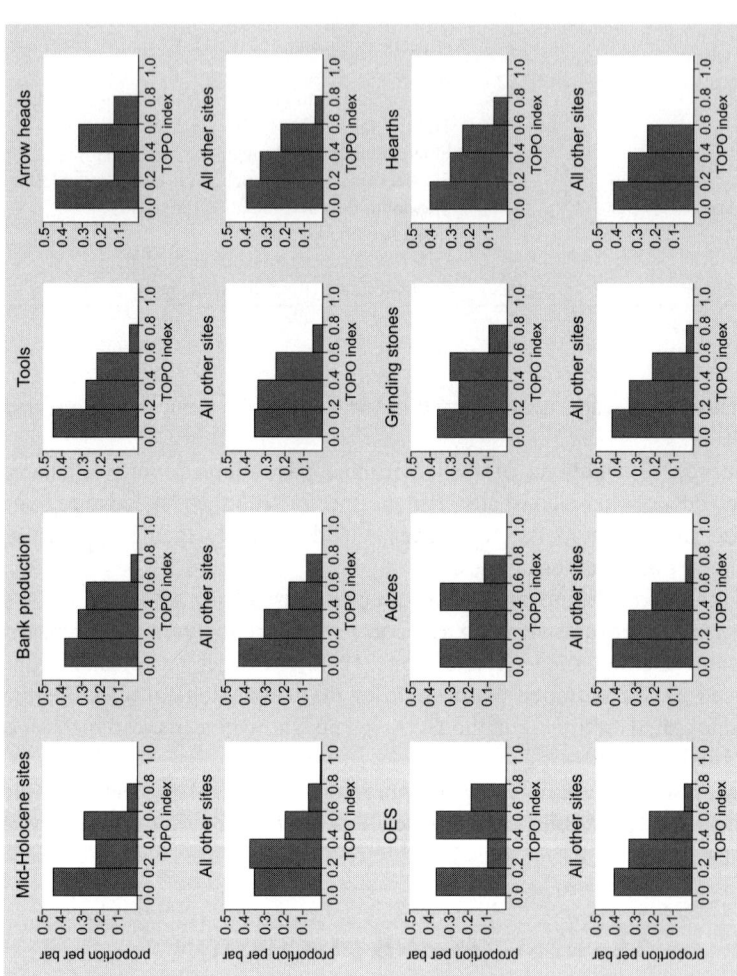

Figure 2.4. Histograms showing the distribution of the Topo-Index (see Table 2.2) values among sites with a certain age, or presence/absence of artefact types, against their background. OES – ostrich egg shell artefacts

Many assemblages could be found near hydrologically favored positions, however, there are also sites located on the higher and drier end of the spectrum. These sites seem to contain preferably signs of blank production and/or hearths. On the other hand, grinding stones and/or stone tools were found in great numbers near channels and flood plains (Figure 2.5).

The ordination results[3] support the general hypothesis. The CCA triplot shows the first two ordination axes to be correlated with the topographical and hydrological features derived from the digital elevation model (Figure 2.6). Attribute–environment correlations are significant (Monte Carlo test, 99 runs). The Topo-Index correlates with axis 1 (−0.943) and separates the two study regions, suggesting a different occupation pattern in response to landscape morphology. The Hydro-Index is correlated with axis 2 (−0.996), separating the Early from Mid-Holocene sites in Djara. One possible interpretation is that human exploitation in earlier phases of the Holocene optimum concentrated on the plains and highlands, where a diffuse vegetation of steppic character promised optimal hunting grounds. New and probably more profitable resources were found during later chronological units, when temporary wetlands came into existence and attracted people to settle on the deeper and more clayey soils along the shorelines. These were the places where seasonal inundation led to the vigorous growth of wild herbs and grasses, nutrient rich and worth being collected for their cereal products. If this activity was an autochthonous innovation or the result of interregional exchange remains one of the main archaeological problems to be solved for this region.

2.6. CONCLUSION

The study presented here focused on the three-dimensional physical aspects of landscape, emphasising its role as the ultimate basis of human livelihoods and life-supporting system, necessarily leaving aside all the cultural aspects of the environment. Nevertheless we acknowledge that people are more than passive recipients of climatic changes. However, it has to be borne in mind that during the Holocene humid optimum, the ecological situation already provided difficult conditions for the people in the Western Desert of Egypt. Neither vegetation nor fauna were abundant during the Mid-Holocene and therefore these conditions only supported mobile populations of predominant hunter-gatherers that were highly dependent on adequate rain, extremely vulnerable to climatic changes, and so reacted sensitively. That is just why arid zones, especially for the time slice examined, form an almost experimental situation, as key resources are limited factors, and

---

[3] To conclude the presentation of the results, some caution is warranted because the archaeological database is far from statistically ideal. The number of cases within categories varies over a wide range. There is also some ambiguity in the Topo-Index, which creates small values at valley positions, but also within level terrain.

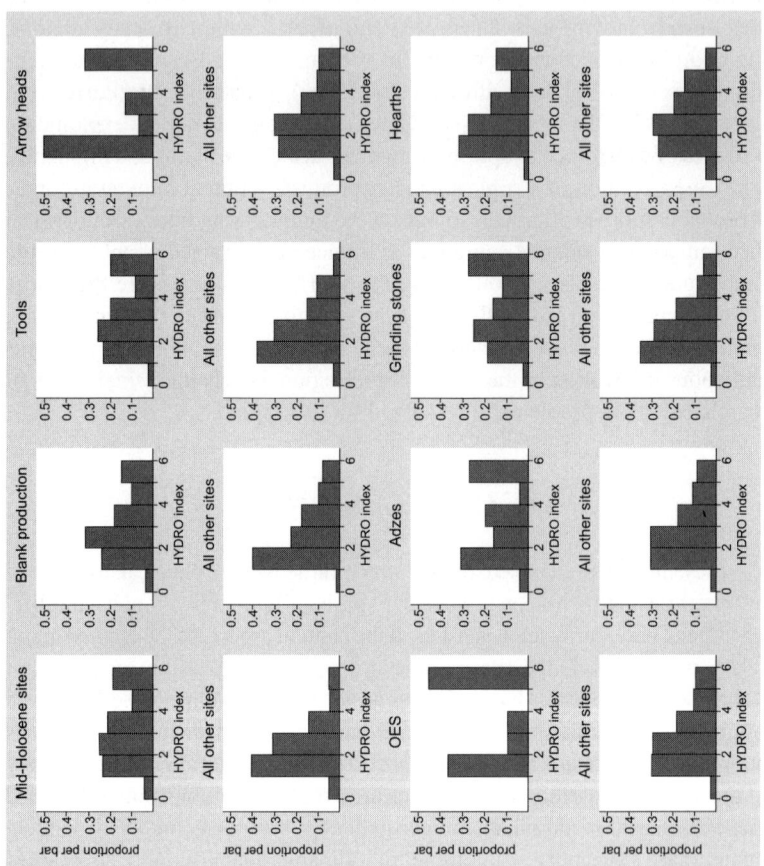

Figure 2.5. Histograms showing the distribution of the Hydro-Index (see Table 2.2) values among sites with a certain age and presence/absence of artefact types, against their background. OES – ostrich egg shell artefacts

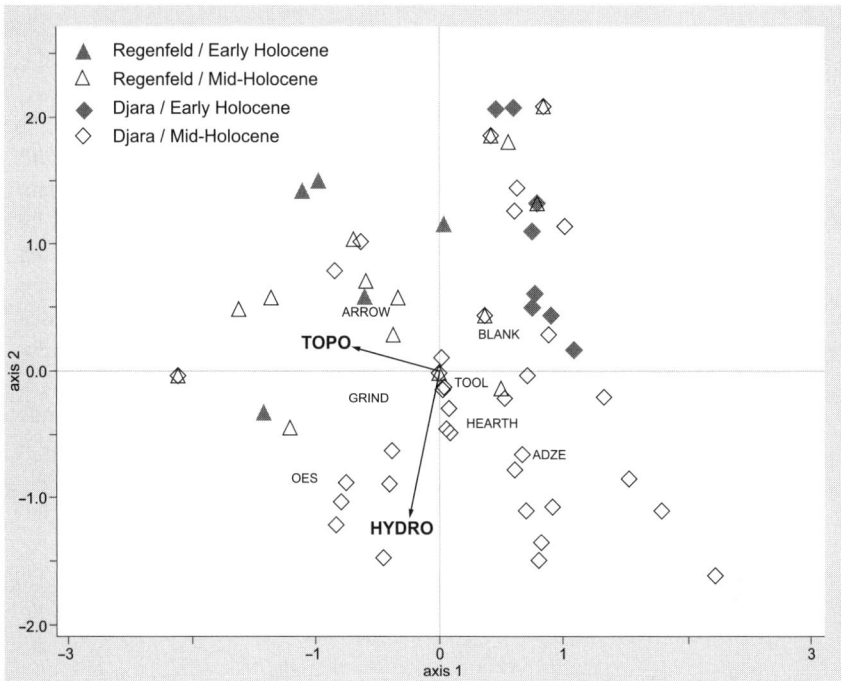

Figure 2.6. Triplot of the CCA, showing the ordination of prehistoric sites with respect to axes 1 and 2 from presence/absence of observations of archaeological remains. Artefact types are placed at their centroids in ordination space. Topo-Index is correlated with axis 1; Hydro-Index is correlated with axis 2. For details and abbreviations, see text

the environmental constraints of the arid landscape can be separated from other nonenvironmental factors

Starting from this point, our analysis of past patterns of adaptation was based mainly on a functional view, which sees landscape as a system composed of compartments and flows of energy and matter. In sum, by the coordination of scattered archaeological evidence and the establishment of relationships with their geographical background, we located places with ecologically favorable conditions for the prehistoric people and found reasonable explanations for site distributions in terms of environmental features.

APPENDIX

Steps of Deriving Environmental Parameters

At the beginning of the project the source of digital elevation model data and the necessary grid resolutions were discussed. Free available elevation data (e.g., GTOPO 30 model) is unsuited to the question described, because its resolution is too coarse. The Advanced Spaceborne Thermal Emission and Reflection Radiometer (ASTER) onboard the National Aeronautics and Space Administration's (NASA's) TERRA spacecraft (launched in 1999) provides digital stereo data at 15 m resolution (Yamaguchi et al., 1998). ASTER's spectral and geometric capabilities include 14 bands in different wavelengths. A comparison of ASTER and the well-known LANDSAT ETM+ spectral bands is given in Figure 2.A1.

Figure 2.A1. ASTER spectral bands (top) compared to Landsat ETM+ (bottom). The rectangular boxes indicate the sensor channel with the respective spatial resolution on the top of the boxes. VNIR – visible and near infrared; SWIR – short-wave infrared; TIR – thermal infrared.

For generating DEMs the L1A raw data (available at the EOS DG, 2007) with the two stereo bands (3b and n) are in use (Fujisada, 1998; Abrams, 2000). The orientation is set by the same tie points because of the lack of exact ground control points in the unsurveyed region. It is for this reason that only relative DEMs can be extracted. In very smooth terrain, as well as under problematic conditions (cloud cover) the generation and the quality of DEMs are imprecise. However, the comparison between generated DEMs by the ASTER team and the generated DEM of the same scene by our project shows particular agreement. Another problem concerns the mosaicing of several generated elevation models, because artefacts might arise from small absolute elevation differences in the models between to track rows. However, the problems described are quite rare and most often attributed to errors in the raw data (Bolten & Bubenzer, 2006; Bubenzer & Bolten, 2008). Remarkable is the enormous calculation time for a DEM extraction (about 30 min calculation time without setup work for one dataset), so a generation of a wide-range elevation model could only be done with up-to-date computer power.

(continued)

Figure 2.A2 shows a flowchart starting with the extracted DEM. Thence-forward there are two directions of analysis.

Firstly the determination of surface data with a GIS and secondly the extraction of hydrologic parameters in a watershed modelling software (TOPAZ, Garbrecht, 2000). Then, secondary combinational parameters (e.g., a topographic valley index [Myburgh, 1974]) in combination with the external data could be extracted as an origin for a statistical processing.

The part 'surface data' means the distribution of altitude, slope, and aspect for the entire DEM area. After this the data of specific points – defined by archaeological sites (external data) – can be extracted and transferred into a database. The determination of slope and aspect, as simple primary geomorphometric parameters (Schmidt & Dikau, 1999), are part of most raster-based GIS systems.

The part 'hydrology data' contains the determination of flow-direction, flow-accumulation, and watershed basins, called complex primary geomorphometric parameters (Schmidt & Dikau, 1999). With these parameters a quantitative calculation of a hypothetical drainage system is possible. The quality of the drainage system is checked by comparison with a ground-checked drainage system analysis by Jäkel and Rückert (1998). The extracted drainage system shows good correspondence, except in flat playa areas, when the error noise of the DEM is more important and higher than the elevation variety (Bubenzer et al., 2007c).

Another derived vector parameter is the favorable path to the drainage system in dependency of the surface slope.

The derived parameters from the statistical software can be used to identify different landscape characteristics regarding the land use potential. At a sufficient number of investigated regions an areal interpolation is possible.

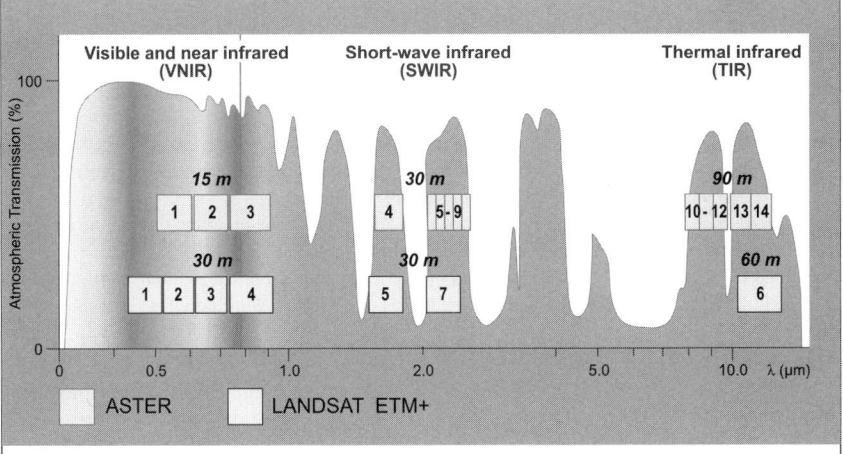

Figure 2.A2. Flow chart for further calculation (*See also Color Plates*)

(continued)

REFERENCES

Abrams, M. (2000). The Advanced Spaceborne Thermal Emission and Reflection Radiometer (ASTER): Data products for the high spatial resolution imager on NASA's Terra platform. *International Journal of Remote Sensing*, 21, 847–859.

Adams, J.M. & Faure, H. (Eds.) (1997). QEN members. Review and atlas of palaeovegetation: Preliminary land ecosystem maps of the world since the Last Glacial Maximum. Oak Ridge National Laboratory, TN, 1999 (http://www.esd.ornl.gov/projects/qen/adams1.html).

Alaily, F. (1993). Soil association and land suitability maps of the Western Desert, SW Egypt. *Catena Suppl.*, 26, 123–153.

Anhuf, D. (1997). Paleovegetation in West Africa for 18.000 B.P. and 8.500 B.P. *Eiszeitalter und Gegenwart*, 47, 112–119.

Anhuf, D. & Frankenberg, P. (2000). Die mittelholozäne Feuchtphase 5000 BP – Eine Vegetationsrekonstruktion für Afrika. *Regensburger Geogr. Schriften*, 33, 99–126.

Barbour, M.G., Burk, J.H., Pitts, W.D., Gilliam, F.& Schwartz, M. (1999). *Terrestrial Plant Ecology*. Menlo Park, CA: Addison Wesley/Longman.

Bastian, O. (1999). Landschaftsbewertung und Leitbildentwicklung auf der Basis von Mikrogeochoren. In U. Steinhardt & M. Volk (Eds.), *Regionalisierung in der Landschaftsökologie* (pp. 287–298). Stuttgart, Leipzig: Teubner.

Bastian, O. & Steinhardt, U. (Eds.) (2002). *Development and Perspectives of Landscape Ecology*. Boston: Kluwer.

Besler, H. (2002). The Great Sand (Egypt) during the late Pleistocene and the Holocene. *Zeitschrift für Geomorphologie*, Suppl.-Bd. 127, 1–19.

Bolten, A. & Bubenzer, O. (2006). New elevation data (SRTM/ASTER) for geomorphological and geoarchaeological research in arid regions. *Zeitschrift für Geomorphologie*, Suppl. 142, 265–279.

Bornkamm, R. & Darius, F. (1999). Probleme der Landnutzungsplanung in der Extremwüste Süd-Ägypteus. *TU International*, 46/47, 39–42.

ter Braak, C.J.F. (1986). Canonical correspondence analysis: A new eigenvector technique for multivariate direct gradient analysis. *Ecology*, 67, 1167–1179.

ter Braak, C.J.F. (1994). Canonical community ordination. Part I: Basic theory and linear methods. *Ecoscience*, 1, 127–140.

ter Braak, C.J.F. & Smilauer, P. (1998). *CANOCO Reference Manual and User's Guide to CANOCO for Windows: Software for Canonical Community Ordination (Version 4)*. Wageningen: Centre for Biometry.

Bubenzer, O. & Besler, H. (2005). Human occupation of Sand Seas during the Early and Mid-Holocene. Examples from Egypt. *Zeitschrift für Geomorphologie*, Suppl. 138, 153–165.

Bubenzer, O. & Bolten, A. (2003). Detecting areas with different land use potential in Egypt, Sudan, and Namibia, GIS helps compare human strategies of coping with arid habitats. *ArcNews*, 25, 2.

Bubenzer, O. & Bolten, A. (2008): The use of new elevation data (SRTM/ASTER) for the detection and morphometric quantification of Pleistocene megadunes (draa) in the eastern Sahara and the southern Namib. *Geomorphology* (in press, available online).

Bubenzer, O. & Hilgers, A. (2003). Luminescence dating of Holocene playa deposits of the Egyptian Plateau, Western Desert, Egypt. *Quaternary Science Reviews*, 22, 1077–1084.

Bubenzer, O. & Riemer, H. (2007). Holocene climatic change and human settlement between the central Sahara and the Nile Valley: Archaeological and geomorphological results. *Geoarchaeology*, 22, 607–620.

Bubenzer, O., Besler, H. & Hilgers, H. (2007b). Filling the gap: OSL data expanding ^{14}C chronologies of Late Quaternary environmental change in the Libyan Desert. *Quaternary International*, 175, 41–52.

Bubenzer, O., Bolten, A. & Darius, F. (Eds.) (2007a). *Atlas of Cultural and Environmental Change in Arid Africa. Africa Praehistorica*, 21, 240 p. Cologne: Heinrich-Barth-Institut e.V.

Bubenzer, O., Bolten, A. & Ritter, M. (2007c). Scale-specific geomorphometry of arid regions – Examples from the eastern Sahara. *Proceedings of an International ACACIA Conference* held at Königswinter, Germany, October 1–3, 2003. *Colloquium Africanum*, 2: 17–34. Cologne: Heinrich-Barth-Institut e.V.

Bubenzer, O., Hilgers, A. & Riemer, H. (2006). Luminescence dating and archaeology of Holocene flu-vio-lacustrine sediments of Abu Tartur, Eastern Sahara. *Quaternary Geochronology*, 2, 314–321.

El Kady, H.F., Ayyad, M.A. & Bornkamm, R. (1995). Vegetation and recent land-use history in the desert of Maktala, Egypt. In H.-P. Blume & S. Berkowicz (Eds.), *Arid Ecosystems* (pp. 109–123). Cremlingen-Destedt: Catena.

EOS DG (2007): Earth Observing System Data Gateway (http://edcimswww.cr.usgs.gov/pub/imswelcome/).

Farshad, A. (2001). Reconstruction of the evolution of the past agrarian landscape as a clue for assessing sustainability: A case study of Iran. In D. Van der Zee & I.S. Zonneveld (Eds.), *Landscape Ecology Applied in Land Evaluation, Development and Conservation. Some Worldwide Selected Examples* (pp. 1–48.) ITC publication, 81. International Institute for Aerospace Survey and Earth Sciences, Enschede.

Fujisada, H. (1998). ASTER level-1 data processing algorithm. *IEEE Transactions and Remote Sensing*, 36, 1101–1112.

Garbrecht, J. (2000). *TOPAZ User Manual* (http://grl.ars.usda.gov/topaz/USERMAN/PAGE1.htm).

Gardner, I.S. (1977). *Physical Geography*. New York: Harper's College Press.

Gehlen, B., Kindermann, K., Linstädter, J. & Riemer, H. (2002). The Holocene occupation of the Eastern Sahara: Regional chronologies and supra-regional developments in four areas of the Absolute Desert. In:Jennerstr. 8 (Eds.), *Tides of the Desert – Gezeiten der Wüste. Contributions to the Archaeology and Environmental History of Africa in Honour of Rudolph Kuper. Africa Praehistorica* (Vol. 14, pp. 85–116). Cologne: Heinrich-Barth-Institut.

Haase, G. (1996). Geotopologie und geochorologie – Die Leipzig-Dresdener Schule der Landschaftsökologie. In G. Haase & E. Eichler (Eds.), *Wege und Fortschritte der Wissenschaft; Sächsische Akademie der Wissenschaften zu Leipzig* (pp. 201–229). Berlin: Akademie-Verlag Berlin.

Haynes, V.C. (2001). Geochronology and climate change of the Pleistocene-Holocene transition in the Darb el Arba´in Desert, Eastern Sahara. *Geoarchaeology*, 16, 119–141.

Jäkel, D. & Rückert, H. (1998). Recent rainfall distribution patterns of the Republic of Sudan as a model for rainfall variations in the past and climate induced geomorphological processes in Sahelian countries. *Paleoecology of Africa*, 25, 101–120.

Kääb, A., Huggel, C., Paul, F., Wessels, R., Raup, B., Kieffer, H. & Kargel, J. (2002). Glacier monitoring from ASTER imagery: Accuracy and applications. *Proceedings LIS-SIG* Workshop Berne, March 11–13, 2002.

Kehl, H. & Bornkamm, R. (1993). Landscape ecology and vegetation units of the Western Desert of Egypt. *Catena*, 26, 155–178.

Kindermann, K. (2003). Djara: Prehistoric links between the Desert and the Nile. In Z. Hawass & L. Pich Brock (Eds.), *Egyptology at the Dawn of the Twenty-First Century. Proceedings of the Eighth International Congress of Egyptologists* (pp. 272–279). Cairo.

Kindermann, K. (2004). Djara: Excavations and surveys of the 1998–2002 seasons. *Archéo-Nil*, 14, 31–50.

Kindermann, K. (2006). Djara. Zur mittelholozänen Besiedlungsgeschichte zwischen Niltal und Oasen (Abu-Muhariq-Plateau, Ägypten). (Unpublished Ph.D. thesis, Köln 2006).

Kindermann, K. & Bubenzer, O. (2007). Djara – Humans and their environment on the Egyptian limestone plateau around 8,000 years ago. In O. Bubenzer, A. Bolten & F. Darius (Eds.), Atlas of Cultural and Environmental Change in Arid Africa *Africa Praehistorica* (Vol. 21, pp. 26–29). Cologne: Heinrich-Barth-Institut e.V.

Kindermann, K., Bubenzer, O., Nussbaum, S., Riemer, H., Darius, F., Pöllath, N. & Smettan, U. (2006). Palaeoenvironment and Holocene land use of Djara, Western Desert of Egypt. *Quaternary Science Reviews*, 25, 1619–1637.

Kröpelin, S. (1993). Geomorphology, landscape evolution and paleoclimates of Southwest Egypt. *Catena*, Suppl. 26, 31–65.

Kuper, R. (1989). The Eastern Sahara from north to south: Data and dates from the B.O.S. Project. In L. Krzyżaniak & M. Kobusiewicz (Eds.), *Late Prehistory of the Nile Basin and the Sahara. Studies in African Archaeology* (pp. 197–203). Poznan: Poznan Archaological Museum.

Kuper, R. (2002). Routes and roots in Egypt's Western Desert: The Early Holocene resettlement of the Eastern Sahara. In R. Friedman (Ed.), *Egypt and Nubia. Gifts of the Desert* (pp. 1–12 (pl. 1–24)). London: The British Museum Press.

Kuper, R. (2006). After 5000 BC: The Libyan desert in transition. *C.R. Palevol*, 5, 409–419.

Kuper, R. & Kröpelin, S. (2006). Climate-controlled Holocene occupation in the Sahara: Motor of Africa's evolution. *Science*, 313, 803–807.

Kutter, A., Nachtergaele, F.O. & Verheye, W.H. (1997). The new FAO approach to land use planning and management, and its application in Sierra Leone. *ITC Journal*, 3/4, 278–283.

Leser, H. (1997). *Landschaftsökologie. Ansatz, Modelle, Methodik, Anwendung*. Stuttgart: Ulmer.

McCune, B .& Mefford, M.J. (1999). *PC-ORD. Multivariate Analysis of Ecological Data.Version 4.0*. Gleneden Beach, OR: MjM Software.

Myburgh, J. (1974). An index to relate local topography to mean minimum temperatures. *Agrochemophysica*, 6, 73–78.

Naveh, Z. (1995). Interactions of landscapes and cultures. *Landscape and Urban Planning*, 32, 43–54.

Neumann, K. (1989). Holocene vegetation of the Eastern Sahara: Charcoals from prehistoric sites. *African Archaeological Review*, 7, 97–116.

de Noblet-Ducoudré, N., Claussen, M. & Prentice, C. (2000). Mid-Holocene greening of the Sahara: First results of the GAIM 6000 year BP experiment with two asynchronously coupled atmosphere/biome models. *Climate Dynamics*, 16, 643–659.

Pachur, H.-J. & Altmann, N. (2006). *Die Ostsahara im Spätquartär. Ökosystemwandel im größten hyperariden Raum der Erde*. Berlin: Springer.

Penck, A. (1894). *Morphologie der Erdoberfläche*. Stuttgart: J. Engelhorn.

Prentice, I.C., Jolly, D. & BIOME 6000 participants (2000). Mid-Holocene and glacial-maximum vegetation geography of the northern continents and Africa. *Journal of Biogeography*, 27, 507–519.

Richthofen, F.v. (1877). Die heutigen Aufgaben der wissenschaftlichen Geographie. In F.v. Richthofen (Ed.), *China, Bd. 1* (pp. 729–733). Berlin: D. Reimer.

Riemer, H. (2000). Regenfeld 96/1 – Great Sand Sea and the question of human settlement on whaleback dunes. In L. Krzyżaniak, K. Kroeper & M. Kobusiewicz (Eds.), *Recent Research into the Stone Age of Northeastern Africa. Studies in African Archaeology* (pp. 21–31). Poznan: Poznan Archaological Museum.

Riemer, H. (2006). Archaeology and environment of the Western Desert of Egypt: ^{14}C-based human occupation history as an archive for Holocene palaeoclimatic reconstruction. In Youssef & El-Sayed, A.A. (Eds.), *Geology of the Tethys. Proceedings of the First International Conference on the Geology of the Tethys* (pp. 553–564). Cairo: Cairo University 2005, The Tethys Geological Society.

Rohdenburg, H. (1989). *Landschaftsökologie – Geomorphologie*. Cremlingen: Catena Verlag.

Schild, R. & Wendorf, F. (2001). Geoarchaeology of the Holocene climatic optimum at Nabta Playa, Southwest Desert, Egypt. *Geoarchaeology*, 16, 7–28.

Schmidt, J. & Dikau, R. (1999). Extracting geomorphometric attributes and objects from digital elevation models – Semantics, methods, future needs. In R. Dikau & H. Saurer (Eds.), *GIS for Earth Surface Systems* (pp. 153–174). Stuttgart: Gebrüder Borntraeger.

Schulz, E., Akhtar-Schuster, M., Agwu, Ch., Beck, C., Dupont, L., Jahns, S., Niedermeyer, M., Ousseini, I. & Salzmann, U. (2001). *The Holocene Landscape and Vegetation History of Northern and Western Africa – A Palaeoecological Atlas* (http://www.uni-wuerzburg.de/geographie/fachi/pal_atlas_afrika/index_atlas.htm).

Spikins, P. (2000). GIS models of past vegetation: An example from Northern England, 10000–15000 BP. *Journal of Archaeological Science*, 27, 219–234.

Syrbe, R.-U. (2002). Entwicklung von Struktur und Funktionsweise in der Landschaft. In G. Haase & K. Mannsfeld (Hrsg.), *Naturraumeinheiten, Landschaftsfunktionen und Leitbilder am Beispiel von Sachsen. Forschungen zur deutschen Landeskunde*, 250 (pp. 19–26). Flensburg: Academy Publication.

Tress, B., Tress, G. & Fry, G. (2003). Potential and limitations of interdisciplinary and transdisciplinary landscape studies. In B. Tress, G. Tress, A. van der Valk & G. Fry (Eds.), *Interdisciplinary and Transdisciplinary Landscape Studies: Potential and Limitations. Delta series*, 2 (pp. 182–192). Wageningen: Alterra Green World Research.

Troll, C. (1939). Luftbildplan und ökologische Bodenforschung (Aerial photography and ecological studies of the earth). *Zeitschrift der Gesellschaft für Erdkunde*, 241–298.

Troll, C. (1966). Landschaftsökologie als geographisch-synoptische Naturbetrachtung. In: ökologische Landschaftsforschung und vergleichende Hochgebirgsforschung. *Erdkundliches Wissen*, 11, 1–13.

Turner, M.G. & Gardner, R.H. (Eds.) (1990). *Quantitative Methods in Landscape Ecology. Ecological Studies*, 82. New York: Springer.

Wendorf, F., Schild, R. & Associates (Eds.) (2001). *Holocene Settlement of the Egyptian Sahara. The Archaeology of Nabta Playa, 1*. New York: Kluwer/Plenum.

Yamaguchi, Y., Kahle, A.B., Tsu, H., Kawakami, T. & Pniel, M. (1998). Overview of Advanced Spaceborn Thermal Emission and Reflection Radiometer (ASTER). *IEEE Transactions on Geoscience and Remote Sensing*, 36, 1062–1071.

Chapter 3

Landscape Ecology of Savannas: From Disturbance Regime to Management Strategies

ANJA LINSTÄDTER

In this contribution, the sustainability of range management in savannas is related to dynamics within a savanna landscape. Concepts of landscape ecology may help to identify crucial aspects of a sustainable management, such as a specific disturbance regime. Some important goals for future research activities are identified, either related to the process and implementation of research, or to the content of research.

The transdisciplinary science of landscape ecology studies the structure, function, and development of landscapes. It is still passing through a process of self-discovery and does not yet offer a unified theory. As its roots are deep in geography as well as in geobotany and land management, self-discovery comprises the search for the unification of landscape ecology as a discipline, the relation between basic research and application, and between sectoral and holistic approaches.

The key landscape types for landscape ecology have been the cultural land-scapes in Europe and North America. Some of the concepts developed for these landscapes are difficult to transfer to African savannas. Nevertheless, the dynamics in a savanna are best dealt with on the spatial and conceptual level of a landscape. Here the different disturbances can be integrated into a disturbance regime, and their functional aspects can be discussed. It is argued that the sustainable use of

M. Bollig, O. Bubenzer (eds.), *African Landscapes*,
doi: 10.1007/978-0-387-78682-7_3, © Springer Science+Business Media, LLC 2009

a savanna landscape by nomadic herders implies a disturbance regime roughly similar to the regime occurring under natural conditions. A sustainable use has to ensure fodder reserves as an ecological buffer against temporal variability, and it has to be adapted to spatial variability on a landscape level.

3.1. THE SAVANNA LANDSCAPE: THEORETICAL BACKGROUND

3.1.1. What Are Savannas?

Savannas are tropical ecosystems with a continuous layer of grasses and with a discontinuous layer of shrubs and/or trees (Solbrig, 1996, p. 1). The continuous lower vegetation layer consists mainly of grasses, usually interspersed with a certain number of forbs. Savannas are the most common vegetation type of the tropics and the subtropics (Solbrig, 1996, p. 31). Savannas cover more than 40% of the African continent. They are the African landscape type with the largest human population and with the fastest growing population. Nevertheless, this important landscape type is increasingly endangered by maladaptive human land use (Ayoub, 1998; Darkoh, 1998). For centuries many African savanna landscapes were utilised by nomadic pastoralists and by agropastoralists. The majority of these grazing systems were changed when they came in contact with 'western' civilisation during the past 200 years. Nevertheless, a few such societies have survived until today (Mainguet & Da Silva, 1998, p. 377). They practise a sustainable land management, whereas other ways of range management (such as a sedentary communal land management) often lead to degradation and desertification.

Up to now, there is very little known about the mechanisms of a sustainable land use in savannas. One reason for this remarkable discrepancy between economical importance and ecological understanding is that the 'good practice' case of pastoral-nomadism has been poorly investigated up until now (Tapson, 1993; Behnke & Scoones, 1993). Secondly, the spatiotemporal dynamics of savannas used as rangeland are particularly complex (Ellis, 1995). In order to understand these dynamics, the relationships between landscape patterns and ecological processes have to be examined. This is the main goal of landscape ecology (Turner et al., 1989; Tischendorf, 2001).

3.1.2. Landscape Ecology: A Transdisciplinary Science

Landscape ecology is a discipline currently considered as a bridge between basic and applied ecology. It explicitly includes humans causing functional changes on the landscape (Sanderson & Harris, 2000). Bastian (2001, p. 759) quotes two 'concise and comprehensive' definitions of the term landscape ecology. The definition by Forman, 1981: 'Landscape ecology . . . studies the structure, function, and development of landscapes' appears to be clearer than the definition by Leser, 1997: 'Landscape ecology deals with the interrelations of all functional and visible factors representing the landscape ecosystem.' The IALE Executive Committee (in Moss, 2000) gives a narrower definition. Here landscape ecology is understood as 'the study of spatial variation in landscapes at a variety of scales' and 'includes the biophysical and societal causes and consequences of landscape heterogeneity.'

Landscape ecology is one of the youngest branches of ecology (Farina, 1998, p. 1) and was born as a human-related science (Naveh, 1989). Recently, landscape ecology has become a unique and dynamic global science. Nevertheless, it is still passing through a process of self-discovery (Bastian, 2001, p. 757). As its roots are deep in geography as well as in geobotany and land management, this process contains the search for the unification of landscape ecology as a discipline, the relation between basic research and application, and between sectoral and holistic approaches and methods.

Due to its youth, landscape ecology does not offer unique definitions or a unified theory. There are still many divergent definitions of landscape, and even in recent publications on landscape ecology these definitions are merely compiled, often without a clear statement of which definition to favour (e.g. Bastian, 2001, p. 758), or retreating on a broad but highly imprecise definition (such as Farina, 1998, p. 2). The scientific term 'landscape' was shaped by geographers, most importantly by Alexander von Humboldt, 200 years ago (Table 3.1). A functional definition of landscape (or, as he calls it, landscape ecosystem) is given by Leser (1997, p. 25). His broad but clear definition integrates other, narrower, views.

Table 3.1. Definitions of the term landscape

"the total character of a region"	Von Humboldt
"landscapes dealt with in their totallty as physical, ecological and geographical entities, integrating all natural and human (caused) patterns and processes"	Naveh, 1987
"a heterogeneous land area composed of a cluster of interacting ecosystems that is repeated in similar form throughout"	Forman & Godron, 1986
"a particular configuration of topography, vegetation cover, land use, and settlement pattern which delimits some coherence of natural and cultural processes and activities"	Green et al., 1996
"a highly complex system (...) where abiotic, biotic and anthropogenic components and their interrelations form the structure and functioning of a generic entity (...)"[a]	Leser, 1997

[a]This is an English translation, which also reduces Leser's long work definition to its essential aspects. Leser's original definition is: "Das Landschaftsökosystem ist ein in der Realität hochkomplexes Wirkungsgefüge von physiogenen, biotischen und anthropogenen Faktoren, die mit direkten und indirekten Beziehungen untereinander einen Öbergeordneten Funktlionszusammenhang bilden, dessen räumlicher Repräsentant die 'Landschaft' ist."

Because of the various aspects of landscape (components, processes, and relations), landscape ecology should be regarded as a 'multidisciplinary' (Leser, 1997, p. 191), an 'interdisciplinary' (IALE Executive committee, 1998), or even better, a 'transdisciplinary' science. Transdisciplinary exists 'where interaction involves not only the scientific and technological disciplines in stated goals, but also where planners and administrators become involved in the processes' (Moss, 2000, p. 305), and 'where different approaches and views are involved in a holistic manner' (Bastian, 2001, p. 757). Other authors even consider a transdisciplinary approach as not far-reaching enough, such as Moss (2000, p. 303), who demands that 'landscape ecology must develop more as a discipline with its own theoretical bases and foci.'

New impulses may come from disturbance ecology and metapopulation theory. Even with its current lack of a strong theoretical and methodological base, holistic landscape ecology renders important tools to cope with increasing environmental problems and to achieve the sustainability of land use, at the same time reducing the high risk of 'considering landscape ecology from an exclusively anthropocentric viewpoint' (Farina, 1998, p. 7). The main strength of landscape ecology is its ability to transfer information across different families of processes occurring on different spatial and temporal scales. The greatest danger is that 'the term landscape ecology may become wishy-washy like terms such as ecological equilibrium, ecological stability . . . or sustainability' (Bastian, 2001, p. 759).

3.1.3. Savanna Landscapes: Scale and Heterogeneity

Landscapes are complex ecological systems that operate over broad spatiotemporal scales (O'Neill et al., 1989). Hierarchy theory provides guidelines for the scientific study of such complex systems (Allen & Starr, 1982). It predicts that landscapes will often develop hierarchical structure (O'Neill et al., 1989, 1992). A basic principle of hierarchy theory concerns how higher levels of organisation act to constrain lower levels. Higher levels of organisation act on larger spatial and temporal scales and constitute their context (Urban et al., 1987). Levels of organisation in ecology are: (1) cell, (2) organism, (3) population, (4) community, (5) ecosystem, (6) landscape, (7) biome, and (8) biosphere. Each level is composed of the subsystems on the next lower level and is controlled by the level above it. The hierarchy of these eight levels is often presented as a straight tower, but there is not necessarily a time- or space-dependent difference between the classes. These ecological levels of organisation are not scale-dependent, but are criteria for telling foreground from background, or the object from its context (Allen & Hoekstra, 1990).

3.1.4. Patch Dynamics in Savanna Landscapes

Apart from a simple ecological classification of biotic entities where the savanna landscape is part of a hierarchy reaching from cell level to biosphere level, landscape can also be classified in view of its components (Leser, 1997, p. 118 ff.).

The basic entity of this approach is a 'patch'. Because a landscape is intrinsically heterogeneous, the landscape components can be perceived as 'individual patches inserted into a matrix, by which we mean the dominant cover' (Farina, 1998, p. 11). Each landscape is a mosaic of discrete patches, which have different sizes, compositions, shapes, and longevities (Forman & Godron, 1986; Urban et al., 1987).

The borders of the patches can either be natural such as valley margins or anthropogenic such as forest margins following political territories. Patch boundaries are artificially imposed and are only meaningful when referenced to a particular scale (i.e., grain size and extent). Farina (1998, p. 12) compiles five different approaches for classifying a landscape and its component patches: (1) the structural patch composed of a soil type overlapped by vegetation types; (2) the functional patch which means an area homogeneous for a function or a physical trait; (3) the resource patch for the description of an animal's home range; (4) the habitat patch which may be defined as distinct plant community types generally larger than an individual home range; and (5) the corridor patch which is a functional portion of the land mosaic that is used by an organism to move, explore, disperse, and migrate.

In the so-called patch-corridor-matrix model (Forman & Godron, 1986), the resource or habitat patches of a landscape are connected by corridor patches. Here corridor patches are not defined by their function, but on the basis of their structure. Corridors are merely linear landscape elements which may function as habitat patches or functional patches, for example, dispersal conduits or barriers (Figure 3.1).

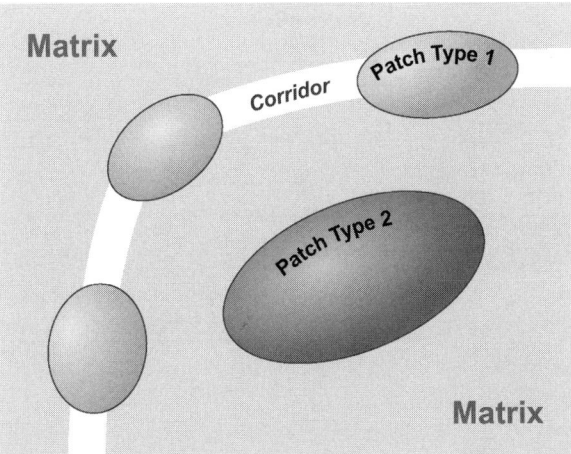

Figure 3.1. A visualization of the patch-corridor-matrix model. The matrix is the most extensive and most connected landscape component (Forman & Godron, 1986). Corridors are linear landscape elements, and patches are nonlinear landscape elements inserted into the matrix. In this visualization, patches of type 1 are connected by corridors, whereas patches of type 2 are not

The patch-corridor-matrix model is convenient for conceptualizing and representing landscape elements in a categorical map and has gained a broad acceptance in landscape ecology. It is particularly useful to classify fragmented cultural landscapes and to develop management plans for the maintenance of species diversity. The model as well as Farina's patch types 3–5 (see above) show a dominating biological view. This is criticised by some landscape ecologists with a geographic background or a holistic approach (such as Moss, 2000; Bastian, 2001). In order to understand the structure and function of a landscape system as a whole, the functional patch concept seems most appropriate. It is crucial to define patches relative to the phenomenon under investigation or management.

Urban et al. (1987) suggest that the patches within a landscape, and the patterns they create, be analysed along continua of spatial and temporal scales. With this technique, Belsky (1989) categorised the visually obvious vegetation patches in the savanna landscape of the Serengeti as being mostly relatively small, ranging from 1 to 20–30 m in diameter (Figure 3.2). A categorisation of the vegetation patches similar to the approach of Belsky (1989) is yet to be done for most other savanna landscapes.

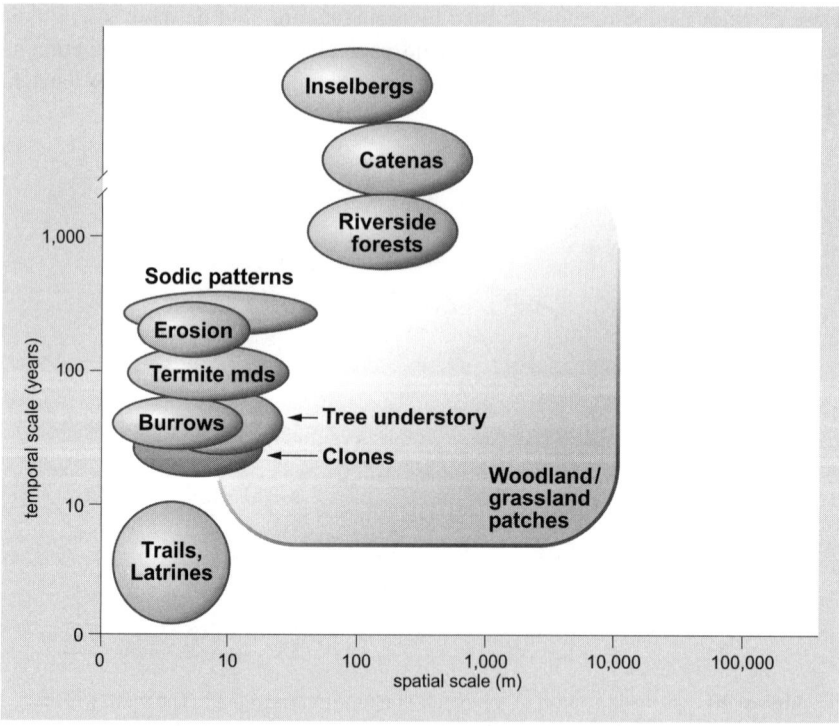

Figure 3.2. Sizes and longevities of vegetation patch types in the savanna landscape of Serengeti National Park, Tanzania (After Belsky, 1989, p. 267)

3.2. SAVANNA LANDSCAPES AND THE SIGNIFICANCE OF DISTURBANCE

3.2.1. What Characterises a Savanna Landscape?

Patterns and processes in savanna landscapes differ considerably from those in mesic regions. Early models of savanna ecology have focussed on a competitive equilibrium between tree and grass layer. The models were based on water availability as the principal limiting factor (Walter, 1954, 1971; Walker & Noy-Meir, 1982). Later the ecological understanding of savanna ecology was determined by four main variables: (1) the amount of water available for plants, (2) the amount of nutrients available for plants, (3) fire, and (4) herbivory (Frost et al., 1986; Walker, 1987). These determinants were thought to interact on all ecological levels (from landscape to microhabitat), but their relative significance was thought to vary between individual levels (Solbrig, 1991; Medina, 1996; Solbrig et al., 1996). In the past decade, there has been a paradigm shift away from equilibrium models (Figure 3.3). Recent models focus on the importance of environmental variability and disturbance (Justice et al., 1994; Jeltsch et al., 1996, 1998; Gómez Sal et al., 1999; Weber & Jeltsch, 2000). Savannas are seen as event-driven systems in which no equilibrium between grass layer and tree layer exists. The dominance between the woody and the grass component constantly changes in correlation with environmental conditions. In this context, Belsky (1989, p. 267) points out the great importance of patch-producing processes for savannas, and Jeltsch et al. (1999) compile a list of buffering mechanisms (Figure 3.3, bottom). These are ecological factors that either impede the transition from savanna to woodland (e.g., fire effects, browsing, or elephant damage of the tree layer) or to grassland (e.g., grazing or patch formation by ants or termites). Both patch-producing processes and buffering mechanisms, which create and maintain the typical patch dynamics of a savanna landscape, are mostly disturbances. A disturbance is defined as 'any relatively discrete event in time that disrupts ecosystem, community or population structure and changes resources, substrate availability, or the physical environment' (Picket & White, 1985). Landscape structure may change dramatically over time due to these disturbances (Baker, 1989). A savanna landscape is therefore characterised by its unique disturbance regime.

3.2.2 Savanna Landscapes and Disturbance Regime

Disturbances can alter the age, size, and spatial structure of landscape patches (Picket & White, 1985; White & Jentsch, 1991). Much theoretical research in landscape ecology has focussed on understanding the effect of disturbance processes, both natural and anthropogenic, on the vegetation (see Gardner & O'Neill, 1991; Turner et al., 1991 for reviews).

To be able to distinguish between 'natural' vegetation dynamics and disturbance-driven dynamics, one requires a thorough knowledge of the mechanisms

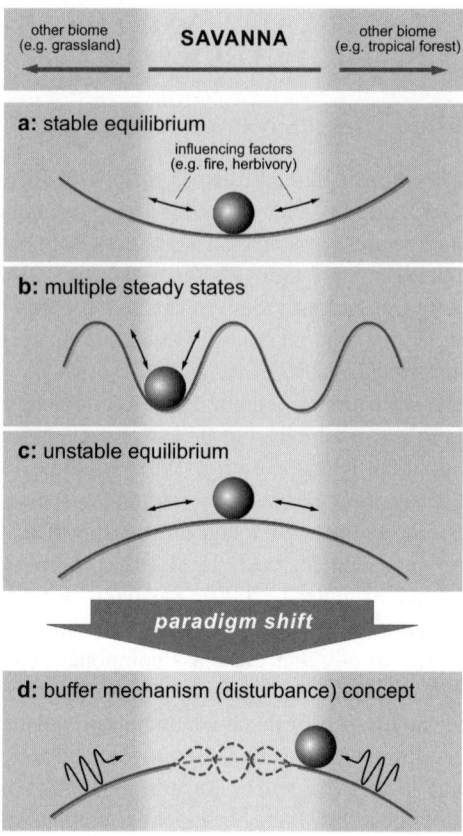

Figure 3.3. The paradigm shift in savanna ecology between equilibrium/disequilibrium concepts (a)–(c) and the concept of ecological buffering mechanisms (d). The first three concepts focus on the system's behaviour within its boundaries: (a) stable equilibrium, (b) multiple steady states, and (c) unstable equilibrium. The new ecological buffering concept (d) focuses on the system's boundaries, in particular the mechanisms that allow a savanna to persist in critical situations where the system is driven to its boundaries (After Jeltsch et al., 2000, p. 162)

involved and of the natural constraints of the landscape under investigation. In his conceptual model based on the hierarchy theory, Acker (1990) regards vegetation as part of a nonnested hierarchical system. Two components are mentioned as natural constraints: firstly the inherent constraints that cannot be changed by the vegetation, such as macroclimate and topography, and secondly the maximum available flora (Acker, 1990). Abrupt changes to the inherent constraints of the system manifest themselves as disturbances. When distinguishing between inherent and variable environmental factors, their scale in time and space is significant. A fire in a certain vegetation patch may be regarded as a disturbance, whereas seen on a landscape level, it might be regarded as part of the natural fire regime and thus as part of the inherent dynamics of a landscape. In this context not only fire, but also the highly variable climate of semi-arid savannas (which comprises extremely dry as well as extremely wet years) is part of the inherent constraints of this landscape. Similarly, herbivory by large mammals is not a disturbance of the total system, but should be seen as a 'disturbance regime' within its inherent constraints. A change in a landscape[1] is thus a change in landscape structure associated with considerable changes in the system's disturbance regime. Thus the disturbances are no longer the (original) inherent constraints of the landscape.

3.2.3. Disturbances and Land Use

Many ecologists have investigated savanna vegetation with regard to their specific responses to disturbances caused by land use (e.g. Shackleton et al., 1994; Fuhlendorf & Smeins, 1997; Landsberg et al., 1999). The best-known attempt to classify savanna species on the basis of their response to grazing led to the development of the so-called ecological status (Dyksterhuis, 1949; Trollope et al., 1990). Numerous South African research projects on range evaluation make use of this classification system. In principle the concept of ecological status is based on the notion of a linear succession proceeding from a (disturbed) 'disclimax' to the climax vegetation of a certain site (Clements, 1916). Dyksterhuis (1949, pp.108 ff.) observed that species along a grazing gradient either decrease or increase in abundance (Figure 3.4). Thus he divided the species of the disclimax into increasers and decreasers, defining three groups: species whose relative coverage decreases and which are eventually eliminated under heavy grazing ('decreaser'), species that increase for a certain period and then decrease (called 'increaser', and later 'increaser I'), as well as species whose coverage monotonously increases ('invaders', later called 'increaser II'; the term 'invader' now being restricted to exotic species, cf. Trollope et al., 1990).

Meanwhile the concept of a linear succession has largely been replaced by other theories of vegetation dynamics in arid and semi-arid landscapes (see above). Consequently Dyksterhuis' classification system from 1949 has forfeited its original theoretical foundation. However, because in principle it is based on a similar response of grass species to disturbances, it implicitly determines response groups.[2] Therefore the ecological status can be regarded as an attempt at functional classification which may retain its values as disturbance indicator. However, functional groups should be objectified by correlating plant abundance

---

[1] It is difficult to give an adequate English translation of the German term *Landschaftswandel*.

[2] According to Gitay and Noble (1997, p. 6), the term 'response group' should be used to describe groups of organisms on the basis of their response to a certain disturbance. Response patterns may be identified according to each species' individual responses along a disturbance gradient. Species may then be functionally classified according to similarities in their responses to disturbances, which manifest themselves in similar trait syndromes.

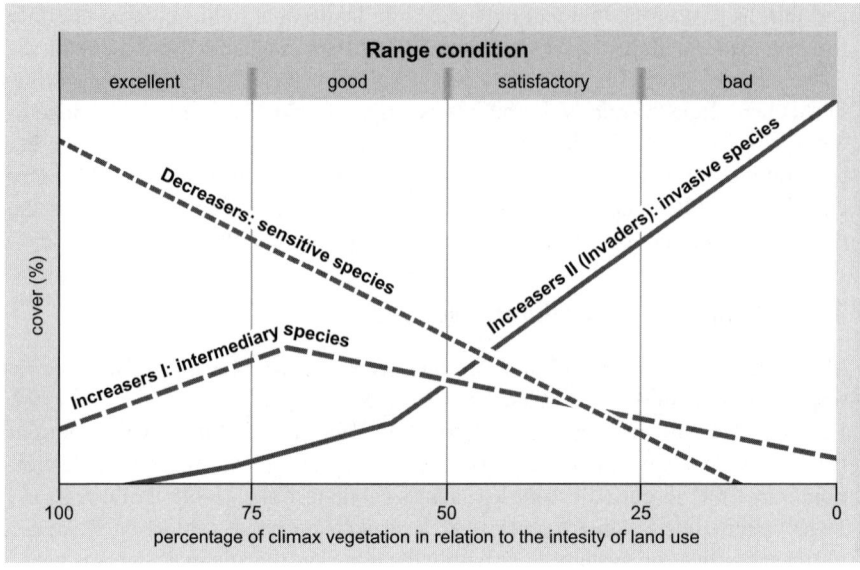

Figure 3.4. The ecological status: reaction of species along a grazing gradient
(Modified from Dyksterhuis, 1949, p. 109)

along a disturbance gradient with disturbance-related plant traits. Suitable sets of traits are, for example, compiled by McIntyre et al. (1995) and Weiher et al. (1999). To develop a set of fundamental traits, Weiher et al. (1999) began with the primary challenges faced by plants: dispersal, establishment, and persistence. As most of these traits are hard to measure, they suggest a number of easy to measure analogues such as seed mass, dispersal mode, specific leaf area, leaf water content, height, above-ground biomass, life history, and resprouting ability.

Response groups assist in classifying savanna landscape patches with respect to their grazing disturbance regime. Thus they connect spatial aspects of land use (i.e., patch size) with land use intensity and frequency.

3.3. HYPOTHESIS OF DISTURBANCE REGIME SIMILARITY

3.3.1. Hypothesis

In this chapter I hypothesize that a sustainable range management in a savanna landscape implies a disturbance regime that resembles the regime occurring under natural conditions (Schulte, 2001, p. 180). A similar pattern has already been observed for rainforest farming, where human land use implicitly copies the natural mosaic cycles of the vegetation.

3.3.2. Case Study: Kaokoland

In order to discuss this assumption, the landscape of Kaokoland in northwestern Namibia is compared to the near-natural savanna landscape in Etosha National Park (serving as a benchmark landscape[3]), and to other mopane savannas in southern Africa under a different land management. Kaokoland is inhabited by the Himba people, who have been living in this area as pastoral nomads for at least 200 years (Bollig, 1997, 2006, this volume). As a result of the remoteness of the area and due to politically conditioned isolation (Bollig, 1998), the Himba have preserved most aspects of local range management (Sander et al., 1998; Behnke, 1999; Bollig, 2006). Therefore Kaokoland provides ideal conditions for investigating the effects of pastoral-nomadism on a mopane savanna landscape.

In order to determine the influence of environmental and anthropogenic factors in vegetation dynamics, a paired site study was conducted of grazed and adjacent ungrazed plots in several settlement areas with different land use history (Schulte, 2001). The advantage of grazing exclosure experiments is that they allow a determination of the rate and extent of vegetation change following the removal of livestock grazing pressure. However, they do not render information about the natural vegetation grazed by large wild herbivores and burnt by frequent bush fires. For the natural disturbance regime imposed by wild herbivores and for the natural fire regime, information from the mopane savanna in adjacent Etosha National Park is extrapolated.

Covarying with rainfall, the abundance of predominant species fluctuated substantially on all grazed and ungrazed plots. In the area with traditional pasture management, considerable changes in vegetation structure and composition occurred within the first five years of protection from grazing. Most important was an increase of perennial grasses (Schulte, 2001, 2002). This may be interpreted as grass layer recovery after heavy disturbance by livestock grazing. If the traditional ways of nomadic land use have been abandoned due to the impact of state politics and due to a bundle of internal factors (e.g., demographic growth, institutional decline) on the local grazing regime, overexploitation may trigger a transition of the savanna landscape. In this case, biomass production decreases significantly, as only short-lived annual grasses are able to complete their life cycle. Usually this process is accompanied by severe soil erosion and a massive decline of the tree cover.

---

[3] The desired condition of a technical, ecological, or social-ecological system is known as a benchmark. For the social-ecological system 'rangeland,' it represents the optimal state or condition of the range with regard to a certain mode of land use. Therefore benchmarks vary depending upon the ecology of the area assessed and the demands of the relevant user group (Aucamp et al., 1992, p. 9). In the past benchmarks in range ecology were not frequently regarded as being relative to a certain context but were understood as being either 'the percentage of the present vegetation which is original vegetation for the site' (Dyksterhuis 1949, p. 105) or reduced to the ?'condition under normal climate and best practicable management' (Hawley 1944, quoted according to Dyksterhuis, 1949, p. 104), or they were borrowed from other studies.

Grazing exclosure experiments allow a reconstruction of the main vegetation changes after the onset of pastoral-nomadic use, with the restriction that grazing exclosure data are not fully congruent to data from a savanna under a natural disturbance regime (see above). However, they give valuable information on the recovery after a severe disturbance. In the landscape before the onset of pastoral-nomadic land use, the grass layer was dominated by perennial grasses. Grazing by domesticated herbivores led to an adaptation of the grass layer to high-frequency disturbances. The perennial grasses were mostly replaced by productive annuals. Nevertheless, the production of palatable biomass remained relatively constant. The main driving force for short-term vegetation dynamics in this secondary savanna is rainfall variability. In this sense, the savanna in Kaokoland is a typical arid rangeland. The opportunistic range management of the OvaHimba is well-adapted to the spatial and temporal variation of plant production and can be regarded as sustainable (Bollig & Schulte, 1999; Schulte, 2001, 2002). This system is a human-shaped landscape. Only if a certain threshold is crossed, is the system altered in the sense of degradation, leading to a destroyed landscape.

3.3.3. Disturbance Regimes in Savanna Landscapes

3.3.3.1. Spatial Extent

Does the disturbance regime imposed by pastoral-nomadic land use diverge considerably from the natural regime? In other words: Are savannas used by pastoral nomads 'man-made landscapes'? To answer these questions (1) the spatial dimension and (2) the temporal dimension of patch-creating disturbance events have to be compared.

Turner et al. (1993) propose to use a ratio to deal with spatial aspects of a disturbance event. They relate the size of a disturbance patch to the total area of the landscape. In this contribution, this ratio is named the *spatial extent* of a disturbance event (Figure 3.5, *x*-axis). It expresses the proportion of the savanna landscape affected by a particular type of disturbance: 'local' indicates a small proportion and 'extensive' implies a large proportion of disturbed area.

3.3.3.2. Recovery Potential

As regards the temporal dimensions of a disturbance event, an approach following Turner et al. (1993) is used. Here, the disturbance interval is related to the recovery interval. Recovery is complete when the predisturbance status has been re-established. Thus the recovery interval depends upon the intensity of the particular disturbance (Böhmer & Richter, 1997) and upon the system's ability to regenerate. The ratio between disturbance interval and recovery interval can be expressed as the *recovery potential* (Figure 3.5, *y*-axis).

The more localised the disturbance and the higher the recovery potential of an ecosystem, the sooner a system returns to a dynamic state of equilibrium. Where the recovery potential is low and the disturbance extensive, the ecosystem's state of equilibrium is unstable and its catastrophic breakdown imminent.

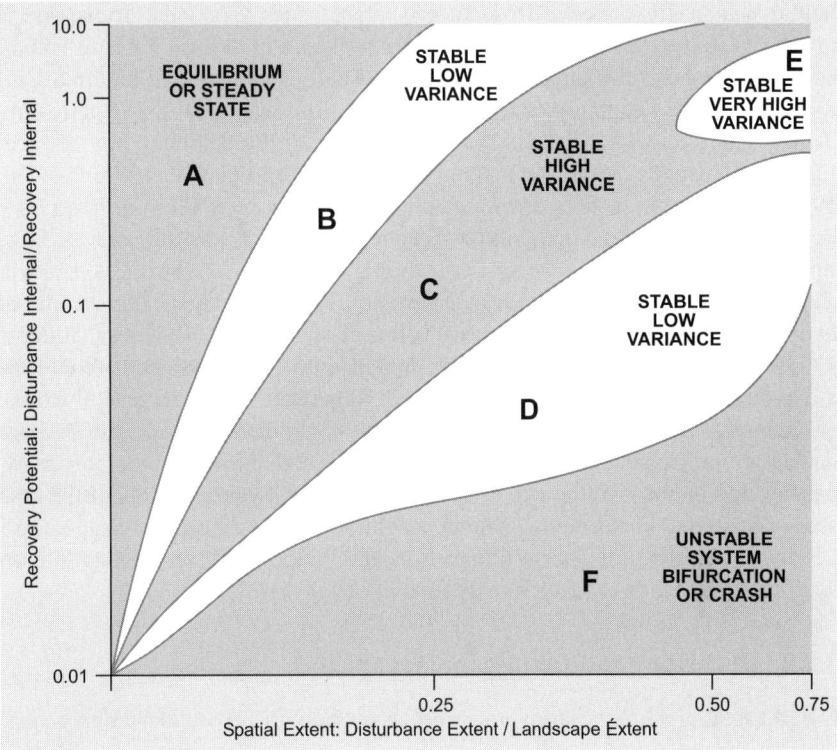

Figure 3.5. Stability and variance of landscape dynamics as a function of two ratios. On the *x*-axis, the spatial extent, which is the spatial relation between disturbance patch size and landscape area is given, on the *y*-axis, the recovery potential (i.e., the temporal relation between disturbance interval and recovery interval). The lines separate landscape types that display qualitatively different dynamics. These types are derived from regions of high or low standard deviation (SD) in the proportion of the landscape occupied by the mature stage of vegetation (Graph after Turner et al., 1993, pp. 219, 220)

3.4. DISTURBANCE REGIMES IN SAVANNA LANDSCAPES

Large-scale disturbances in natural savannas are either caused by drought (as a result of the high temporal and spatial rainfall variability), by fire, by flooding, by herbivory, or by large mammals which have the ability to serve as landscape architects.

3.4.1. Fire

Fire is a critical factor in savanna landscapes. Intensity, frequency, and the area affected essentially depend upon environmental conditions. Whereas dry savannas only burn in years of extraordinarily high biomass production in the grass layer, humid savannas experience frequent and intense fires unless the dry season is too short to allow the grass layer to dry out sufficiently to burn. Fires in natural savannas are generally caused by lightning, usually during one of the first thunderstorms at the beginning of the rainy season. Fire events at this stage mainly damage the tree layer, whereas the dormant grasses are distinctly less harmed. Early fires usually stimulate perennial grasses to sprout (Mentis & Tainton, 1984). Because these fires are soon followed by the rainy season, the grasses experience a full vegetation period and can replenish their nutrient reserves. Most fires started by humans, however, occur during the dry season (Edwards, 1984) at a time when both trees and grasses are dormant. The new growth of grass induced by the fire dies, due to a lack of water, or it is consumed by herbivores (Trollope, 1992). Thus, the grasses cannot replenish their nutrient reserves and are eventually replaced by ephemeral species (Frost & Robertson, 1987). In general woody species – in spite of protective adaptations such as a thick corky bark – suffer more damage from fire than grasses. If savannas are protected from burning, diversity, density, and biomass of the tree layer increases (Braithwaite, 1996). For the case study Kaokoland in Northern Namibia, a natural fire frequency between 5 and 15 years is estimated. In adjacent Etosha National Park, it is recommended to burn landscape patches every 5 to 9 years (Stander et al., 1993).

3.4.2. Herbivory and Landscape Architects

Out of a total of 92 large herbivorous species on the African continent, rarely more than 20 species occur together in one type of savanna (Cumming, 1982). With the exception of national parks and areas infested with tsetse-fly, wild herbivores have generally been replaced by domesticated animals. Both the indigenous and the domesticated herbivores vary considerably in size, digestive system, and grazing habits. A distinction is usually made between grazers and browsers: browsers feed predominantly on the leaves of trees, shrubs, and forbs, whereas grazers eat mainly grasses and herbs. Some of the large herbivores also serve as landscape architects. In a natural mopane savanna such as in Namibia's Etosha National Park, elephants are a prominent example, causing frequent damage to the tree layer, sometimes coined 'bulldozing' (Frost et al., 1986). Damage is often localised and is related to areas of high elephant concentration (Timberlake, 1995, p. 11). The African bush elephant (*Loxodonta africana*) used to be common in Kaokoland until recently, but has become rare in the past decades (Bollig, 2006). Abundant numbers of other large herbivores have been reported in Kaokoland until the 1990s, including grazers such as Hartmann's mountain zebra (*Equus zebra hartmannae*) and blue wildebeest (*Connochaetus taurinus*), and browsers such as giraffe (*Giraffa camelopardalis*) and greater kudu (*Tragelaphus strepsiceros*). Today most of these species are locally extinct or extremely rare in Kaokoland, but are still found in adjacent Etosha National Park (Legget et al., 2003).

In contrast to the large herbivores, relatively little is known about the role of other herbivorous species in savanna landscapes. It is assumed that insects are significant herbivores.

In nutrient-rich savannas, trees in particular are exposed to episodic denudation by caterpillars (Scholes & Walker, 1993). This also holds true for the mopane savanna in the study area Kaokoland and adjacent Etosha National Park. Here the dominant tree species *Colophospermum mopane* is regularly affected by mopane worms (caterpillars of the mopane emperor moth *Imbrasia belina*; Styles, 1994). Mopane worms are a regular disturbance for tree individuals, causing large patches of the tree layer to be defoliated at the end of the rainy season (personal observation in the years 1995–2006).

3.4.3. Small-Scale Disturbances

The effects of small-scale disturbances affecting soil properties may be comparatively long-lasting due to the slow weathering of soil (Belsky, 1989, p. 267 ff). This preserves small-scale patches created by digging animals such as ants, termites, aardvarks (*Orycteropus afer*), and warthogs (*Phacochoerus africanus*). The subsurface particles and salts that are brought aboveground by the animals are not quickly weathered or leached. This implies a low recovery potential of the savanna landscape with respect to soil disturbances. On the other hand, natural soil erosion may overlay patch dynamics created by digging, leading to a different disturbance regime.

When comparing the temporal and spatial extent of individual disturbance types in a near-natural savanna and a savanna managed by pastoral nomads (Table 3.2), it becomes evident that differences among the frequency, intensity, and the spatial dimension of disturbances are small. With the exception of the factor drought (in the sense of a sequence of very dry years), which cannot be controlled by man, both systems are characterised by extensive, highly frequent disturbances of the grass layer (grazing) and the tree layer (browsing) as well as localised and intensive but infrequent disturbances of the tree layer. In the natural system, it is tree damage caused by elephants and bush fires, in the pastoral-nomadic system, tree damage caused by woodcutting and other settlement activities. These disturbances are concentrated around water places. In both systems, small-scale soil disturbances are caused by insects and small herbivores, which show a trade-off with the intensity of herbivory.

However, grazing as practised by pastoral nomads due to its higher frequency and intensity reduces the recovery potential of the grass layer to a greater degree than grazing by wild herbivores. Consequently perennial grasses are replaced by annual grasses. Due to their short life cycle, annual grasses are well adapted to survive frequent and intensive disturbances (r-strategists, cf. Oksanen & Ranta, 1992). The tree layer in large areas under pastoral-nomadic land use, due to the absence of fire and elephants, experiences fewer disturbances than it would in a natural ecosystem. Only in the vicinity of human settlements does human wood consumption produce a roughly similar disturbance pattern. Accelerated soil erosion may well occur under pastoral-nomadic land management (Schulte, 2001, pp. 154, 155), and may overlay small-scale patch dynamics created by digging animals.

Table 3.2. Comparison of the disturbance regime (1) in a near-natural savanna landscape (Etosha National Park), (2) in a landscape utilised according to pastoral-nomadic practice (Kaokoland), and (3) in an overutilised landscape (Kaokoland, close to permanent settlements). All landscapes are situated in northwestern Namibia. Some types of disturbances affect the tree layer (TL) and the grass layer (GL) to different degrees. Consequently, the recovery potential (i.e., the time required to return to the original state) also differs.

Landscape status	Source of disturbance	Disturbance event	Spatial extent	Frequency	Intensity	Recovery potential
(1) Natural landscape	Rainfall variability	Drought	Extensive	Rare	TL: high GL: very high	TL: medium GL: medium
	Zebra etc.	Grazing	Extensive	±annual	GL: medium	GL: good
	Giraffe etc.	Browsing	Extensive	Frequent	TL: low	TL: good
	Mopane worms, etc.	Herbivory by insects	Local to extensive	Frequent to annually	TL: very high GL: very high	TL: good GL: good
	Elephant	"Bulldozing"	Local to extensive	Rare	TL: very high	TL: medium
	Fire	Fire	Local to extensive	Rare (ca. 5-10 years)	TL: very high GL: medium	TL: medium GL: good
	Digging animals	Soil disturbance	Small-scale	Frequent	GL: high Soil: high	GL: good Soil: low
	Rainfall events	Soil erosion	Local to extensive	Rare	Soil: high	Soil: good
(2) Cultural landscape (Pastoral nomadic land use)	Rainfall variability	Drought	Extensive	Rare	TL: high GL: very high	TL: medium GL: medium
	Cattle	Grazing	Extensive	Annually	GL: very high	GL: **low**
	Goats	Browsing	Extensive	Annually	TL: low	TL: good
	Mopane worms etc.	Herbivory by insects	Local to **extensive**	Frequent to annually	TL: very high GL: very high	TL: good GL: medium
	Settlement	Cutting, grazing	Local	Rare	TL: very high GL: very high	TL: medium GL: good
	Digging animals	Soil disturbance	Small-scale	Frequent	GL: high Soil: high	GL: good (?) Soil: **low**
	Rainfall events	Soil erosion	**Extensive**	Frequent	Soil: high	Soil: **low**

Landscape status	Source of disturbance	Disturbance event	Spatial extent	Frequency	Intensity	Recovery potential
(3) Degrading landscape (Over-utilisation due to high density of population)	Rainfall variability	Drought	Extensive	Rare	TL: high GL: very high	TL: medium GL: medium
	Cattle	Grazing	Extensive	Annually	GL: very high	GL: low
	Goats	Browsing	extensive	Frequent	TL: medium	TL: medium
	Mopane worms etc.	Herbivory by insects	Local to extensive	Frequent to annually	TL: very high GL: very high	TL: low (?) GL: low (?)
	Settlement	Cutting, browsing, grazing	Local	Frequent	TL: very high GL: very high	TL: low GL: low
	Digging animals	Soil disturbance	Local	Frequent	GL: high Soil: high	GL: low (?) Soil: low
	Rainfall events, wind	Soil erosion	Extensive	Frequent high	Soil: very	Soil: very low

Pastoral-nomadic land use does indeed seem to be characterised by a disturbance regime, which in many aspects is similar to the disturbance regime occurring under natural conditions. It is a form of land use to which flora and vegetation are, to a certain degree, pre-adapted. For instance, the tree species *C. mopane,* which dominates the tree layer of the savannas in northwestern Namibia, is exceptionally tolerant to being cut due to its tolerance of fire and elephant damage. However, there are specific differences between the two disturbance regimes, especially with regard to the shorter recovery intervals of the grass layer under pastoral-nomadic land use. Perennial grasses adapted to moderate grazing due to their ability of a compensatory growth ('Decreaser', see Figure 3.4), are replaced by productive annual species with the ability to opportunistically cover bare ground ('Increaser I'). Because the bulk of the grass biomass is consumed, fires are eliminated (cf. Skarpe, 1995). Projected on a landscape level, both the natural and the anthropogenic disturbance regimes result in a dynamic equilibrium, due to interaction between the disturbed areas.

3.5. DISTURBANCE REGIME IN A DEGRADING LANDSCAPE

When the population and the stocking rate increase, the disturbance regime changes dramatically when compared to that of a natural landscape (Table 3.2, bottom). Initially this is only the case on a relatively small 'sacrifice area', that is, the immediate vicinity of a settlement. Because the settlements are no longer abandoned after a certain period of time (about 20 years) neither the tree layer nor the grass layer have enough time to recover. This manifests itself in more frequent and more intensive damage (disturbance) to the tree layer as a result of browsing and woodcutting. Furthermore, highly accelerated soil erosion occurs. Due to the low recovery potential of the vegetation in the vicinity of such settlements and due to the extremely low recovery potential of soils in general, the ecosystem eventually collapses (Sander et al., 1998).

3.6. USER-SPECIFIC DISTURBANCE REGIMES

The divergent economic objectives of different user groups lead to considerable ecological differences between the various types of land use in southern African savanna landscapes. This fact is clearly reflected by specific benchmarks and the management strategies applied by the user groups (Table 3.3).

3.7. PASTORAL-NOMADIC MANAGEMENT STRATEGIES

The ecological characteristics of a disturbance regime shaped by traditional pastoral-nomadic land use have been discussed in the previous paragraph. Here the economic motivation for the implicit preference of this system is investigated.

Table 3.3. Divergent economic objectives of different user groups in southern African savannas in relation to the ecological status. The 'desired' disturbance regime of a landscape is realised by specific management practices

User group	Economic objective	Implicit ecological objective	Desired disturbance regime of savanna landscape
Pastoral nomads (OvaHimba)	Maximum annual production of palatable biomass	Maintaining the dominance of annual grasses (Increaser I species)	Long, intensive grazing during the dry season
Farmer: conservative	Consistent production of fodder biomass	Maintaining the perennial grass populations (Decreaser species)	High frequency and extended grazing periods
Farmer: Holistic Resource Management	Consistent production of palatable fodder biomass	Maintaining the perennial grass populations with high fodder value	Low frequency and intensive, short grazing periods (imitating the natural disturbance regime)
Nature conservators	Variable (diversity, tourism, etc.)	Maintaining natural or desired (e.g. diverse) systems	Natural disturbance regime or regime of the desired system

The herdsmen of Kaokoland are primarily interested in a maximum annual production of palatable fodder biomass (Table 3.3, top). They regard the fodder value of certain annual grasses as distinctly higher than that of the perennial species, which would dominate under a natural disturbance regime. Consequently the benchmark for local range management is not a natural landscape dominated by perennial grasses. On the contrary, the perennial grasses should, if possible, be suppressed in favour of productive annual grasses. This goal is reached by long intensive grazing during the dry season. The availability of large tracts of fodder biomass during periods of drought is apparently not the main objective of their opportunistic range management (Bollig & Schulte, 1999; Linstädter & Bolten, 2007). Likewise, the slightly reduced productivity of the grass layer as compared to that of the natural vegetation is tolerated.

3.8. FARMERS' MANAGEMENT STRATEGIES

Land use practised by farmers in southern Africa is characterized by a very low user density compared to user groups with communal management strategies. Commercial farming may be separated into conservative and holistic management (Table 3.3). The benchmark for both land management systems is the natural condition of the vegetation, and the maintenance of perennial grass populations is a major objective. It is mainly argued that only perennial grasses will provide sufficient fodder on the farm during dry years. As farmers are much less flexible horizontally than pastoral nomads, it is an economic necessity to maintain local standing biomass for periods of drought (Müller et al., 2007b). Survival of the perennial grasses is achieved by conservative management and by a relatively unnatural disturbance regime: fenced-off camps are grazed for relatively long periods at low stocking densities.

The 'Holistic Resource Management' (Savory, 1988) attempts to imitate the disturbance regime of large herds of wild herbivores by keeping livestock at high stocking rates in small camps for short periods. The objectives are: (1) to get an intense but short disturbance impact by trampling, and (2) to prevent the animals from selectively grazing the particularly palatable grasses, inasmuch as long-term selective grazing would lead to a reduction in pasture quality. All palatable species are heavily grazed to a roughly equal degree and experience long recovery phases. Similar to conservative pasture management, holistic management focuses on the conservation of perennial grasses (Squires et al., 1992) ensuring standing crop for drought situations, on a local level.

3.9. SPACE MATTERS: LESSONS OF LANDSCAPE ECOLOGY FOR A SUSTAINABLE RANGE MANAGEMENT IN SAVANNAS

The landscape-level approach to examine and conceptualise the disturbance regimes and the implied ecological objectives of savanna management has helped to identify crucial aspects of a sustainable range management: It not only has to be adapted to temporal variability, but also to spatial variability on a landscape level (Linstädter & Bolten, 2007).

These findings have important implications for the scale of research. The analysis of land use patterns and disturbances has to be done at different spatial scales. Plot-scale research is not necessarily relevant to the scales at which savannas and other (semi-) arid landscapes are managed. It might be misleading if the properties of the broader-scale landscape override findings at the small homogeneous plot level (Fisher et al., 1999, p. 167). A hierarchically structured sampling design, where patterns of vegetation, management, and disturbances are documented at several spatial scales eliminates shortcomings of the plot scale. A major focus of research needs to be on understanding how different landscape units can provide for and respond to different levels of use (Fisher et al., 1999, pp. 167, 168). For the case study region of Kaokoland, this has been done within the framework of a multidisciplinary research project.

A sustainable grazing management in savanna landscapes has to ensure fodder reserves as an ecological buffer against temporal variability (Müller et al., 2007a, b), and it has to be adapted to spatial variability, in particular spatial differences in vegetation productivity and recovery potential. A fodder reserve may either be ensured on a local level as biomass of perennial species, and/or as protected reserves for drought. It may also be obtained on the landscape level through external (nonlocal) resources, either by moving livestock to external pastures, or by fodder supply. Focusing on the recovery potential of vegetation patches, vegetation patches with a comparably high recovery potential after disturbances such as drought or heavy grazing, will be used more intensely and more regularly than less resilient vegetation patches (Müller et al., 2007b). Ecological buffers are crucial for two key situations: firstly for scarce times within the annual grazing cycle, and secondly for drought times. Different user groups have alternative strategies ensuring the availability of fodder reserves. Whereas farmers tend to rely on internal biomass built up by perennial grasses (Table 3.3), pastoral nomads make more use of external fodder resources. These differences have already been described as the horizontally flexible strategy of mobile pastoralists. Thus spatial flexibility on a landscape level is not the only strategy to cope with temporal variability of resources, but it is crucial in cases where fodder shortages in key situations may not be buffered by local fodder resources.

Acknowledgements Financial support by the Deutsche Forschungsgemeinschaft (SFB 389) and by the Volkswagen Foundation (AZ: II/79 041) is gratefully acknowledged.

REFERENCES

Acker, S.A. (1990). Vegetation as a component of a non-nested hierarchy: A conceptual model. *Journal of Vegetation Science*, 1, 683–690.

Allen, T.F.H. & Starr, T.B. (1982). *Hierarchy: Perspectives for Ecological Complexity*. Chicago: University of Chicago Press.

Allen, T.F.H. & Hoekstra, T.W. (1990). The confusion between scale-defined levels and conventional levels of organization in ecology. *Journal of Vegetation Science*, 1, 5–12.

Arnold, G.W. (1995). Incorporating landscape pattern into conservation programs. In L. Hansson, L. Fahrig & G. Merriam (Eds.), *Mosaic Landscapes and Ecological Processes* (pp. 309–337). London: Chapman & Hall.

Aucamp, A.J., Danckwerts, J.E. & Tainton, N.M. (1992). Range monitoring in South Africa: A broad perspective. *Journal of the Grassland Society of South Africa*, 9, 8–10.

Ayoub, A.T. (1998). Extent, severity and causative factors of land degradation in the Sudan. *Journal of Arid Environments*, 38, 397–409.

Baker, W.L. (1989). A review of models of landscape change. *Landscape Ecology*, 2, 111–133.

Bastian, O. (2001). Landscape ecology – Towards a unified discipline? *Landscape Ecology*, 16, 757–766.

Behnke, R.H. & Scoones, I. (1993). Rethinking range ecology: Implications for rangeland management in Africa. In Behnke, R.H. Scoones & I. Kerven (Eds.), C. *Range Ecology at Disequilibrium: New Models of Natural Variability and Pastoral Adaptation in African Savannas* (pp. 1–30). London: Overseas Development Institute.

Behnke Jr, R.H. (1999). Stock-movement and range-management in a Himba community in north-western Namibia. In M. Niama-Fuller (Ed.), *Managing Mobility in African Rangelands: The Legitimization of Transhumance* (pp. 184–216). London: Intermediate Technology.

Belsky, A.J. (1989). Landscape patterns in a semi-arid ecosystem in East Africa. *Journal of Arid Environments*, 17, 265–270.

Böhmer, H.J. & Richter, M. (1997). Regeneration of plant communities – An attempt to establish a typology and a zonal system. *Plant Research and Development*, 45, 74–88.

Bollig, M. (1997). Risk and risk minimisation among Himba pastoralists in northwestern Namibia. *Nomadic Peoples*, 1, 66–89.

Bollig, M. (1998). The colonial encapsulation of the north-western Namibian pastoral economy. *Africa*, 68, 506–536.

Bollig, M. (2002). Problems of resource management in Namibia's rural communities: transformations of land tenure between state and local community. *Die Erde*, 133, 155–182.

Bollig, M. (2006). *Risk Management in a Hazardous Environment – A Comparative Study of Two Pastoral Societies*. Berlin: Springer.

Bollig, M. & Schulte, A. (1999). Environmental change and pastoral perceptions: Degradation and indigenous knowledge on two African pastoral communities. *Human Ecology*, 27, 493–514.

Braithwaite, R. (1996). Biodiversity and fire in the savanna landscape. In O.T. Solbring, E. Medina & J.F. Silva (Eds.), *Biodiversity and Savanna Ecosystem Processes. A Global Perspective* (pp. 121–140). Berlin: Springer.

Clements, F.E. (1916). *Plant succession: An analysis of the development of vegetation*. Washington, DC: Carnegie Institution of Washington, publication No. 242.

Cumming, D.H.M. (1982). The influence of large herbivores on savanna structure in Africa. In Huntley & B.J. Walker (Eds.), B.H. *Ecology of Tropical Savannas* (pp. 217–245). Berlin: Springer.

Darkoh, M.B.K. (1998). The nature, causes and consequences of desertification in the drylands of Africa. *Land Degradation and Development*, 9, 1–20.

Dyksterhuis, E.J. (1949). Condition and management of range land based on quantitative ecology. *Journal of Range Management*, 2, 104–115.

Edwards, P.J. (1984). The use of fire as a management tool. In P. De v. Booysen & N.M. Tainton (Eds.), *Ecological Effects of Fire in South African Ecosystems* (pp. 349–362). Berlin: Springer.

Ellis, J.E. (1995). Climate variability and complex ecosystem dynamics: Implications for pastoral development. In I. Scoones (Ed.), *Living with Uncertainty: New Directions in Pastoral Development in Africa* (pp. 37–46). London: Intermediate Technology.

Ellision, L. (1960). Influence of grazing on plant succession of rangelands. *Botanical Revue*, 26, 1–78.

Farina, A. (1998). *Principles and Methods in Landscape Ecology*. London: Chapman & Hall.

Fisher, J.T., Stafford Smith, M., Cavazos, R., Manzanilla, H., Ffolliott, P.F., Saltz, D., Irwin, M., Sammis, T.W., Swietlik, D., Moshe, I. & Sachs, M. (1999). Land use and management: research implications from three arid and semi-arid regions of the world. In T.W. Hoekstra & M. Shachak (Eds.), *Arid Lands Management Towards Ecological Sustainability* (pp. 143–170). Chicago: University of Illinois Press.

Foran, B.D., Tainton, N.M. & Booysen, P. De V. (1978). The development of a method for assessing veld condition in three grassveld types in Natal. *Proceedings of the Grassland Society of Southern Africa*, 13, 27–33.

Forman, R.T.T. (1981). Interaction among landscape elements: A core of landscape ecology. *Proceedings of the International Congress organized by the Netherlands Society for Landscape Ecology*, Veldhoven, The Netherlands, April 6–11, 1981. Pudoc, Wageningen, pp. 35–48.

Forman, R.T.T. & Godron, M. (1986). *Landscape Ecology*. New York: Wiley.

Frost, P.E., Medina, P.E., Menault, J.-C., Solbrig, O., Swift, M. & Walker, W. (1986). Responses of savannas to stress and disturbances. *Biology International*, 10, 1–82.

Frost, P.E. & Robertson, F. (1987). The ecological effects of fire in savannas. In B.H. Walker (Ed.), *Determinants of Tropical Savannas* (pp. 93–140) Miami: ISCU Press.

Fuhlendorf, S.D. & Smeins, F.E. (1997). Long-term vegetation dynamics mediated by herbivores, weather and fire in a Juniperus-Quercus savanna. *Journal of Vegetation Science*, 8, 819–828.

Gardner, R.H. & O'Neill, R.V. (1991). Pattern, process, and predictability: The use of neutral models for landscape analysis. In M.G. Turner & R.H. Gardner (Eds.), *Quantitative Methods in Landscape Ecology* (pp. 289–307). New York: Springer-Verlag.

Gitay, H. & Noble, I.R. (1997). What are functional types and how should we seek them? In T.M. Smith, H.H. Shugart & F.I. Woodward (Eds.), *Plant Functional Types: Their Relevance to Ecosystem Properties and Global Change* (pp. 3–19). Cambridge: Cambridge University Press.

Gómez Sal, A., Rey Benayas, J.M., López-Pintor, A. & Rebollo, S. (1999). Role of disturbance in maintaining a savanna-like pattern in Mediterranean Retama sphaerocarpa shrubland. *Journal of Vegetation Science*, 10, 365–370.

Hurt, C.R., Hardy, M.B. & Tainton, N.M. (1993). Identification of key grass species under razing in the Highland Sourveld of Natal. *African Journal of Range and Forage Science*, 10, 96–102.

IALE Executive Committee (1998). IALE mission statement. Bulletin, International Association for Landscape Ecology, 16, 1.

Jeltsch, F., Milton, S.J., Dean, W.R.J. & Van Rooyen, N. (1996). Tree spacing and coexistence in semiarid savannas. *Journal of Ecology*, 84, 583–595.

Jeltsch, F., Milton, S.J., Dean, W.R.J., Van Rooyen, N. & Moloney, K.A. (1998). Modelling the impact of small-scale heterogeneities on tree-grass coexistence in semi-arid savannas: A modelling study. *Journal of Ecology*, 86, 780–794.

Jeltsch, F., Moloney, K., & Milton, S.J. (1999). Detecting process from snapshot pattern: lessons from tree spacing in the southern Kalahari. *Oikos* 85, 451–466.

Jeltsch, F., Weber, G.E., & Grimm, V. (2000). Ecological buffering mechanisms in savannas: A unifying theory of long-term tree-grass coexistence. *Plant Ecol.* 150, 161–171.

Justice, C., Scholes, R.J. & Frost, P. (Eds.) (1994). *African Savannas and the Global Atmosphere. Global Change Report 31*. Miami: ISCU Press.

Landsberg, J., Lavorel, S. & Stol, J. (1999). Grazing response groups among understorey plants in arid rangelands. *Journal of Vegetation Science*, 10, 683–696.

Leggett, K., Fennessy, K. & Schneider, S. (2003). Does land use matter in arid environments? A case study from the Hoanib River catchment, north-western Namibia. *Journal of Arid Environments*, 53, 529–543.

Leser, H. (1997). *Landschaftsökologie*. Stuttgart: Verlag Eugen Ulmer.

Linstädter, A. & Bolten, A. (2006). Learning from the Himba nomads. Investigation of a land use strategy in the savannas of Namibia. *Geographische Rundschau, International Edition*, 2, 21–27.

Linstädter, A. & Bolten, A. (2007). Space matters – Sustainable range management in a highly variable environment. In Bubenzer. et al. (Eds.), O. *Atlas of Cultural and Environmental Change in Arid Africa* (pp. 104–107). Institut für Ur- und Frühgeschichte der Universität zu Köln, Forschungsstelle Afrika.

Mainguet, M. & Da Silva, G.G. (1998). Desertification and drylands development: What can be done?. *Land Degradation and Development*, 9, 375–382.

McIntyre, S., Lavorel, S. & Tremont, R.M. (1995). Plant life-history attributes: Their relevance to disturbance response in herbaceous vegetation. *Journal of Ecology*, 83, 31–44.

Medina, E. (1996). Biodiversity and nutrient relations in savanna ecosystems: Interactions between primary producers, soil microorganisms, and soils. In O.T. Solbring, E. Medina & J.F. Silva (Eds.), *Biodiversity and Savanna Ecosystem Processes. A Global Perspective* (pp. 45–57). Berlin: Springer.

Mentis, M.T. & Tainton, N.M. (1984). The effect of fire on forage production and quality. In P.De v. Booysen & N.M. Tainton (Eds.), *Ecological Effects of Fire in South African Ecosystems* (pp. 245–254). Berlin: Springer.

Moss, M.R. (2000). Interdisciplinarity, landscape ecology and the 'Transformation of Agricultural Landscapes'. *Landscape Ecology*, 15, 303–311.

Müller, B., Frank, K. & Wissel, C. (2007b). Relevance of rest periods in non-equilibrium rangeland systems – A modelling analysis. *Agricultural Systems*, 92, 295–317.

Müller, B., Linstädter, A., Frank, K., Bollig, M. & Wissel, C. (2007a). Learning from local knowledge: Modeling the pastoral-nomadic range management of the Himba, Namibia. *Ecological Applications*, 17, 1857–1875.

Naveh, Z. (1987). Biocybernetic and thermodynamic perspectives of landscape functions and land use patterns. *Landscape Ecology*, 1, 75–83.

Naveh, Z. (1989). The challenges of desert landscape ecology as a transdisciplinary problem-solving oriented science. *Journal of Arid Environments*, 17, 245–253.

Oksanen, L. & Ranta, E. (1992). Plant strategies along mountain vegetation gradients: A test of two theories. *Journal of Vegetation Science*, 3, 175–186.

O'Neill, R.V., Gardner, R.H. & Turner, M.G. (1992). A hierarchical neutral model for landscape analysis. *Landscape Ecology*, 7, 55–61.

O'Neill, R.V., Johnson, A.R. & King, A.W. (1989). A hierarchical framework for the analysis of scale. *Landscape Ecology*, 3, 193–205.

Picket, S.T.A. & White, P.S. (1985). Natural disturbance and patch dynamics: An introduction. In Picket, & S.T.A.White, (Eds.), P.S. *The Ecology of Natural Disturbance and Patch Dynamics* (pp. 3–13). Orlando, FL: Academic Press.

Sander, H., Bollig, M. & Schulte, A. (1998). Himba paradise lost – Stability, degradation, and pastoralist management of the Omuhonga Basin (Namibia). *Die Erde*, 129, 301–315.

Sanderson, J. & Harris, L.D. (Eds.) (2000). *Landscape Ecology: A Top-Down Approach.* Boca Raton, FL: Lewis.

Savory, A. (1988). *Holistic Resource Management.* Harare, Zimbabwe: Jongwe.

Scholes, R.J. & Walker, B.H. (1993). *An African Savanna – Synthesis of the Nylsvley Study.* Cambridge: Cambridge University Press.

Schulte, A. (2001). *Weideökologie des Kaokolandes. Struktur und Dynamik einer Mopane-Savanne unter pastoralnomadischer Nutzung.* Ph.D. thesis, Cologne. URN: urn:nbn:de:hbz:38–112312740; URL: http://kups.ub.uni-koeln.de/volltexte/2003/432/

Schulte, A. (2002). Stabilität oder Zerstörung? Veränderungen der Vegetation des Kaokolandes unter pastoralnomadischer Nutzung. *Kölner Geographische Arbeiten,* 77, 101–118.

Shackleton, C.M., Griffin, N.J., Banks, D.I., Mavrandonis, J.M. & Shackleton, S.E. (1994). Community structure and species composition along a disturbance gradient in a communally managed South African savanna. *Vegetatio,* 115, 157–167.

Skarpe, C. (1995). Vegetation ecology in African savanna. *Verhandlungen der Gesellschaft für Ökologie,* 24, 11–16.

Solbrig, O.T. (1991). Savanna modelling for global change. *Biology International,* 24, 1–47.

Solbrig, O.T. (1996). The diversity of the savanna ecosystem. In O.T. Solbrig, E. Medina & J.F. Silva (Eds.), *Biodiversity and Savanna Ecosystem Processes. A Global Perspective* (pp. 1–27). Berlin: Springer.

Solbrig, O.T., Medina, E. & Silva, J.F. (1996). Determinants of tropical savannas. In O.T. Solbrig, E. Medina & J.F. Silva (Eds.), *Biodiversity and Savanna Ecosystem Processes. A Global Perspective* (pp. 31–41). Berlin: Springer.

Squires, V.R., Mann, T.L. & Andrew, M.H. (1992). Problems in implementing improved range management on common lands in Africa: An Australian perspective. *Journal of the Grassland Society of South Africa,* 9, 1–7.

Stander, P.E., Nott, T.B. & Mentis, M.T., (1993). Proposed burning strategy for a semi-arid African savanna. *African Journal of Ecology,* 31, 282–289.

Styles, C. (1994). Mopane worms: More important than elephants? *Farmer's Weekly,* July 29, 14–19.

Tapson, D. (1993). Biological sustainability in pastoral systems: The Kwazulu case. In R.H. Behnke, I. Scoones & C. Kerven (Eds.), *Range Ecology at Disequilibrium. New Models of Natural Variability and Pastoral Adaptation in African Savannas* (pp. 118–135). London: Overseas Development Institute.

Timberlake, J.R. (1995). *Colophospermum mopane.* Annotated bibliography and review. *The Zimbabwe Bulletin of Forestry Research,* 11, 1–49.

Tischendorf, L. (2001). Can landscape indices predict ecological processes consistently? *Landscape Ecology,* 16, 235–254.

Trollope, W.S.W. (1992). Veld management in grassland and savanna areas. In F.P. Van Oudtshoorn (Ed.), *Grasses of South Africa* (pp. 45–56). Arcadia, Pretoria: Briza Publikasies.

Trollope, W.S.W., Trollope, L. & Bosch, O.J.H. (1990). Veld and pasture management terminology in southern Africa. *Journal of the Grassland Society of South Africa,* 7, 52–61.

Turner, M.G., O'Neill, R.V., Gardner, R.H. & Milne, B.T. (1989). Effects of changing spatial scale on the analysis of landscape pattern. *Landscape Ecology,* 3, 153–162.

Turner, M.G., Romme, W.H., Gardner, R.H., O'Neill, R.V. & Kratz, T.K. (1993). A revised concept of landscape equilibrium: Disturbance and stability on scaled landscapes. *Landscape Ecology,* 8, 213–227.

Turner, S.J., O'Neill, R.V., Conley, W., Conley, M.R. & Humphries, H.C. (1991). Pattern and scale: Statistics for landscape ecology. In M.G. Turner & R.H. Gardner (Eds.), *Quantitative Methods in Landscape Ecology* (p. 536). New York: Springer-Verlag.

Urban, D.L., O'Neill, R.V. & Shugart, H.H. (1987). Landscape ecology. *Bioscience,* 37, 119–127.

Walker, B.H. (1987). A general model of savanna structure and function. In B.H. Walker (Ed.), *Determinants of Tropical Savannas* (pp. 1–12). Miami: ISCU Press.

Walker, B.H. & Noy-Meir, I. (1982). Aspects of the stability and resilience of savanna ecosystems. In B.J. Huntley & B.H. Walker (Eds.), *Ecology of Tropical Savannas* (pp. 556–590). Berlin: Springer

Walter, H. (1954). Die Verbuschung, eine Erscheinung der subtropischen Savannengebiete, und ihre ökologischen Ursachen. *Vegetatio,* 5/6, 6–10.

Walter, H. (1971). *Ecology of Tropical and Subtropical Vegetation*. Edinburgh: Oliver & Boyd.

Weber, G.E. & Jeltsch, F. (2000). Long-term impacts of livestock herbivory on herbaceous and woody vegetation in semiarid savannas. *Basic and Applied Ecology*, 1, 13–23.

Weiher, E., Van der Warf, A., Thompson, K., Roderick, M., Garnier, E. & Eriksson, O. (1999). Challenging Theophrastus: A common core list of plant traits for functional ecology. *Journal of Vegetation Science*, 10, 609–620.

White, P.S. & Jentsch, A. (2001). The search for generality in studies of disturbance and ecosystem dynamics. *Progress in Botany*, 62, 399–449.

Chapter 4

Quantitative Classification of Landscapes in Northern Namibia Using an ASTER Digital Elevation Model

GUNTER MENZ AND JOCHEN RICHTERS

In quantitative landscape ecology and environmental modelling, the spatial delineation of a landscape and its differentiation into landscape units is often based on subjective criteria. At the same time, the precise spatial definition of a landscape as a study area is often crucial for deriving accurate model outputs. In this study we present a promising solution to this problem by quantitatively assessing the complex term 'landscape' using an empirical statistical approach. This analysis is based principally on a digital elevation model (DEM) derived from stereoscopic ASTER sensor data. DEMs were produced for varying spatial subsets of two ACACIA landscapes in Northern Namibia: (1) Lower Kunene Hills (LKH) and (2) Upper Kunene Hills (UKH).

These subsets were converted into a multilayer dataset, which was then statistically transformed using a principal component analysis (PCA). We attempted to interpret the resulting landscape elements (which we term 'landform types') and to utilize calculated variances to explain their geomorphological contribution to establishing the individual landscape.

For the Upper Kunene Hills (UKH) landscape, the highest variance was obtained with a quadratic subset size of 50×50 raster elements and a spatial coverage of 50%. For the Lower Kunene Hills (LKH) landscape, the highest variance was obtained with a subset size of 25×25 raster elements and coverage of 25%.

M. Bollig, O. Bubenzer (eds.), *African Landscapes*,
doi: 10.1007/978-0-387-78682-7_4, © Springer Science+Business Media, LLC 2009

This difference can be explained by the different dominant landforms of the two study areas. Whereas the UKH is principally characterized by large planar geomorphological features, the LKH is dominated by smaller-scale features such as mountain ridges intersected with steep valleys. These differences in dominant landforms among the two landscapes make it necessary to establish site-specific variance thresholds in order to operationally implement this methodology and produce a final classification of the study areas.

4.1. INTRODUCTION

The abstract term 'landscape' is difficult to scientifically define as a unit for classifying landforms (Bubenzer, this volume). When studying landscapes in detail, it can be extremely difficult to account for the structural variety within a landscape, as well as to accurately delimit a landscape from neighboring ones. Difficulty may occur with the location of spatially representative information in clearly defined subspaces, as these subspaces usually follow political or other 'artificial' boundaries. However, traditional regionally based ecological models typically attempt to quantify geobiophysical data and processes in the spatial unit of a landscape, and require a spatial definition and clear delineation of the study units. For example, the quantitative estimate of the water balance in a watershed or the quantification of vegetation degradation using remote sensing data require a spatial delineation and the direct or indirect incorporation of elevation information from digital elevation models (DEMs) (Chumura et al., 1992). It is also important that model results may be reproduced, compared, and if desired, integrated with other studies. Our goal in this study is to allow improved quantification of the often subjective or imprecise term 'landscape' and thus contribute to an objective definition of the term 'landscape.'

The hilly to mountainous savanna region of the Central Kaokoveld in Northern Namibia includes various important geomorphic features. Ecologically relevant landscapes in this region are determined principally by geomorphic features and their internal structure. Distribution and development of soils and plants, as well as climatological regimes, closely follow specific landform characteristics of the region. In this study, these landforms are examined as an example of a spatially distributed ecosystem variable, with the goal of finding a new approach to delineating and internally differentiating these landscapes.

Since the year 2000, digital elevation models with adequate spatial resolution have been available to researchers. Two such datasets were available for this study: a radar-based DEM produced by the Space Radar Topography Mission (SRTM), and stereo image pairs produced by the ASTER (Advanced Spaceborne Thermal Emission and Reflection Radiometer) sensor onboard the TERRA satellite (Bolten et al., this volume). For this study of the Kaokoveld region, we utilized a DEM (termed here an 'ASTER-DEM') derived from ASTER sensor data.

The delineation of three different landscapes and their internal differentiation was conducted using a well-known empirical statistical approach called principal component analysis (PCA) (Benichou et al., 1987). The results of the principal component transformation are referred to here as 'landform types,' characteristic for the three selected landscape units: Upper Kunene Hills (UKH), Escarpment (ESC), and Lower Kunene Hills (LKH).

4.2. THE TERM 'LANDSCAPE' AND ITS USE

The term landscape traditionally has been understood by geographers and landscape ecologists as the sum of the features of a three-dimensional section of the earth's surface, consisting of atmosphere, biosphere, pedosphere, and lithosphere, including direct and indirect factors (compare Bubenzer, this volume). This general definition of landscape was developed by geographic researchers at the onset of the twentieth century (e.g., Passarge, 1933). With this approach, a landscape is subdivided both spatially and contextually into different assortments of geographic factors and 'landscape cells.' The research is transformed from purely descriptive to methodological studies.

The guiding approach to the present study is based on the theory of the 'geographic dimension' as defined by Barsch (1990), Haase (1991), Leser (1991), and Mosimann et al. (1992). This assumes that within the spatially differentiated continuum of the landscape there are so-called continuity areas in the value distribution of individual properties, which are delineated by borders with discontinuous and rapidly changing value distributions. These homogeneous landscape units form the basic elements of the landscape. According to the theory of the 'geographic dimension', Mosimann and Duttmann (1992) developed the methodology of ecological complexity analysis of landscapes, which is a research approach geared toward a general overview. This approach uses map-based information and other measurements (including remote sensing) to study the structure and function of so-called elemental landscape units within the complex landscape ecosystem.

4.3. STUDY AREA AND UNDERLYING DATA

4.3.1. Central Kaokoveld

The Central Kaokoveld study area is a representative section of the transitional region in Northern Namibia between the topographically higher Upper Kunene Hills (UKH) in the north and northeast (average elevation 1100 m ASL), the northern escarpment (ESC), and the Lower Kunene Hills (LKH) in the south and southwest (average elevation 700 m ASL) (Figure 4.1). There is a prono-unced east to west annual mean precipitation gradient within the study area, decreasing from 350 mm in the east to 150 mm in the west (Sander & Becker, 2002). This has a corresponding pronounced change in vegetation from a relatively closed mopane tree savanna to a very dispersed dwarf-brush savanna or to woodland areas, respectively (Linstädter, this volume). The spatial dimensions of the Central Kaokoveld study area measure approximately 27 km east to west, and

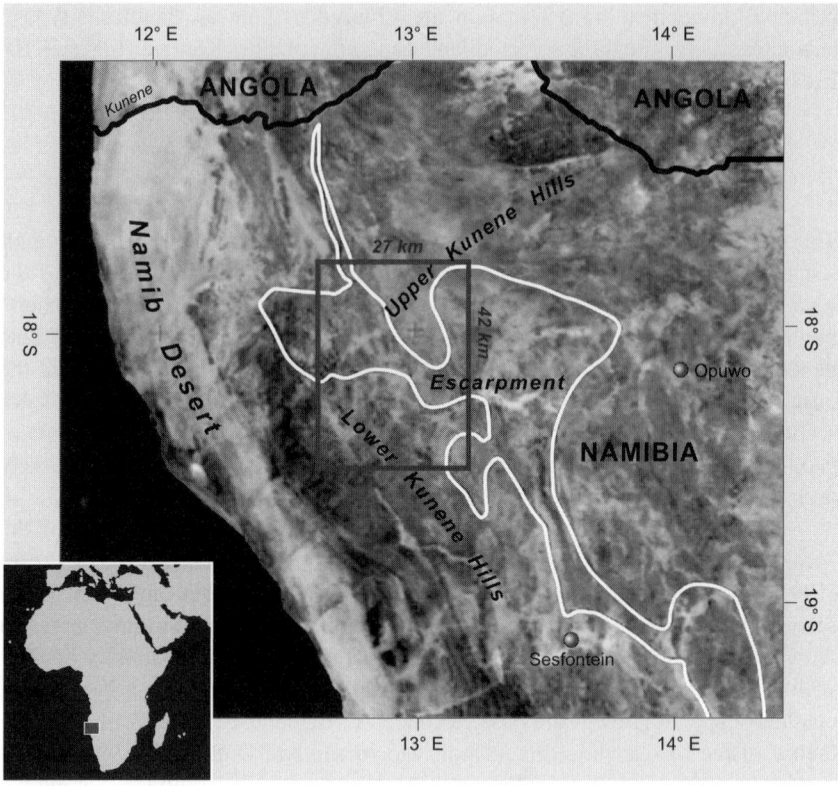

Figure 4.1. MODIS satellite image mosaic of NW Namibia. True color composite compiled
from images acquired between June and July 2001. Overlaid with geomorphologic data
from the Digital Atlas of Namibia (2001) and the 'Central Kaokoveld' study area (red)
(See also Color Plates)

42 km north to south. Study area vertical relief varies from 550 m ASL at the Lower
Kunene Hills in the southwest part of the study area to more than 1700 m ASL at
the Upper Kunene Hills (Jürgens & Bähr, 2002). Overall the Central Kaokoveld
study area comprises a broad variety of ecosystem properties and dynamics.

The MODIS satellite image mosaic (Figure 4.1), compiled from images
obtained between June and July 2001, shows northwestern Namibia, including
the study area in the center, in a true colour composite, combining MODIS
channels 1, 2, and 3. The spatial resolution of this image is 500 m. Clearly visible
are the green shades of various vegetated surfaces, the dark brown metamorphic
bedrock, and the beige shades of the extensive sand fields and of the gravel
surfaces at the transition to the Namib Desert. In Figure 4.1, the satellite data were
overlaid with the 2001 geomorphologic classification of Namibia (Mendlesohn
et al., 2002), with the northern escarpment in the centre. This classification into
the topographic infrastructure provides the basis for the selection of sites for
detailed study.

4.3.2. Generating a Digital Elevation Model from ASTER Satellite Data

The digital elevation model for the Central Kaokoveld site was calculated within the ACACIA-subproject E1 (Bolten et al., this volume) using the stereo image data of the TERRA satellite's ASTER sensor acquired during 2001. ASTER data has a spatial resolution of 15 m. Using the two infrared sensor channels, 3N (pointing at nadir) and 3B (oriented backwards), the relative elevation of any point on the image can be calculated from the parallax difference. When combined with a minimum of three points of known elevation, the elevation model can be converted to absolute topography (m ASL). The resulting DEM, like the underlying data, has a spatial resolution of 15 m.

The shaded relief (Figure 4.2) clearly shows how the study area is divided into two principal parts. The northeast portion of the study area is a part of the African shield high plateau, whereas the southwest area is characterized by pronounced topographic incision and individual valley structures with associated mountain ridges. Bolten (*ibid.*) performed a field verification to analyze the accuracy of this DEM dataset and document a high level of accuracy, with RMS errors in x- and y-directions of approximately 8.5 m.

4.3.3. Selection of Study Sites and Establishment of a Multi-Layer Structure

Based on a regionalisation study by Benichou (1987), three Central Kaokoveld landscape units were selected within the ASTER DEM dataset as characteristic for the site: Upper Kunene Hills (UKH), Escarpment (ESC), and Lower Kunene Hills (LKH). These landscape units contain representative landscape features and are distinctly different from each other (Figure 4.2). Three systematic samples with varying subset sizes were then selected within each study area. Square subsets of 25×25 pixels (375×375 m^2), 50×50 pixels (or 750×750 m^2), and 100×100 pixels (1500×1500 m^2) were chosen with the aim of most effectively representing the typical landforms (valleys, mountains, etc.) found within each landscape unit. Subsets that are too small do not sufficiently represent large landforms (e.g., the Tönnesen Mountains), whereas subsets that are too large cannot characterize smaller landforms (e.g., the valley system of the Khumib River). To more thoroughly evaluate the influence of the amount of coverage present in each subset in delineating the three landscape units, we analysed coverage for each subset at 25%, 50%, and 100% of the study area (Table 4.1).

Elevation values of a grid-based subset have been shown to be highly correlated and adjacent elevation values are often redundant (Yue-Hong et al., 1999). To remove this redundancy within a single grid-based subset and within all grid-based subsets of an area, as well as to identify the characteristic landforms within the data, the subsets were transformed using a principal component analysis (described in detail in Section 4.4.2). In preparation for this statistical

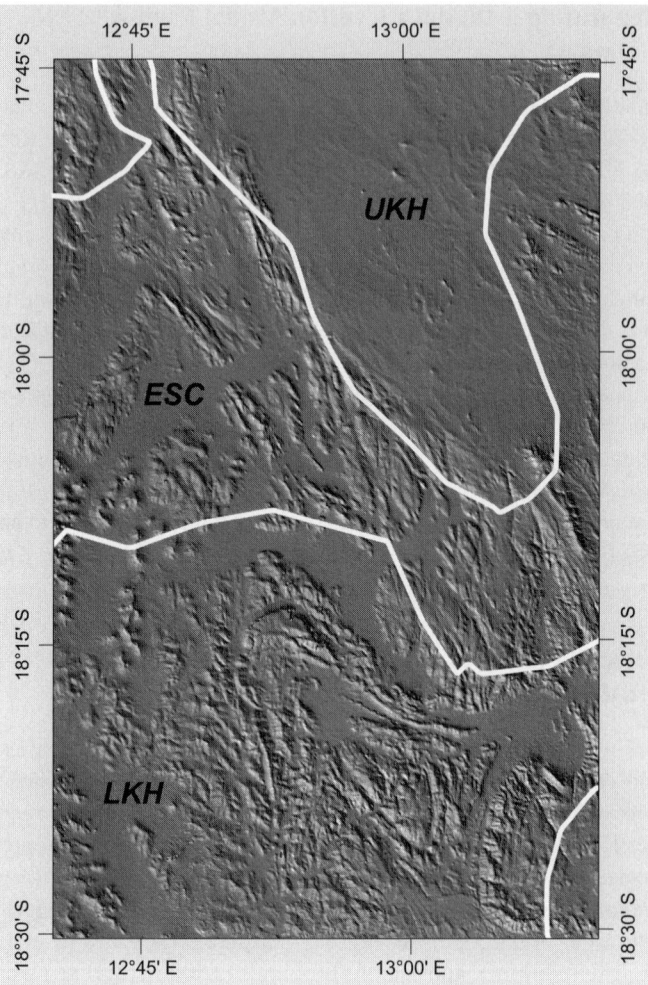

Figure 4.2. ASTER-based Digital Elevation Model (DEM) of the Central Kaokoveld as represented as a shaded relief image. Geomorphic boundaries of the three study areas UKH, ESC, and LKH are shown as modified from the *Digital Atlas of Namibia* (2001)

analysis, all subsets of a sample were transformed to a multi-layer dataset. As an example, the UKH landscape unit contains 855 'subsets' when analyzed at a 25 × 25 matrix and 50% coverage.

This multi-layer dataset is derived from the single layer DEM by automatically selecting samples using the systematic sampling method. Samples having the same subset size in the *x*- and *y*-dimensions (e.g., a 25 × 25 matrix) are input sequentially as individual layers into a multilayer grid-based dataset. Through this sampling process, 18 individual datasets were created for the UKH and LKH regions together (Table 4.1).

Table 4.1. Number of samples for various subset sizes (e.g., 50 × 50 matrix) and percentage coverages for the three landscape units (UKH, ESC, and LKH)

	Upper Kunene Hills (240.49 km²)	Escarpment (398.33 km²)	Lower Kunene Hills (429.51 km²)
SET I *(25×25 Matrix)*			
25%	427	708	758
50%	855	1,416	1,516
100%	1,710	2,832	3,032
SET II *(51×51 Matrix)*			
25%	102	170	182
50%	205	340	364
100%	410	680	728
SET III *(101×101 Matrix)*			
25%	26	43	46
50%	52	86	92
100%	104	173	185

Preparation of multi-layer dataset

N = 30

Principal Component Analysis (PCA)

4.4. METHODOLOGY AND ANALYSIS

4.4.1. Determination of Landform Types

The determination of landform types is based on the multi-layer grid-based structure generated as described in Section 4.3.3 above. This procedure is based on the fundamental knowledge that the values of all grid points within a certain perimeter correlate strongly with the mean elevation and thus with the absolute elevation (m ASL) of the centrepoint of the image. It is advantageous for the calculation of the principal components to first calculate only relative elevations (Bahrenberg et al., 1990); this is done by subtracting the mean value of all points on the grid-based surface from the individual elevations. The mean of all grid points is the reference surface for the individual grid points. If absolute elevation values were used for the principal component analysis, little more than a principal axis could be extracted, depending on the subsets. Güßefeld (1991) showed that deriving relative elevations is an essential precondition for the application of a principal component analysis even for noncontiguous landscape samples with differing elevations, as in our study.

4.4.2. Principal Component Analysis

Employing the methodology developed by Benichou (1987), the multilayer landscape dataset was transformed through principal component analysis. This allows for the reduction of the highly correlated, multidimensional variable statistical image space (identical with the respective sample size) to its most important, linearly independent principal axis and principal components (Kreyszig, 1979). The principal component values then provide the desired quantitative description of the landform types present in the grid. In a statistical evaluation of the results, the respective data structure within the three landscape units was compressed (e.g., based on the intrinsic value) to obtain a number of significant principal components or landform types.

The transformations were calculated for the selected Lower Kunene Hills (LKH) and Upper Kunene Hills (UKH) study areas using the 'Principal Component Analysis' module within the ENVI/IDL image processing software package (version 3.5, RSI, 1999). The ESC subregion was not subject to further analysis, because, as a transition zone, it is comprised of surface features present within the LKH and UKH study areas. Using a preliminary sample size of 30 samples per study area, we were able to reduce the 30-dimensional variable space to the first five principal axes (Table 4.2), thereby greatly reducing the spatial extent of the study areas: the LKH study area was reduced to 9% of the original extent, and the UKH area to 16%.

For the Upper Kunene Hills study area, the first principal component can explain more than 53% of the variance (50 × 50 matrix) in the overall landscape. This is shown by the associated landform type, which is basically a SW to NE downward sloping surface. Field measurements corroborate this plateau surface inclined to the east (Figure 4.4). To correlate the standardised landform types derived from the PCA to the elevations of the original digital elevation model as well as to absolute topography, it is necessary to retransform the attributes of each principal component, using a linear equation:

$$T = A*s + x$$

where:

Table 4.2. Variance of the first principal component (PCA_1) and the cumulative values of principal components 1 through 5 ($\sum PCA_i$, i = 1 through 5) for varying subset sizes for the Upper Kunene Hills (UKH) and Lower Kunene Hills (LKH) study area (sample size is 30)

Subset size		25 × 25 Matrix%	50 × 50 Matrix%	100 × 100 Matrix%
UKH	PCA_1	39.34	53.30	41.47
	$\sum_{i=1}^{5} PCA_j$	86.37	83.87	79.27
LKH	PCA_i	38.31	36.17	26.82
	$\sum_{i=1}^{5} PCA_j$	91.21	83.60	71.62

T = matrix of the absolute elevations (size: 30 × 30 grid elements),

A = matrix of the principal component attributes (size: 30 × 30 grid elements),

s = scalar quantity that represents the standard deviation of the absolute elevation of all grid points,

x = offset which represents the arithmetic mean of the subset.

Preliminary evaluation of this retransformation procedure has demonstrated that it has little effect on the spatial landform patterns present in the LKH and UKH study areas, and may be disregarded in this study.

4.5. RESULTS AND DISCUSSION

The results of the 18 principal component analyses for the Upper Kunene Hills (UKH) and Lower Kunene Hills (LKH) study areas are described in the following summary.

4.5.1. Influence of the Subset Size

Optimal representation of the specific landforms present in both study areas requires different subset sizes for the principal components analysis. For UKH, the highest statistical correlation in the first principal component is achieved with a subset size of 50 × 50 pixels (750 × 750 m²), yielding a variance of 53.3%. In comparison, the highest correlation for LKH occurs with a smaller 25 × 25 pixel (375 × 375 m²) subset, yielding a 38.31% variance (Table 4.2).

The statistical analyses confirm two important geomorphologic findings:

–   The LKH study area represents a very heterogeneous geomorphic environment with small-scale transitions between valleys and ridges of varying spatial dimension and arrangement. The UKH study area consists of a more uniform plateau geomorphology.
–   The significant differences in the variance of the two first principal components also can be explained by the differences in study area geomorphology. The first principal component can explain significantly more landscape detail for the UKH study area due to the less complex landforms present there.

4.5.2. Influence of Varying Coverage Fractions

Following determination of the optimal subset sizes for each test area, it was necessary to calculate the fractional portion of the sample squares (as part of each of the study areas) that provided the highest information content for the first principal component (PCA_1) and for the first five principal components ($\sum PCA_i$, i = 1 through 5).

The variance of the first five principal components ($\sum PCA_j$, $i = 1$ through 5) changes only slightly with the transition from 30 samples (7.3% of entire study area) to samples including 25%, 50%, and 100% of the study area (Table 4.2). In contrast, the variance of PCA_1 for the UKH study area shows a decrease from 53.30% to 38.25% (approximately 15%); whereas an increase in variance from 38.31% to 53.13% was found for the LKH study area (Table 4.2 and Table 4.3). As is the case with varying subset size, these results are clearly related to the geomorphological landforms of the two study areas (Table 4.3). They show that an increase in sample size does not result in a significantly better description of the landforms for the relatively homogeneous UKH, whereas a larger sample size does lead to a better description of the landforms for the more complex LKH.

4.5.3. Influence of Landscape Representation

Principally due to their geomorphic differences, PCA results vary significantly for the two different landscape study areas (LKH and UKH). Each PCA represents the weight of the principal components as standardised values and thus characterises a specific landform type present within each study area. The first five principal components can explain more than 80% of the total variance of the landscape found in both the LKH and UKH study areas; although different subset sizes and coverage fractions are required in each study area. The grey-scale PCA images (Figure 4.3 and Figure 4.4) show the various landform types for each study area with higher elevations represented in brighter tones.

4.5.3.1. Lower Kunene Hills

The dominant landform type in the LKH study area is a SW–NE sloping plain with little relief (PCA_1; Figure 4.3). This plain is associated with a NW to SE sloping surface (PCA_2). The third landform type identified in the study area consists of isolated SE–NW oriented ridges (PCA_3). Finally, the entire landscape is characterized by small-scale topographical depressions (PCA_4) as well as an N–S

Table 4.3. Variance for the first principal component (PCA_1) and cumulative values of principal components 1 through 5 ($\sum PCA_j$, $i = 1$ through 5) when varying coverage fractions for the optimized subset sizes (from Table 4.2) for the study areas

Coverage		25 % coverage	50 % coverage	100 % coverage
UKH 50×50	PCA_1	34.09	38.25	37.31
Matrix	$\sum_{j=1}^{5} PCA_j$	80.45	82.68	80.24
LKH 25×25	PCA_1	53.13	49.92	42.74
Matrix	$\sum_{j=1}^{5} PCA_j$	89.79	87.73	82.62

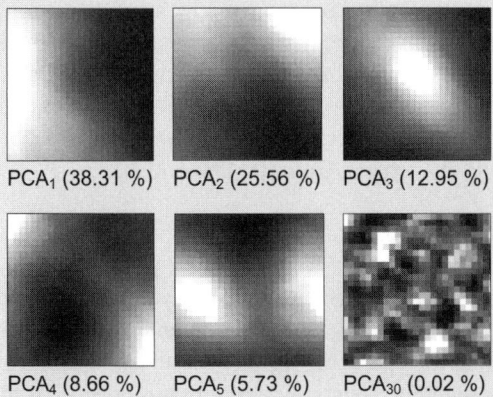

Figure 4.3. Expression of the landform types for the Lower Kunene Hills (LKH) as a result of the PCA analysis (size: 25 × 25 matrix and coverage: 25%)

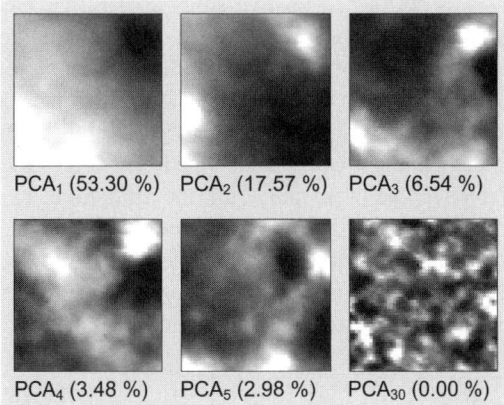

Figure 4.4. Expression of the landform types for the Upper Kunene Hills (UKH) as a result of the PCA analysis (size: 50 × 50 matrix and coverage: 50%)

transverse saddle (PCA_5). The highest principal component (PCA_{30}) represents the smallest variance within the dataset (0.02%) and displays undefined image background noise.

4.5.3.2. Upper Kunene Hills

The dominant landform of the UKH study area is a surface inclined from SW–NE (PCA_1; Figure 4.4). Small depressions are integrated into this surface in the northeastern portion of the study area. A secondary landform is a NW–SE oriented longitudinal saddle (PCA_2). Beginning with the third principal component

(PCA$_3$), cumulative covariances are generally high (above 70%); this is also shown by the more heterogeneous grey values present in the LKH images (Figure 4.3). Interpretation of landform types in the UKH study area becomes more difficult for PCA$_4$ and PCA$_5$. For example, PCA$_4$ may be broadly interpreted as a NW–SE oriented ridge, which is in the process of being transformed into smaller valleys by erosion from the SW and NE.

4.5.4. Overall Conclusions

The landform types for the UKH and LKH study areas which appear in the PCA data correspond to the general geological, tectonic, and topographical setting of the Kaokoveld study region. They reflect the clear bedrock orientation of the region (which parallels the coast), although more incision of the lower bedrock layers is evident in the LKH study area compared to the relatively undisturbed plateau surface of the UKH. These results clearly demonstrate the potential of principal component analysis for the delineation and internal differentiation of landscapes.

This analysis was performed for two study areas in the Central Kaokoveld region. We believe that future research related to this methodology should include three principal topics: (1) further development and refinement of the dynamic sample selection, (2) automation of the principal component analysis process, and (3) objective determination of maximum variance and threshold values.

By applying the principal component transformation to the high-resolution digital elevation model, we were able, for the first time, to objectively and reproducibly derive landscape types for the Central Kaokoveld region. In this approach, the homogeneity of landscape units (as shown by their topography) is used as a delineating factor and is reflected in the variance of the individual principal components. This study extends the work of Güßefeld (1991) and Goßmann et al. (1993) in two principal ways: model inputs are simplified to a single-layer DEM dataset and variably sized samples – in terms of subset spatial size as well as sample coverage portions – are used.

These calculated landscape types are especially useful for the selection of study areas for modelling and can provide an important foundation for future local and regional environmental modelling efforts.

Acknowledgements The digital elevation model of the central Kaokoveld region was calculated within the ACACIA-subproject E1.

REFERENCES

Bahrenberg, G., Giese, E., & Nipper, J. (1990). *Statistische Methoden in der Geographie. Univariate und bivariate Statistik.* Stuttgart: Teubner Verlag.

Benichou, P. (1987). Annual and interannual variability of statistical relationships between precipitation and topography in a mountainous area. In *10th Conference on Probability and Statistics in Atmospheric Science*, 273–278. Canada: Edmonton.

Benichou, P. & Le Breton, O. (1987). Prise en compte de la topographie pour la cartographie des champs pluviométrique statistiques. *La Météorologie*, 7(19), 23–34, Paris.

Chumura, G.L., Costanza, R., & Kosters, E.C. (1992). Modelling coastal marsh stability in response to sea level rise: A case study in coastal Louisiana, USA. *Ecological Modelling*, 64(1), 47–64.

DEA (2001). Digital Atlas of Namibia. (http://www.dea.met.gov.na/website/NamibiaAtlas/ – 6.10.2001).

Gossmann, H., Banzhaf, E., & Klein, G. (1993). Regionalisierung Ökologischer Daten – alte Aufgaben, neue Lösungswege. Das Freiburger Regionalisierungsmodell FREIM. *Würzburger Geographische Arbeiten*, 87, 399–418.

Güssefeldt, J. (1991). *Quantitative Relieftypen zur Parametrisierung geoökologischer Modelle*, Manuskript (Freiburg).

Haase, G. (1991). Theoretisch-methodologische Schlussfolgerungen zur Landschaftsforschung. In Nova acta Leopoldina. *Abhandlungen der Deutschen Akademie der Naturforscher Leopoldina, Neue Folge, Nr. 276, Bd.* 64, 173–186.

Jürgens, U. & Bähr, J. (2002). *Das südliche Afrika*. Perthes Regionalprofile. Gotha, Stuttgart: Klett-Perthes Verlag.

Kreyszig, E. (1979). *Statistische Methoden und ihre Anwendungen*. Göttingen: Vandenhoeck & Ruprecht.

Leser, H. (1991). *Landschaftsökologie. Ansatz, Modelle, Methodik, Anwendung: mit einem Beitrag zum Prozeß-Korrelations-Systemmodell von Thomas Mosimann*. Stuttgart: Ulmer Verlag.

Mendlesohn, J., Jarvis, R., & Robertson, T. (2002). *Atlas of Namibia: A Portrait of the Land and Its People*. Cape Town: David Philip.

Mosimann, T. & Duttmann, R. (1992). Die digitale Geoökologische Karte als Ergebnis einer pro-zeßorientierten Landschaftsanalyse am Beispiel der Nienburger Geest. *Berichte zur Deutschen Landeskunde*, 66(2), 335–336.

Passarge, S. (1933). *Einführung in die Landschaftskunde*. Leipzig: B.G. Teubner.

RSI, Research Systems Inc. (1999). ENVI User Guide. RSI.

Sander, H. & Becker, T. (2002). Klimatologie des Kaokolandes. In: M. Bollig, E. Brunotte, & Th. Becker (Eds.) Interdisziplinäre Perspektiven zu Kultur- und Landschaftswandel im ariden und semiariden Nordwest Namibia, Kölner Geographische Arbeiten, H. 77, Selbstverlag des Geographischen Instituts der Universität zu Köln, pp. 57–68.

Yue-Hong, C., Pin-Shuo, L., & Dezzani, R. (1999). Terrain complexity and reduction of topographic data. *Journal of Geographical Systems*, 1(2), 179–198.

Chapter 5

Risks and Resources in an Arid Landscape: An Archaeological Case Study from the Great Sand Sea, Egypt

Heiko Riemer

Although conventional site archaeology tends to ignore the potential of an analysis of the wider spatial setting, in this contribution landscape archaeology defines an approach which is concerned with the understanding of human behaviour and interaction with resources, topography, and environment of the whole landscape, using a multiperiod systematic approach. Between 1996 and 2000 archaeological excavations and regional surveys were conducted in the southern Great Sand Sea of Egypt by interdisciplinary teams from the ACACIA project of the University of Cologne. A study region, named 'Regenfeld', was selected for detailed analysis of the geophysical units, past environmental conditions, and prehistoric settlement patterns. The chronological affiliation of surveyed sites shows temporary activities of prehistoric desert dwellers during the Holocene pluvial. Although the palaeoenvironmental archives clearly indicate a Holocene 'wet-phase' during that period, the southern Great Sand Sea was an arid environment with scarce and highly variable key resources, such as surface water.

The holistic approach as proposed in landscape archaeology entailing a comprehensive analysis of a great number of sites and assemblages indicates that

M. Bollig, O. Bubenzer (eds.), *African Landscapes*,
doi: 10.1007/978-0-387-78682-7_5, © Springer Science + Business Media, LLC 2009

the dwellers were highly mobile hunters and gatherers, covering hundreds of kilo-metres during their episodic rounds through the desert. The settings of sites within the landscape clearly indicate campsite nucleation around sources of open water. These adaptive and flexible subsistence strategies were developed to cope with the unpredictable climate and the risks of resource shortage in an arid landscape. Landscape as an integrative concept becomes a useful tool to analyse the long-term interaction between humans and their environment under changing conditions.

5.1. INTRODUCTION

Since the beginning of the ACACIA project in 1995, supraregional research has been conducted in five different areas of the Western Desert of Egypt to gather information about human subsistence during the last 10,000 years in contrasting types of landscapes following a landscape approach that integrates various disciplines from archaeology to environmental studies and geo-sciences (Kuper & Kröpelin, 2006; Bubenzer et al., 2007; Bubenzer & Riemer, 2007; Bubenzer et al., this volume).[1] As a case study, the following chapter focuses on one of the regions situated in the southern Great Sand Sea in the most remote West of Egypt (Figure 5.1).

Landscape archaeology can hardly find a more appropriate field for its stud-ies than the Great Sand Sea. By the unparalleled uniformity of its giant dunes (Figures 5.2–5.4), rising up to 100 m high and running more than 500 km from north to south, as well as by its complete lack of vegetation, it provides an ideal natural laboratory for the study of different environmental and cultural influences during the so-called Holocene 'humid phase' lasting approximately from 9500 to 6000 BP (c. 8900–4900 BC) when the eastern Sahara received somewhat higher rainfall than today. The harsh environmental conditions of the Great Sand Sea and its spatial dimension concern its role as a possible cultural barrier between Egypt and the central Sahara, as well as the fact that it is crossed by the frontier between winter and summer rains, whose movements during the Holocene can be seen reflected in the remains of the human occupation in the area.

---

[1] Exceptional preconditions for landscape archaeology, for instance, are provided by the extended open-air coal mines in central Europe, that have been used within the frame of the outstanding archaeological project 'Siedlungsgeschichte der Aldenhovener Platte' (settlement history of the Aldenhovener Platte) west of Cologne since the mid-1960s for continuous excavation of an entire microlandscape (Lüning, 1983). The SAP project was dedicated to a landscape approach, and this tradition continued and developed during the B.O.S. project of the early 1980s to the ACACIA project starting in 1995/96, although the climatic and cultural frame changed. Already in the early 1980s archaeological research had been carried out by the B.O.S. project of the University of Cologne at the western and eastern fringes of the Sand Sea, in the Glass Area, and the area west of Abu Minqar respec-tively. Since 1995/96 this research has been continued within the frame of the ACACIA project, then focussing on the core area of the Sand Sea where climatic influences were expected to be more clearly detectable in prehistoric sites than in the marginal zones.

Figure 0.1. Albrecht Altdorfer, Danube Landscape with Castle near Wörth (Courtesy Bildarchiv Preußischer Kulturbesitz)

Figure 0.2. Pieter Brueghel; The Fall of the Icarus (Courtesy Bildarchiv Preußischer Kulturbesitz)

Figure 0.3. Humboldt, 'Geographie der Pflanzen in den Tropenländern' (geography of plants in the tropic countries; Courtesy Bildarchiv Preußischer Kulturbesitz)

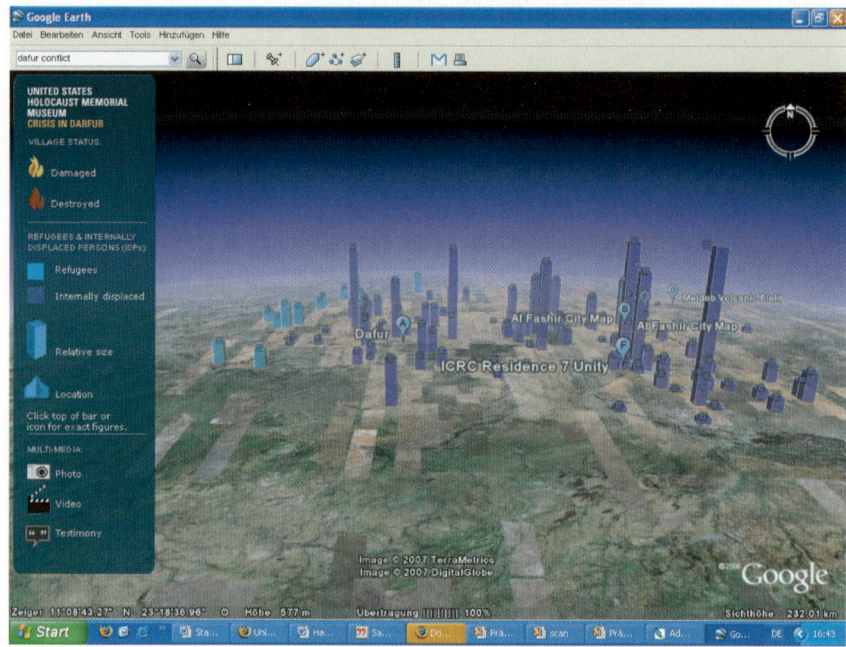

Figure 0.7. Google-Earth presentation of the Darfur conflict (Courtesy Google-Earth in cooperation with the Holocaust Memorial Museum in Washington)

Figure 2.1. Archaeological sites within the Western Desert of Egypt recorded during archaeological surveys by B.O.S. and ACACIA (1980–2002). Displayed are three different spatial scale factors according to the scale discussion. 1: Landsat 5 image (30 m resolution); 2: ASTER image (15 m resolution); 3: Quickbird image (0.61 m resolution)

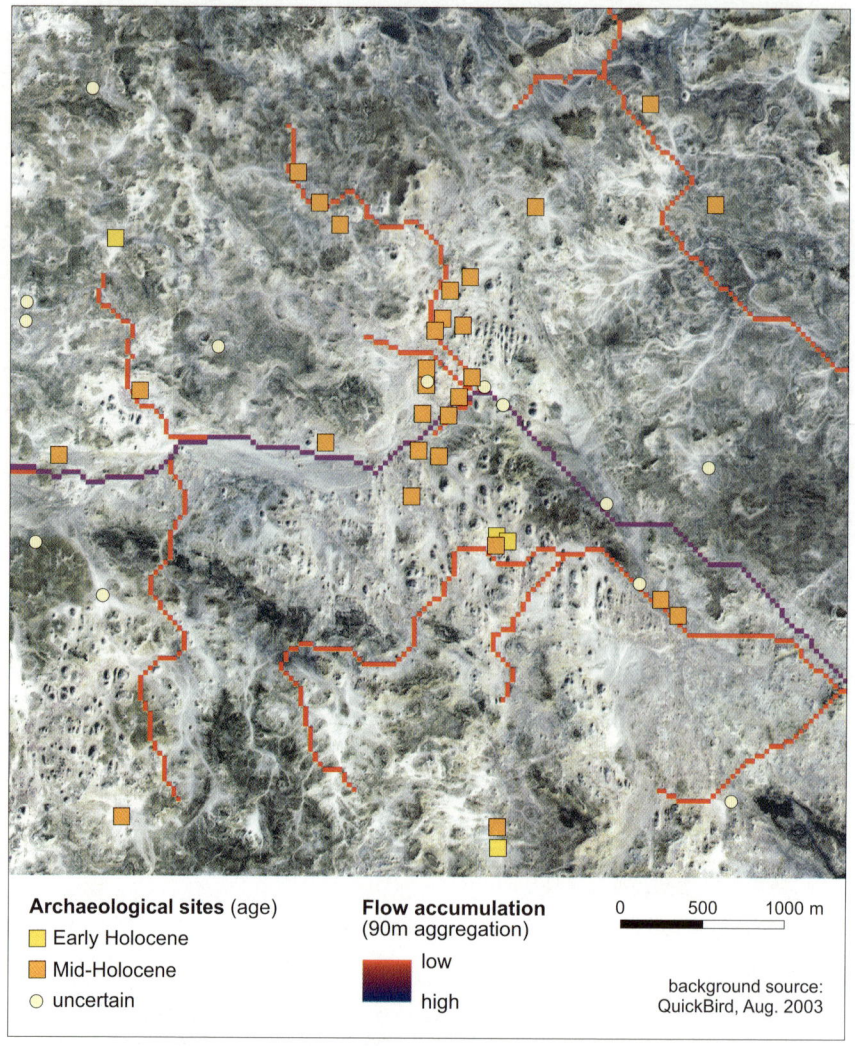

Archaeological sites (age)
- 🟨 Early Holocene
- 🟧 Mid-Holocene
- ⚪ uncertain

Flow accumulation
(90m aggregation)

low

high

0   500   1000 m

background source:
QuickBird, Aug. 2003

Figure 2.2. Archaeological sites and flow-accumulation in the Djara area. The flow-accumulation is derived from the digital elevation model. Clearly visible is the accumulation of sites along the main drainage line in the centre of the figure

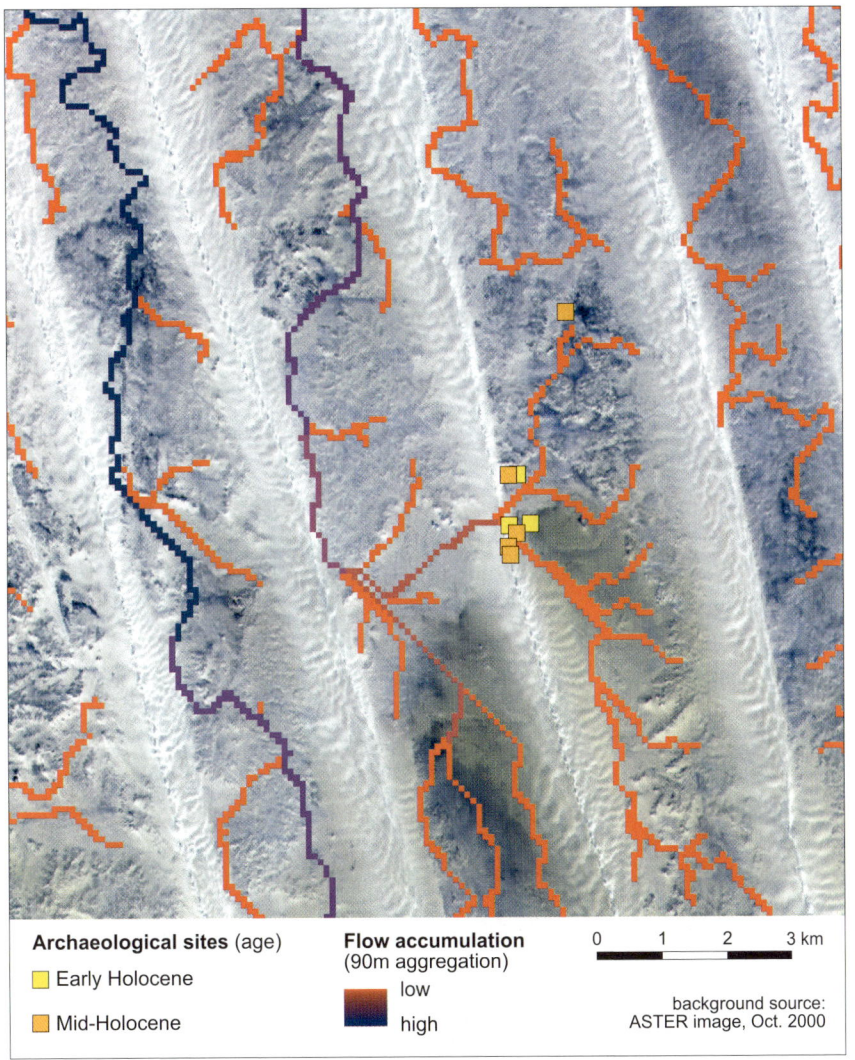

Archaeological sites (age)

☐ Early Holocene

☐ Mid-Holocene

Flow accumulation
(90m aggregation)

low

high

0   1   2   3 km

background source:
ASTER image, Oct. 2000

Figure 2.3. Archaeological sites and flow-accumulation in the Regenfeld area.
The flow-accumulation is derived from the digital elevation model. For better understanding the flow-accumulation raster is aggregated from 30 to 90 m pixel resolution. Clearly
visible is the incidence of archaeological sites at the calculated dune penetration point

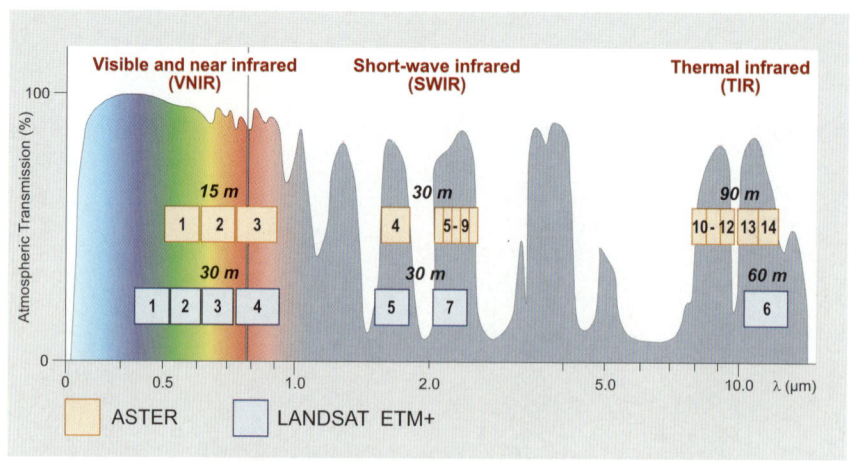

Figure 2.A2. Flow chart for further calculation

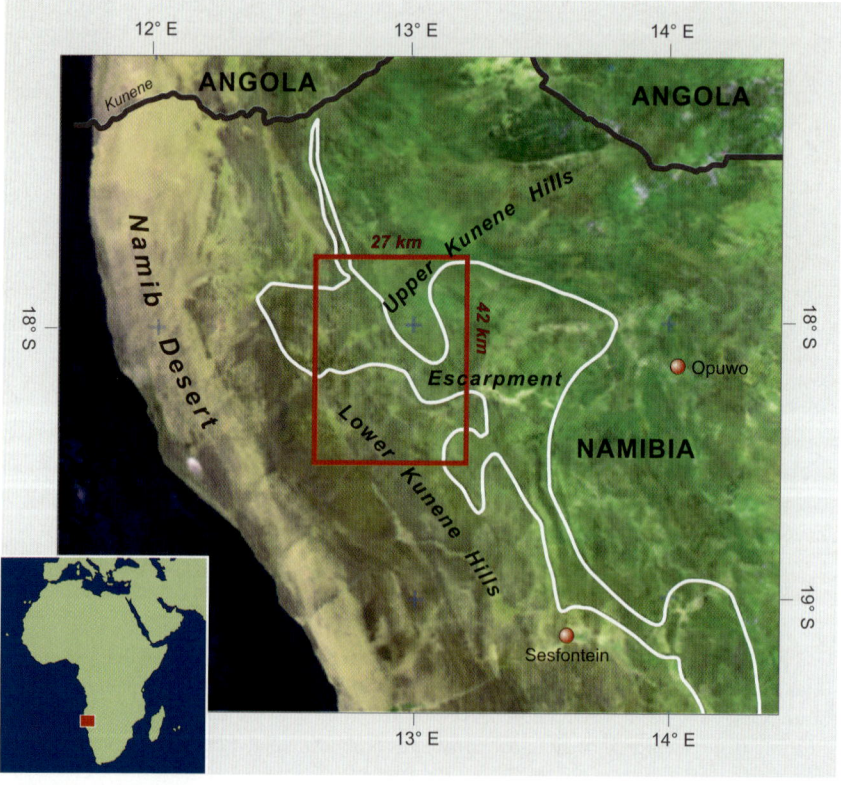

Figure 4.1. MODIS satellite image mosaic of NW Namibia. True color composite compiled from images acquired between June and July 2001. Overlaid with geomorphologic data from the Digital Atlas of Namibia (2001) and the 'Central Kaokoveld' study area (red)

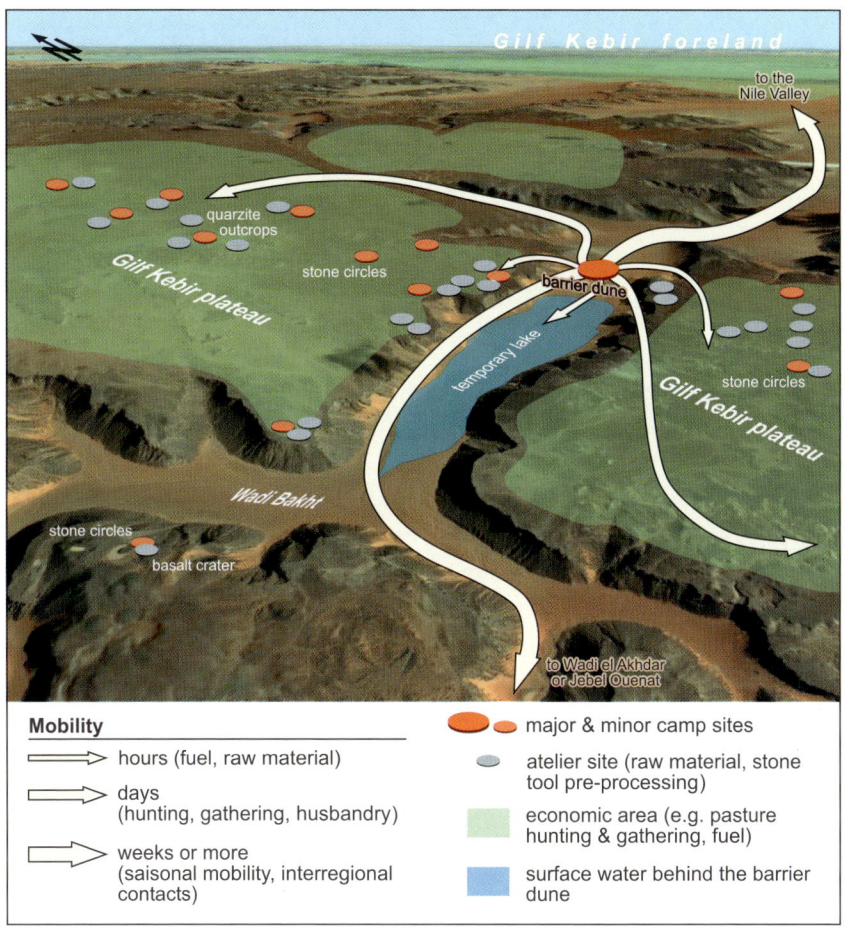

Figure 6.9. A section of the Later Stone Age Brandberg/Daureb landscape indicating resources and reconstructing the use patterns in a schematic representation. Note that the eco-tope did not undergo a change comparable to that of the Sahara (Photo: courtesy of H. Mooser)

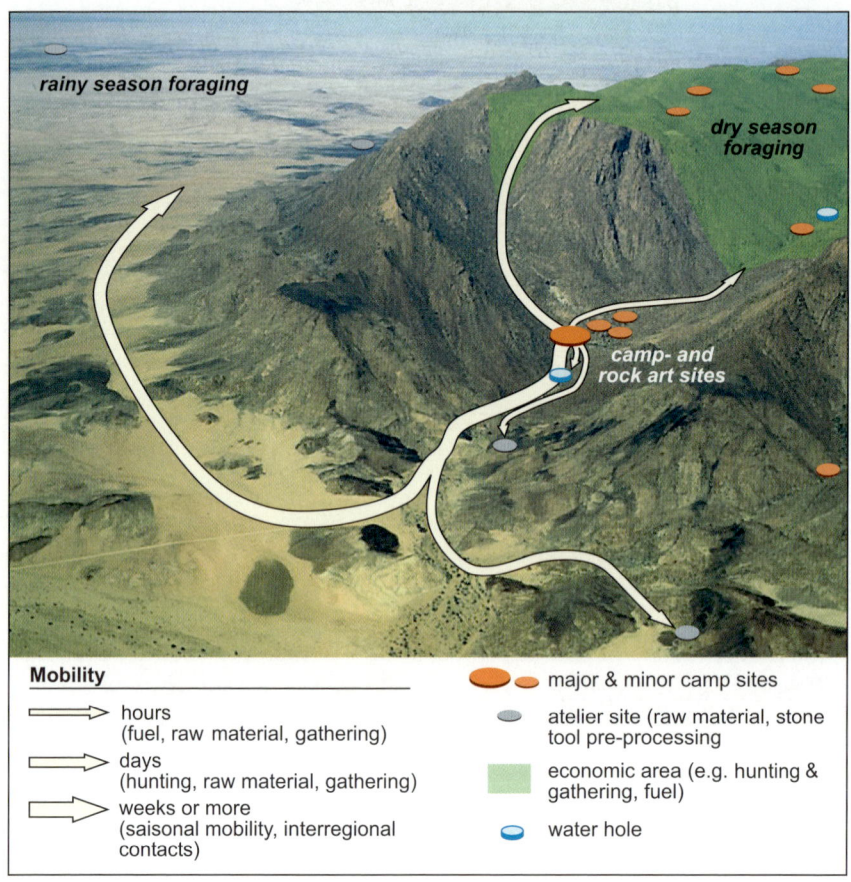

Figure 6.10. Analysis of relations between use of resources and human needs for the Chad case study

Figure 7.8. Inner façade of the temple of Esna, offering scene (Photograph by Dagmar Budde, Mainz)

Figure 7.9. Decoration over the gate of the inner façade of the temple of Esna, showing the nocturnal sun personified by the god Khnum (Photograph by Dagmar Budde, Mainz)

Figure 8.4. Postcolonial landscape. Visit to the graves of the Kwanyama kings near Namakunde in southern Angola, officially designated as national patrimony, 1999 (Photographs Patricia Hayes)

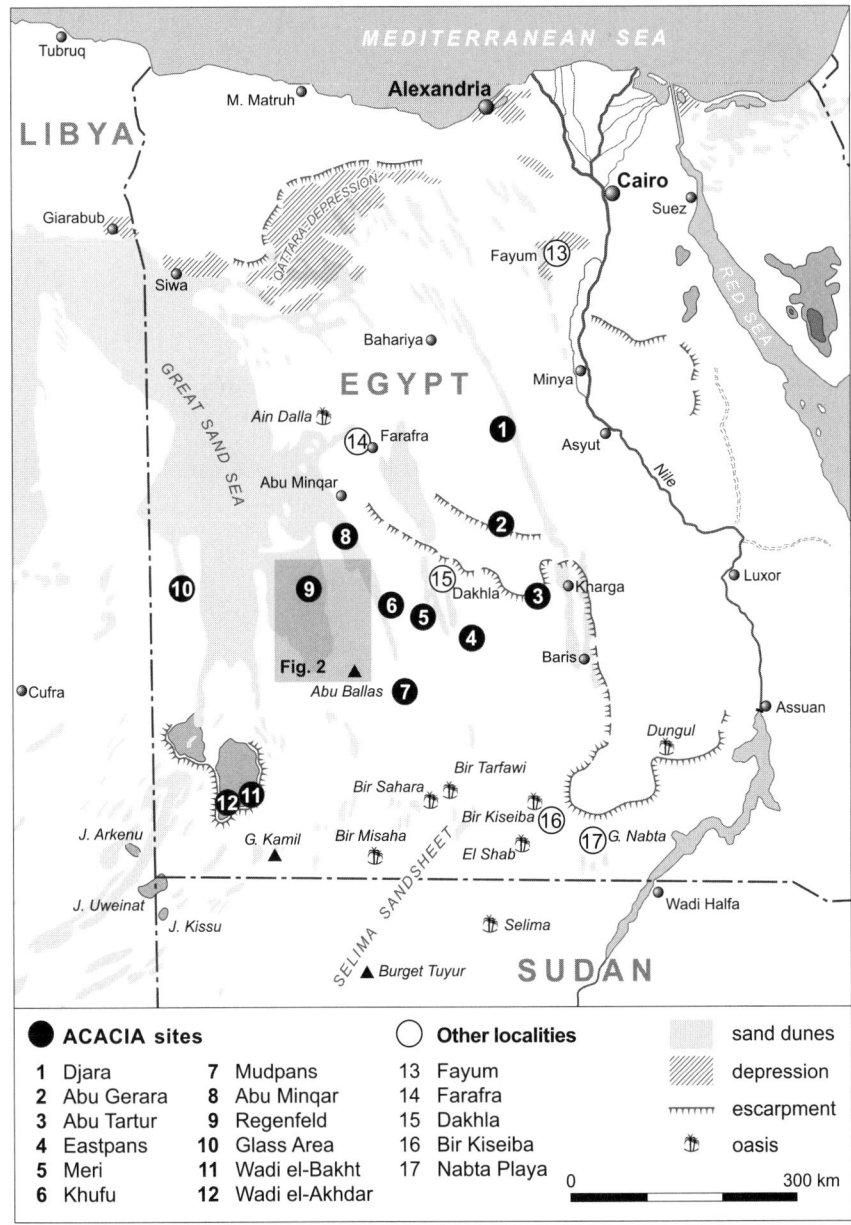

Figure 5.1. Map of northeast Africa showing the ACACIA sites (1–12) and other localities (13–17) mentioned in the text: 1, Djara; 2, Abu Gerara; 3, Abu Tartur; 4, Eastpans; 5, Meri; 6, Khufu; 7, Mudpans; 8, Abu Minqar; 9, Regenfeld; 10, Glass Area; 11, Wadi el-Bakht; 12, Wadi el-Akhdar; 13, Fayum; 14, Farafra; 15, Dakhla; 16, Bir Kiseiba; 17, Nabta Playa

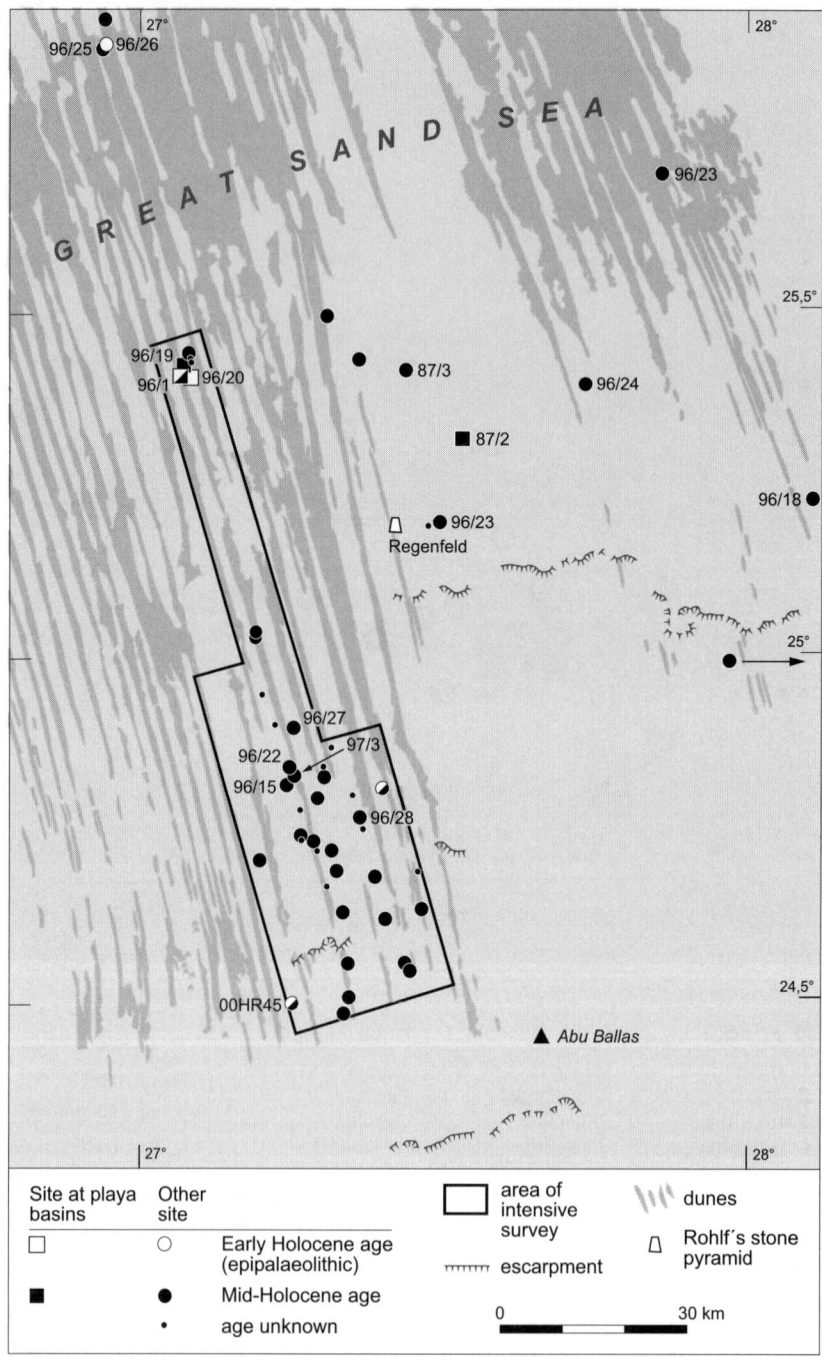

Figure 5.2. Southeastern Great Sand Sea showing the study area of Regenfeld and distribution of excavated or surveyed sites (map outlined after Klitzsch et al., 1987)

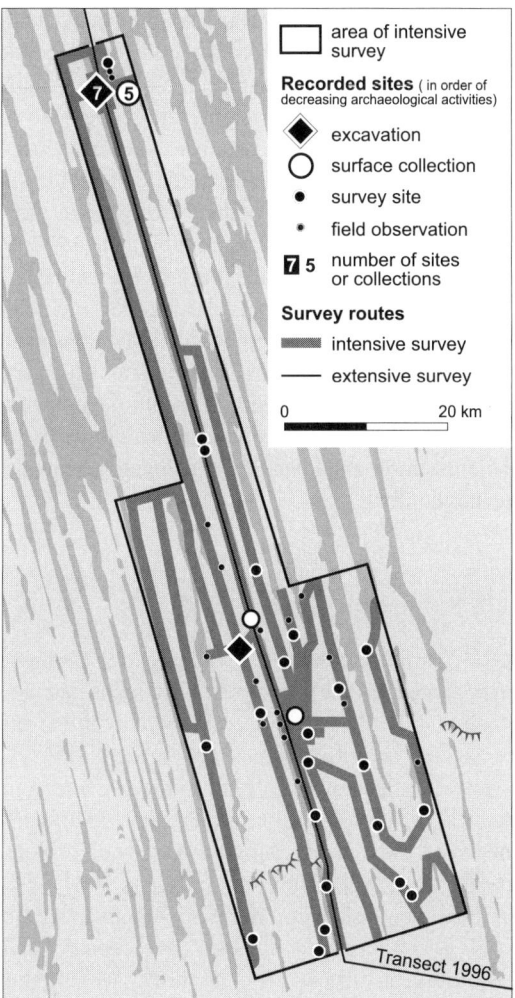

Figure 5.3. Study area of Regenfeld showing the survey routes and the recorded sites in order of decreasing archaeological activities.

5.1.1. Landscape in Archaeology

Site and landscape are different spatial units. Individual sites are embedded into a wider landscape with a set of potential resources (Lenssen-Erz & Linstädter, this volume). Archaeology traditionally concentrates on the study of individual sites by excavation and the analysis of artefacts, artificial structures, and their intrasite relations. Although a site may contain information about its context, conventional archaeological approaches tend to ignore the potential of an analysis of the wider spatial setting. Contextual evidence was developed by the analysis of the site's catchment area or exploitation territory (Vita-Finzi & Higgs, 1970) which includes the study of the site's contents and their sources in the surroundings of a site. But this methodological tool again evolves from the study of an individual site in order to understand the past activities on-site within their wider spatial context.

It is, therefore, wise to combine site and landscape analysis with the more general techniques of regional survey to analyse sites in their cultural and environmental setting. A regional view can help to understand the site's function and interaction within a complex system of other sites, and to estimate how representative the conditions at individual sites are (Johnson, 1977; Plog et al., 1978; Dunnell & Dancey, 1983). Moreover, the study of the entire landscape includes the 'off-site' evidence, as single artefacts, landmarks, old roads, and the like. This approach is particularly useful when people led a nonsedentary way of life, and only a little accumulation of artefacts was on site. Thus, appropriate survey methods were developed from being simply a preliminary stage to detect sites for excavation to an independent research tool in archaeology. In this way, the landscape provides the epistemological base for the development of methods capable of integrating site and context.

Spatial Units in Landscape Archaeology

Environment may denominate either the social environs ('milieu') or the natural surrounding (in particular, biotic) with which humans interact, or both. The present chapter dominantly focuses on the natural conditions, inasmuch as they dominate hunter-gatherer activity in zones of extreme aridity.

Landscape refers to the configuration of space on a regional scale mainly based on geophysical attributes. Important aspects are how humans perceive the landscape in which they live ('cognitive landscape'; 'mental maps'), and how they modify the landscape ('cultural landscape').

Site is a basic archaeological unit defined as a place with some significant evidence of human activity. Practically, sites have been defined by clusters of cultural material (artefacts, structures, or features) which contrast in density with the surrounding. Whereas 'intrasite analysis' combines methods to study the relationship between things found inside the boundary of a site, 'intersite analysis' focuses on the interaction between sites.

Territory is the total area covered by a nomadic group during its seasonal cycle ('annual home range').

Exploitation territory or *catchment area* demarcates the daily or micro-moves around a camp to exploit resources.

5.1.2. Deserts – Landscapes of Extremes

Archaeological investigations on arid zones naturally emphasise the palaeoecological aspects of landscape archaeology, as natural key resources are thought to be extremely scarce and unpredictable, and to be the major constraint for cultural evolution. A prime interest is to reconstruct the environmental conditions which define how and where people lived (Butzer, 1971, 1982; Vita-Finzi, 1978). This includes the reconstruction of the general climatic background as well as regional landforms, vegetation, and animal life. Climate, relief, and sediments determined the precipitation, run-off, and evaporation of surface water which in turn affected plant cover and animal life. The natural factors deeply influenced subsistence strategies or, at least, determined the presence or absence of humans in a certain landscape.

In arid zones the availability of water plays a critical role. Therefore, arid zone archaeology focuses on hydrological features as specific markers of landscapes. Water may occur as a permanent aquifer in oases or as temporary sub-surface water, but in the desert it is mostly available as surface water created by episodic rain events. The most significant hydroclimatic aspect of arid zones is the small amount of annual precipitation (<100–150 mm/a; Western Desert <2 mm/a) combined with the high rate of evaporation (potential evaporation about 5 m/a in the Eastern Sahara, Haynes 1982) and the high spatial and temporal variability of rainfall (relative variability >150% in the Western Desert, Besler, 1992). These factors are constraints for humans when they design adequate strategies of movement and resource management. Consequently, the second focus of this study is on how people attempted to create adequate subsistence strategies with a minimum of effort and risk. Plant remains and animal bones recorded from excavations may provide not only environmental data, but show the prime aspects of subsistence. Artefact assemblages of campsites can help to reconstruct the range of activities performed on site.

The Holocene wet phase, which is the period in question here, lasted for about 4000 years and was framed by two profound climatic changes. The initial change at the beginning of the Early Holocene is the northward shift of tropical summer rains following the hyperarid Pleistocene after 10,000 BP (9500 BC). The fundamental impact of this climatic change can be seen in the reoccupation of the depopulated desert. The final event is the drying up of the desert in the Mid-Holocene about 6000 BP (c. 4900 BC) which effected the depopulation of most areas in Egypt's Western (or Libyan Desert). On the base of typochronological dating, the wet phase in the study area can be grouped into two chronological units, the Epipalaeolithic or Early Holocene, c. 9500–7800 BP (c. 8900–6600 BC) and the Mid-Holocene phase, c. 7700–6000 BP (c. 6500–4900 BC). Further chronological subdivisions are possible (Gehlen et al., 2002), but reduce the number of dated sites below the limit of fruitful analysis.

5.1.3. Survey Methods at Regenfeld

Research strategies based on a landscape approach combine in-depth site research and extensive surveys. Whereas excavations aim at exhaustive or very detailed collection of data, surveys focus on sampling information over a wider area. In many cases, it is not possible to collect artefacts or samples for further analysis during a survey, so artefacts have to be documented and classified on site. In other words, there is no possibility of controlling or re-examining the primary data, nor in changing the focus of investigation after the fieldwork has been conducted.

Therefore a detailed mode of recording was worked out before starting into the field. It was helpful to concentrate investigations on specific attributes of surface sites with a standardised survey form, completed by sample collections of diagnostic finds or test excavations. The survey form defines key variables of site location and artefact composition such as geomorphological units (playa pan, dune, etc.), site's extension, artefact density, and artefact classes.

The different spatial levels of investigations addressed during the field work in the Great Sand Sea and in adjacent study areas can be listed in order of decreasing complexity as follows (Figure 5.3).

1. Excavating and sediment sieving of activity areas of the site including a full collection of organic material, debitage products, and so on.
2. Collection and measurement of the surface assemblage, including potsherds, stone tools, and cores, however, ignoring the fine-grained material which could only be gathered in an excavation.
3. Intensive areal survey as walking tours or with motor vehicles in low-speed gear within a microregional scale. Information was recorded on the survey form.
4. Extensive transect surveys with motor vehicles and random site stops. Sites were recorded on the survey form. This type of survey is mostly conducted as an accompanying activity during ground reconnaissance or other field trips.
5. As a complementary part, field observations – geographical, botanical, as well as archaeological – were recorded by Global Positioning System (GPS) within the logbook of each vehicle. The latter forms the fundament of a digital database used for further analysis of the landscape and site distribution.

In the Great Sand Sea three long-distance surveys from south to north along individual dune trains were realised during three geoarchaeological transects on palaeodune formation processes (Besler, 2002). The southern part of the eastern transect was selected for a more intensive survey because of a number of sites with well-preserved organic material and artefact scatters. The area intensively investigated is called 'Regenfeld' after a location where a stone cairn was erected by the German explorer Gerhard Rohlfs in 1873 (1875, 166 f) (Figures 5.2 and 5.3). Excavations and survey activities were conducted in 1996, 1997,2000 and 2000 (Riemer, , 2003). The majority of the collected material has been analysed in the meantime and has produced detailed information on environmental and cultural development.

The survey area includes a 100 km long strip covering two neighbouring interdune corridors. The Regenfeld playa (96/1) forms the northern edge of the survey area. In the south two neighbouring corridors were additionally surveyed. Although this survey unit was artificially set in the absence of natural limits, it represents a typical landscape of the southern Sand Sea.

Although the total number of sites increased during the surveys, their number was still not satisfactory, and sites discovered during other field reconnaissance conducted between Regenfeld and the eastern border of the Sand Sea have been incorporated into the analysis of settlement patterns as far as possible.

Investigations into the past require answers to questions of chronology in order to understand long-term developments of climate, landscapes, and cultural processes. Although the establishment of chronology is a fundamental aim in archaeology, the dating of sites is frequently problematic in regional surveys because dating must be carried out in the field, and on the base of small sample collections or individual artefacts. Absolute dating by the ^{14}C-method[2] and stratigraphical evidence, form the backbone of dating methods on excavations in the Western Desert, whereas survey sites were rather tentatively dated by the typological comparison of stone tools.

5.2. RECONSTRUCTING THE PAST ENVIRONMENT

Climatic conditions constrain human activities in arid landscapes in an essential way. However, evidence of past climatic conditions, the amount of rainfall, temperatures, or seasonal variation to name but a few variables cannot be measured directly, but have to be deduced from environmental archives. Using comparable data from recent environments helps to shed some light on past climatic processes and variables.

The reconstruction of the climatic development in the eastern Sahara has been outlined by a number of environmental studies over the past decades (Wendorf & Schild, 1980; Wendorf et al., 1984; Kuper, 1989; Wendorf et al., 2001). The most informative arch ives studied were as follows.

- Sediments of perennial lakes and episodic mud pans (playas) as a result of higher rainfall and runoff.
- Archaeobotanical and -zoological studies led to the reconstruction of plant cover and wild life.[3]

---

[2] All dates mentioned are approximate, based on the 14C-method. They are given as uncalibrated dates in BP and calibrated calendar years in BC. As to the calibration, the dates were calculated and plotted by the 2-D Dispersion Calibration Program Version Cologne 2003 of CALPAL (Cologne Radiocarbon Calibration & Palaeoclimate Research Package) by Bernhard Weninger, Radiocarbon Laboratory, University of Cologne.

[3] The paleoenvironmental data outlined by charcoal and bone analyses give only poor evidence on the absolute limits of the Holocene wet phase. Despite the valuable role played by archaeobotanical and archaeozoological archives to determine past environmental conditions, it has to be recognised that most organic macroremains constituting the paleofauna and paleoflora come from archaeological excavations deposited as a result of human subsistence and consumption activity. Accumulation on campsites and preservation in campfires are the most important sources for plant and animal remains. A complete break-off of the human occupation in the desert regions, or the absence of people during the dry intervals, means negative evidence for any environmental study that relies on botanical or zoological data. This connection of archaeological sites and faunal or floral sources gives rise to a serious problem for all the periods which followed the depopulation of the deserts after 6000 BP (c. 4900 BC). Many plants and animals in dry lands can still survive under extreme arid conditions although humans can no longer exist. A continuation of well-adapted arid species can truly be suggested, however, when the occupation of the deserts ended, nothing remained to deduce the past animal and plant life.

- The intensity of settlement activities (^{14}C chronologies) led to the reconstruction of more favoured climatic phases or phases of climatic deterioration.

5.2.1. Dating the Holocene Wet Phase

With the onset of humid conditions after the hyperarid Pleistocene, the tropical wet front moved 700–1000 km northwards (Neumann, 1989a; Haynes, 1987). Rainfall increased from less than 10 mm to a maximum of 50 or 100 mm per year (Neumann, 1989a, b). The compilation of initial playa and lacustrine deposits from Egypt and northern Sudan (Gabriel, 1986; Haynes, 1987, 2001; Kröpelin, 1993; Hoelzmann, 2002) clearly shows an onset of rains and fluvial activity in the early Holocene between 9800 and 9500 BP (c. 9200–8900 BC).

Although the onset of the rains in the Early Holocene is supposed to have been a rapid climatic process, the Mid-Holocene trend towards hyperaridity cannot be dated so well. The end of fluvial activity is difficult to determine from playa stratigraphies because their upper parts are often obscured by deflation. In the case of the Bahr playa stratigraphy near Farafra (Ashour, 1995; Hassan et al., 2001), the uppermost silt deposits at 6000 BP (c. 4900 BC) are covered by rubble which most probably prevented an extensive deflation. The upper playa layer of Djara 98/20 is roughly dated around 6000 BP (c. 4900 BC) (Bubenzer & Hilgers, 2003; Kindermann et al., 2006). It is covered by a dense scatter of lithics which only show little effects of deflation. A *terminus ante quem* for wadi sedimentation about 6500 BP (c. 5400 BC) is suggested by a ^{14}C-date from 'Locality 165' in Dakhla Oasis (Kleindienst et al., 1999, p. 13); some proof might also be given from Bashendi B phase artefacts (6500–5200 BP, c. 5400–4000 BC) (McDonald, 2001) covering the surface. However, artefact scatters could also be affected by deflation, and it is often hard to say whether the artefacts were once embedded into the sediments. The extended playa deposits of Kharga have so far received only minimal attention. Most sites excavated by the Combined Prehistoric Expedition (CPE) were connected to lacustrine or spring mound sediments (Wendorf & Schild, 1980) which were effected by artesian water.

Although all this indicates a beginning of the drying trend around 6000 BP (4900 BC) the data from southern Egypt and Sudan point to a differentiated picture. The silts of Nabta Playa were also lacustrine sediments (Wendorf et al., 2001). Here, no 'playa' silts postdate the Early Neolithic (10,000–7300 BP, c. 9500–6100 BC), and the authors suggest the pluvial maximum for the El-Nabta/El-Jarar Early Neolithic phase (8050–7300 BP, c. 7000–6100 BC). The deposits are capped by gravel wash or dunes containing artefacts from the Middle Neolithic to the Final Neolithic (7300–4500 BP, c. 6100–3200 BC). On the other hand, in the neighbouring Bir Kiseiba area the final silt sedimentation (so-called Playa III) lasted until about 5400 BP (c. 4300 BC) (Wendorf et al., 1984, p. 404).

Playas and human occupation in the Gilf Kebir (Kröpelin, 1987, 1989; Linstädter & Kröpelin, 2004) lasted longer, to about 4500 BP (c. 3200 BC). Lake sediments from northern Sudan (Gabriel, 1986; Haynes, 1987; Kröpelin, 1987; Pachur & Hoelzmann, 1991; Hoelzmann et al., 2001; Hoelzmann, 2002) show an onset of predominant aeolian activity between 4300 and 3600 BP (c. 3000–2000 BC). In the Laqiya region, northern Sudan, nomadic occupation stopped between 3600 and 3200 BP (c. 2000–1500 BC) (Lange, 2006; Schuck, 1989). These dates indicate a slowly progressing aridisation; the monsoonal summer rain belt shifted to the south with an average time gradient of about 35 km/100 years (Kröpelin, 1993; Haynes, 1987). The Wadi Howar shows continuous occupation until 3200 BP (c. 1500 BC) (Keding, 2000), and nomads still gather temporary groundwater from the wadi sediments nowadays. In the case of the Gilf Kebir, an increasing influence of the Mediterranean rains in the most western part of Egypt during the Mid-Holocene is suggested (Kröpelin, 1987; Linstädter & Kröpelin, 2004) as an additional factor, whereas most desert areas in Egypt dried up completely after 6000 BP (c. 4900 BC).

Frequencies of ^{14}C-dates from archaeological excavations can help to reconstruct phases of human activities in the desert (Nicoll, 2001; Gehlen et al., 2002; Kuper & Kröpelin, 2006; Riemer, 2006). The dates collected by the B.O.S. and ACACIA projects over the last two decades clearly show the time limits of the Holocene wet phase in the Eastern Sahara which lasted approximately from 9500 to 6000 BP (c. 8900–4900 BC) in this area. Among the data derived between Dakhla Oasis and the high plateau of the Gilf Kebir, which embraces the southern Great Sand Sea and the Abu Ballas scarp land, the earliest date indicating the reoccupation of the desert territory after the hyperarid Pleistocene is that of Regenfeld 96/1 falling at about 9400 BP (c. 8700 BC). The final drying up is set about 6000 BP (c. 4900 BC) as shown by the drop-off of ^{14}C-curves in most regions of the eastern Sahara (Figure 5.4). This general picture corresponds well with other compilations of data from the eastern Sahara (Nicoll, 2001; Vermeersch, 2002). However, the histograms of ^{14}C-dates from

Figure 5.4. Excavation 96/1–2 on the eastern fossil dune slope at Regenfeld playa in progress (view northeast). The playa is silted up to the lorry's position, but covered with a thin layer of windblown sand. The lower edge of the shadow in the background approximately marks the foot of the steep active dune crest rising up to 50 m above the playa's shoreline, and the underlying fossil dune shows a rounded body (taken December 1997)

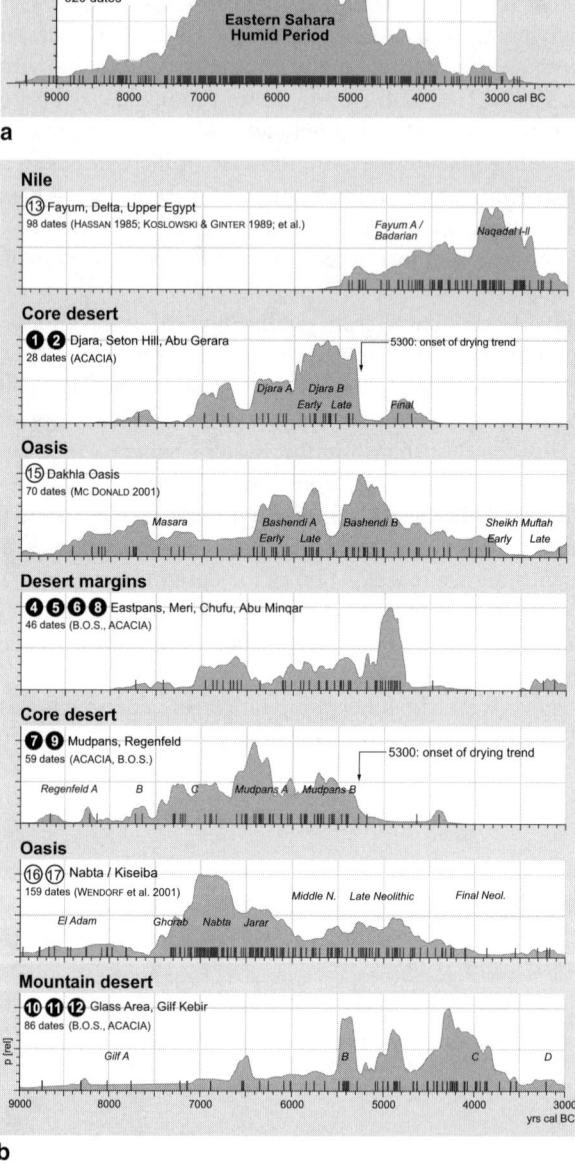

Figure 5.5. Cumulative histograms of calibrated ^{14}C-dates from the Eastern Sahara illustrating the occupational development and related climate change: A, Egyptian Western Desert curve; B, Regional curves from the Nile Valley and the Western Desert including the core desert of Mudpans and Regenfeld. Ticks on the *y*-scale indicate the mean values of individual dates

the Great Sand Sea and the Abu Ballas scarp land suggest a somewhat earlier onset of the drying trend at about 6300 BP (c. 5300 BC). This observation is linked to the fact that this area is – and most probably was – not only the driest part of the Sahara but also most sensitive for low-graded climatic fluctuations (Kuper & Kröpelin, 2006; Riemer, 2006).

5.2.2. Palaeovegetation

The reconstruction of plant spectra in the desert areas of Egypt's Western Desert by anthropological analysis of charcoal remains from campfires was developed by Neumann (1989a, b). On the base of her studies, she suggested a maximum rainfall for the Holocene wet phase which never exceeded 50 mm/a in the Early Holocene, and 100 mm/a in the Mid-Holocene of the Western Desert.

A close relationship exists between plant cover, the topography, and the substrate which control the water availability of an area (Kehl & Bornkamm, 1993). Comparison with recent environments helps to reconstruct the plant communities and their response to soil or substrate, precipitation, and topographic patterns. Thus, Neumann (1989a) reconstructed the palaeovegetation cover on the sand dunes and sand sheets of the Great Sand Sea which can be classified as (semi-) desert with diffuse perennial vegetation of grasses and shrubs (Neumann, 1989a, p. 110). Unfortunately grasses, and often shrubs as well, are difficult to ascertain among the charcoals of prehistoric campfires. The interdune valleys were characterised by desert vegetation contracted on the wadi channels and playas. The depressions between the dunes were probably covered with dwarf shrubs, tuft grasses, and small trees such as tamarisks or tarfa (*Tamarix* sp.), and probably also *Acacia* sp. (Neumann, 1989a, p. 111). During long-lasting water stands, there was an azonal growing of reed or cattail (*Phragmites* sp. or *Typha* sp.) in the basins (Neumann, 1989a, p. 112).

The Regenfeld sequence suggests acacia trees and tamarisks as the dominant plant species during the Holocene wet phase (Figure 5.5). As to the climatic development, the charcoal remains found on sites in Mudpans south of the Regenfeld area show an increasing number of plant taxa during the Mid-Holocene interpreted by Neumann as the phase of maximum rainfall during the Holocene wet phase. The Regenfeld sequence does not verify this interpretation, but shows a continuation of the Early Holocene desert flora throughout the Mid-Holocene

Neumann also summarises the complementary advantages of the Sand Sea and sand sheets on the one hand, and the playa pools (Neumann, 1989a, p. 113): dunes and sand sheets most likely had a dense (or diffuse) vegetation cover due to the advantage of water storage in the sandy substrate. Therefore, they provided good grazing areas for herbivores, such as gazelles, antelopes, and ostriches as well as goats, sheep, and cattle. On the other hand, the dune fields did not provide open water because of the less-pronounced relief in the interdune valleys, and the infiltration of rain water on sandy grounds. Areas which combine sand sheets or dunes with developed playa deposits were generally favourable places for

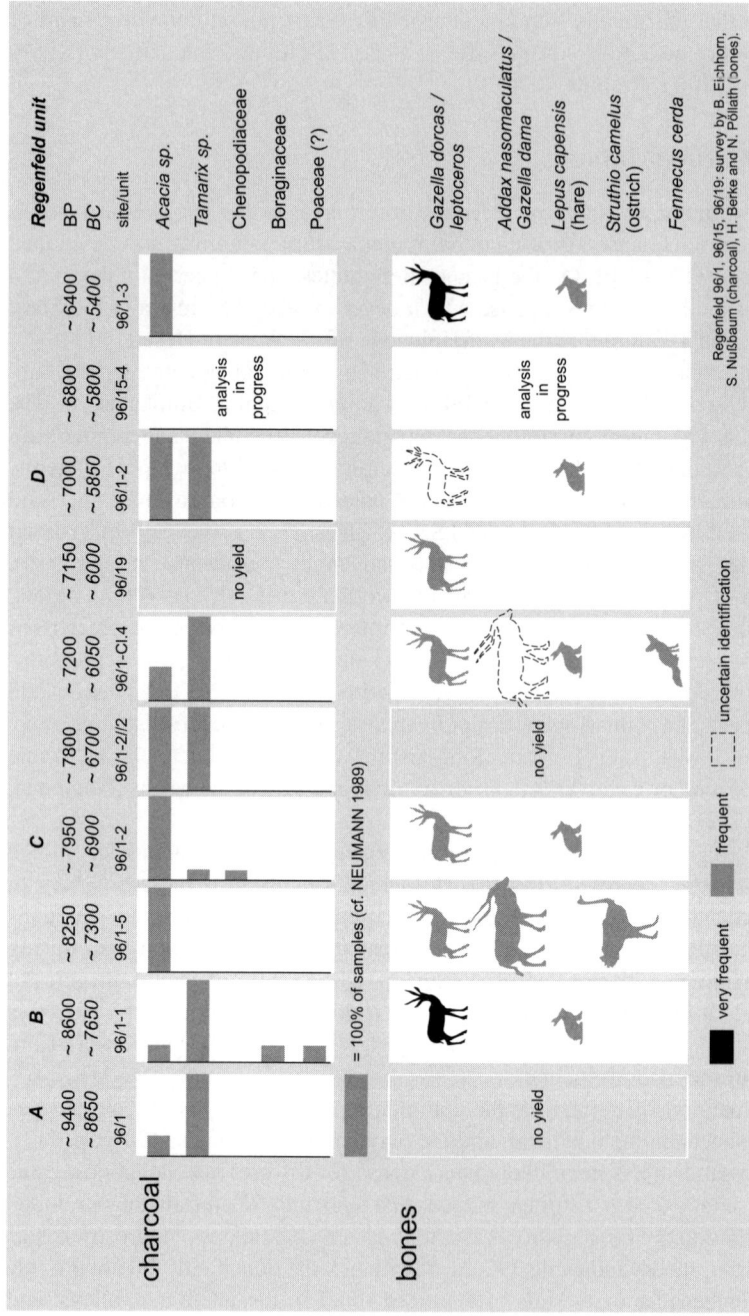

Figure 5.6. Regenfeld sequence showing the results of archaeobotanical and archaeozoological studies of the Regenfeld assemblages

long-lasting campsites because they provided enough water and food resources for a few months. The Regenfeld playa also served as such a favoured place, but it was only an isolated playa pond far away from any other surface water. Within the survey area of Regenfeld, this playa was the only basin to hold surface water for a longer period.

5.2.3. Palaeofauna

The study of animal bones points to a similar climatic picture (Figure 5.5). The Regenfeld sites frequently produced remains of small gazelles (*Gazella leptoceros/dorcas*), hare (*Lepus capenis*), and fennek (*Fennecus cerda*) which are the main animals in all the desert sites. Larger gazelles or antelopes are less frequent; only two determinations of *Addax nasomaculatus* or *Gazella dama* are known from Regenfeld (Berke, 2001, p. 241). The mentioned taxa are characteristic for an arid environment, because they do not (or not frequently) need to drink from open water (Dorst & Dandelot, 1972; Haltenorth & Diller, 1977; Estes, 1992). Their water demand is provided by their plant diet. The very small dorcas (*G. dorcas*) and Loder's gazelles (*G. leptoceros*) (body weight 45–50 kg) are best adapted to desert conditions because they exploit food resources which are not only beyond the range of water-dependent herbivores, but too meagre to support larger competitors such as addax or oryx. They penetrate the deepest into the desert, and indeed they are the only bovids in the Western Desert where a minimum of vegetation exists. *A. nasomaculatus* (weight 180–270 kg) and *G. dama* (160 kg) are not water-dependent, but have slightly higher demands.

A small number of bones from Regenfeld were identified as ostrich (*Struthio camelus*), which are restricted to sandy regions of the semidesert or desert, and can survive without water for longer periods of time (Sauer & Sauer, 1980, p. 90). Although ostrich bones are rarely found within archaeological sites, the eggshells or eggshell beads are typical remains found on nearly all desert sites. Bones were only reported from Regenfeld (Berke, 2001, p. 241), Mudpans, and Wadi el-Akhdar (Van Neer & Uerpmann, 1989, p. 328).

Scimitar-horned oryx (*Oryx gazella dammah*; weight 450 kg) have not been observed in Regenfeld, although it is another desert-adapted mammal, and also known from prehistoric desert sites elsewhere (Mudpans, Glass Area, Van Neer & Uerpmann, 1989). Semiarid species, for example, hartebeest (*Alcelaphus buselaphus*) or giraffe (*Giraffa camelopardalis*) are rarely present in desert inventories (Van Neer & Uerpmann, 1989). Two bone samples of hartebeest were determined within the new collection of Glass Area 81/61 ('Willmann's Camp'; N. Pöllath, personal communication), however, it was not observed on other desert sites. The hartebeest is a semiarid grassland or dry savanna antelope. It normally drinks daily, but in waterless countries it can compensate water intake by using melons, roots, or tubers for days or months (Estes, 1992, p. 139; Haltenorth & Diller, 1977, p. 86).

Giraffes form a major component of the bone assemblages on a few sites in the Gilf Kebir wadis, and a bone fragment as well as depictions in rock art were found in Mudpans (Van Neer & Uerpmann, 1989, p. 328). The giraffe is distributed throughout the savanna, but it also penetrates into desert zones wherever trees grow because it can feed on the crowns of trees. Acacia trees mostly form the bulk of their diet, especially in dry lands. Giraffes drink at intervals of three days or less when water is available, but can also substitute part of their demand with green leaves (Estes, 1992, p. 202 f).

The elephant (*Loxodonta africana*) can only subsist in arid zones where trees provide shade and browse and surface water is available (Estes, 1992, p. 260). Elephant bones were found in Dakhla Oasis (Churcher, 1999, p. 143) but never in the desert. The curious fragment of elephant ivory in Mudpans is clearly worked and may be a result of transport by people (Van Neer & Uerpmann, 1989, p. 328).

5.2.4. Water Resources

The most important factor influencing human activity in desert areas is the availability of water. The Egyptian oases have a permanent water supply charged by the fossil water of the Nubian Aquifer (Figure 5.6; Ball, 1927; Thorweihe, 1990; Heinl & Thorweihe, 1993). In contrast to the rains, the artesian flow was relatively stable during the Holocene.[4]

In the absolute desert of the Great Sand Sea and the Abu Ballas scarp land the fossil groundwater table was by far too deep (Thorweihe & Heinl, 1993; see Figure 5.6). Water was only temporarily available in open pools (playas or mud pans). Locally fluctuating subsurface reservoirs fed by infiltrated rainwater in the sediments of larger wadi streams or on impermeable subsurface layers may predominantly occur in mountainous areas such as the Gilf Kebir or the southern Egyptian Limestone Plateau. After a downpour the water infiltrates into porous sediments or cracks. Impermeable beds below the surface stop the downward progress of the water. It only comes to the surface at escarpments or slopes of deep wadi channels, often higher than the depression's or wadi's floor. For instance, the well of 'Ain Amur on the northern edge of the Abu Tartur Plateau stands about 300 m above the depression (King, 1925, p. 316). For the western Gilf Kebir, it was reported that Tubu nomads used wells in the Wadis Talh and Wadi Abd el-Melik (Almásy, 1939, pp. 157–164).

Subsurface water is not suggested for the Great Sand Sea and the Abu Ballas scarp land which are mostly characterised by a less-developed relief. There, water was only available as surface water in open pools where extended drainage systems guaranteed effective run-off after heavy rainfall. Large drainage areas,

---

[4] In spite of only very limited dates, a continuity of settlement activities lasting over the millennia between the drying up of the desert and the early Pharaonic occupation is documented only for the oases (McDonald, 1999, 2001). This is believed to be a result of the high reliability of the groundwater supply.

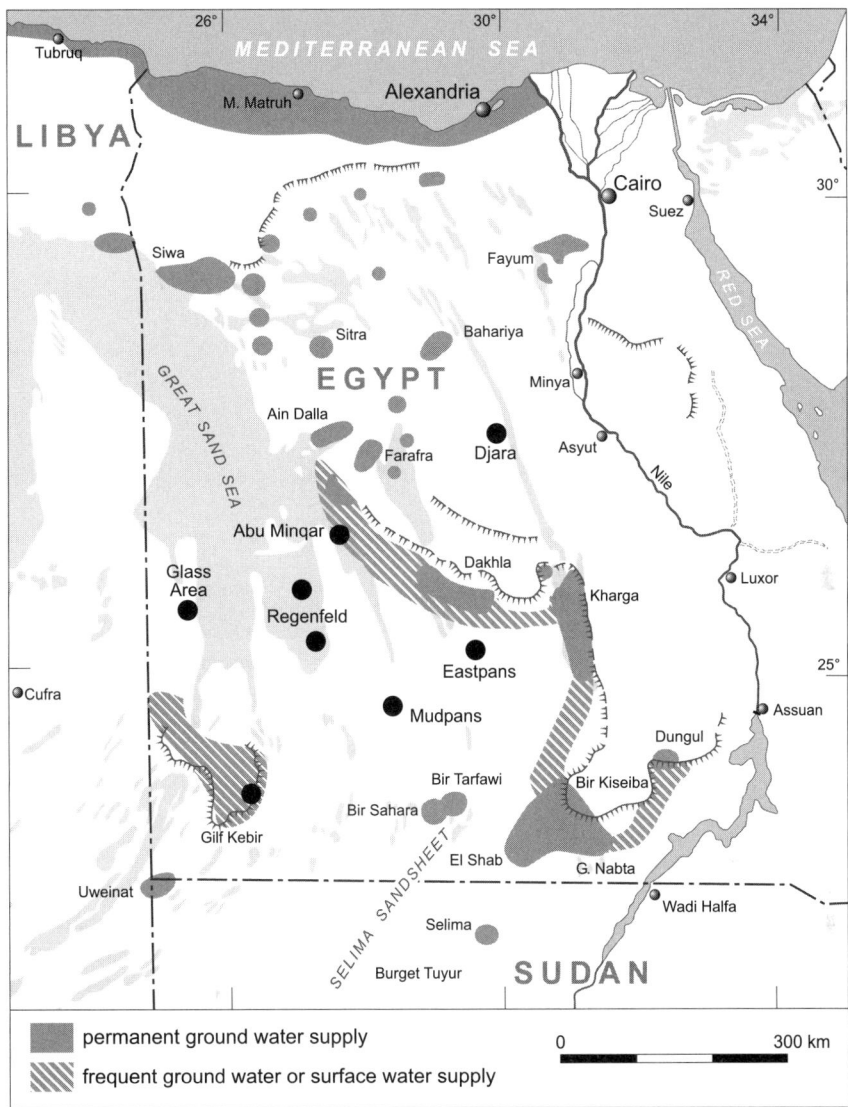

Figure 5.7. Map of northeast Africa showing areas with permanent availability of groundwater and with relief favourable to frequent local groundwater during the Holocene

hollows, and depressions, and blocked drainage systems provided run-off and high-water stands. The pools were characterised by water-repellent fine-grained sediments deposited in still water areas during inundation. Although these ephemeral drainage basins (also called mud pans or playas) are found in most desert areas (Embabi, 1999; Pachur, 1999, 2001; Pachur & Hoelzmann, 2000), they are rare and usually small in extent.

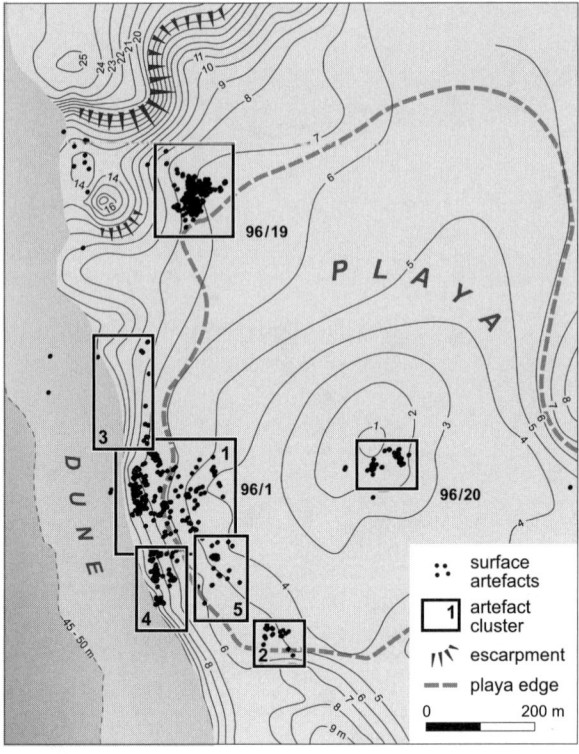

Figure 5.8. Map of Regenfeld playa showing the major physiographic units (cf. Figure 5.8)

A playa pool was discovered in the Regenfeld area indicated by mud deposits which were dated into the Holocene wet phase by artefacts and intercalation of Holocene dune layers. The onset of Early Holocene playa deposition is generally assigned to the horizon mentioned, around 9800–9500 BP (c. 9200–8900 BC) which might also be a realistic age for the Regenfeld playa. The Regenfeld playa extends about 800 m in a north–south direction, forming a shallow basin of 6 m depth (Figures 5.7 and 5.8). Taking into account that the higher terraces of the shoreline indicate the maximum high-water level, it is possible to estimate the period of inundation after the pool was filled. Of course, the original altitude of the basin's floor or the rate of deflation during the last 7000 years is not easy to ascertain, but should not extend beyond 1 m, based on observations of artefact scatters and hearth mounds on the playa. Available data on measured evaporation are scanty and often unreliable, because of numerous phenomena affecting the results, such as the 'oasis effect' or changes in atmospheric moisture or air temperature.

To get a crude image of the time needed to evaporate the playa pool, these phenomena were ignored. Data on actual potential evaporation range from about 3 m/a for Kharga (Simons, 1973, p. 462) and Farafra Oasis (Barich et al., 1991, p. 42), about 5 m/a for the central Sahara (Haynes, 1982), to 5.6–5.9 m/a for Dakhla and Kharga (after Griffiths, 1972, p. 90). On the base of an average evaporation rate of 5 m/a,

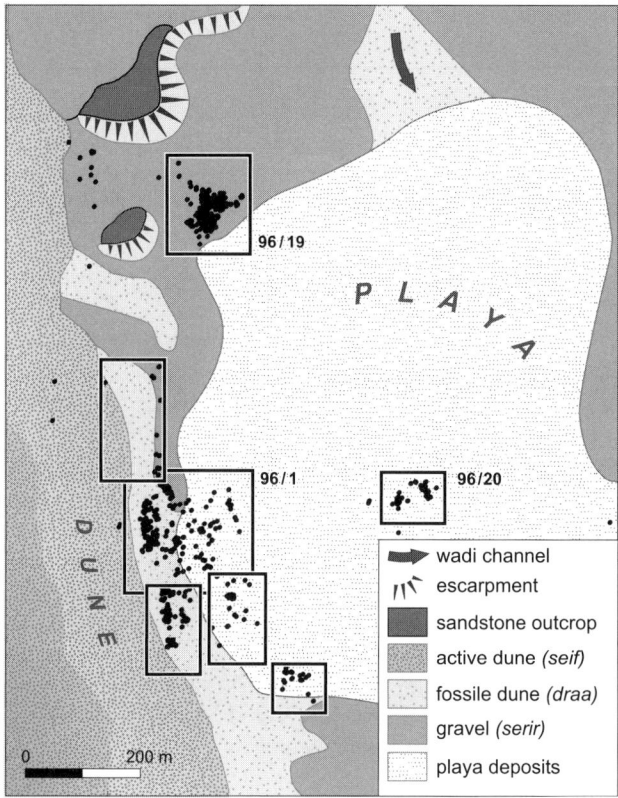

Figure 5.9. Topographic map of Regenfeld playa showing the approximate edge of the playa and the distribution of sites and surface finds (based on tachymetric measurement in December 1996)

surface water may have been available for more than a year. In the southern Great Sand Sea the rainy season most probably was in the summer when temperatures were high and the potential evaporation aggregated three times the winter values (Griffiths, 1972, p. 90). This reduces the period of inundation to approximately ten months (Table 5.1). Moreover, such a high-water stand was undoubtedly an exceptional case. Probably, the periods of surface water were much shorter. For instance, after an inundation of 0.5 m the water only lasted for three weeks in the summer months.

The study of the Regenfeld playa sediments didn't produce any remains of any bioactivity of aquatic fauna or flora, with the exception of stem or root casts of 1 cm diameter identified as reed or cattail (*Phragmites* sp. or *Typha* sp.), visible on the playa surfaces. The ecology of the Regenfeld playa is comparable with playas, for example, in the eastern Gilf Kebir, which were analysed by Kröpelin (1989). He suggests a low-graded salinity for the playa sediments in Wadi el-Bakht and Wadi el-Akhdar on the base of electric conductivity logging (1989, pp. 231, 278). The low-grade salinity provided good conditions for aquatic or semiaquatic life. However, the lack of bioremains such as other plant elements or remains of aquatic molluscs confirms the assumption of very rare and irregular water stands.

Table 5.1. Cumulative evaporation rates per month after rainfall at the beginning of May (calculated on values from Dakhla; source of daily potential evaporation: Griffiths, 1972

Amount of evaporation (mm)					
Month	Daily	Cumulative	Month	Daily	Cumulative
1 May	21.8	675.8	10 February	10.0	4,915.1
2 June	23.8	1,413.6	11 March	13.4	5,330.5
3 July	22.4	2,108.0	12 April	17.7	5,861.5
4 August	21.3	2,768.3	13 May	21.8	6,537.3
5 September	19.5	3,353.3	14 June	23.8	7,275.1
6 October	15.4	3,830.7	15 July	22.4	7,969.5
7 November	10.9	4,157.7	16 August	21.3	8,629.8
8 December	7.7	4,396.4	17 September	19.5	9,214.8
9 January	7.7	4,635.1			

With the reconstruction of landscape features, one can assume that in the absolute desert neither a sedentary way of life nor a long-lasting or regular seasonal presence of desert dwellers was possible. After the surface water dried up the people had to return to more favourable areas such as the oases or mountainous refuges.

5.3. ASPECTS OF LAND USE RESPONDING TO THE LANDSCAPE

5.3.1. Mobility

The availability of water conditioned the mobility of the desert dwellers. The prehistoric sites of the southern Sand Sea are situated between 150 and 250 km away from permanent water sources. These distances, at least, had to be covered by desert dwellers during their seasonal rounds between the oases and the episodic water pools of the desert. The study of artefacts and animal remains found on sites in the desert areas can add further details to the movement of people.

'Exotic' goods, for example, the Nile bivalves *Spathopsis* (*Aspatharia*) sp., and gastropods from the Red Sea or eastern Mediterranean such as cowry (*Cypraea* sp.), point to the 'embeddedness' of localized prehistoric forager communities into wider exchange systems. Good examples of both species were found in Mudpans (Van Neer & Uerpmann, 1989, p. 328). However, exotic molluscs were only rarely found in the deserts west of the oases.

Raw material procurement for lithic production as well as traditions of tool production can help to establish a more detailed picture of the territory covered by the desert dwellers during their seasonal rounds. The fact that the southern Great Sand Sea only holds rare resources of good quality raw material for lithic production creates an almost experimental situation: most stone artefacts from Regenfeld were made of material originating from outside the area. The Regenfeld sites yielded a large amount of artefacts made out of flint, Libyan

Desert Silica Glass (LDSG), and quartzitic sandstone. None of it was gathered within the daily micromoves of about 10 or 15 km around the campsites.

Most probably generated by a meteoritic impact, LDSG originates at the western margins of the Great Sand Sea (Riemer, 2007; Figure 5.9). This area,

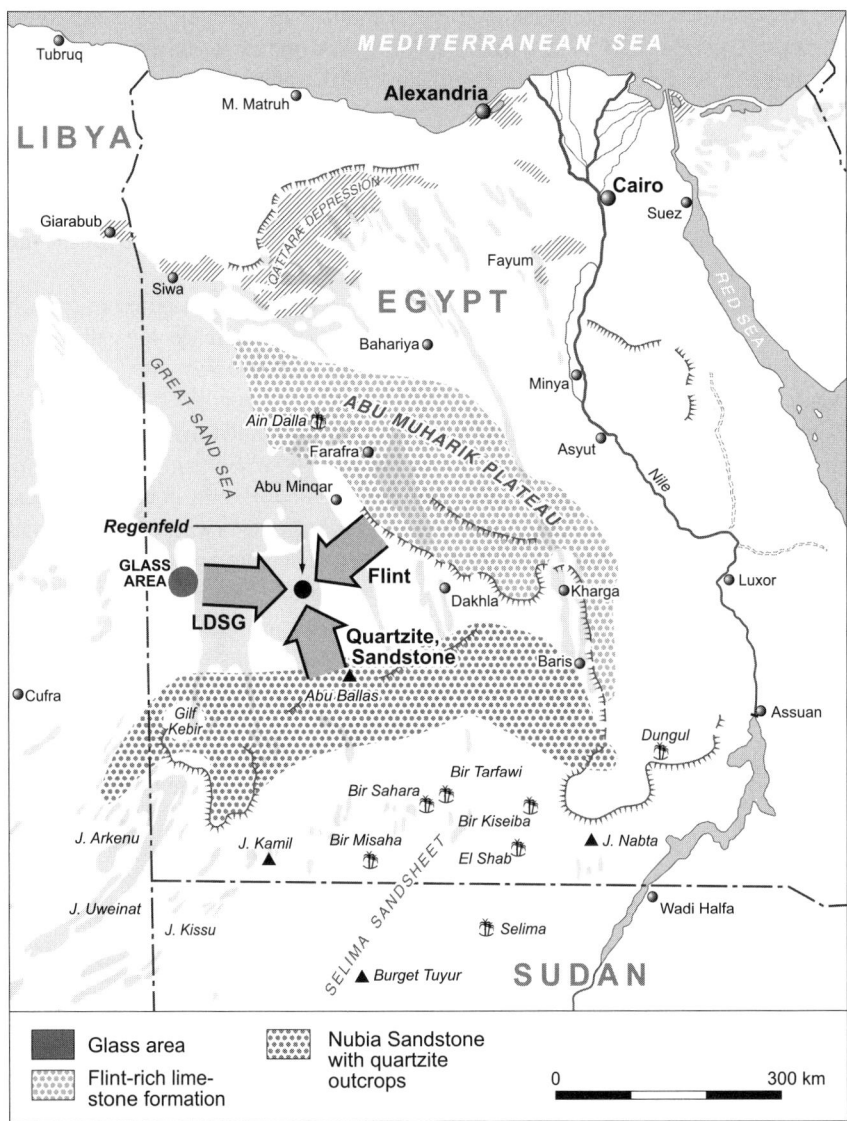

Figure 5.10. Provenience of raw material classes for stone artefact production found on the northern Regenfeld sites. Flint originated on the Eocene limestone formation back of the oases, quartzite and quartzitic sandstone most probably came from the south, and Libyan Desert Silica Glass (LDSG) provenienced in the Glass Area at the western margins of the Great Sand Sea (Klitzsch et al., and after a sketch by F. Klees)

which is named 'Glass Area' is more than 150 km away from the Regenfeld sites. Flint stone is available on the Egyptian Limestone Plateau within air line distances of more than 120 km northeast of Regenfeld.

Although Regenfeld is situated on the geological formation of Nubian Sandstone, good quality quartzite or quartzitic sandstone most likely came from the Abu Ballas scarp land about 80–100 km south of the Regenfeld playa. Cores, debitage, and retouched tools made out of these raw materials indicate blank production and modification processes on-site, and therefore a large distance transport of raw material or tested pieces. The percentages of quartzitic sandstone and flint are significantly higher than those of LDSG. Cores and blanks made of flint or LDSG are generally smaller than those of quartzitic sandstone. Initial stages of the production sequence, such as primary or cortical flakes, are rare among the LDSG and flint artefacts. This roughly correlates with the distances between the sites and the sources.

The central question is whether the raw material was distributed over such large distances as a result of direct access or of exchange. In archaeological studies, raw material procurement among hunter-gatherers is generally interpreted as a result of direct access during the seasonal rounds (Weniger, 1991; Floss, 1994), following the model of 'embedded procurement' as defined by Binford (1978). Raw material procurement over distances of up to 200 or 300 km was obviously usual, whereas larger distances are rarely documented. Exotic stones, molluscs, jewellery, and other items were exchanged over large distances, however, their number in archaeological assemblages is rather small. The proportion of LDSG artefacts related to the distance from its source in case of the assemblages of the Sand Sea and neighbouring regions (Figure 5.10) shows a fall-off curve with a rapid decrease at 200 km from the source. Gradual decrease is characteristic for a distribution as a function of geographical distance without concentration effects through centralised exchange (Renfrew, 1984, pp. 119–128), such as markets or meeting points. The fall-off may be interpreted as the transition from the 'supply zone' with predominating direct access to the 'contact zone' where material was distributed by simple modes of exchange (Torrence, 1986, pp. 13–15).

On the other hand, distribution of artefacts and raw material resulting from exchange processes is documented in ethnographic contexts, for instance, by a number of studies of Australian hunter-gatherers (McCarthy, 1939–1940; Gould, 1978; Gould & Saggers, 1985). This has provoked some critique in archaeology, but should not be overestimated. Although tools were exchanged, raw material exchange does not play a major role, neither in recent nor in past hunter-gatherer societies. Assemblages are generally dominated by ad hoc procurement of raw material, and the number of exotic pieces is relatively small.

Exchange is a good explanation regarding the 'exotic' tools located very far from their sources. The oases may have served as central places and meeting points where the exchange of 'exotic' material was predominantly realised (Hassan & Gross, 1987, p. 99 f.) on the occasion of episodic population agglomerations. This seems a useful hypothesis regarding goods such as the molluscs mentioned which apparently passed through the oases on their way from the Nile Valley or

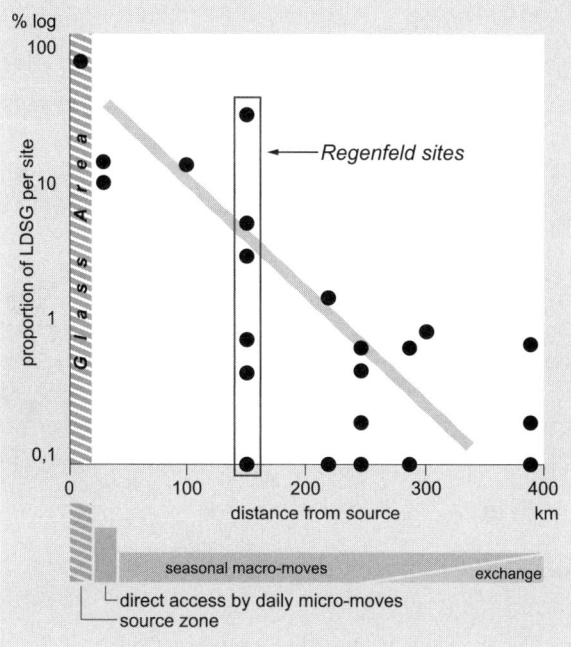

Figure 5.11. The fall-off in artefacts made of Libyan Desert Silica Glass (LDSG) with increasing distance from its source

the Red Sea towards the west. Moreover, there are a few examples from sites in the Western Desert where individual objects made of exotic stone material were found, such as polished planes of granite, 'green stone', and others.

There is also the ecological argument featuring the direct access model. As suggested for the oases of Siwa and Farafra, the population density in the area close to the oases was not more than 0.05 or 0.03 persons per km² as a consequence of the poor carrying capacity (Hassan & Gross, 1987, p. 99; Barich et al., 1991, p. 45). The situation in the desert itself most likely allowed a population density in the rainy season of far less than the mentioned ratios. Patchy distribution and a highly temporal variance of rainfall minimised the potential of regular contacts in the desert. The ecological circumstances thus did not favour the deve-lopment of exchange localities in the desert. It is a truism that desert areas exclude regular modes of exchange, but they may virtually be regarded as barriers. The percentage of raw material within a site's assemblage accumulated during a number of occupations indicates the degree of direct contact between the site and the source. It shows that far-ranging contacts were established. The rapid drop-off of the LDSG curve around 200 km indicates the limits of direct access. Single pieces of LDSG found up to 370 km away from the Glass Area most likely hint at simple modes of exchange (see Figure 5.10).

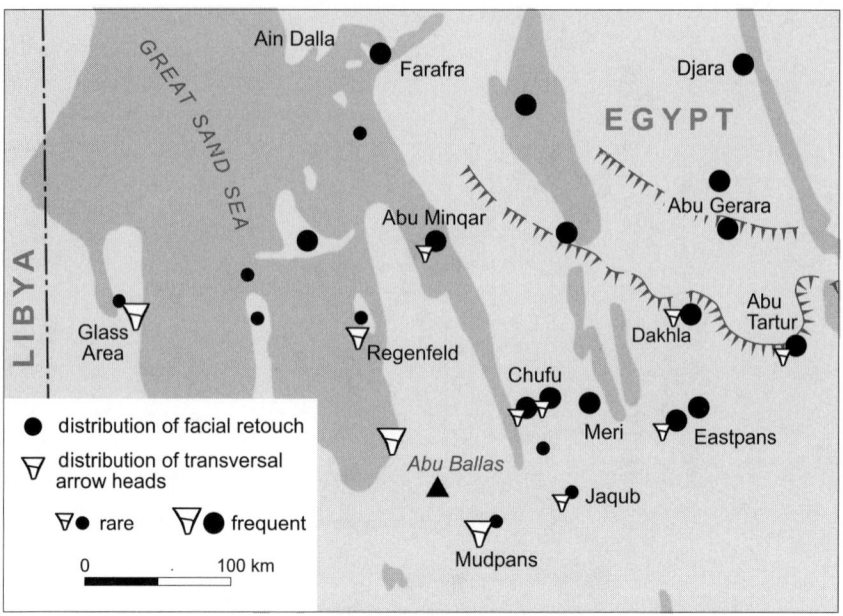

Figure 5.12. Lithic flake tool traditions in the southern Great Sand Sea and vicinities. Small arrowheads or dots indicate existence (not numbered) on survey sites or rare occurrence. Individual signature can incorporate several sites

Although raw material for stone tool production tentatively draws attention to an ad hoc acquisition, the tool kit found on sites illustrates the lithic traditions and the cultural identity and provenience from which the toolmaker originated. The Early Holocene microlithic tradition is rather uniform all over the Sahara featuring a blade industry and elongated triangles and backed points (Figure 5.12 1–9). The Mid-Holocene assemblages of Regenfeld indicate two different traditions (Figures 5.11 and 5.12, 10–15): the tool kit of the south which features the transversal arrow head as characteristic (Figure 5.12, 12–14), and the northeastern tradition which has the facial and bifacial technology (Figure 5.12, 10 and 11) as well as a number of diagnostic tools as side-blow flakes, knives, and leaf-shaped or stemmed points (Figure 5.12, 15), to name but a few.

The work on sites south of Abu Minqar (Klees, 1989a), in Eastpans, south of Dakhla (Gehlen et al., 2002), and current studies conducted within the area between the eastern margins of the Great Sand Sea and Dakhla Oasis (Figure 5.11: study areas of Chufu and Meri) clearly underline the dominance of the bifacial technocomplex in the assemblages between Regenfeld and Dakhla Oasis. Also survey sites in the north and the northwest of the Regenfeld playa (west of Ain Dalla and Ammonite Scarp) produced a typical bifacial tool kit. In contrast, the southern Regenfeld sites (99/15 and 99/22) as well as Mudpans and the Gilf Kebir show transversal arrowheads without or with only little

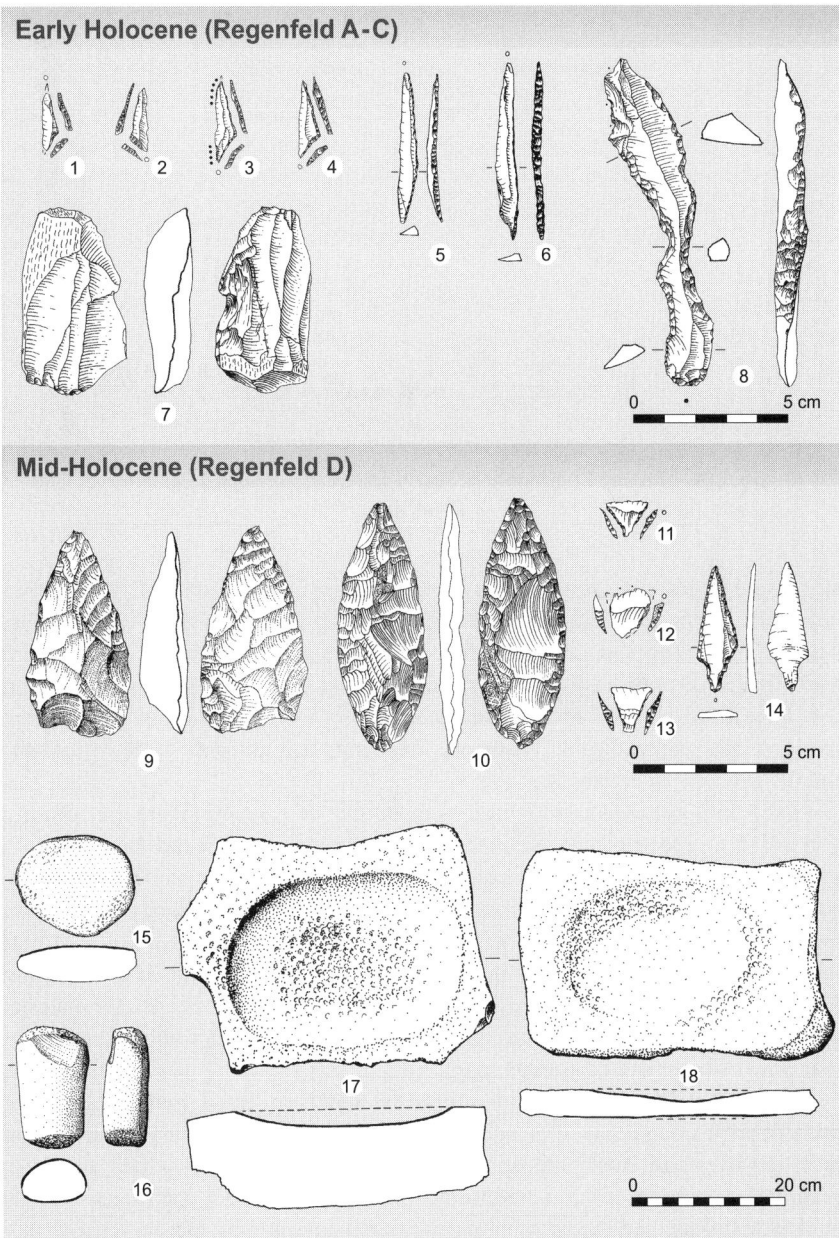

Figure 5.13. Characteristic stone tools from Regenfeld: 1–7, elongated triangles and backed points; 8, blade core; 9, strangulated blade; 10–11, bifacial tools; 12–14, transversal arrowheads; 15, stemmed point; 16–17, upper grinders; 18–19, lower grinders

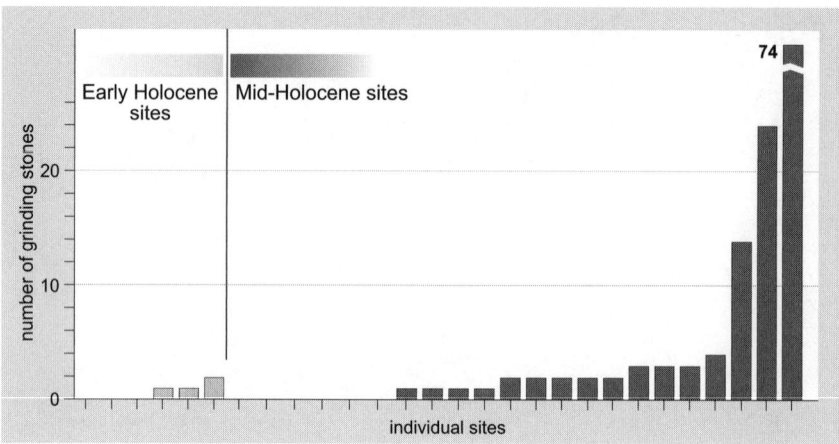

Figure 5.14. Number of grinding stones (upper and lower grinders) in Regenfeld assemblages

admixture of bifacial tools. However, mixing is observed on nearly all larger sites, and every region holds a certain percentage of sites with exotic tools. This points to the fact that the larger sites, situated at the water pools, were attractive for groups of different origins. Smaller transitory sites are not mixed up. This indicates that tools used by members of an individual group were made within one tradition. Therefore, exchange did not play a major role for the distribution of artefacts.

The tool types and tool traditions found at Regenfeld are not strictly connected to specific raw material types. Although the bifacial technology is commonly related to the flint-rich areas of the Egyptian Limestone Plateau, the bifacial tools of Regenfeld are also made of sublocal quartzitic sandstone (Figure 5.12, 10). Likewise, the transversal arrow heads at Regenfeld are not predominantly made out of quartzite or quartzitic sandstone, but of flint as well (Figure 5.12, 12–14). Indeed, this could be an indicator of a certain rate of raw material exchange, but could also point to the long-distance movements and an opportunistic way of raw material procurement.

Taking all this into consideration, the southern Great Sand Sea cannot be seen as a completely impenetrable natural barrier for cultural distribution, although it was the most inhospitable core zone of the desert which generally separated diverging cultural traditions. The overlap of tool traditions with its shifting margins, and the far-ranging contacts suggested by the analysis of raw materials show that the southern Sand Sea and the Abu Ballas scarp land were episodically visited by highly mobile groups of hunters and gatherers coming from the refuges around. Although not impenetrable, the area between the Gilf Kebir and Dakhla Oasis was an area of hundreds of kilometres without a permanent water supply, and obviously supported the evolution of individual cultural traditions in the north and the south.

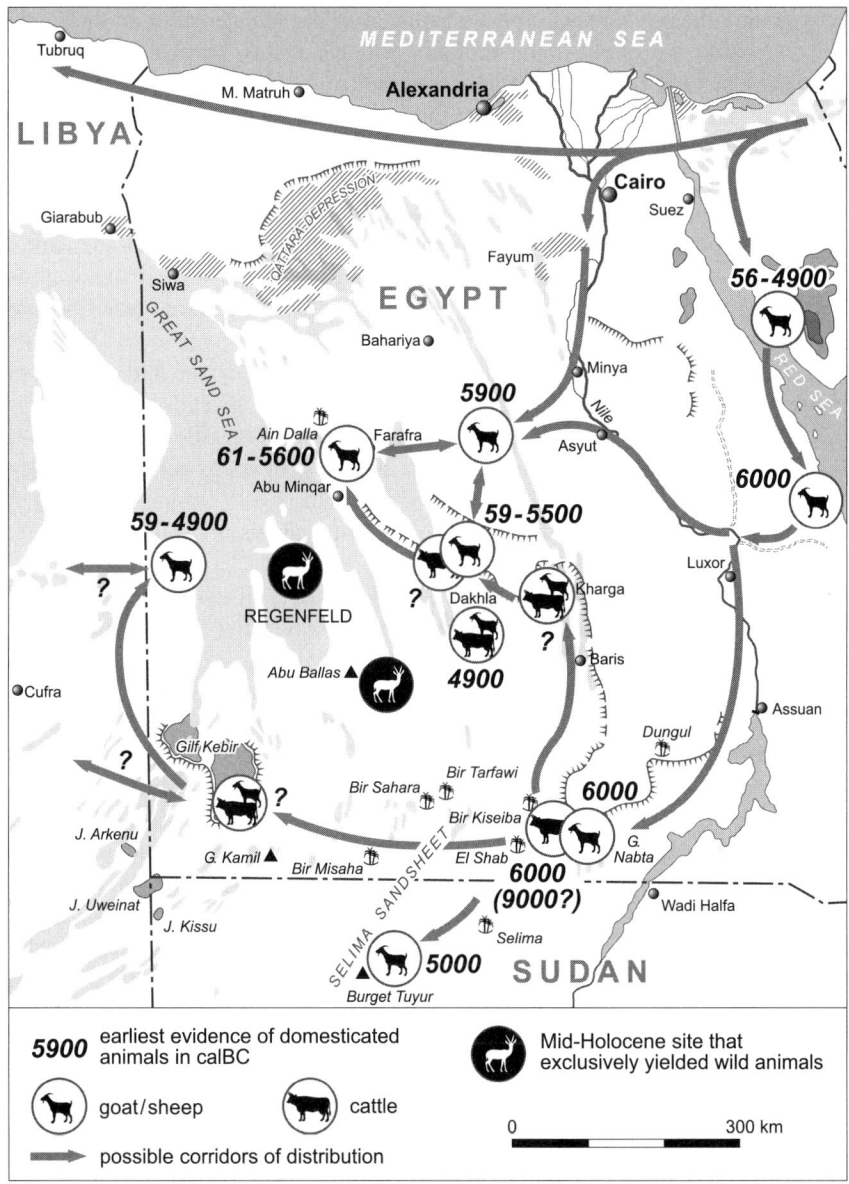

Figure 5.15. Map of northeastern Africa showing the distribution and age of early domesticated animals

5.3.2. Subsistence

The Mid-Holocene sites of the southern Sand Sea are often characterised by a mass of grinding equipment (Figure 5.12, 16–19), which were most likely used for grinding the seeds of wild plants gathered from the vegetation on the sandy soils. Grinders are rarely found on Early Holocene sites, but it seems that the increasing number of grinding implements in the Mid-Holocene period indicates a general trend towards an extensive utilisation of wild plant resources (Figure 5.13). Domesticated plants, as firstly seen in Egypt for the Fayum A Culture (starting about 6500–6000 BP, c. 5400–4900 BC; Stemler, 1980), were not found in the context of desert occupation during the Holocene pluvial.

The bones show that wild animals were hunted throughout the entire Holocene pluvial (Figure 5.5). Arrowheads or insets in arrow shafts as well as butchering places found from all periods of the desert occupation underline that hunting was a major element of subsistence.

The introduction of domesticated cattle, sheep, and goats is indicated from a number of oases and other favoured places during the seventh millennium BP (c. sixth millennium BC; Figure 5.14). Although the earliest cattle is suggested for the sites of Nabta Playa and Bir Kiseiba in southeast Egypt as early as the Early Holocene (Wendorf et al., 1984; Wendorf & Schild, 1994; Gautier, 1984, 001), there is no evidence of a distribution of domestic animals in Egypt before 7000 BP (c. 5900 BC; Gautier, 1987; Close, 2002).

In the wider vicinity of the Great Sand Sea, bones of domesticated animals have been identified in the Wadi el-Bakht, Gilf Kebir (Gautier, 1980, pp. 340–342; Wendorf & Schild, 1980, pp. 217–222, with a problematic date), in the Glass Area at the western margins of the Great Sand Sea (site Glass Area 81/61 'Willmann's Camp', Mid-Holocene context, N. Pöllath, personal communication 2003), from Dakhla (6900 BP, c. 5800 BC; McDonald, 1998; Churcher, 1999), Farafra (6700 BP, c. 5600 BC; Barich & Hassan, 2000), and Eastpans about 80 km south of Dakhla (6000 BP, c. 4900 BC; Berke, 2001). In all cases the domestic animals were found within hunter-gatherer contexts, according to the dominance of wild animals in all bone assemblages, and high proportions of arrow tips. Although domesticated animals became established after 7000 BP (c. 5900 BC) in the eastern Sahara, they are missing at the sites of the southern Great Sand Sea and the central Abu Ballas scarp land. The sites of Mudpans and Regenfeld did not yield bones of domesticates (Van Neer & Uerpmann, 1989; Berke, 2001).

During the hot summer, which is the rainy season in this area, domestic animals can survive for two or three days without water. Within the survey region of Regenfeld, an area of about 100 km in length, there is only one large playa basin like the Regenfeld playa. Other playa pools recorded during the survey are too shallow to keep water for a long period of time. The nearest large basins are approximately 100 km to the south. The patchy distribution of playas suggests that the meagre resources and the enormous distances between the water pools made it impossible for the hunter-gatherer-pastoralists to enter the core zone of the Sand Sea and the Abu Ballas scarp land with cattle, goats, or sheep. Until hyperaridity set in between 6300 and 6000 BP (c. 5300–4900 BC), these regions were obviously occupied by hunter-gatherers only. The site of Eastpans about 80 km south of Dakhla Oasis might indicate the range of the hunter-gatherer-pastoralists who entered the desert margins in favourable years.

5.3.3. Settlement Pattern

The initial aim was to investigate the functions of different sites and their setting in the landscape. The study of the physical landscape of Regenfeld indicates a relief interrupted by longitudinal dunes up to 100 m high, running from north-northwest to south-southeast. The dunes consist of an old Pleistocene body, called whaleback or *draa*, crowned by active *seif* dunes which developed during the last 5000 years (Besler, 2002; Figure 5.4). The fossil dunes were stabilised by formation processes and vegetation during the Holocene pluvial. In the southern Great Sand Sea the interdune corridors, extending between 3 and 10 km, are characterised by sand sheets and plains or rolling gravel surfaces rarely intercepted by small hills, ridges, or shallow wadis. Playa basins occur at different places as large depressions, but they are few (Figure 5.15).

During the archaeological survey it was noticed that there are sites which clearly differ in the number of artefacts. To measure these differences the number of artefacts, tools, or other distinctive artefact classes should have been counted. However, such an effort could never have been realised in a large-scale survey. Because of that we tested the value of the site's extension and the artefact density as possible substitutes for the number of artefacts. Whereas the extension of a site was estimated in metres, the density was classified in three categories, as 'dense', 'light', and 'isolated artefacts'. To test the relationship between the artefact number and the possible substitutes, the number of lithic tools was recorded on a restricted number of sites diagrammed in Figure 5.16. This indicated that the number of tools does not significantly correlate with the site's extension but with the density classes. The dense scatters always showed many tools whereas the light scatters and those with isolated artefacts yielded small amounts of tools. Therefore, density classes were established for further analysis of settlement patterns.

Although many sites were discovered on the plains far away from the playas, 'dense' sites were predominantly situated on or at playa pools (Figure 5.17). Undoubtedly, these locations were the most attractive settings in the landscape.

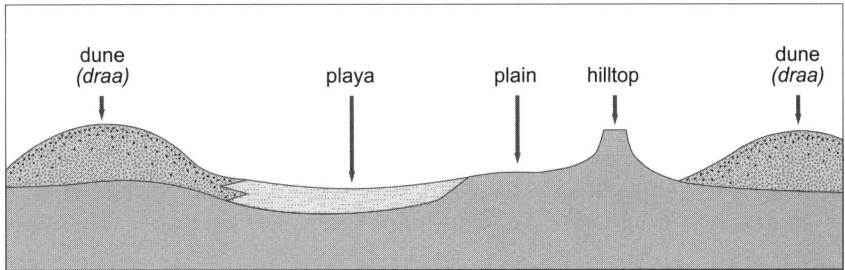

Figure 5.16. Schematic profile of the southern Great Sand Sea landscape showing the possible landscape elements (physiographic units)

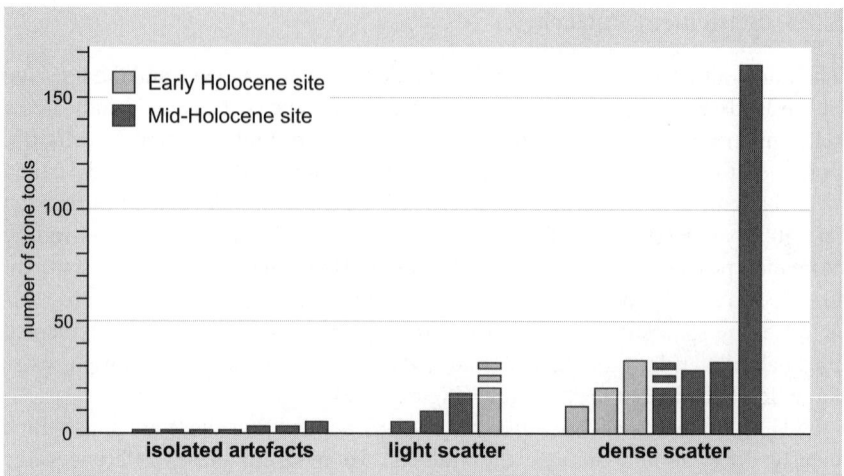

Figure 5.17. Campsite pattern: Number of stone tools (flake tools and grinders) in individual Regenfeld assemblages classified by different classes of artefact density

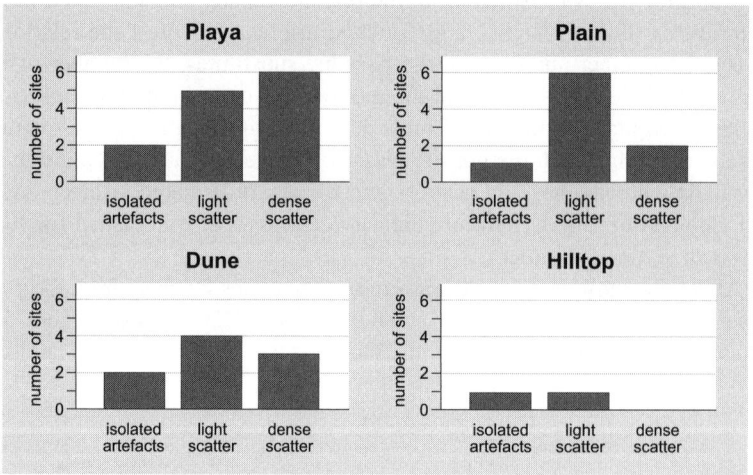

Figure 5.18. Campsite pattern: Setting of sites in different physiographic units in the Regenfeld area

In the Mid-Holocene, these sites with dense scatters were characterised by a diversified activity spectrum including primary and tool production, use of grinding equipment, production of jewellery (ostrich egg shell beads), butchering, and use of campfires (Figure 5.18). The sites were clearly used as long-term transitory campsites, repeatedly occupied over many years. The slightly rounded bodies of the *draa* dunes were also often used for larger campsites as indicated by a number

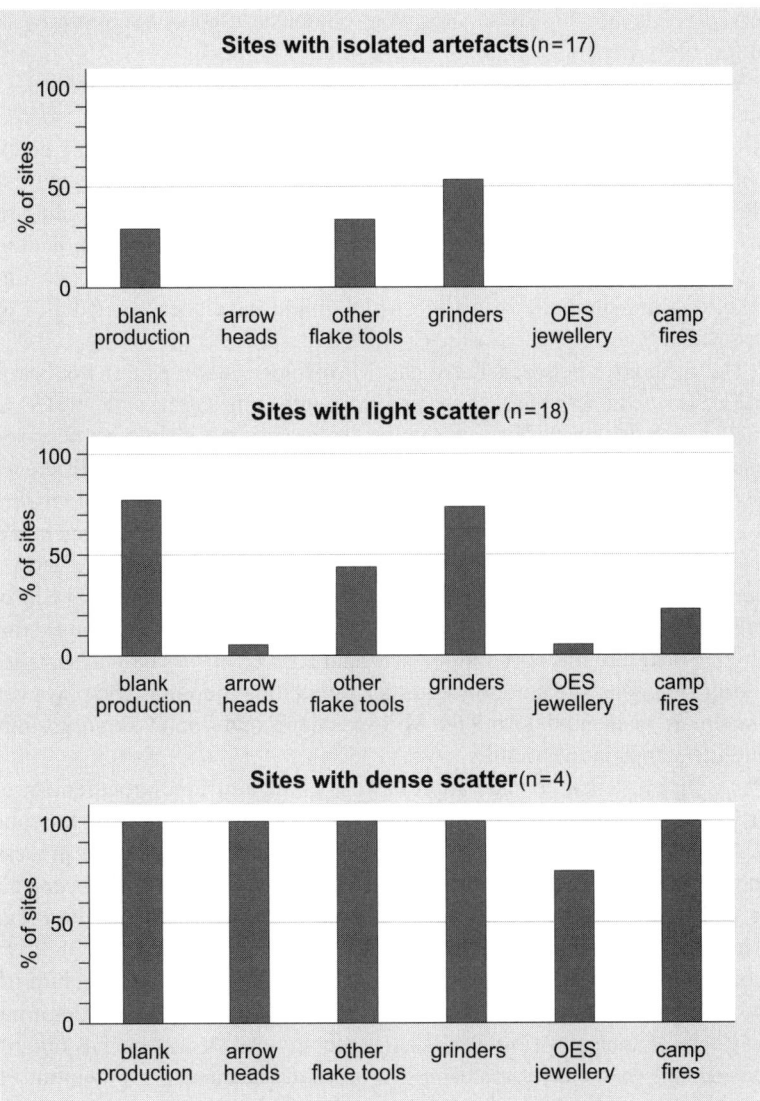

Figure 5.19. Campsite pattern: Artefact classes found on Mid-Holocene sites of Regenfeld. Blank production includes production debitage and cores (OES = Ostrich Egg Shells)

of dense and light scatters placed on the dune's slopes (Figure 5.17). But most of them existed near to playas.

The plains produced sites with isolated artefacts, designated as 'small' sites (Figure 5.17). They show a restricted tool kit which may include blanks, tools, or grinders. Intensive working places for jewellery, tools, and other items were not observed (Figure 5.18). These sites were obviously used as overnight or short-term transitory camps.

Two sites – one was determined as dense, the other as light scatter – produced a high proportion of arrowheads. The densely scattered site, moreover, showed blank and tool production which probably served for repairing and hafting of arrow tips ('retooling'). The second site is a hilltop covered with stone circles which may have served as watch posts for the game hunt. Again arrowheads were produced, and wild animals butchered or cut on-site. Both sites are designated as special-task sites for hunting. They are situated in the vicinity of a 4 km long human-made stone line which might have been erected as a game drive or trap.

The difference between Early and Mid-Holocene site patterns is characterised as follows. In both phases sites predominantly concentrate at the water pools, but in the Early Holocene sites are smaller suggesting only short-term occupations and less repeated visits at a specific site. There is no significant difference in the variation of Early Holocene sites at the playa and those in the plains. In the Mid-Holocene sites became varied in size and function. There are such special task sites as hunting camps; and very large, repeatedly occupied long-term sites at the water pools which indicate a great range of activities. In terms of Binford's mobility categories (1980) the Mid-Holocene sites indicate a radiating mobility (high logistical mobility) around the seasonal base camps at the playas, whereas the Early Holocene site pattern suggests a circulating mobility. The playa camps became more residential during the Mid-Holocene, and small task groups moved to satellite camps in the vicinity.

It is an important consideration that the change of food procurement coincides with the changing of settlement patterns. In the Early Holocene, the subsistence seems to have been predominantly based on hunting, and plant processing did not increase before the end of the Early Holocene. In the Mid-Holocene plants were gathered in on the dunes and were mostly processed in the base camps indicated by the grinding equipment at these sites (Figure 5.12, 16–19). It is reasonable to suggest that many grinders were left on-site to await the return of the group in the next season (Figure 5.12, 18), although some of the smaller grinders show traces of transport (Figure 5.12, 19). In any case it was useful to increase the residential character of the camps in order to minimise the amount of effort to be spent for the production or transport of grinders.

In turn specialised hunting and butchering sites were situated at a considerable distance to the base camps during the Mid-Holocene. This can be seen as an effect of the residential camps at the playas. As Gautier (2001, p. 633) suggests for the Nabta/Bir Kiseiba region, gazelle and antelope have visited the playas less frequently when humans were present at the water. This is also a plausible scenario for the Regenfeld area. Seasonal residential camps scared off the game from the playa. As a consequence hunters had to cover larger distances to meet gazelle and antelope resulting in the establishment of satellite hunting camps.

5.4. DISCUSSION AND OUTLOOK

The reconstruction of the environment of the southern Great Sand Sea indicates fully arid conditions during the Holocene wet phase with episodic surface water and rainfall that correlates with other sites at Mudpans and elsewhere in the Western Desert. Although precipitation was significantly higher during the humid phase than during the Pleistocene and after c. 5000 BC, oscillations between moister phases and dry interphases during the humid phase have not yet been reconstructed satisfactorily. In contrast to other sequences at Nabta/Bir Kiseiba (Wendorf et al., 2001) or at Mudpans (Neumann, 1989b) the Regenfeld sequence does not indicate any significant change during the Holocene humid phase.

Despite the general problems of reconstructing the climatic development during the wet phase, the southern Great Sand Sea may play a key role in examining when and under what circumstances human existence in arid zones tended to fail. This desert area forms the most arid core zone of the Sahara and therefore can be seen as a landscape with only small environmental use-potential. One consequence is perhaps the lack of evidence for domesticated animals. Moreover, the Regenfeld area has a signalling position for the onset of climate deterioration earlier than in other places, because the depopulation started in Regenfeld at 6300 BP (c. 5300 BC).

Although the archaeological evidence reflects a complex system of natural factors forcing or preventing human activities, none except the scarcity or absence of water completely ruled out human existence. Animal and plant life existed under conditions of a fully arid desert without fossil groundwater supply and with only a minimum of rain per year. All the animals examined within the assemblages of the absolute desert do not need to drink from open water. Especially the Great Sand Sea has a great percentage of sand-dune cover which is a favourable substrate for vegetation, and in turn for the animals. Therefore, plant and animal life were not the limiting factors in the desert, but surface water. Open water in playa basins was rare and scattered over a large area with enormous distances in between.

Oases served as refuges for the people banished from the desert when it dried up. In the oases water from the Nubian Aquifer was permanently available, but the agglomeration of people in an area with only a minimum expanse may have led to an overexploitation of other bioresources such as game and vegetation. In the case of severe and extended droughts, a long-lasting overpopulation of the oases exceeded their natural carrying capacities and apparently evoked the danger of food shortage. As a consequence, the danger of serious land degradation in the oases was generally high (Neumann, 1994) whereas the potential human impact on the desert vegetation and on game was rather small. Although the oases and the desert are contrasting environments, they can be seen as complementary elements within the human subsistence of nonsedentary groups.

The studies on settlement patterns and mobility in the Regenfeld area have shed some light on the seasonal rounds and the distances covered by prehistoric hunter-gatherers. The rainy season in this area was the summer when temperatures and evaporation were high. The pans were filled after rainfall for some days, weeks, or even months. That was the time when prehistoric groups started to occupy the deserts and to move over hundreds of kilometres. The Early Holocene hunters had a high residential mobility, however, the Mid-Holocene hunter-gatherers lived in larger base camps near the playa pools. Satellite camps were used during the stay, such as hunting or killing sites. When surface water ran out, moisture from roots of tubers could be obtained for a short while. Then the desert dwellers returned to the oases where water was permanently available.

The investigations into interrelation between human beings and arid land-scape in the Great Sand Sea have yielded a number of valuable results with the potential of application to other studies in arid landscape archaeology. They can be summed up as follows.

1. An unpredictable water supply as a consequence of the spatiotemporal variability and uncertainty of rains was apparently the key constraint. Local environmental conditions differed in a highly dynamic subsystem. They often have complementary potential, as in the case of the azonal oases versus the desert.
2. The sporadic availability of key resources and the variability of environ-mental conditions led to flexible adaptation patterns, which in turn meant that routes and locations used over the year were often changing.
3. The high mobility and spatial flexibility of the desert dwellers was the basic device to minimise the risks of food and water shortage.
4. Multiresource management based on gathering wild cereals, hunting gazelles, hare, and fennek (and most likely other smaller animals), and a small proportion of pastoral subsistence in the Mid-Holocene were another adaptational consequence of scarce resources and incalculable risks.
5. Sedentary elements and agriculture were not adequate strategies in this highly arid climate. Pastoralism, too, was not very common in fully arid zones with precipitation less than 100 mm/a, if there was no hinterland or azonal refuge which offered better and more stable conditions.

When the desert dried up and the landscape changed between 6300 and 6000 BP (c. 5300–4900 BC), certain strategies and techniques failed. The desert changed into an inhospitable environment and the oases and the Nile appeared as the only niches where people could settle.[5]

---

[5] It has been said that there was no revival of the rains, nor a reoccupation of the deserts since hyperarid conditions set in. This is basically true, following the definition of 'dwell-ers' as people who, although episodically, lived within the desert lands over longer periods of time utilising the scarce desert potential for their subsistence. However, recent research has shown evidence for the existence of people in the desert after the drying up. Pottery was found in the deserts – the so-called 'Clayton rings' – which fall into the time of the late Predynastic/Early Dynastic (about 4500 BP, c. 3200 BC) to the end of the Old Kingdom (Riemer & Kuper, 2000). But as yet there is no meaningful explanation regarding the use of this pottery. This development stands outside the scope of this study; it most probably reflects the rise of desert travelling and exchange due to the introduction of the domesti-cated donkey. It therefore indicates a completely different land-use system adapted on a fundamentally changed environment.

The landscape approach, as used here in the case study of Regenfeld, offers the opportunity to understand long-term regularities and variability of past land use as a response to arid environmental conditions. Excavations of individual sites and regional surveys provide valuable archaeological instruments if other disciplines such as geosciences, botany, and zoology are incorporated. A comparative synthesis covering various microenvironments (Bolten et al., this volume) will allow us to move towards more diverse and better grounded assessments of the prehistoric development, and to patterns and variations in the land use systems which can be examined in a single case study.

Acknowledgements The research has been carried out within the frame of ACACIA's subproject A1 'Climatic Change and Human Settlement between the Nile Valley and the Central Sahara' directed by Rudolph Kuper and Helga Besler. I wish to thank the colleagues who participated with their valuable studies, comments, and help, in particular Hubert Berke, Helga Besler, Andreas Bolten, Olaf Bubenzer, Frank Darius, Barbara Eichhorn, Ursula Eisenhauer, Karin Kindermann, Stefan Kröpelin, Rudolph Kuper, Jörg Linstädter, Stefanie Nussbaum, and Nadja Pöllath.

REFERENCES

Almásy, L.E. (1939). *Unbekannte Sahara*. Leipzig: Brockhaus.

Ashour, M.M. (1995). A report on the field trip in the Western Desert of Egypt. *Bulletin de la Société de Géographie d'Égypte*, 8, 5–31.

Ball, J. (1927). Problems of the Libyan Desert. *Geographical Journal*, 70, 21–38; 105–128; 209–224.

Barich, B. & Hassan, F.A. (2000). A stratified sequence from Wadi el-Obeiyd, Farafra: New data on subsistence and chronology of the Egyptian Western Desert. In L. Krzyzaniak, K. Kroeper & M. Kobusiewicz (Eds.), *Recent Research into the Stone Age of Northeastern Africa* (pp. 33–46). Studies in African Archaeology, 7. Poznan: Archaeological Museum.

Barich, B., Hassan, F.A. & Mahmoud, A.A. (1991). From settlement to site: Formation and transformation of archaeological traces. *Scienze dell'Antichita*, 5, 33–62.

Berke, H. (2001). Gunsträume und Grenzbereiche. Archäozoologische Beobachtungen in der Libyschen Wüste, Sudan und Ägypten. In B. Gehlen, M. Heinen & A. Tillmann (Eds.), *Zeit-Räume. Gedenkschrift für Wolfgang Taute.* Archäologische Berichte, 14. Bonn: Deutsche Gesellschaft für Ur- und Frühgeschichte. pp. 257–282.

Besler, H. (1992). *Geomorphologie der ariden Gebiete*. Erträge der Forschung 280. Darmstadt: Wissenschaftliche Buchgesellschaft.

Besler, H. (2002). The Great Sand Sea (Egypt) during the late Pleistocene and the Holocene. *Zeitschrift für Geomorphologie N.F.*, 127, 1–19.

Binford, L.R. (1978). *Nunamiut Ethnoarchaeology*. New York: Academic.

Binford, L.R. (1980). Willow smoke and dog's tails: Hunter-gatherer settlement systems and archaeological site formation. *American Antiquity*, 45, 4–20.

Bubenzer, O. & Hilgers, A. (2003). Luminescence dating of Holocene playa sediments of the Egyptian Plateau, Western Desert, Egypt. *Quaternary Science Reviews*, 22, 1077–1084.

Bubenzer, O. & Riemer, H. (2007). Holocene climatic change and human settlement between the Central Sahara and the Nile Valley: Archaeological and geomorphological results. *Geoarchaeology*, 22, 607–620.

Bubenzer, O., Bolten, A. & Darius, F. (Eds.) (2007). *Atlas of Cultural and Environmental Change in Arid Africa*. Africa Praehistorica, 21. Köln: Heinrich-Barth-Institut.

Butzer, K.W. (1971). *Environment and Archaeology* (2nd ed.). Chicago: Aldine.

Butzer, K.W. (1982). *Archaeology as Human Ecology*. Cambridge: Cambridge University Press.

Churcher, C.R. (1999). Holocene faunas of the Dakhleh Oasis. In C.S. Churcher & A.J. Mills (Eds.), *Reports from the Survey of the Dakhleh Oasis, Western Desert of Egypt, 1977–1987* (pp. 109–115). Oxford: Oxbow.

Close, A.E., (2002). Sinai, Sahara, Sahel: The introduction of domestic caprines to Africa. In Heinrich Barth Institut (Ed.), *Tides of the Desert – Gezeiten der Wüste. Contributions to the Archaeology and Environmental History of Africa in Honour of Rudolph Kuper* (pp. 459–469). Africa Praehistorica, 14. Köln: Heinrich-Barth-Institut.

Dorst, J. & Dandelot, P. (1972). *The Larger Mammals of Africa* (2nd ed.). London: Collins.

Dunnell, R.C. & Dancey, W.S. (1983). The siteless survey: A regional scale data collection strategy. In M.B. Schiffer (Ed.), *Advances in Archaeological Method and Theory 6* (pp. 267–287). New York/London: Academic.

Embabi, N.S. (1999). Playas of the Western Desert, Egypt. In J. Donner (Ed.), *Studies of Playas in the Western Desert of Egypt* (pp. 5–47). Annales Academiae Scientiarum Fennicae, Geologica-geographica, 160. Helsinki: Academia Scientiarum Fennica.

Estes, R.D. (1992). *The Behavior Guide to African Mammals.* Berkeley: University of California.

Floss, H. (1994). *Rohmaterialversorgung im Paläolithikum des Mittelrheingebietes.* RGZM Monographien, 21. Bonn: Habelt.

Gabriel, B. (1986). *Die östliche Libysche Wüste im Jungquartär.* Berliner geographische Studien, 19. Berlin: Institut für Geographie der Technischen Universität Berlin.

Gautier, A. (1980). Contributions to the archaeozoology of Egypt. In F. Wendorf & R. Schild (Eds.), *Prehistory of the Eastern Sahara* (pp. 317–344). New York: Academic.

Gautier, A. (1984). Archaeozoology of the Bir Kiseiba Region, Eastern Sahara In Wendorf et al. (Eds.) *Cattle-Keepers of the Eastern Sahara: The Neolithic of Bir Kiseiba* (pp. 49–72). Dallas: Southern Methodist University Press.

Gautier, A. (1987). Prehistoric men and cattle in North Africa: A dearth of data and a surfeit of models. In A. Close (Ed.), *Prehistory of Arid North Africa. Essays in Honor of Fred Wendorf* (pp. 163–187). Dallas: Southern Methodist University Press.

Gautier, A. (2001). The early to late neolithic archaeofaunas from Nabta and Bir Kiseiba. In F. Wendorf, R. Schild, & A. Close (Eds.), *Cattle-Keepers of the Eastern Sahara* (pp. 609–635). Dallas: Southern Methodist University Press.

Gehlen, B., Kindermann, K., Linstädter, J. & Riemer, H. (2002). The Holocene occupation of the Eastern Sahara: Regional chronologies and supra-regional developments in four areas of the absolute desert. In Heinrich Barth Institut (Ed.), *Tides of the Desert – Gezeiten der Wüste. Contributions to the Archaeology and Environmental History of Africa in Honour of Rudolph Kuper* (pp. 85–116). Africa Praehistorica, 14. Köln: Heinrich-Barth-Institut.

Gould, R.A. (1978). The anthropology of human residues. *American Anhtropologist*, 80, 815–835.

Gould, R.A. & Saggers, S. (1985). Lithic procurement in Central Australia: A closer look at Binford's idea of embeddedness in archaeology. *American Antiquity*, 50, 117–136.

Griffiths, J.F. (1972). The climate of the United Arabic Republic. In J.F. Griffiths (Ed.), Climates of Africa (pp. 79–92) World Survey of Climatology, 10. Amsterdam, London, New York: Elsevier.

Haltenorth, Th. & Diller, H. (1977). *Säugetiere Afrikas und Madagaskars.* München: BLV.

Hassan, F.A. & Gross, T. (1987). Resources and subsistence during the early Holocene at Siwa Oasis, Northern Egypt. In A.E. Close (Ed.), *Prehistory of Arid North Africa. Essays in Honor of Fred Wendorf* (pp. 85–103). Dallas: Southern Methodist University Press.

Hassan, F.A., Barich, B., Mahmoud, M. & Hemdan, M.A. (2001). Holocene Playa deposits of Farafra Oasis, Egypt, and their palaeoclimatic and geoarchaeological significance. *Geoarchaeology*, 16, 29–46.

Haynes, C.V. Jr. (1982). Great Sand Sea and the Selima sand sheet, Eastern Sahara: Geochronology of desertification. *Science*, 217, 629–633.

Haynes, C.V. Jr. (1987). Holocene migration rates of the Sudano-Sahelian wetting front, Arba'in Desert, eastern Sahara. In A.E. Close (Ed.), *Prehistory of Arid North Africa. Essays in Honor of Fred Wendorf* (pp. 69–84). Dallas: Southern Methodist University Press.

Haynes, C.V. Jr. (2001). Geochronology and climate change of the Pleistocene – Holocene transition in the Darb el Arba'in Desert, Eastern Sahara. *Geoarchaeology*, 16, 119–141.

Heinl, M. & Thorweihe, U. (1993). Groundwater resources and management in SW Egypt. *Catena*, 26, 99–121.

Hoelzmann, P. (2002). Lacustrine sediments as indicators of climate change during the Late Quaternary in Western Nubia (Eastern Sahara). In Heinrich Barth Institut (Ed.), *Tides of the Desert – Gezeiten der Wüste. Contributions to the Archaeology and Environmental History of Africa in Honour of Rudolph Kuper* (pp. 375–388). Africa Praehistorica, 14. Köln: Heinrich-Barth-Institut.

Hoelzmann, P., Keding, B., Berke, H. & Kröpelin, S. (2001). Environmental change and archaeology: Lake evolution and human occupation in the Eastern Sahara during the Holocene. *Palaeogeography, Palaeoclimatology, Palaeoecology*, 169, 193–217.

Johnson, G.A. (1977). Aspects of regional analysis in archaeology. *Annual Review of Anthropology*, 6, 479–508.

Keding, B. (2000). New data on the Holocene occupation of the Wadi Howar region (Eastern Sahara/Sudan). In L. Krzyzaniak, K. Kroeper & M. Kobusiewicz (Eds.), *Recent Research into the Stone Age of Northeastern Africa* (pp. 89–104). Studies in African Archaeology, 7. Poznan: Archaeological Museum.

Kehl, H. & Bornkamm, R. (1993). Landscape ecology and vegetation units of the Western Desert of Egypt. *Catena*, 26, 155–178.

Kindermann, K., Bubenzer, O., Nussbaum, S., Riemer, H., Darius, F., Pöllath, N. & Smettan, U. (2006). Palaeoenvironment and Holocene land use of Djara, Western Desert of Egypt. *Quaternary Science Reviews*, 25, 1619–1637.

King, W.J.H. (1925). *Mysteries of the Libyan Desert*. London: Seeley, Service & Co.

Klees, F. (1989a). Lobo: A contribution to the prehistory of the eastern Sand Sea and the Egyptian oases. In L. Krzyzaniak, K. Kroeper & M. Kobusiewicz (Eds.), *Recent Research into the Stone Age of Northeastern Africa* (pp. 223–231). Studies in African Archaeology, 7. Poznan: Archaeological Museum.

Klees, F. (1989b). Die Große Sandsee. Wurzeln des ägyptischen Neolithikums in der Wüste. *Archäologie in Deutschland*, 2, 15–18.

Kleindienst, M.R., Churcher, C.S., MsDonald, M.M.A. & Schwarcz, H.P. (1999). Geography, geology, geochronology and geoarchaeology of the Dakhleh Oasis region: An interim report. In C.S. Churcher & A.J. Mills (Eds.), *Reports from the Survey of the Dakhleh Oasis, Western Desert of Egypt* (pp. 1–54). Dakhleh Oasis Project, Monograph, 2. Oxford: Oxbow.

Klitzsch, E., List, F.K. & Pöhlmann, G. (Eds.) (1987). *Geological Map of Egypt 1:500000*. Cairo: Technische Fachhochschule Berlin.

Kröpelin, S. (1987). Palaeoclimatic evidence from Early to Mid-Holocene playas in the Gilf Kebir (Southwest-Egypt). *Palaeoeology of Africa*, 18, 189–208.

Kröpelin, S. (1989). Untersuchungen zum Sedimentationsmilieu von Playas im Gilf Kebir (Südwest-Ägypten) In. R. Kuper (Ed.), *Forschungen zur Umweltgeschichte der Ostsahara* (pp. 183–305) Africa Praehistorica, 2. Köln: Heinrich-Barth-Institut.

Kröpelin, S. (1993). Geomorphology, landscape evolution and paleoclimates of southwest Egypt. *Catena*, 26, 31–65.

Kuper, R. (Ed.) (1989). *Forschungen zur Umweltgeschichte der Ostsahara*. Africa Praehistorica, 2. Köln: Heinrich-Barth-Institut.

Kuper, R. & Kröpelin, S. (2006). Climate-controlled Holocene occupation in the Sahara: Motor of Africa's evolution. *Science*, 313, 803–807.

Lange, M. (2006). *Wadi Shaw – Wadi Sahal. Studien zur holozänen Besiedlung der Laqiya-Region (Nordsudan)*. Africa Praehistorica, 19. Köln: Heinrich-Barth-Institut.

Linstädter, J. (2003). Prehistoric land use systems in the Gilf Kebir. In Z. Hawass & L. Pinch Brock (Eds.), *Egyptology at the Dawn of the Twenty-First Century. Proceedings of the Eighth International Congress of Egyptologists* (pp. 381–388). Cairo: American University in Cairo Press.

Linstädter, J. & Kröpelin, S. (2004). Wadi Bakht revisited: New insights on Holocene climate change and prehistoric occupation in the Gilf Kebir region (Central Eastern Sahara, SW Egypt). *Geoarchaeology*, 19, 753–778.

Lüning, J. (1983). *Stand und Aufgaben der siedlungsarchäologischen Erforschung des Neolithikums im Rheinischen Braunkohlenrevier. Archäologie in den Rheinischen Lößbörden* (pp. 34–46).

Beiträge zur Siedlungsarchäologie im Rheinland [=Rheinische Ausgrabungen, 24]. Köln, Bonn: Rheinland Verlag.

McCarthy, F.D. (1939–1940). 'Trade' in aboriginal Australia and 'Trade' relationships with Torres Strait, New Guinea and Malaya. *Oceania*, 9(4), 405–438; 10(1), 80–104; 10(2), 171–195.

McDonald, M.M.A. (1998). Early African pastoralism: View from Dakhleh Oasis (South Central Egypt). *Journal of Anthropological Archaeology*, 17, 93–99.

McDonald, M.M.A. (1999). Neolithic cultural units and adaptations in the Dakhleh Oasis In. C.S. Churcher & A.J. Mills (Eds.), *Reports from the Survey of the Dakhleh Oasis, Western Desert of Egypt* (pp. 117–132). Dakhleh Oasis Project, Monograph, 2. Oxford: Oxbow.

McDonald, M.M.A. (2001). Late prehistoric radiocarbon chronology for Dakhla Oasis within the wider environmental and cultural settings of the Egyptian Western Desert In. M. Marlow (Ed.), *The Oasis Papers 1: Proceedings of the First International Symposium of the Dakhleh Oasis Project*. Dakhleh Oasis Project, Monograph, 6. Oxford: Oxbow.

Neumann, K. (1989a). Zur Vegetationsgeschichte der Ostsahara im Holozän. Holzkohlen aus prähistorischen Fundstellen In. R. Kuper (Ed.), *Forschungen zur Umweltgeschichte der Ostsahara* (pp. 13–181). Africa Praehistorica, 2. Köln: Heinrich-Barth-Institut.

Neumann, K. (1989b). Holocene vegetation of the eastern Sahara: Charcoals from prehistoric sites. *African Archaeological Review*, 7, 97–116.

Neumann, K. (1994). Wirtschaftsweisen im Neolithikum der Ostsahara und ihr Einfluß auf die Vegetation. In M. Bollig & F. Klees (Eds.), *Überlebensstrategien in Afrika* (pp. 47–65). Colloquium Africanum, 1. Köln: Heinrich-Barth-Institut.

Nicoll, K. (2001). Radiocarbon chronologies for prehistoric human occupation and hydroclimatic change in Egypt and Northern Sudan. *Geoarchaeology*, 16, 47–64.

Pachur, H.-J. (2001). Holozäne Klimawechsel in den nördlichen Subtropen. *Nova Acta Leopolina*, 88, 109–131.

Pachur, H.-J. & Hoelzmann, P. (1991). Palaeoclimatic implications of late quaternary lacustrine sediments in Western Nubia, Sudan. *Quaternary Research*, 36, 257–276.

Pachur, H.-J. & Hoelzmann, P. (2000). Late Quaternary palaeoecology and palaeoclimates of the eastern Sahara. *Journal of African Earth Sciences*, 30, 929–939.

Plog, S., Plog, F. & Wait, W. (1978). Decision making in modern surveys. In M.B. Schiffer (Ed.), *Advances in Archaeological Method and Theory 1* (pp. 384–421). New York/London: Academic.

Renfrew, C. (1984). *Approaches to Social Archaeology*. Cambridge: Harvard University Press.

Riemer, H. (2000). Regenfeld 96/1-Great Sand Sea and the question of human settlement on whaleback dunes. In L. Krzyzaniak, K. Kroeper & M. Kobusiewicz (Eds.), *Recent Research into the Stone Age of Northeastern Africa* (pp. 21–31). Studies in African Archaeology, 7. Poznan: Archaeological Museum.

Riemer, H. (2003). The 're-conquest' of the Great Sand Sea. In Z. Hawass & L. Pinch Brock (Eds.), *Egyptology at the Dawn of the Twenty-First Century. Proceedings of the Eighth International Congress of Egyptologists* (pp. 408–415). Cairo: American University in Cairo Press.

Riemer, H. (2006). Archaeology and environment of the Western Desert of Egypt: 14C-Based human occupation history as an archive for Holocene palaeoclimatic reconstruction. In A.A. Youssef El-Sayed (Ed.), *Geology of the Tethys. Proceedings of The First International Conference on the Geology of the Tethys* (pp. 553–564), Cairo. Tethys Geological Society.

Riemer, H. (2007). Mapping the movement of pastro-foragers: The spread of desert glass and other objects in the eastern Sahara during the Holocene 'humid phase'. In O. Bubenzer, A. Bolten & F. Darius (Eds.), *Atlas of Cultural and Environmental Change in Arid Africa* (pp. 30–33). Africa Praehistorica, 21. Köln: Heinrich-Barth-Institut.

Riemer, H. & Kuper, R. (2000). 'Clayton rings': Enigmatic ancient pottery in the eastern Sahara. *Sahara*, 12, 91–100.

Rohlfs, G., Rohlfs, G. (1875). *Drei Monate in der libyschen Wüste*. Cassel: Theodor Fischer [reprinted 1996. Köln: Heinrich-Barth-Institut.].

Sauer, F. & Sauer, E. (1980). Die Lebensweise der Strauße. In B. Grzimek (Ed.), *Grzimeks Tierleben. Enzyklopädie des Tierreichs 7, Vögel 1* (pp. 89–91). München: dtv.

Schuck, W. (1989). From lake to well: 5,000 years of settlement in Wadi Shaw (Northern Sudan). In L. Krzyzaniak & M. Kobusiewicz (Eds.), *Late Prehistory of the Nile Basin and the Sahara* (pp. 421–429). Studies in African Archaeology, 2. Poznan: Archaeological Museum.

Simons, P. (1973). Der Osten der Sahara. In H. Schiffers (Ed.), *Die Sahara und ihre Randgebiete. Vol. 3: Regionalgeographie* (pp. 433–535). München: Weltforum.

Stemler, A.B.L. (1980). Origins of plant domestication in the Sahara and the Nile Valley. In M.A.J. Willis & H. Faure (Eds.), *The Sahara and the Nile* (pp. 503–526). Rotterdam: Balkema.

Thorweihe, U. (1990). Nubian aquifer system. In R. Said (Ed.), *The Geology of Egypt* (pp. 601–611). Rotterdam: Balkema.

Thorweihe, U. & Heinl, M. (1993). Hydrogeology. In B. Meissner & P. Wyeisk (Eds.), Geopotential maps of the Western Desert of Egypt. South-West Egypt 1:1000000. *Catena*, 26, supplement.

Torrence, R. (1986). *Production and Exchange of Stone Tools. Prehistoric Obsidian in the Aegean. New Studies in Archaeology*. Cambridge: Cambridge University Press.

Van Neer, W. & Uerpmann, H.-P. (1989). Palaeoecological significance of the Holocene faunal remains of the B.O.S.-missions In. R. Kuper (Ed.), *Forschungen zur Umweltgeschichte der Ostsahara* (pp. 307–341). Africa Praehistorica, 2. Köln: Heinrich-Barth-Institut.

Vermeersch, P.M. (2002). The Egyptian Nile valley during the early Holocene. In Heinrich Barth Institut (Ed.), *Tides of the Desert – Gezeiten der Wüste. Contributions to the Archaeology and Environmental History of Africa in Honour of Rudolph Kuper* (pp. 27–49). Africa Praehistorica, 14. Köln: Heinrich-Barth-Institut.

Vita-Finzi, C. (1978). *Archaeological Sites in Their Setting. Ancient People and Places*, 90. London: Thames & Hudson.

Vita-Finzi, C. & Higgs, E.S. (1970). Prehistoric economy in the Mount Carmel area of Palestine: Site catchment analysis. *Proceedings of the Prehistoric Society*, 36, 1–37.

Wendorf, F. & Schild, R. (1980). *Prehistory of the Eastern Sahara. Studies in Archaeology*. New York: Academic Press.

Wendorf, F. & Schild, R. (1994). Nabta Playa and its role in northeastern African prehistory. *Journal of Anthropological Archaeology*, 17, 97–123.

Wendorf, F., et al. (2001). *The Archaeology of Nabta Playa. Holocene Settlement of the Egyptian Sahara*, 1. New York: Kluwer/Plenum.

Wendorf, F., Schild, R.&Close, A. (Eds.) (1984). *Cattle-Keepers of the Eastern Sahara*. Dallas: Southern Methodist University Press.

Weniger, G.-C. (1991). Überlegungen zur Mobilität jägerischer Gruppen im Jungpaläolithikum. *Saeculum*, 42, 82–103.

Chapter 6

Resources, Use Potential, and Basic Needs: A Methodological Framework for Landscape Archaeology

Tilman Lenssen-Erz and Jörg Linstädter

Landscape archaeology as an analytical concept is not really new. Compilations of publications on this issue list several hundred references. Although they are far from being unanimous in their understanding of landscape archaeology, there is a common theme among almost all approaches, that is, the use of the term "landscape" as an analytical concept for a comprehensive understanding of the relation of prehistoric people to their environs, how they acted upon it, and, for a lesser number of papers, how they were cognizant of it. To clarifywits full potential the term "landscape" is defined here and an epistemological frame developed for its implementation. The aim is to grasp the complex network of relations between resources, use of resources, and basic human needs in one comprehensive representation. This approach helps to work through all relevant issues in a checklist and facilitates comparisons between different case studies. Case studies from northern and southern Africa demonstrate the value of such an approach.

M. Bollig, O. Bubenzer (eds.), *African Landscapes*,
doi: 10.1007/978-0-387-78682-7_6, © Springer Science+Business Media, LLC 2009

6.1. INTRODUCTION

This chapter discusses a method to systematically record the archaeological remains of a certain landscape at a certain time in its full range. The aim of this approach is to present a checklist of possible human–environment interactions and to propose a procedure for making these interactions comparable across different case studies.

Another innovation of this method is its strict deduction from a definition of the concept "landscape." This definition results from pertinent investigations in archaeology as well as in other disciplines. The authors understand the concept of landscape as a notion that develops at the interface of natural assets and human agency, use as well as cognition. These four columns have to be translated into categories that can be investigated by archaeological means. A "natural asset" is a resource such as water, food, or raw materials and "human agency" subsumes all traces of how humans shape their surroundings (*Gestaltung*). To investigate the relations between the parameters all usable resources are listed and their use potential discussed. Additionally, all human needs from nourishment up to spirituality are accounted for. The way in which these needs are satisfied on the basis of which resources are available shows the complexity of human land use at a certain point in time and space. Finally, cognition would seem to be an ephemeral phenomenon in the archaeological record but through the occurrence of rock art in a region we find stable, symbolically loaded markers of locations of relevance for the prehistoric people in the landscape and with indicators for the linkages of empirical and imaginary space.

The archaeology of landscapes—the landscape of archaeologies: a brief overview

Concepts of landscape archaeology are usually more clearly discernible in the archaeological practice than being the demonstration of a previously defined body of procedures, working concepts, rules, and postulates. A short scan of the relevant literature shows the different foci which can be the basis for approaches in this vein (cf. Anschuetz et al., 2001, pp. 164ff.). The scale reaches from unabridged positivist concepts to fully fledged hermeneutic narratives, which also express the discrepancies between processual and post-processual approaches. There are, however, further conceptual positions along this continuum, such as the pragmatic position, the position giving priority to emic views or the position trying to consolidate the benefits of processual and postprocessual approaches.

The *empirical approach* is rooted in the 1960s and 1970s (e.g., Hodder & Orton, 1976; Vita-Finzi, 1978) being very close to the natural sciences and trying to grasp the complex information of the human existence in a landscape through empirical verification, quantifiable and measurable in artefacts and other material evidence. Very clearly in this approach the landscape is mainly seen as a resource apt for human exploitation, being a set of assets conditioning human livelihoods.

The *pragmatic approach* to a landscape grows out of a rather intense preoccupation with an archaeological corpus inseparably linked to a spatial environment. It is in rock art studies where the link between artefacts and the natural surroundings is particularly stable (e.g., Bradley, 1994; Bradley et al., 1994; Swartz & Hurlbutt, 1994) because the sites and the landscape setting in most cases remain more or less unchanged in their large-scale properties such as geology or topography.

Of course *emic views* depend on indigenous voices which still have a chance to be heard. They can help to challenge the western conceptualization of landscape in which often pristine nature plays an important role (e.g., Schama, 1995; Luig & von Oppen, 1997). It has long been known that landscapes can be perceived in very different ways (e.g., Littlejohn, 1963; see also Rössler as well as Dieckmann, this volume) and this realization has led to concepts that move away from the Western bias when using the term landscape. In order to emphasize this shift, new concepts have been introduced, such as "taskscape" (Ingold, 1993) or "mindscape" (Ouzman, 1998a,b, 2002). Taskscape sets the focus on activities and consequently on time which is inseparably linked to action (Ingold, 1993, pp. 157ff.; see also Widlok, this volume). Backed by the views of nineteenth century San hunter-gatherers and their relation to rock art, the mindscape approach emphasizes the cultural specificity of every individual mind in landscape perception (Ouzman, 2002, p. 101).

The *postmodern* turn archaeology has recently taken is clearly visible in a lot of papers which are exercises in writing up histories in the sense of ever-new stories that lie behind the perceived (e.g., Tilley, 1994). Postmodern researchers expressly link up with a phenomenological approach (e.g., Bender, 2002, p. 108) and Tilley's study is based entirely on a phenomenology founded on Heidegger and Merleau-Ponty (Tilley, 1994, p. 14), providing a highly theoretical and abstract background. In its essence this approach is based on the notion that experiencing a landscape—if it is only intensive enough—bears some trustworthiness even if today we are the ones who want to understand a Mesolithic landscape (ibid.: 74f.).

The attempts of consolidating processual and postprocessual approaches in archaeology obviously grew from the opinion that neither of these approaches is completely obsolete and that relevant information can be gathered either way. Thus R. Layton and P. Ucko concede that it "has become impossible to deny that our explanations are culturally constructed; even if they refer to an independent reality, they enable knowledge of the world not as it is, but merely as we represent it to ourselves" (Layton & Ucko, 1999, p. 3). But they also see that "Reading the landscape as an expression of meanings negotiated in past or

(continued)

(continued)

present cultures will depend on identifying a community's reference to external features that we also perceive" (ibid., 11). At this juncture they bring into play the more pragmatic phenomenology of A. Schütz (following E. Husserl). In its essence this phenomenology is an approach that focuses on those issues in a perceived environment which are liable to intersubjective understanding because this environment, which he termed *Lebenswelt* (= life-world, cf. "lifeworld" in Hodder et al., 1995, p. 239), comprises components that are undisputed and unproblematic, belonging to a world of common experience and interpretation (Schütz & Luckmann, 1975).

Another consolidating strand is manifest in the teaching of landscape archaeology at universities. A short survey on the Internet (e.g., www. bristol.ac.uk 2007, www.exeter.ac.uk 2007, www.oxford.ac.uk 2007, www. sheffield.ac.uk 2007) reveals that the aim is the reconstruction of human interaction with a landscape through time, encompassing the earliest hunter-gatherers as well as recent historical times. What makes this understanding of landscape archaeology new and establishes its broadness is the search for analogies in human geography, anthropology, and art history, as well as in philosophy. On the other hand, these studies embrace an interpretative framework that links this kind of landscape archaeology to postprocessual archaeology by looking at emotional and political values. The interpretive framework includes issues such as ritual and cognitive landscapes, sacred geography and the political dimension of the past in the present.

Landscape archaeology in Germany

Landscape archaeology in Germany is rooted mainly in the concept of settlement archaeology. A detailed discussion of the history of the term and of the sources and methods of settlement archaeology can be found in Jahnkuhn (1977). He defines settlement archaeology as a field of research that, first of all, strives to study questions relating to settlements on the basis of archaeologically comprehensible and explainable material sources without considering phylogenetic or ethnic aspects. Janhkuhn emphasizes the closeness of settlement archaeology to settlement geography and its two branches, physiogeography and anthropogeography. In this context he applies the term settlement archaeology not only to the examination of the records of a settlement but encloses the economic w pertaining to the settlement as well as raw materials and burial sites (Jahnkuhn, 1977, p. 6ff.; cf. Schade, 2000, p. 140).

Shortly after Jankuhn's paper had appeared the term "landscape" was used in the German archaeological literature. In his article of 1982, "Siedlung in bandkeramischer und Rössener Zeit," J. Lüning remarks that one should preferably speak of an archaeology of prehistoric cultural landscape if the archaeology of settlement is combined with the examination of agricultural areas, cult, political, and military works and networks of traffic (Lüning, 1982, p. 9). His paper, "Landschaftsarchäologie in Deutschland – Ein Programm," (Lüning, 1997) deals explicitly with the concept of landscape archaeology, defining it on an abstract theoretical level:

> The term landscape archaeology . . . describes mainly an overarching view. With the help of this view older approaches of research, namely settlement-, economic-, social-, and eco-archaeology, each with its own focus, can be combined to form a strand of questionings. The classical settlement archaeology is closest to the term landscape archaeology. . . . (Lüning, 1997, p. 277, translation by J.L.)

Another German publication that tries to substantiate the concept of landscape archaeology is C. Schade`s Masters thesis: "Landschaftsarchäologie – eine inhaltliche Begriffsbestimmung" (Schade, 2000). Although he does not contradict Lüning's position he suggests a definition of landscape archaeology that is closer to practice:

> The term landscape archaeology denotes the systematic examination of regions settled in prehistoric times (Leser, 1997, p. 690), which usually aims at the reconstruction of settlement structures and diachronic settlement processes. The structure of a settlement system and of economic areas provides clues for the type and extent of land use and via this strand also for the communal system of a concrete historic and cultural-spatial section of a landscape. Changes in society and economic conditions can be discerned in the choice of different locations and also in number, size, and function of settlements. (Schade; 2000, p. 182, translation by J.L.)

Based on this definition a synthesis of different archaeologies is emphasized, thus stressing an intradisciplinary interlacing in addition to the interdisciplinary teamwork with the natural sciences.

In general the reconstruction of the cultural landscape and its mutual effects with the surrounding natural sphere is the objective of landscape archaeology. Therefore the archaeological sites of the area under investigation have to be recorded as comprehensively as other elements of the landscape (Lüning, 1997, p. 227f.; Schade, 2000, p. 184). The core thought of this concept is that, in a given landscape, there is no isolated site but every trace of human activity is part of a settlement system that has to be recorded (Schade, 2000, p. 160). Because it is impossible and also of little use to dig up whole landscapes, apart from excavation, surveys combining natural scientific and archaeological methods are indispensable (Lüning, 1997, p. 281f.; Schade, 2000, p. 172ff.).

With the help of statistical methods the representativity of the surveyed and excavated areas (Zimmermann, 2001) and of the examined random samples of the inventories (Linstädter et al., 2002) can be checked. Mappings with geographic information systems (GIS) show the distribution of sites in space

(continued)

(continued)

and on a timescale as well as the relationship between the archaeological sources and the natural factors. Apart from a two- or three-dimensional representation of the information, GIS also enables their statistical interpretation.

An excellent example of landscape archaeological work is the examination of the Linearband-ceramic settlements of the Aldenhovener Platte (North Rhine-Westphalia, Germany). In the project, Settlement Archaeology of the Aldenhovener Platte (SAP), five settlements were almost entirely excavated during the years 1972–1973. On the basis of this record the area became a main focus of research. The SAP project continued into the 1980s. In the meantime in an area of approximately 300 km?, 34 sites are known and all the work concerning the structure of settlements (Boelicke, 1982), ceramic typology (Stehli, 1973), and the supply of raw materials (Zimmermann, 1995) have formed the basis for further archaeological research. In addition the settlement archaeological examinations were always connected to palaeoecological and especially palaeobotanical research (Kalis, 1988; Kalis & Zimmermann, 1988). Thus evidence was presented concerning vegetation history, agriculture, and anthropogenic environment change in the settlement's surroundings.

Since 1998 settlement and environmental archaeological work has been continued in the project Landschaftsarchäologie des Neolithikums im Rheinischen Braunkohlerevier (LAN). This project is based on an explicit landscape archaeological concept. The objective of the project is to investigate the settlement corridors of the Rhenish Lössbörde in representative sectors (small regions). Both the internal structure of single settlement clusters and the connecting economic and social networks are the focus of this project (Frank & Wendt, 2003). To compare or complete data from these different levels of scale, methods of upscaling and downscaling have been developed (Zimmermann et al., 2004). All in all, empirical approaches dominate German landscape archaeology.

6.2. LANDSCAPE-ARCHAEOLOGICAL DEFINITIONS AND METHODOLOGY

6.2.1. Definition of the Term "Landscape"

Different scientific disciplines foster their own definitions of the term "landscape," as the articles of this volume show, and within archaeology itself there is no consistent definition. Archaeologists often make use of the dichotomy of "cultural landscape" versus "natural landscape" (Schade, 2000, p. 156). The term "cultural landscape" is used to describe the human impact on the environment. Opposed to that, "natural landscape" is used for a system which is barely or not influenced by humans at all, which is the state people encounter when first colonizing an area.

In landscape ecology the term landscape is synonymous to the concept of "landscape ecosystem." This system is characterized as part of the earth's biogeosphere (ecosphere), realized as a highly complex, substantial, and energetic system of natural influences to which anthropogenic factors and processes stand in direct or indirect relation (Leser, 1997, p. 187). The relationship of natural and anthropogenic factors builds the center of this definition, termed the society-milieu relationship (Hirsch, 1995, p. 9). Even geographical definitions of the term "landscape" such as Sauer's (1963, p. 343) follow this understanding: "The cultural landscape is fashioned from a natural landscape by a cultural group. Culture is the agent, the natural area is the medium[;] the cultural landscape is the result."

According to these landscape-ecological and geographical definitions and in synthesizing the terms cultural and natural landscape frequently used in an archaeological context, we apply landscape as a concept in between human cognition and action (subsuming both as *Gestaltung*) on the one hand and independently existing material resources. We postulate that there is an independent natural complement to human agency. It is well established in the humanities that a distinction of nature versus culture does not correspond to the views of non-Western societies (e.g., Heyd, 2002; Dowson, 2007). However, in a Western scholarly context there are useful aspects in the existence and use of these terms if the incongruence with emic categories is always kept in mind. But the denial of such distinguishing categorization inevitably leads to an unsystematic mingling of two epistemologically different corpora.

In order to avoid getting caught up in a lasting discussion on the validity of these two categories we suggest the analyzing of landscape with a categorization which is not particularly biased, although the analogy to other concepts is evident (Figure 6.1).

6.2.2. A General Procedure for Landscape-Archaeological Practice

According to our understanding of the term landscape it incorporates methodological properties that enhance the strength of landscape archaeology as an integrative tool. Natural resources, subdivided according to their parameters, are correlated with basic human needs. Both are linked by the use potential that transforms a resource into an asset.

Through the configuration of both antipodes a lot of possible relationships between use potential and the satisfaction of human needs can be generated. The pattern that emerges through the satisfaction of a specific need by making use of particular options produces the individual fingerprint of a community utilizing a certain landscape at a certain time. This fingerprint is termed in the following the general pattern of use.

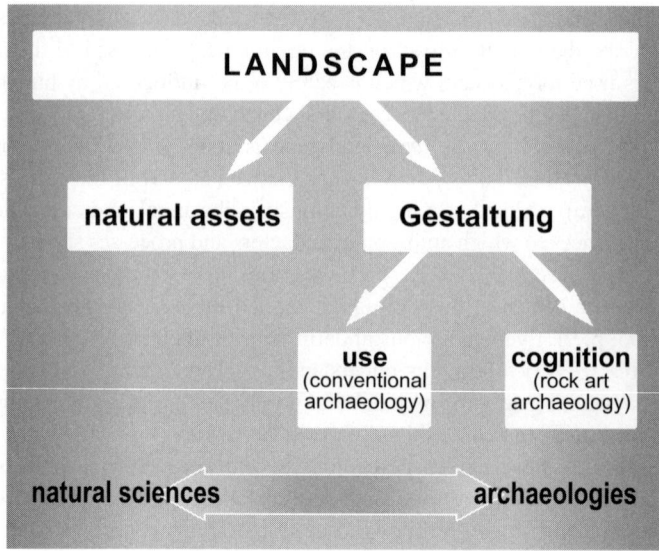

Figure 6.1. Categorization of landscape relevant for landscape archaeology

Although resources of a prehistoric landscape are often still detectable today, the former needs can only be assumed. However, these assumptions are founded on an empirical base. According to the landscape-archaeological concept no isolated site exists without being part of a settlement or use system. Likewise no isolated human action exists beyond the common human behavior of nourishment, settling, tool production, mobility, or interaction with the environment and his or her own species (identity, communication, symbolism).

A general procedure for landscape-archaeological projects has already been described elsewhere (Schade, 2000, p. 184ff.). Modified to our methodological approach, the procedure is divided into six steps (Figure 6.2). The first step is the generation of archaeological data. This includes the definition of the research goal as well as the research area, followed by surveys, excavation, and documentation of all sites and finds. The second step follows the determination and the ranking of human needs (a step which has to be done only once, unlike the generation of archaeological data). Step number three in our *chaîne opératoire* is the identification and mapping of all potential resources. The resource survey takes place most appropriately with the search for the materials recorded on the sites. Moreover, the knowledge and understanding of resources that were not used provide interesting insights as well. Step four results automatically from step three, that is, the definition of the use potential of all resources and the qualitative assessment of the raw materials. Finally this step shows what the landscape actually offers to its inhabitants. In the fifth step the sites are analyzed and combined to temporal and spatial units. Here archaeological entities are evaluated with their natural resources and

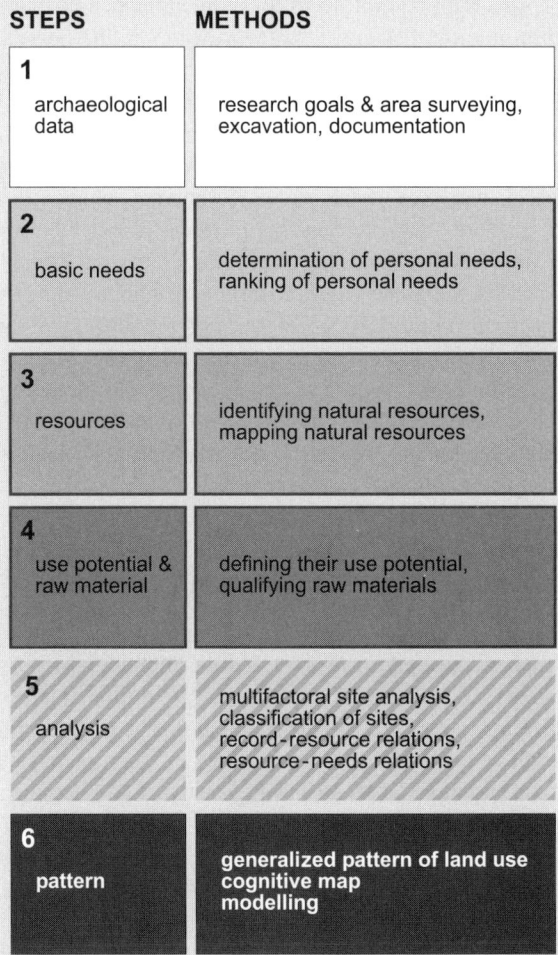

Figure 6.2. *Chaîne opératoire* for landscape-archaeological research

their use potential for the consequential comparison. At the final step generalized patterns of land use can be formulated, land use maps or cognitive maps can be produced, and particular questions can be investigated by modeling methods.

6.2.3. Concepts of Landscape Description: More Definitions and their Methodological Implications

6.2.3.1. *Nature and Natural Assets*

Nature, we maintain, is the empirical landscape that can be examined by the natural sciences in order to supply evidence that provides explanations and understanding for us but may not do so for indigenous people. The advantage of describing nature in natural scientific categories lies in the possibility of developing comparisons and analogies between different landscapes, procedures that do not lie in the scope of emic knowledge systems. Accordingly nature retains relevant information for outsiders without indigenous insights. We may yet be potentially able to communicate parts of our notion of nature to indigenous people because it refers to that part of the world (following Schütz' understanding of phenomenology) where they can meet us physically, we, who are not cultural insiders of their world. Nature holds the resources of interest for human exploitation or symbolization; it is the arena where natural assets are available or negotiable. The choice of the term "natural assets" is based on the aspect that through its connotation of being useful or beneficial to someone, it implies human involvement, thus strengthening the view that a landscape concept is futile without the human role in it, even if it were manifest only in perception. Moreover, the term natural assets should be understood as a heuristic descriptive tool by which the resources that help to satisfy human needs can be registered.

6.2.3.2. Gestaltung

The complementary second component in this approach to landscape is termed *Gestaltung*, which means giving a *gestalt* to something. More specifically it denotes the process by which a structure or configuration is given to physical, biological, or psychological phenomena so integrated as to constitute a functional unit with properties not derivable from its parts in summation (after *Webster's*, 1993, p. 952). In part this refers to the physical act of shaping or processing given assets (e.g., fitting a stone tool or seeding crops) and using the resources whereas another part of the concept *Gestaltung* refers to cognition, that is, the "knowledge, purposes, practices, and skills of the people" (Segal, 1994, p. 22) who have interacted with the landscape. These two broad aspects of *Gestaltung* are grasped here with the terms "use" (see below), for which field archaeology is the relevant research tool, and "cognition" where cognitive archaeology is the appropriate instrument.

Cognition is listed here as a means of *Gestaltung* because it denotes an active procedure by which perception is processed in the mind in order eventually to be uttered as behavior and action. Among the cognitive means of *Gestaltung* rock art takes a salient position inasmuch as it is a sign system often with universally readable elements. They may become understandable to a certain extent even without indigenous comments through the employment of information of intercultural knowledge, such as animal behavior (e.g., Lenssen-Erz, 1997, 2000; Hollmann, 2003). Further advantages of rock art are the restrictedness of taphonomic processes (usually weathering, erosion, and/or repatination only) and, as a consequence thereof, the reliability of the spatial context in which rock art is found which, better than in any other artefact class, enables considerations as to the original, intended spatial arrangement.

6.2.3.3. Basic Needs

Use of resources is a behavior that is inseparably linked to the basic needs which people must satisfy if they want to lead a decent human life. Our list of basic needs follows conventional needs after Abraham Maslow (Figure 6.3, left; Maslow, 1970, 1981; see also Lenssen-Erz, forthcoming, for a detailed discussion) such as food, protection, and so on. The hierarchy established with this pyramid was subdivided by Maslow into the four lower D-needs = deficiency needs and the three higher B-needs = being needs, indicating that the lower needs necessarily have to be satisfied whereas higher needs may not even turn up in every person (Maslow, 1981, pp. 102, 128f.). Once someone has reached the upper levels, he or she may—at least temporarily—dispose of the satisfaction of the lower ones (ibid., 79, 102). This model of needs should therefore be understood as flexible with permeable levels that provide a framework for the motivations under which people may act in any given situation.[1]

As with Maslow's model, needs are ranked differently according to their priority. Our list includes issues that are normally not registered in the many variations of Maslow's pyramid of needs, such as tool production or mobility (Figure 6.3, right). The reason for including such needs lies in the fact that, first, they are means and sometimes preconditions for satisfying the more basic needs and, second, the ubiquity of the respective items (e.g., tools) indicates that people everywhere and at any time display the behavior of producing these traces. Tool production defines humanness and is the major diagnostic evidence of human activity. There is a cogent link of human life with tool production that consequentially is conceptualized here as one of the more or less basic needs.

In order to adapt Maslow's understanding of needs and its psychology-loaded terminology to the conventions of archaeology, Figure 6.3 provides a correlation of Maslow's levels with terms which are current in archaeology.

6.2.3.4. Resources

According to the dictionary, "resource may refer to any asset or means benefiting or assisting one, often to an additional, new, previously unused, or reserve asset" (*Webster's*, 1993, p. 1934). On purpose there is little specification in this definition but there are characteristics that can still be contextualized archaeologically. Accordingly resources are those assets which

---

[1] Recently Malsow"s model has been developed into a paradigm based upon another metaphor, i.e.that is, the "spiral dynamics" model (Beck & Cowan, 1996). This, however, does not invalidate the Maslow pyramid which would seem to be more down to earth and therefore more adequate for archaeological appropriation even though the spiral dynamics model has been put into political practice.

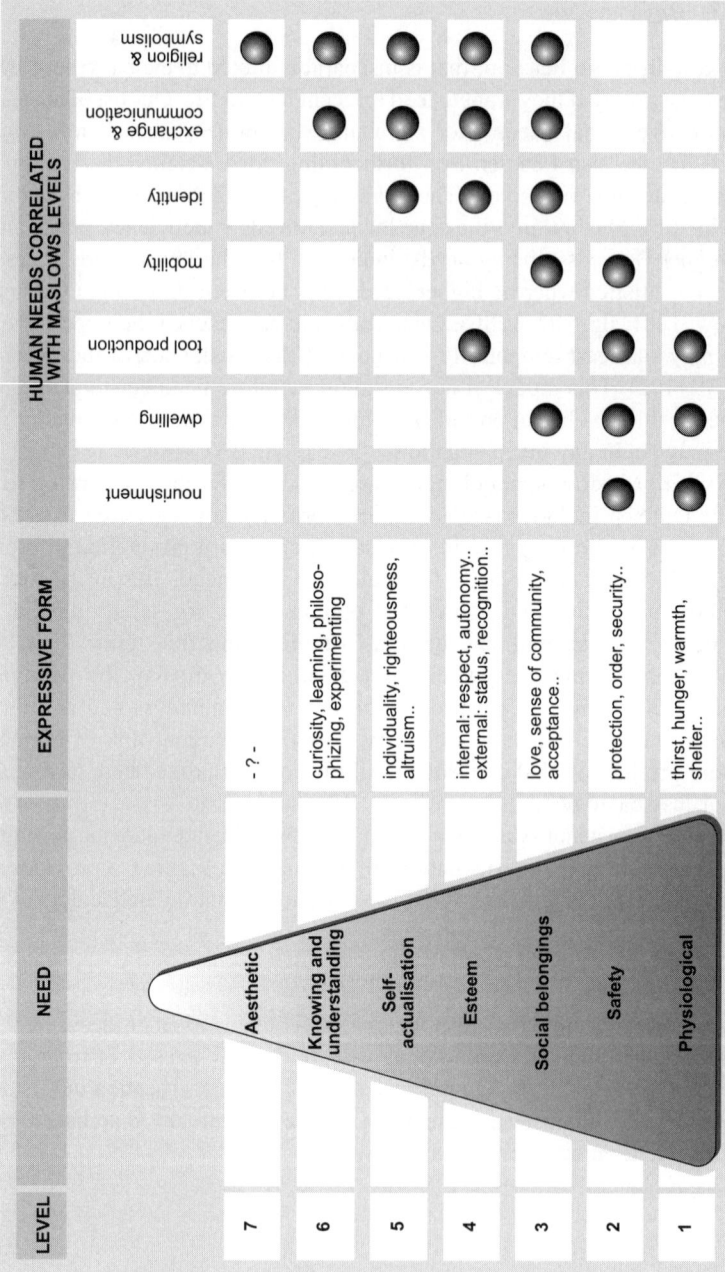

Figure 6.3. Human needs as defined in the present study, correlated with the levels of Maslow's pyramid of human needs (Maslow, 1970)

- Cover all human needs (cf. Figure 6.3)
- Have always been used by humans since the early hominids
- Are accessible to and can be made use of by every able person
- Can be grasped empirically to a large extent, thus being described in formats that enable comparisons and analogies

Such a holistic view on resources sets up a close connection with the concept of lifeworld as established by Schütz (Schütz & Luckmann, 1975).[2] The purpose of linking lifeworld to resources is to establish a descriptive unit that could be an autarchic entity in which the entire life of a population could possibly take place without the necessity to leave this territory in order to gain access to vital outlying resources.

In the context of landscape archaeology, as it is suggested here, five main resources can be established: water, abiotic raw materials, plants, animals, and space but the inclusion of further resources may be possible in future, for example, time or human.[3] The first four resources follow a universal rationality and underlie universal causalities. Here the forces of physical laws, evolution, or geology are at work and restrict the options for human intervention. They are therefore prone to be dealt with by natural sciences if a general first overview of the availability and richness of these resources is needed in a case study. Space, by contrast, is a resource that, to a large extent, underlies culture-specific rationality and causalities (e.g., Dünne & Günzel, 2006) with use being guided by cognition to a much higher degree than that of the other resources.

Each resource is apportioned into several parameters. These parameters define the resource precisely and have to be checked as to whether they are available in the research area. The resource "water," for example, is characterized by

---

[2] Hannah Arendt, following A. Schütz, developed and summarized his concept for a modern world context, but nonetheless her interpretation matches all premodern lifeworlds very well sincebecause "the world of common experiences and interpretation (Lebenswelt) is taken to be primary and theoretical knowledge is dependent on that common experience in the form of a thematization or extrapolation from what is primordially and pre-reflectively present in everyday experience" (after Yar, 2001).

[3] The parameters of time range from day/night over seasons and lifetime to generations and also past and future may be listed here. Time has a potential to be used for labor, leisure, recreation, or movement, as well as social and religious management. Traces of these kinds of use are either too ephemeral or too variant to be analyzed systematically. The human resource certainly awaits a clear definition but it may comprise parameters such as ratio, language, symbolic thinking, reproduction control control, and skilled movements. These specific abilities can potentially be used for working power, *Gestaltung*, innovation, abstraction, imagination, communication, or social differentiation, and the like. Finally, it is only through the human resource that the satisfaction of a basic need can be accommodated which undoubtedly is of universal character, namely sexual activity. Although the relevance of time in landscape archaeology has been emphasized recently by, for example, Ingold and Bender (Ingold, 1993; Bender, 2002), the human as a resource has not received similar attention.

the parameters surface or groundwater (in addition to rain). Of special interest are, in this case, questions such as how much water is available (annual precipitation or groundwater) and at what distance. Furthermore it should be clarified whether water is available permanently or just temporarily. The other parameters are listed in Figure 6.4 in the column "resources."

6.2.3.5. *Use and Use Potential*

For the interaction with resources one may resort to a rather simple, everyday term, namely "use" (German: *Nutzung*). It encompasses exploitation, consumption, curation, nurture, development, occupation, and symbolization. Usually the use of resources leaves traces in the ground. As a rule of thumb one might say that, to a large extent, in the physicality of the findings of field archaeology one has to deal with the results of productive targeted activities—mainly aimed at tangible addressees—that are based on a cognitive spectrum of everyday with the causalities and rationalities of the physical world. Such issues have always been dealt with by conventional archaeology concerned with settlements, economy, social structure, or ecology.

In contrast, the cognitive elements are rather tokens of symbolic actions whose cognitive spectrum may not be linked to our real world thus having its own causalities and rationalities. They are not necessarily derivable by logic reasoning and may partially be aimed at intangible addressees. Yet, cognition is not entirely arbitrary and retains many elements that are accessible from the inevitably ethical perspective of a prehistorian (Zubrow, 1994a, b, p. 110f.).

For each resource parameter there is a potential for use. The potential is the maximum of what can be extracted as an asset from a resource but may never have been managed to the full extent in prehistoric small-scale societies (such as gaining energy from water). In this potential there are options of use that are dependent on temporal and spatial circumstances inasmuch as not every landscape will inevitably provide the full range of resource parameters. Therefore, from the options available, every society makes its choices thus producing patterns of use. It may, however, happen that a society does not exploit the full use potential at its availability and in such a case it should be worked out why this is so. Maybe it was conscious choice, inadequate technology or knowledge, and so on, or maybe relevant needs could be satisfied with another resource or access to the respective resource was blocked by a competitor.

The three strands of resources, use potential, and basic needs form the pillars of any livelihood and pattern of land use. It is the archaeological data that hold the information for the understanding of how one pillar is connected to the other with use potential attaining a central position (Figure 6.4). This scheme enables the researcher to make valid statements concerning the way of living for any society and to format the knowledge about it in a layout that enables comparison to any other society.

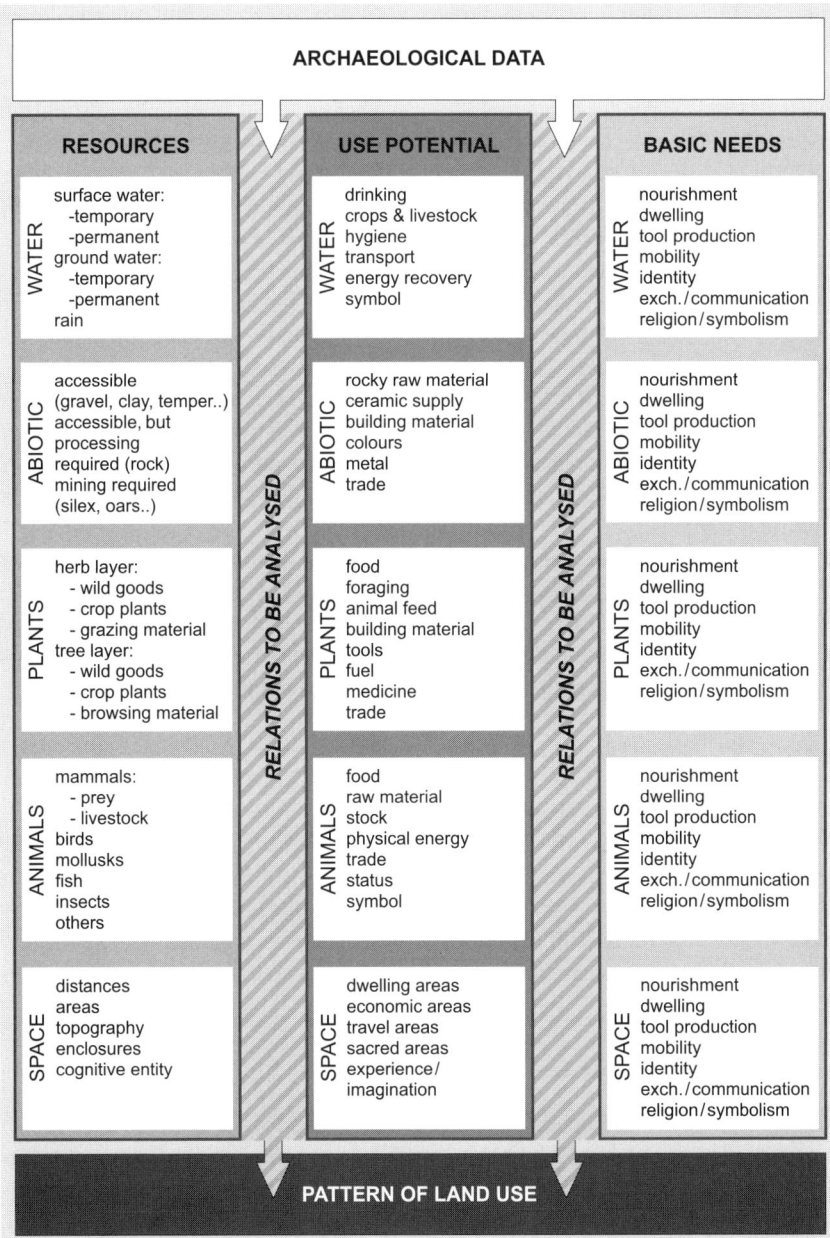

Figure 6.4. Analytical scheme of the relationship among resources, use potential, and human needs

The Millennium Ecosystem Assessment (2005, p. 28) has presented a comparable matrix of relationships working with two basic poles: First "ecosystem services", encompassing "life on earth – biodiversity" which combines components that our model lists under resources but on the other hand are clearly culturally embedded needs. This pole is complemented in the Millennium Ecosystem Assessment (MA) by the second pole of "constituents of well-being" in which basic needs are combined with qualified resources (e.g., "sufficient nutritious food"). Because the use of this model is aimed at political decision making it also contains issues such as "freedom of choice and action" among the constituents of well-being. Both poles are linked with arrows assessing the "intensity of linkages between ecosystem services and human well-being" on the one hand and—here again the political purpose becomes tangible—"potential for mediation by socioeconomic factors."

Both models, the MA as well as ours, are attempts at finding a representation for the complex social–ecological systems' relationships. The necessary differences between both are based on the dissimilar purposes, where our model aims at bringing together all kinds of archaeological finds and data in a consistent empirical framework.

Three case studies presenting archaeological landscapes (the Gilf Kebir in southwest Egypt, the Brandberg/Daureb in central Namibia, and the Ennedi Highlands in eastern Chad) exemplify how the two poles of resource and need can be represented in a matrix showing their interrelation with the use potential thus producing the fingerprints of livelihood of these cultures.

6.2.3.6. Case Studies

The case studies in this chapter serve to exemplify the method proposed here. They are not at all complete descriptions of the discussed sites, cultures, or phenomena. Each of the presented studies is able to fill monographs (Linstädter, 2005) or even book series (Pager, 1989–2006). But the case studies show that by using the relation module of resource/use/basic needs as a guideline the presentation of land use studies can be structured in such a way that different case studies become comparable.

6.2.4. The Gilf Kebir Case Study (Southwest Egypt)[4]

The Gilf Kebir is a sandstone plateau situated 650 km west of the Nile valley on the same geographical latitude as the Aswan Lake. In the north it disappears under the Great Sand Sea, and in the southeast its cliffs rise about 150 m above the surrounding plains. Here the plateau is intersected by broad wadis. Some of them, such as the Wadi Bakht, the Wadi Maftuh, or the Wadi el Akhdar, are of special interest for geographers and archaeologists because of their unique geomorphological

---

[4] The Gilf Kebir case study is based on fieldwork carried out in the framework of several DFG-sponsored projects at the University of Cologne. The Great Sand Sea case study (Riemer, this volume) was undertaken with the same background.

situations: the so-called barrier dunes or terraces (Kröpelin, 1989). Over a period of several millennia, the sediments of playas accumulated behind these barriers. These sediments are the result of seasonal or episodic rainfall which has produced temporary water reservoirs used by prehistoric inhabitants.[5]

Comparable to the SAP-project in Germany described above, in the Gilf Kebir stone tool and ceramic technologies, settlement structures, and subsistence strategies were investigated according to the settlement-archaeological approach. At the end of the twentieth century many of the excavated sites were published and a chronological sequence was established (Figure 6.5) (Hallier, 1996; Schön, 1996; Linstädter, 1999; Gehlen et al., 2002).

The investigation of the archaeological sites close to the playa lakes formed the basis of a conceptual model on settlement activities in the upper reaches of the valleys. Crucial questions of research were the extent of the economic area used by the prehistoric population and the source of raw material supply for the production of lithic tools found in the valleys. It was obvious that the area inside the

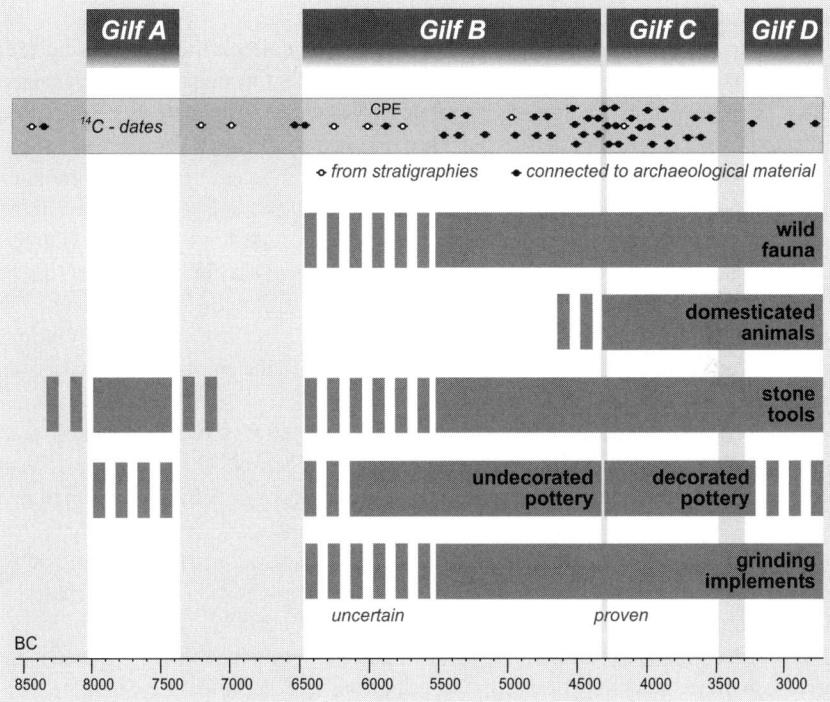

Fig. 6.5. Chronological sequence of the Gilf Kebir

---

[5] The upper reaches of Wadi Bakht, Wadi Maftuh, and Wadi el Akhdar have been subject to archaeological investigation by researchers of the University of Cologne during the past 20 years (Kuper, 1995; Schön, 1994).

wadis and close to the dwelling sites could not have sufficed either for the needs of hunter-gatherers nor for a pastoral-nomadic way of life. Moreover, no quartzite or chalcedony deposits were found close to the barrier dunes. There was also only little evidence of material or cultural exchange with other occupation areas investigated in the Eastern Sahara.

As a first step a research area in the southeastern Gilf Kebir was selected (Step 1). The research area extends from the upper reaches of the Wadi Maftuh and the Wadi Bakht in the west to the eastern plains of the Gilf Kebir in the east and has an extension of about 450 km². It includes three different landforms: plains, valleys, and the plateau surface (Figure 6.7). The vast forelands were surveyed by car, the barely accessible plateau area by foot. As a result 134 sites were documented and parts of them were excavated.

After defining the human needs (Step 2) the natural resources had to be checked (Step 3). In the research area these are mainly surface water behind the barrier dunes, and the quartzite outcrops on top of the plateau. These two resources have a high use potential (Step 4). As a water source there is no alternative to the water reserve of the wadis anywhere in the entire region. Of the stone tools more than 90% are made from the local quartzite.

After standardized analyses in order to obtain data on the material culture, the internal structure, and the age of each site, the findings were categorized (Step 5). One of these categories relates to raw material deposits (so-called outcrops) with diameters between 10 and 60 m, which in almost all cases show evidence of extensive human exploitation. A second category covers campsites which suggest extended stays due to evidence of stone hearths or stone circles or material such as grinding stones, pottery, bones, or ostrich egg shells. Isolated workshops are the third category, indicating short-term stays to renew the supply of stone tools or blanks. In the next step analysis followed that examined the relationships between the natural resources available in the Gilf Kebir including their use options, correlated with the basic needs of the prehistoric inhabitants of the region.

In order to investigate the general land use patterns of the research area all sites were mapped and their relations defined (Step 6). On the base of these data land use models on a local and regional scale were developed (Linstädter & Kröpelin, 2004; Linstädter, 2007).

The mapping of the three different categories shows a very different use of the landforms plain and plateau. Only 12% of the plain sites are raw material deposits and were used for raw material supply. About 88% of the sites (workshops and campsites) indicate short- or medium-term stays. On the plateau the reverse pattern is to be observed. More than half of the sites are quartzite outcrops, used in prehistoric times. The research area appears as a cultural landscape in which special land use systems developed as a function of raw material occurrences and geomorphologic factors.

One of the main aims of the landscape-archaeological concept is to show the change in land use practices at different times. In contrast to the Brandberg/ Daureb case study (see below) where land use patterns of a single phase were examined, two phases with different patterns were compared in the Gilf Kebir case study. The determination of a specific phase is achieved by radiocarbon dates or typological investigations. However, not every site provides datable material because of heavy erosion and deflation in most of the desert landforms. The area with the most dated sites is the plateau region. Here the different patterns of land use in the two main phases, Gilf B (6500–4300 bc) and Gilf C (4300–3500 bc; Figure 6.6) can be clearly distinguished.

The different resources yielded the following evidence.

6.2.4.1. Water

The availability of the resource water is the most likely factor that influenced the land use pattern at all times. According to our current knowledge there was no groundwater available in Neolithic time (Kröpelin, 1989, p. 232). Geological and archaeobotanical research show that precipitation was probably lower than 150 (Kröpelin, 1989, p. 286) or 200 mm/a for the entire period of the Neolithic wet phase. These rainfalls were episodical. In which season they were to be expected depended on the influence of the prevailing climate regime. It is assumed that for the time up till approximately 4300 bc summer rain and from 4300 on winter rain influence was predominant (Linstädter & Kröpelin, 2004). Phase Gilf B therefore falls into the period in which the eastern Sahara was under the influence of the summer rain regime. The wadi barriers in some of the valleys of the Gilf Kebir enabled the water from the brief rainfalls to remain available over a longer period during the entire phase of occupation (Figure 6.7). Because a large part of the heavy summer rains in phase B drained away on the surface, the locations close to the wadi barriers were of special importance.

Settlements during phase Gilf C were under the influence of the winter rains. The change of the rainfall regime had an effect on the water supply but also on flora and fauna, as well as on the economic and settlement system of the people. Despite the same quantity of rainfall as before, species were detected that indicate a more favorable water supply. Moderate rainfalls are better suited to the vegetation on the spot. From time to time it likely generated a grass covering on the plateau, which then could be used as a meadow. At the same time the run-off was diminished so that the settlement sites at the barriers became less attractive and campsites on the plateau confirm its usability.

6.2.4.2. Abiotic Raw Material

The most important aspect for stone tool production is the question concerning raw materials. Through surveys (Linstädter, 2003, p. 385), information on rich deposits of different quartzite varieties on the Gilf Kebir Plateau was gathered. The quartzites occur in outcrops in the vicinity of the valleys and are easily accessible. Eighty to ninety percent of all stone artefacts are made of this material (Schön, 1996; Linstädter, 1999). A *chaîne opératoire* reconstructs the quartzite quarrying on the plateau and its working in the valleys in this phase

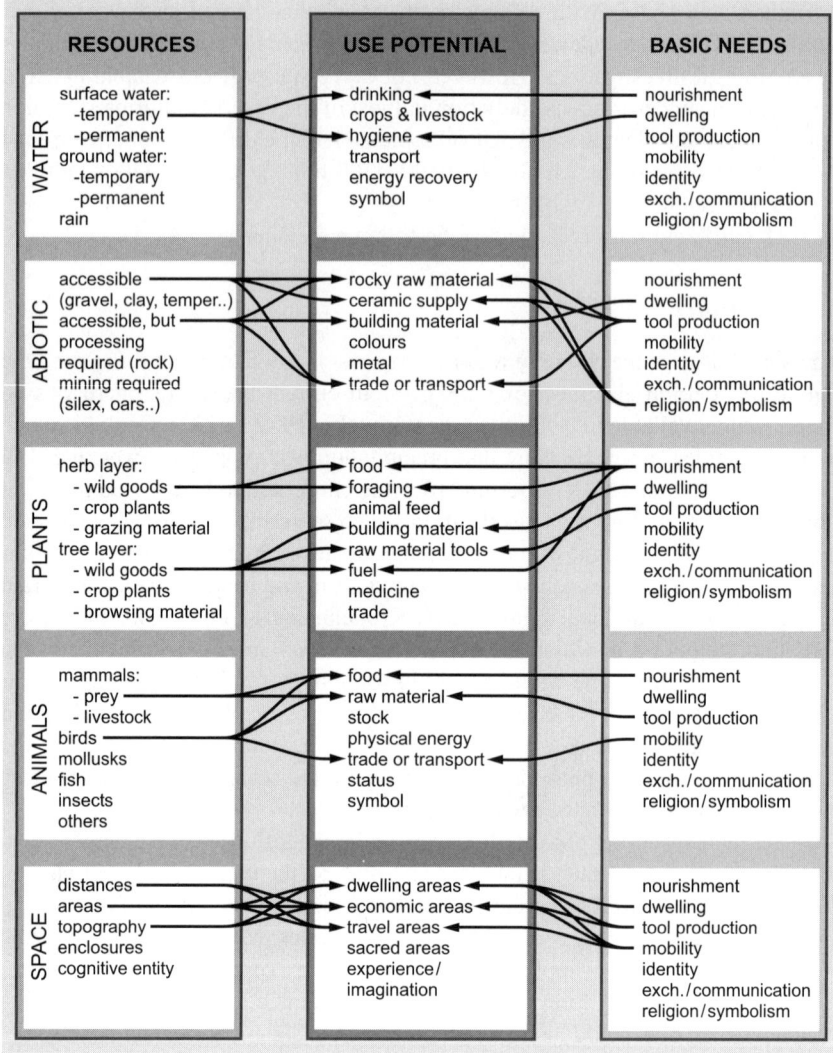

a

Figure 6.6. Analysis of relations between use of resources and human needs for the archaeological phases Gilf B and Gilf C

(Linstädter, 2003, Figure 2). The other artefacts were made of sand- or siltstone, several chalcedony varieties, quartz, basalt, or desert glass (Schön, 1996, p. 353). The source of the quartz and chalcedony have not yet been identified, whereas the origin of the desert glass is precisely known. It originates in the southeastern Sand Sea, directly to the north of the Gilf Kebir and clearly proves contact with this region. Apart from the raw material for the stone tool production, raw material

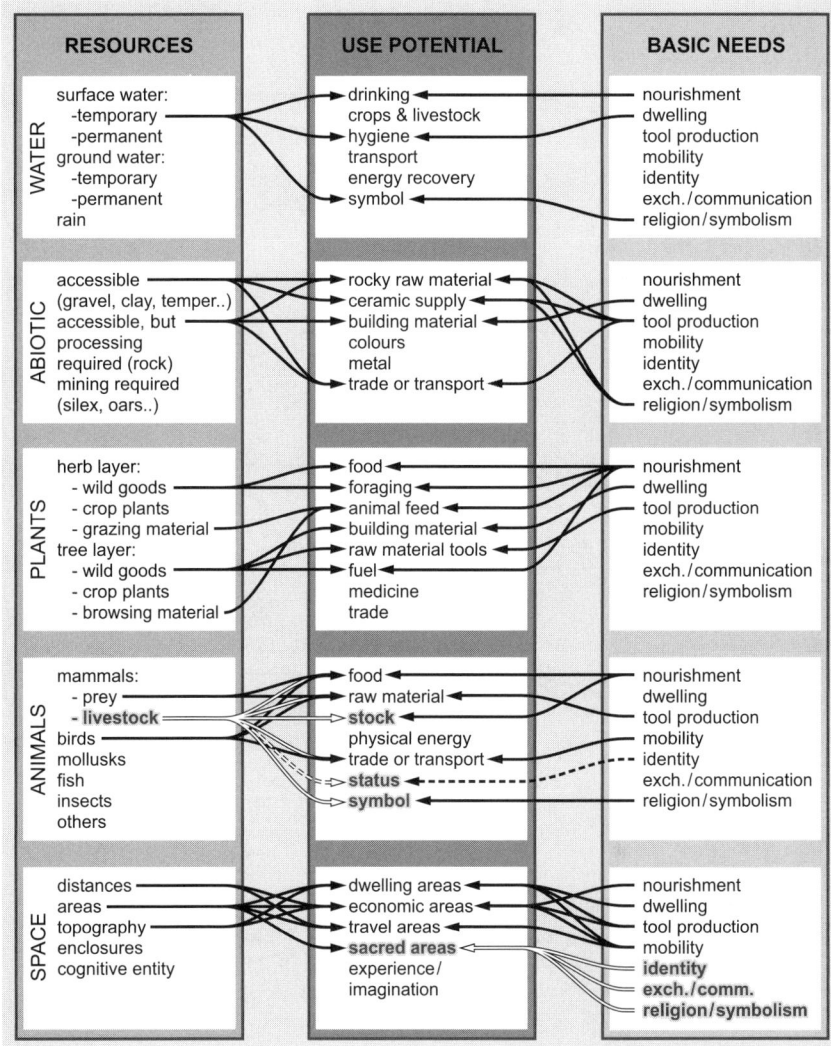

b

Figure 6.6. (continued)

for ceramic production such as clay (playa sediments) or temper material (sand) is available although there has been no indication of local ceramic production in the eastern Sahara up till now.

6.2.4.3. Flora

Information on the tree structure is provided by the research of Neumann (1989, pp. 116ff.) and Nußbaum (Linstädter & Kröpelin, 2004).[1]*The identified species*

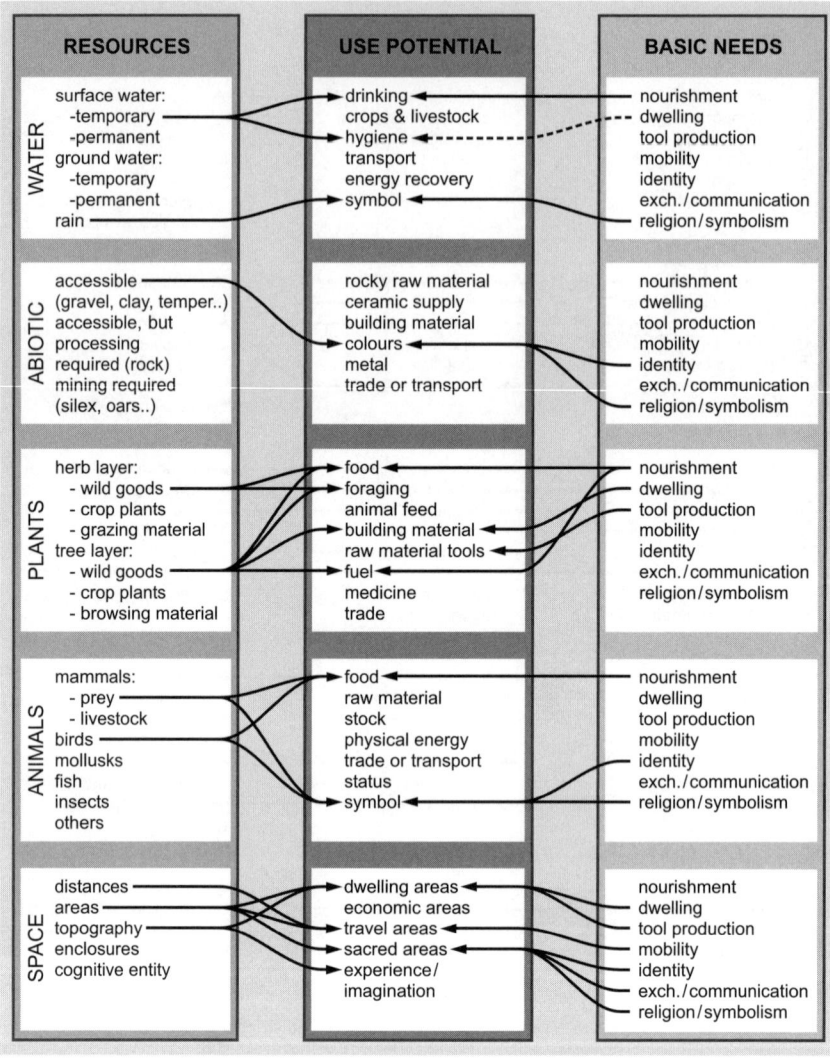

Figure 6.7. Reconstruction of the Wadi Bakht landscape focusing on resources and use patterns during the phase Gilf B (c. 6500–4500 bce)

(*Tamarix* sp., *Acacia* sp., *Maerua crassifolia, Balanitis aegyptiaca, Faidherbia albida*—in the case of Neumann still *Acacia albida*—and *Ziziphus* sp.) are still existent in the Sahara and indicate an arid to semiarid climate. For the early Holocene before 6500 bc only *Tamarix* sp. was detected. This species forms the so-called gallery forests and requires 50–100 mm/a precipitation. From 6500 BC (Gilf B) *Acacia* sp., *Ziziphus* sp., and *Feidherbia albia* can be found, where at least for *Ziziphus* sp. more favorable conditions must be assumed (Neumann, 1989, p. 123).

The occurrence of *F. albida,* which is only present for a period around 5000 BC, indicates a monsoonal influenced summer rain regime for it is clearly a savanna species. Evidence of *Ziziphus* sp. lasts until 3500 BC and includes the phase Gilf C. Therefore it can be assumed that at least the species *Ziziphus* sp. indicates favorable conditions with annual rainfalls from 150 or 200 mm in the phases Gilf B and C. The influence of the summer rain (*F. albida*), however, is only to be expected for the Phase Gilf B. From 3500 BC on again only *Tamarix sp.* can be detected.

Tamarix sp. as well as *Acacia* sp., *M. crassifolia*, and *B. aegyptiaca* provide good firewood and are therefore possibly rather frequent in charcoals. With certain restrictions they can also be used for constructing simple housings.

Evidence of the herb structure is far more complicated to obtain. On the shores of open water *Phragmitis communis,* as well as *Typha* and *Juncus* sp. can be assumed. From the tree structure *Acacia–Panicum* societies can be reconstructed for depressions, wadis, and alluvial plains.

6.2.4.4. Fauna

The different faunal inventories from the Gilf Kebir area have been published by Van Neer and Uerpmann (1989), Peters (1987), and Gautier (1982). The following species were identified: *Giraffa camelopardalis, Oryx damah, Addax nasomaculatus, Gazella dorcas, Gazelle dama, Canis vulpes, Crocuta crocuta, Struthio camelius,* smaller birds, larger bovides (possibly *Bos primigenius* f. taurus), as well as *Ovis aries* and *Capra hircus.* On the site Wadi Bakht 82/21, H. Berke (personal communication) further identified *Lepus* sp., *Rana perezi* (a frog), turtle, and a bird species. For the phase Gilf A no identifications of fauna are available whereas phase Gilf B only yielded wild animals. In this phase the resource fauna only supplied the potential meat and material such as bones and leather. In the following phases Gilf C and also to a lesser degree phase D in addition to wild animals there are also domestic animals (cattle, sheep, goats) present. In what way their occurrence is connected with the changed climatic conditions or whether it is due to cultural development cannot be decided thus far. But it highly increases the usable potential of the resource fauna. In addition to meat, blood, and milk also the physical energy of the animals, for example, as pack animals, may have been used. Furthermore there is the possibility of animal husbandry offering not only the accumulation of food reserves but also a trade and status potential. The symbolic potential that wild and domestic animals had for settlers of the Gilf Kebir are demonstrated by the rock art sites, as, for example, Mogharet el Kantara (Shaw's Cave, Shaw, 1936). Rock art as well as open-air sites at the Gilf Kebir show that the resource animal did not only fulfill diet and tool production needs. Likewise it served to conceptualize the surroundings (symbolism) and the keeper's own role (identity).

6.2.4.5. Space

The spatial extension of the phenomena typical for the prehistoric settlements of the Gilf Kebir is hard to determine. The thoroughly investigated and, in the sense of style and technique, fully comparable sites of the Wadi Bakht and the Wadi el Akhdar are separated by roughly 25 km. If the sites in closer vicinity (predominantly raw material sources) are included they cover an area of about 2000 km^2.

The epipalaeolithic microliths of the phase Gilf A are spread throughout the entire Eastern Sahara, a region of about 1,500,000 km^2. Typical for the phase Gilf B are likewise special microlithic forms but also mostly undecorated pottery, which occasionally exhibits notched rims. The microlithic forms (Linstädter, 1999, Figure 5, pp. 1–7) can also be found in the south of the Great Sand Sea (Wilmanns Camp), and the pottery (Linstädter, 1999, Figure 5, pp. 20–21) is dispersed as far as the northern Sudan (Wadi Shaw). Therefore the area under consideration in this phase probably extends over 300,000 km^2. Phase Gilf C does not consist of any significant stone technology and tool types. The mostly impressed pottery has hardly any parallels outside the Gilf Kebir (Hallier, 1996, p. 107). In view of the fact that, as mentioned above, domestic animals first occur in phase Gilf C, all rock art with cattle depictions can be dated in this phase. The cattle depictions of the Gilf Kebir and the Gebel Uweinat are stylistically similar. Although Shaw's cave is close to the Wadi el Akhdar, the area under consideration expands to about 15,000 km^2 when including the rock art sites of the Wadi Sura and the Wadi Hamra, and to about 40,000 km^2 when including those of the Gebel Uweinat as well. Therefore a reduction of the coverage areas can be detected from the early to the later phases, apparently in a process of regionalization.

6.2.5. The Brandberg/Daureb Case Study (Namibia)

The Brandberg/Daureb in Namibia is an inselberg of 30 km diameter located on the fringes of the Namib Desert to the sparse shrubland at a distance of 80 km from the Atlantic coast. As an area that receives an annual precipitation of about 100 mm it is submitted to a desert climate but the vegetation of the mountain is much richer than the precipitation would suggest. Because the southern subcontinent has not suffered from climatic changes during the Holocene in the same way as the north, conditions of today do not differ drastically from that phase in the Later Stone Age between 2000 and 4000 years ago, when the bulk of the rock art in the area was created.

The Brandberg/Daureb is among the best-studied rock art areas worldwide (Pager, 1989–2006) with research aimed at the link between rock art and space from the onset. In fulfillment of general postulates of landscape archaeology the sites as well as their immediate and wider surroundings have experienced very close attention with systematic recording of contextual data (Lenssen-Erz, 2004). These data together with the whole body of rock art that has been recorded in the area, enabled a classification of sites and, derived thereof, a pattern of use of the whole prehistoric lifeworld (Lenssen-Erz, 2001, pp. 254ff.; 2004).

Proceeding from recording to analysis and to interpretation the *chaîne opératoire* as laid out in Figure 6.2 was implemented: all sites in the area were recorded, surveyed, and documented (first step).

In the second step the basic needs were identified, largely by extrapolating from the vast ethnography on southern African hunter-gatherers (e.g., Marshall, 1976; Tanaka, 1980; Silberbauer, 1981; Guenther, 1986). Here mobility and settlement patterns in relation to the natural resources and carrying capacity play an important role (Lenssen-Erz, 2001, pp. 267–270).

The structure of the landscape with its most prolific and most important resources has to be recorded as a third step. Also features focusing on space such as passes, passages, gorges, or natural travel routes are phenomena that were frequently landmarked by prehistoric people (cf. Bradley, 1994; Swartz & Hurlbutt, 1994) and therefore are part of the comprehensive record.

Step four (Figure 6.2) defines the use potential, that is, which were the actual options for the prehistoric hunter-gatherers in the area for their livelihood and which use could they have made of the given assets. With this background seven types of sites were defined (see below; cf. Lenssen-Erz, 2001, p. 285ff., 2004) in a hypothetico-deductive procedure (Bernbeck, 1997, pp. 58ff.) which accommodate all patterns of behavior that are known from the ethnography of southern African hunter-gatherers.

The detailed analysis (Step 5 in Figure 6.2) looks at rock art as the main cognitive source and at the special features of the sites where also aspects of use play a role. Cognitive phenomena such as specifications of motifs, complexity of depictions at a site, as well as patterns of depicted behavior were all included in the analysis (cf. Lenssen-Erz, 2001, pp. 301ff., 2004). Additionally, the physical features such as distribution of pictures, their visibility, or the interrelations with other sites are part of this analysis.

This complex array of data together with the classification of the sites in seven classes (Step 4 in Figure 6.2) produced first of all a pattern of the frequency of sites (Step 6 in Figure 6.2): (class A) waymark/landmark site (13% of all sites are in this class), (class B) short-term living site (14%), (class C) long-term living site (2%), (class D) aggregation camp (2%), (class E) casual ritual site (34%), (class F) planned ritual site (22%), and (class G) sanctuary, hermitage (14%).

In addition to being the basis for a distributional map, the classification with its patterned features for each site class also served to establish an Idealized Elementary Site (IES). This is a hypothetical site comprising those features of size, location, space, natural infrastructure, artefacts, and rock art which are most common among all sites, being the statistical average site, as it were. Such a site is characterized as follows.

–   A small shelter comprising two large boulders, roofing, five sleeping places under a rather low ceiling.
–   The site is located on the side of a minimum 20 × 20 m level open area.
–   Two further sites can be reached over the level area within a 3–6 min walking distance.
–   A seasonally filled waterhole lies at a distance of about 300 m, yet being farther away than the nearest neighboring sites.
–   The site is located within a few minutes' walk from a natural travel route.

–   There are unambiguous signs of occupation: an amount of several dozen
     artefacts, mainly of LSA origin, is scattered in front of the site, comprising
     stone tools and some ostrich eggshell, but also some pottery shards from
     later periods are present.
–   Paintings are low on the ceiling of the shelter, but not in a hidden loca-
     tion.
–   There are some 50 figures, comprising 80% humans and 20% animals;
     among the human Figures 11% are clearly marked male, 8% are marked
     female, the remaining human figures are zero-marked; animals are mainly
     springbok, giraffe, and gemsbok.
–   The scenes mainly show humans commonly moving in one direction; there
     are only very few superimpositions but variations in the states of preserva-
     tion suggest the practice of a long-lasting painting tradition.

Obviously this average site also mirrors part of the most common patterns of use
and behavior. With this information at hand it is possible to analyze the relation
between natural assets and *Gestaltung* thus providing a picture that illustrates
the degree to which the society of a given case study made use of the options
and coped with the restrictions of their lifeworld (Step 6 in Figure 6.2; see also
Figure. 6.8).

6.2.5.1. Water

In the Brandberg/Daureb, water is an ambiguous resource that may be available
in drastically variable quantities. Although the average precipitation is around
100 mm per annum (Breunig, 1990, 2003, pp. 31ff.; Lenssen-Erz, 2001, pp.
27ff.), in a year when the rainy season fails (as occurred repeatedly in the beginning of the 1980s) vast areas of the mountain are without any accessible surface
water. In years with neither a marked shortage nor abundance the Brandberg/
Daureb would seem to have been an area of retreat in prehistoric times when
months after the rains water resources became scarce in the savannas and shrubland extending north- and eastwards from the mountain (Lenssen-Erz, 2001).
     But direct proximity to water was not an important criterion for the choice
of a place to become a rock art site. For those sites near to reliable waterholes no
cogent correlation with a certain painted motif could be established.

6.2.5.2. Abiotic Material

Abiotic material is virtually absent from rock art, with the exception that the pigments used in the art are abiotic by nature and therefore there is an obvious symbolic value to minerals such as hematite, ochre, or manganese which are the basis
for red, yellow, and black, respectively. The sources for these materials seem to
be restricted to some valley outlets on the southern fringe of the mountain. Raw
material for stone tools, on the other hand, is more commonly scattered around
the mountain, only the availability of crystal, which was used for tool making, is

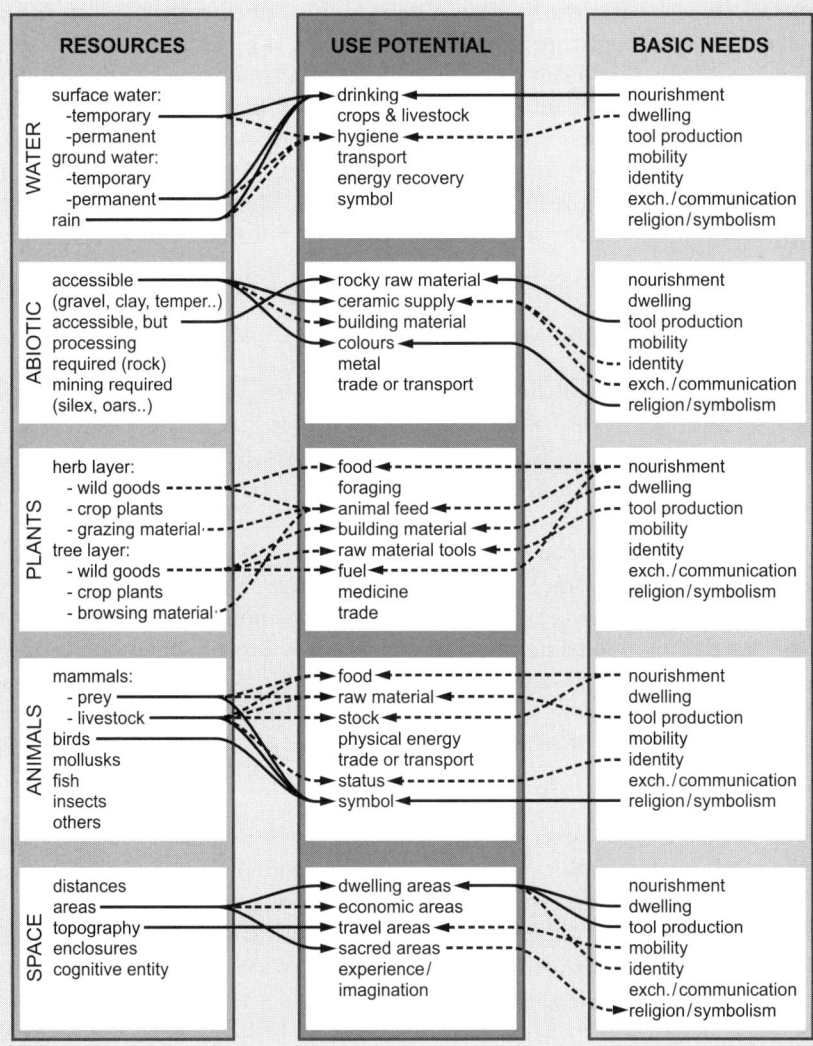

Figure 6.8. Specific matrix of relationships between use of resources and human needs of the prehistoric painters of the Brandberg/Daureb

restricted to a few outcrops in the otherwise relatively homogeneous granite of the Brandberg/Daureb.

6.2.5.3. Plants

Research among the hunter-gatherers of the Kalahari has shown that, besides water, plants are the resource which mainly guides the mobility patterns of nomadic hunter-gatherers (Tanaka, 1980, p. 79; Silberbauer, 1981, p. 202), hence their paramount importance for nourishment as opposed to the unreliable resource of meat. The plant cover of the mountain is rather rich and relatively evenly distributed in the upper regions, thus providing more or less equal conditions for the use of this resource everywhere. In its rich flora the Brandberg/Daureb provides a number of edible plants (Breunig, 1988) but plants play an absolutely negligible role in rock art (less than 0.3% of the motifs). Plant sap may, however, have been used as liquid or binder for colors but there is no evidence of this.

6.2.5.4. Animal Resources

The animal resources of the mountain ranges were of course important as food resources (Van Neer & Breunig, 1999) but they only play a subordinate role in the art (Lenssen-Erz, 2001, pp. 30f.). The bulk of animals on the mountain are small, such as hyrax (*Procavia capensis*) or rabbits (*Pronolagus redensis*). By contrast, the large game animals of the surrounding savannas were a central issue in the symbolic and religious practice of the prehistoric painters thus constituting rather a mental resource during the times people stayed in the mountain area. The focus on large hunting game in the rock art indicates that the mountain area, although it may have had the potential to sustain an autarchic life, did not constitute the entire lifeworld of the people who painted here.

6.2.5.5. Space

Space as such is only in exceptional cases an art motif, for example, in depictions of housings, be they huts or shelters (0.15% of the motifs). But through the spatial relations expressed in the locations and distribution of art the use of this resource is well manifested in the Brandberg/Daureb (Lenssen-Erz, 2001, pp. 254ff., 2003, 2004; Lenssen-Erz & Neubig, 2003) and provides the data that allow us to hypothesize about a cognitive map. The mountain's salient position, which is also supported by being a particularly advantageous biotope, made it the focus of human activities especially during the Later Stone Age (c. 4000–2000 bp).

The matrix of relationships (Figure 6.8) provides clues for the significance of rock art for the early inhabitants of the region. Moreover, it contains clues for the potential of information we can glean from the art. The pattern of relations in Figure 6.8 makes it obvious that space is the resource which can satisfy most basic needs and all resources (except vegetation) share the potential to satisfy the needs of religion and symbolism. It thus appears that vegetation may have been the resource which is nearest to our understanding of nature as the lower needs (corresponding to Maslow's D-needs, Maslow, 1981, pp. 127–130), which have little "other world" connections, are all covered here. Moreover, the use potential of space is best exploited whereas, for example, the faunal resources are less comprehensively utilized.

The combined interpretation of the site distribution map, the idealized elementary site, and the patterns of use of the resources form the basis for a generalized pattern of use for the entire mountain area (Step 6 in Figure 6.2). This again combined with the specialized local patterns of use permits a reconstruction of the cognitive map of the prehistoric people (Figure 6.9; in order not to become too complex the cognitive map outlined below does not take much account of the specialized local patterns of use). What can be gleaned from a cognitive map are most obviously patterns of behavior that to a large extent leave their traces in material remains. Further patterns, like those of perception, are more difficult to grasp, and if so, only by plausible derivations from behavior. Yet, in a framework where comprehensive aspects are taken into consideration even patterns of thought—such as conceiving a certain situation as a crisis—seem to be in reach of our interpretation.

Based on the analysis of the whole rock art corpus from a 135 km^2 segment of the mountain, encompassing all landscape zones from the foot over the slopes up to the high plateau, the mental map of the prehistoric painters was modeled as follows (cf. Lenssen-Erz, 2001, 2003). By reconstructing the use of the resources as listed in Figure 6.4, this mental map evokes a lifeworld where all aspects of human life are accommodated (Figure 6.9). Accordingly the Brandberg/Daureb and its surroundings, about 3000 years ago, were the lifeworld of hunter-gatherers who could rather safely satisfy their basic needs because it provided all resources necessary for an autarchic life. However, these resources were not abundant everywhere in the mountain and in many places could sustain small groups only for a few days. According to the locations where paintings were placed on the rock faces, production as well as consumption of rock art was a public issue enabling all members of the groups to participate in whatever process took place in connection with the art. Consequently, also the reasons for the ritual activities out of which rock art was produced afflicted the whole community. It appears that the ubiquity of rock art is an indicator of a certain critical state of mind in which people repeatedly needed the security and stabilization that can be evoked through communal rituals.

From the characteristics and distribution of 300 rock art sites in the research area one can glean the strategies that were chosen to cope with the crisis, which was probably initially an ecological crisis, such as drought, but with the risk of turning into a social crisis: people limited the size of the groups to about a dozen members and kept up a high level of mobility, changing places every few days. They increased their ritual activities beyond the ordinary frequency of crisis-free times in order to achieve a stabilizing effect through the liturgical repetition of their three major values being community, equality, and mobility. This dominant pattern is superimposed over many other patterns that are expressed in sites that do not fit into the crisis-hypothesis, such as large aggregation camps or sanctuaries.

6.2.6. The Chad Case Study

The third case study also focuses on the rock art of a salient landscape, the Ennedi Highlands in northeastern Chad with the highest peak at 1450 m a.s.l.

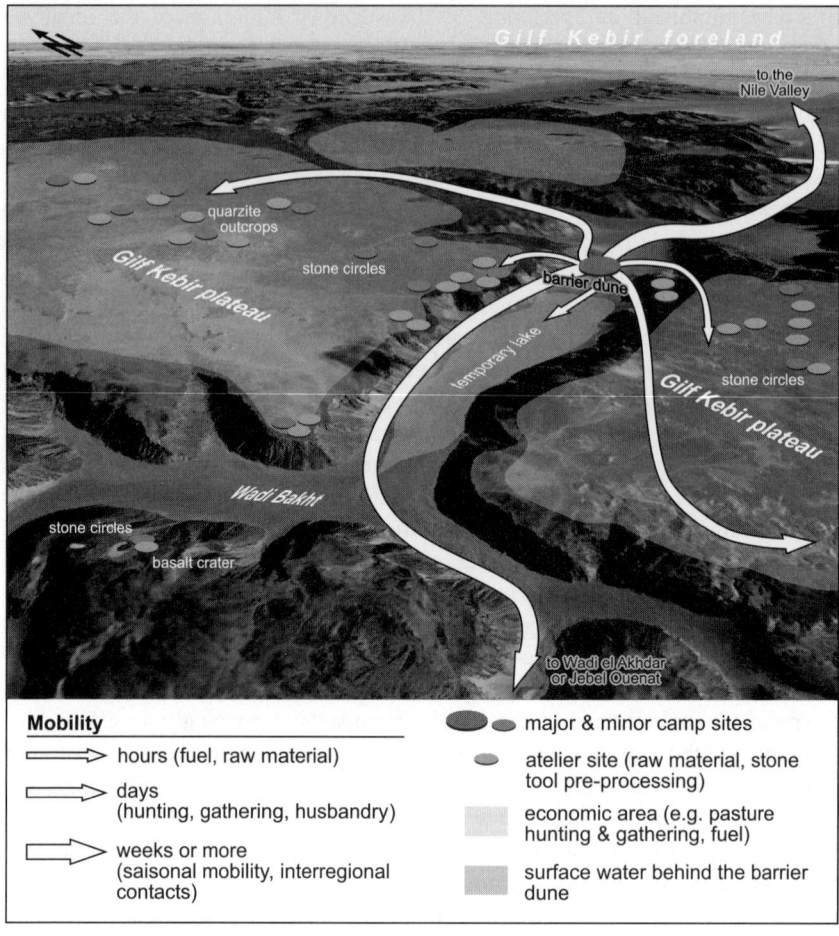

Figure 6.9. A section of the Later Stone Age Brandberg/Daureb landscape indicating resources and reconstructing the use patterns in a schematic representation. Note that the ecotope did not undergo a change comparable to that of the Sahara (Photo: courtesy of H. Mooser) (*See also Color Plates*)

The Ennedi Highlands is a retreat area on the southern margin of the Sahara that is still today settled by camel nomads (Keding et al., 2007). Archaeological research is rudimentary here (Bailloud, 1997) and it is only within recent years that research within the ACACIA project of the University of Cologne has started to unveil the archaeology of the inner highlands (Lenssen-Erz & Czerniewicz, 2005; Lenssen-Erz, 2007; Keding et al., 2007). First intense settlement activities have become archaeologically visible at the time of roughly 3000 bce with the first cattle and small stock being introduced into the region. According to Bailloud (1997) horses and camels simultaneously arrived in the area around the beginning of the ce. Rock art, painted and engraved, is a ubiquitous phenomenon of the region and spans the period from the few early hunter-gatherers through the times of all herders to the present. At a given time subsistence patterns seem to have been rather homogeneous throughout the highlands (Keding et al., 2007), yet stylistic differences and idiosyncrasies of the rock art suggest a rather strong cultural diversity. Research still remains in the early stages, particularly in view of the analysis of the archaeological finds. Yet the investigation of the 148 rock art sites recorded during the research enables interpretations of the use of the prehistoric landscape.

For the demonstration of the landscape archaeological method the focus is on the differences of painted and engraved sites. Between these two groups of sites more divergences seem to manifest than in other selections of sites inasmuch as most of the main motifs turn up in pictures of both techniques. However, because all social groups appear to have shared the same subsistence patterns at a given time, the resource/use/basic need model is unable to detect differences among the various groups of herders. But the model helps to sort out the potential and limitations of the sources at hand.

It is possible to design a network of relations between the three domains of resource/use/basic need but many of the relations are necessarily insinuations based on general knowledge of the livelihood of herders and to some extent on rock art. The latter is particularly true concerning the horse keepers of the Ennedi of whom no archaeological record has been excavated so far but which are very present in the art. In order to exemplify the method on this particular case study, Figure 6.10 shows the matrix of the resource/use/basic needs model for the Neolithic herders of the Ennedi (disregarding Iron Age cattle herders, camel herders of the last 2000 years, and horse keepers probably from the first millennium ad).

6.2.6.1. Water, Plants, Animals

Use of these main resources will have been the same irrespective of whether the people produced paintings or engravings. All shared the same landscape and have produced their art almost equally in all periods. Accordingly painters as well as engravers had to face the gradual shrinking of the water resources and as a consequence thereof a change in plant and animal resources, all of which left no choice other than a nomadic lifestyle with its consequential expansion of the lifeworld.

6.2.6.2. Abiotic Raw Material

The choice of producing either a painting or an engraving was made according to social and/or cultural values because it is the same bedrock on which both techniques can be found thus evincing that it was not a particular texture that attracted either painters or engravers. Yet there is an indirect significance of abiotic material for paintings and engravings because the artists required either pigments or hard stones to produce their artwork. The geological formation of the Ennedi enables rather easy access to both because in particular hematite can be found strewn over wide areas with some interspersed ochre. Also, white mineral

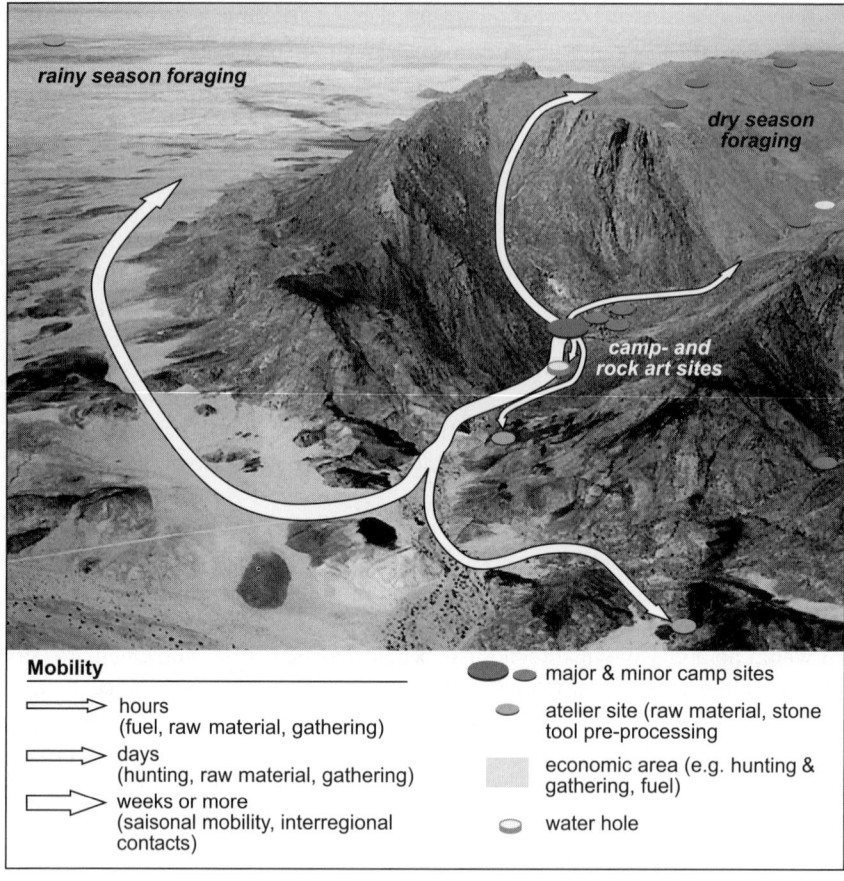

Figure 6.10. Analysis of relations between use of resources and human needs for the Chad case study (*See also Color Plates*)

pigment can be found in many places. For the engravers it would not have been problematic to collect quartz all over the highlands which is one of the handy raw materials that can produce tools hard and sharp enough to engrave the sandstone of the Ennedi. Accordingly it is hard to determine the sources for these main technical media and to draw conclusions from it because they could be attained in many places in the highlands.

6.2.6.3. Space

In the use of space the most manifest differences between painters and engravers can be detected if taking their art sites as points of reference. Whereas painting sites through their location in shelters and through the constant association with other artefacts indicate the use of the sites as dwelling places, engravings are hardly ever in shelters and are less frequently associated with other arte-facts. Accordingly, paintings are found in dwelling or economic areas whereas engravings seem to be located in sacred or in travel areas. Another difference between the two bodies of art is the significantly larger distance between the engraved sites as compared to painted sites. Engravings are scattered more widely without generally allowing visual or acoustic contact between two sites which is frequent among painted sites. Finally, there is a more cogent depend-ency on topography of the location of engraving sites than can be found among painted sites inasmuch as a large percentage of the engravings are located, for example, at the foot of elevations (42% as opposed to 24% of the paintings).

In view of the fact that there are little material differences between the groups in the Ennedi because they shared equal subsistence patterns, the exist-ence of paintings and engravings indicates that cognitive differences may have prevailed between these groups on the side of manifestations of particular identi-ties. In order to grasp these differences the *chaîne opératoire* (Figure 6.2) can be implemented and an idealized elementary site can be established, both for the painters and the engravers by taking the whole landscape setting into account and by drawing as much information as possible from the resource/use/basic needs model. Again the caveat has to be emphasized here that much of this information is based on the readings of rock art and has not yet been supported by the results of archaeological excavations.

6.2.6.4. The Mental Map of the Painters' and the Engravers' Landscapes

The IES of the painters is a small roofed shelter that lies in a group of sites. The next water source is several hundred meters away but an open area of at least 20 × 20 m is just nearby. The site is not located at a particular point of accent in the landscape and it can be easily reached. The sites face either north or south (rarely east or west), providing limited outlook. The relatively high number of paintings (55) is mainly on the ceiling of the shelter and becomes visible only when nearing the site; some pictures are even hidden in "private" locations. Sometimes the surface structure of the rock is incorporated into the paintings.

By contrast, the IES of the engravers is a vertical rockface at the foot of an elevation without a roof and other features of a dwelling site. The next water source and also the next engraving site are many hundred meters away but there is an open field just nearby. The site can be reached easily and it faces either north or south, providing limited outlook. A few of the pictures (14) are occasionally made to achieve far-ranging visibility.

Without going into details on the specific types of sites, these IESs allow us to hypothesize about the use of the landscape by the painters and the engravers.

The painters traveled the land in small groups and accepted almost every upcoming shelter as a place to stay. Even though normally they would not stay for long at a given site, they would rather quickly turn to producing pictures, the comparatively large number of pictures in part also owed to the fact that they would repeatedly visit such a place. Obviously in the same sense as people of the area today stow away their gear in such shelters, in prehistory people would also "stow" their symbolic capital (which were mainly cattle and later camels; pictures of these animals consequently being metasymbolic capital). The painted shelters were part of the everyday lifeworld and the pictures were components of everyday activities (arguably in production and consumption). These characteristics together with the relative density of sites and paintings per site show that the *Gestaltung* of the landscape and its appropriation happened in an everyday context of use of the natural infrastructure (although being ritual activity) and may hence be termed an active approach to the landscape because the painters did not have a particular configuration and symbolism of the landscape in mind for the choice of a site, but rather properties of a place with its suitability for dwelling purposes. If these were fulfilled paintings could be attributed in order to complete the appropriation of the place with ritual means.

The engravers, on the other hand, seem to have had a certain model of the landscape in mind and searched for particular topographical configurations for their engravings. These were independent of the natural infrastructure so that sites were not necessarily linked to everyday life and were more or less unconnected among themselves. Consequently sites were mainly, if not exclusively, used for art production so that the pictures as parts of ritual activities were less connected to everyday activities than the paintings. The entire landscape was symbolically loaded and the marked places remained part of a sacred landscape and did not become part of the everyday lifeworld through mundane activities. This approach to landscape may be termed passive as it is not the people who establish the symbolically charged places through their use of the natural infrastructure, but these places are predetermined by the landscape and the engravings (in part a single picture suffices) are a means to set this sacred status free by making it visible.

Landscape-archaeological reconstructions

Based on the comprehensive mapping and analysis of the archaeological data of a given region the patterns of use and behavior at a certain time become discernible. These two examples demonstrate the complexity of early herder and of hunter-gatherer livelihood in the Gilf Kebir and the Brandberg/Daureb. Based on the given resources, that is, the natural component, it is shown how they are interrelated with culture, that is, the use people made of this landscape.

The framework presented here aims at the reconciliation of two extremes of archaeological work, that is, pragmatism of fieldwork and theoretical foundations. This apart, it is also designed to express any conceivable case study in an all-inclusive format that allows the comparison of it to other case studies.

The methodology is founded on a concept of landscape that puts equal emphasis on the empirical landscape as on the culturally mediated landscape by implementing the subconcepts of infrastructure and *Gestaltung*.

The three case studies presented here focus on the resources of water, abiotic material, vegetation, fauna, and space thus following a strict systematization which is further strengthened through *chaînes opératoires* (Figures 6.2, 6.6, 6.8, and 6.10) that were implemented in both studies. Their origins lie in two rather different archaeologies, namely the cognitive archaeology of two African rock art traditions and the conventional field archaeology of the Eastern Sahara where only relatively few indicators of symbolic behavior were found. Nevertheless both projects had independently developed analogous working procedures. The resource/use/basic needs model, by contrast, grew out of intense discussions on how the everyday terms of resource, use, or needs could be forged into concepts with methodological substance. This article is an attempt to provide working procedures for intradisciplinary cooperation of different archaeologies.

The methodological approach of landscape archaeology forwarded here cannot completely annul differences in the corpora of data that are collected by cognitive and field archaeology. However, it opens a road to analyze data, from whatever origin, in a basic scheme that is concerned with a comprehensive view of a lifeworld wherein, by means of the checklist character of the scheme, all aspects receive the same attention and are being assessed within the same frame of understanding.

In comparing the case studies from Gilf Kebir and the Ennedi Highlands with the Brandberg/Daureb it becomes evident that through domestic animals there seem to be more possibilities to exploit the use potential of the resources. This is manifest in the more numerous arrows of relation (between the two left columns) for the phase Gilf C in Figure 6.6b and the herders of the Ennedi. There is little surprise in the fact that through the innovation of domestication more complexity is added to a society. However, as can be seen from the comparison with the predomesticated animals phase of Gilf B and particularly with the pure hunter-gatherers of the Brandberg/Daureb (Figure 6.8), this growth of use potential does not go together with a likewise significant growth in relations of use potential and human needs; in other words, the novel exploitation does not open many more options for the satisfaction of needs.

Even the evidence from Brandberg/Daureb rock art—which is only peripherally supported by data from excavations (e.g., Breunig, 1989, 2003) and which is therefore scarce in terms of material finds—shows a rather complete coverage of needs through the different resources for this society of comparatively little complexity. This may hypothetically be seen to indicate that, assuming general conditions do not change, the introduction of domesticated animals is not a factor that makes life easier because it might enhance the satisfaction of human needs. Rather the introduction of livestock may be understood as a diversification of use options when general conditions deteriorate, notwithstanding the potential to advertise status and wealth through livestock.

Admittedly, the present form of this approach is not yet a complete method. It still suffers from shortcomings such as that it is unable to accommodate the resource of time adequately and therefore lacks a tool to systematically grasp the phenomenon of change in its dynamics. For the time being the scheme requires us to study two sequential phases (such as the Gilf B and C phases) as two more or less static events. The same counts for the various keepers of domestic animals in the Ennedi Highlands.

Further development of the method will have to make it more flexible and also to mitigate its deterministic character towards the inclusion of a module that accommodates human agency even better without disposing of its universal applicability. The acceptance of this newly proposed method will depend on its capability to serve as a reconciling procedure that provides useful aspects for all views of a landscape, be they empirical, pragmatic, emic, or postmodern.

REFERENCES

Anschuetz, K.F., Wilshusen, R.H. & Scheick, C.L. (2001). An archaeology of landscapes: Perspectives and directions. *Journal of Archaeological Research*, 9(2), 157–211.

Bailloud, G. (1997). *Art rupestre en Ennedi*. Saint-Maur: Éditions Sépia.

Beck, D.E. & Cowan, C.C. (1996). *Spiral Dynamics: Mastering Values, Leadership and Chance*. Cambridge, MA: Blackwell.

Bender, B. (2002). Time and landscape. *Current Anthropology*, 43, 103–112.

Bernbeck, R. (1997). *Theorien in der Archäologie*. Tübingen, Basel: A. Francke Verlag.

Boelicke, U. (1982). Gruben und Häuser: Untersuchungen zur Struktur bandkeramischer Hofplätze. In E. Bakels (Ed.), *Siedlungen der Kultur mit Linearkeramik in Europa* (pp. 17–28). Internationales Kolloquium Nové Vozokany.

Bradley, R. (1994). Symbols and signposts – Understanding the prehistoric petroglyphs of the British Isles. In C. Renfrew & Ezra B.W. Zubrow (Eds.), *The Ancient Mind – Elements of Cognitive Archaeology* (pp. 95–106). Cambridge: Cambridge University Press.

Bradley, R., Criado Boado, F. & Fábregas Valcarce, R. (1994). Rock art research as landscape archaeology: A pilot study in Galicia, north-west Spain. *World Archaeology*, 25(3), 374–390.

Breunig, P. (1988). Botanisch-archäologische Beobachtungen in einem afrikanischen Hochgebirge. Aspekte zur prähistorischen Besiedlung eines ariden Gunstraumes. *Archäologische Informationen* (11)1, 53–73.

Breunig, P. (1989). Archäologische Untersuchungenzur Besiedlungsgeschichte des Brandbergs – Archaeological invetsigations into the settlement history of the Brandberg. In: H. Pager. *The Rock Paintings of the Upper Brandberg, Part I – Amis Gorge* (pp. 17–45). Köln: Heinrich-Barth-Institut.

Breunig, P. (1990). Temperaturen und Niederschläge im Hohen Brandberg. *Journal Namibia Wissenschaftliche Gesellschaft*, 42, 7–24.

Breunig, P. (2003). *Der Brandberg. Untersuchungen zur Besiedlungsgeschichte eines Hochgebirges in Namibia*. Köln: Heinrich-Barth-Institut.

Dowson, T.A. (2007). Debating shamanism in southern African rock art: Time to move on... *South African Archaeological Bulletin*, 62, 185, 46–61.

Dünne, J. & Günzel, S. (Eds.) (2006). *Raumtheorie – Grundlagen aus Philosophie und Kulturwissenschaft*. Frankfurt: Suhrkamp.

Frank, T. & Wendt, K.P. (2003). Up & down. Scaling archaeological data provided by GIS based procedures. Paper presented at the congress *Enter the Past*, Vienna, 8–12 April 2003.

Gautier, A. (1982). Neolithic faunal remains in the Gilf Kebir and the Abu Hussein Dunefield, Western Desert, Egypt. In F. El-Baz & M.A. Maxwell (Eds.), *Desert Landforms in Southwest-Egypt: A Basis for Comparison with Mars* (pp. 335–339). Washington, DC: NASA.

Gehlen, B., Kindermann, K., Linstädter, J. & Riemer, H. (2002). The Holocene occupation of the Eastern Sahara: Regional chronologies and supra-regional developments in four areas in the absolute desert. In Heinrich Barth Institut (Ed.) *Tides of the Desert – Gezeiten der Wüste* (pp. 85–116). Köln: Heinrich-Barth-Institut.

Guenther, M. (1986). *The Nharo Bushmen of Botswana – Tradition and Change.* Hamburg: Helmut Buske.

Hallier, M. (1996). Zwei keramische Fundplätze am Übergang vom 5. zum 4. Jahrtausend vor Christi Geburt in Südwest-Ägypten: Wadi Bakht 82/15 und 82/24. (Magisterarbeit Köln 1996).

Heyd, T. (2002). Natural heritage: Culture in nature. In German Commission for UNESCO/Brandenburg University of Technology at Cottbus (Eds.), *Natur und Kultur – Ambivalente Dimensionen unseres Erbes – Perspektivenwechsel/Nature and culture – Ambivalent dimensions of our heritage – Change of perspective.* Deutsche UNESCO-Kommission (pp. 85–95). Cottbus: UNESCO.

Hirsch, E. (1995). Landscape: Between place and space. In E. Hirsch & M. O'Hanlon (Eds.), *The Anthropology of Landscape. Perspectives on Place and Space* (pp. 1–30). Oxford: Clarendon Press.

Hodder, I., Shanks, M., Alexandri, A., Buchli, V., Carman, J., Last, J. & Lucas, G. (Eds.) (1995). *Interpreting Archaeology.* London/New York: Routledge.

Hodder, I.R. & Orton, C. (1976). *Spatial Analysis in Archaeology.* London/New York: Cambridge University Press.

Hollmann, J. (2003). Indigenous knowledge and paintings of human-animal combinations: Ostriches, swifts and religion in Bushman rock-art, Western Cape Province, Unpubl. Masters Thesis, Johannesburg. .

Ingold, T. (1993). The temporality of the landscape. *World Archaeology*, 25(2), 152–174.

Jahnkuhn, H. (1977). *Einführung in die Siedlungsarchäologie.* Berlin/New York: Walter de Gruyter.

Kalis, A.J. (1988). Zur Umwelt des frühneolithischen Menschen: Ein Beitrag der Pollenanalyse. Forschungen und Berichte zur Vor- und Frühgeschichte Baden-Württembergs 31: 125–137.

Kalis, A.J. & Zimmermann, A. (1988). An integrative model for the use of different landscapes in Linearbandkeramik times. In J.L. Bintliff, D.A. Davidson & E.G. Grant (Eds.), *Conceptual Issues in Environmental Archaeology* (pp. 145–152). Edinburgh: Edinburgh University Press.

Keding, B., Lenssen-Erz, T. & Pastoors, A. (2007). Pictures and pots from pastoralists – Investigations into the prehistory of the Ennedi Highlands, NE Tchad. *Sahara*, 18, 23–45.

Kröpelin, S. (1989). Untersuchungen zu Sedimentationsmilieu von Playas im Gilf Kebir (Südwest-Ägypten). In R. Kuper (Ed.), *Forschungen zur Umweltgeschichte der Ostsahara* (pp. 183–305). Köln: Heinrich-Barth-Institut.

Kuper, R. (1995). Prehistoric research in the Southern Libyan desert. A brief account and some conclusions of the B.O.S. project. *Cahier de Recherches de l'institut de Papyrologie et d'Egyptologie de Lille* 17, 123–140.

Layton, R. & Ucko, P. (1999). Introduction: Gazing on the landscape and encountering the environment. In P. Ucko & R. Layton (Eds.), *The Archaeology and Anthropology of Landscape* (pp. 1–20). London/New York: Routledge.

Lenssen-Erz, T. (1997). Metaphors of intactness of environment in Namibian rock paintings. In P. Faulstich (Ed.), *Rock Art as Visual Ecology* (pp. 43–54). Tucson, AZ: American Rock Art Research Association.

Lenssen-Erz, T. (2001). *Gemeinschaft – Gleichheit – Mobilität. Felsbilder im Brandberg, Namibia, und ihre Bedeutung. Grundlagen einer textuellen Felsbildarchäologie.* Köln: Heinrich-Barth-Institut.

Lenssen-Erz, T. (2003). Mental mapping of arid landscapes in Southern Africa: A cognitive ethnographic-archaeological approach. Paper presented to the session 'Glimpses of a Landscape's Past,' *Fifth World Archaeological Congress*, 21–26 June 2003. Washington, DC.

Lenssen-Erz, T. (2004). The landscape setting of rock-painting sites in the Brandberg, Namibia: Infrastructure, Gestaltung, use and meaning. In C. Chippindale & G. Nash (Eds.): *The Figured Landscapes of Rock Art* (pp. 131–150). Cambridge: Cambridge University Press.

Lenssen-Erz, T. (2007). Rock art in African Highlands: Ennedi Highlands, Chad – Artists and herders in a lifeworld on the Margins. In O. Bubenzer, A. Bolten & F. Darius (Eds.), *Atlas of Environmental Change and Human Adaptation in Arid Africa* (pp. 48–51). Africa Praehistorica 21. Köln: Heinrich-Barth-Institut.

Lenssen-Erz, T. & Czerniewicz, M. von (2005). Résultats préliminaires des recherches archéologiques dans l'Ennedi. *Revue Scientifique du Tchad*, 7(2), 5–18.

Lenssen-Erz, T. & Neubig, J. (2003). Augenblick und Ewigkeit, Raum und Diskurs. Artefakte der prähistorischen Kunst Namibias und die Arteplage in Murten, Expo.02 in der Schweiz. In A. Pastoors & G.-C. Weniger (Eds.), *Höhlenkunst und Raum: Archäologische und architektonische Perspektiven - Cave art and space:* Archaeological and architectural perspectives. 74–90 Mettman: Neanderthal Museum..

Leser, H. (1997). *Landschaftsökologie*, 4th edn. Stuttgart: UTB für Wissenschaft.

Linstädter, J. (1999). Leben auf der Düne. Der mittelneolithische Fundplatz Wadi Bakht 82/21 im Gilf Kebir (Südwest-Ägypten). *Archäologische Informationen*, 22(1), 115–124.

Linstädter, J. (2003). Neolithic land-use systems in the Gilf Kebir, South-West Egypt. In Z. Hawass & L. Pinch Brock (Eds.), *Egyptology at the Dawn of the 21st Century. Proceedings of the Eighth International Congress of Egyptologists in Cairo 2000* (pp. 381–389). American University in Cairo Press.

Linstädter, J. (Ed.) 2005. Wadi Bakht – Landschaftsarchäologie einer Siedlungskammer im Gilf Kebir (SW-Ägypten). *Africa Praehistorica*, 18, 372–387.

Linstädter, J. (2007). Rocky islands within oceans of sand – Archaeology of the Jebel Ouenat/Gilf Kebir region, Eastern Sahara. In O. Bubenzer, A. Bolten & F. Darius (Eds.), Atlas of environmental and cultural change in arid Africa. *Africa Praehistorica*, 21, 34–37.

Linstädter, J. & Kröpelin, S. (2004). Wadi Bakht revisited: New data on Holocene climate and prehistoric occupation in the Gilf Kebir plateau (Central Eastern Sahara, SW Egypt). *Geoarchaeology* 19: 753–778.

Linstädter, J., Richter, J. & Linstädter, A. (2002). Optimale Datenerhebung mit minimalen Aufwand. *Archaeologische Informationen*, 25(1&2), 99–106.

Littlejohn, J. (1963). Temne space. *Anthropological Quarterly*, 36(1), 1–17.

Luig, U. & von Oppen, A. (1997). Landscape in Africa: Process and vision. *Paideuma*, 42, 7–45.

Lüning, J. (1982). Siedlung und Siedlungslandschaft in bandkeramischer und Rössener Zeit. *Offa*, 39, 9–33.

Lüning, J. (1997). Landschaftsarchäologie in Deutschland – Ein Programm. *Archäologisches Nachrich tenblatt*, 3, 277–285.

Marshall, L. (1976). *The !Kung of Nyae Nyae*. Cambridge, MA: Harvard University Press.

Maslow, A. (1970). *Motivation and Personality*, 2nd edn. New York: Harper.

Maslow, A. (1981). *Motivation und Persönlichkeit*. Reinbek: rororo.

Millennium Ecosystem Assessment (2005). *Ecosystem and Human Well-Being: Current State and Trends*. Washington, DC: Island Press.

Neumann, K. (1989). Zur Vegetationsgeschichte der Ostsahara im Holozän. Holzkohlen aus prähistorischen Fundstellen. In R. Kuper (Ed.), *Forschungen zur Umweltgeschichte der Ostsahara* (pp. 13–181). Köln: Heinrich-Barth-Institut.

Ouzman, S. (1998a). Mindscape. In P. Bouissac (Ed.), *Encyclopedia of Semiotics* (pp. 419–421). New York: Oxford University Press.

Ouzman, S. (1998b). Towards a mindscape of landscape: Rock-art as expression of world-understanding. In C. Chippindale & P.S.C. Taçon (Eds.), *The Archaeology of Rock-Art*. Cambridge: Cambridge University Press.

Ouzman, S. (2002). Encountering an encultured nature – Some edifying examples from indigenous Southern Africa. In German Commission for UNESCO/Brandenburg University of Technology at Cottbus (Eds.), *Natur und Kultur – Ambivalente Dimensionen unseres Erbes – Perspektivenwechsel/Nature and Culture – Ambivalent Dimensions of our Heritage – Change of Perspective. Deutsche UNESCO-Kommission* (pp. 99–117). Cottbus: UNESCO.

Pager, H. (1989–2006). *The Rock Paintings of the Upper Brandberg, Part I – VI*. Köln: Heinrich-Barth-Institut.

Peters, J. (1987). The faunal remains collected by the Bagnold-Mond Expedition in the Gilf Kebir and Gebel Uweinat in 1938. *Archéologie du Nil moyen*, 2, 251–264.

Sauer, C. (1963). The morphology of landscape. In J. Leithly (Ed.), *Land and Life: A Selection of Writings of Carl Sauer*. Berkeley, CA: University of California Press.

Schade, C.C.J. (2000). *Landschaftsarchäologie – eine inhaltliche Begriffsbestimmung. Studien zur Siedlungsarchäologie II.* Bonn: Rudolf Habelt.

Schama, S. (1995). *Landscape and Memory.* London: Fontana.

Schön, W. (1994). The late Neolithic of Wadi el Akhdar (Gilf Kebir) and the eastern Sahara. *Archéologie du Nil moyen,* 6, 131–175.

Schön, W. (1996). *Ausgrabungen im Wadi el Akhdar, Gilf Kebir (SW-Ägypten).* Köln: Heinrich-Barth-Institut.

Schütz, A. & Luckmann, T. (1975). *Strukturen der Lebenswelt. Luchterhand, Neuwied. (Engl: The Structures of the Life-World.)* London: Heinemann.

Segal, E.M. (1994). Archaeology and cognitive science. In C. Renfrew & E.B.W. Zubrow (Eds.), *The Ancient Mind – Elements of Cognitive Archaeology* (pp. 22–28). Cambridge: Cambridge University Press.

Shaw, W.B.K. (1936). An expedition to the Southern Libyan Desert. *The Geographical Journal* 87, 193–221.

Silberbauer, G.B. (1981). *Hunter and Habitat in the Central Kalahari Desert.* Cambridge: Cambridge University Press.

Stehli, P. (1973). Keramik. In J.-P. Farruggia, R. Kuper, J. Lüning & P. Stehli Der (Eds.), *Bandkeramische Siedlungsplatz Langweiler* (pp. 57–105) (*Rheinische Ausgrabungen, Band 13*). Bonn: Rheinland Verlag Köln

Swartz, B.K. Jr. & Hurlbutt, T.S. (1994). Space, place and territory in rock art interpretation. An integration of concepts of space and their application to an unusual petroglyph locality in the Great Basin, *USA Rock Art Research,* 11, 1, 13–22.

Tanaka, J. (1980). *The San Hunter-Gatherers of the Kalahari.* Tokyo: University of Tokyo Press.

Tilley, C. (1994). *A Phenomenology of Landscape: Places, Paths and Monuments.* Oxford: Berg.

Van Neer, W. & Breunig, P. (1999). Contribution to the archaeozoology of the Brandberg, Namibia. *Cimbebasia,* 15, 127–140.

Van Neer, W. & Uerpmann, H.-P. (1989). Palaeoecological significance of the Holocene faunal remains of the B.O.S. missions. In R. Kuper (Ed.), *Forschungen zur Umweltgeschichte der Ostsahara* (pp. 307–341). Köln: Heinrich-Barth-Institut.

Vita-Finzi, C. (1978). *Archaeological Sites in Their Setting.* London: Thames and Hudson.

Webster's Third New International Dictionary. (1993). Cologne: Könemann.

www.bristol.ac.uk, 2007. MA Landscape Archaeology. URL: www.bristol.ac.uk/archanth/postgrad/landscape.html Last update: 2007-05-10. Access date: 2007-09-02.

www.exeter.ac.uk, 2007. MA in Landscape Archaeology. URL: www.exeter.ac.uk/postgraduate/degrees/archaeology/landscapema.shtml Last update: 2007-04-25. Access date: 2007-09-02.

www.oxford.ac.uk, 2007. MSc in Applied Landscape Archaeology. URL: www.awardbearing.conted.ox.ac.uk/archaeology/mscala.php Last update: 2007. Access date: 2007-09-02.

www.sheffield.ac.uk, 2007. MA Landscape archaeology. URL: www.shef.ac.uk/archaeology/prospectivepg/masters/landscape.html Last update: 2007. Access date: 2007-09-02.

Yar, M. (2001). Hannah Arendt 1906–1975. In J. Fieser & B. Dowden (Eds.), *The Internet Encyclopedia of Philosophy.* URL: www.utm.edu/research/iep/a/arendt.htm. Access date: 2003-05-30.

Zimmermann, A. (1995). *Austauschsysteme von Silexartefakten in der Bandkeramik Mitteleuropas.* Universitätsforschungen zur prähistorischen Archäologie 26. Bonn: Habelt.

Zimmermann, A. (2001). Ist die politische Frorderung nach der 'beispielhaften Ausgrabung' aus fachlicher Sicht immer unerfüllbar? Zum Aspekt der Repräsentativität von Ausgrabungsergebnissen. *Archäologisches Nachichtenblatt,* 6, 131–137.

Zimmermann, A., Richter, J., Frank, T. & Wendt, K.P. (2004) Landschaftsarchäologie II, Überlegungen zu Prinzipien einer Landschaftsarchäologie. *Bericht der RGK* 85, 37–95.

Zubrow, E.B.W. (1994a). Cognitive archaeology reconsidered. In C. Renfrew & E.B.W. Zubrow (Eds.), *The Ancient Mind – Elements of Cognitive Archaeology* (pp. 187–190). Cambridge: Cambridge University Press.

Zubrow, E.B.W. (1994b). Knowledge representation and archaeology: A cognitive example using GIS. In C. Renfrew & E.B.W. Zubrow (Eds.), *The Ancient Mind – Elements of Cognitive Archaeology* (pp. 107–118). Cambridge University Press: Cambridge.

Part II

State, Power, and Control in Africa's Arid Landscapes: Perspectives from the Historical Sciences

Chapter 7

The 'Landscapes' of Ancient Egypt: Intellectual Reactions to the Environment of the Lower Nile Valley

MICHAEL HERB AND PHILIPPE DERCHAIN

La nature est le vis-à-vis que la culture s'est inventé pour se donner une contenance. Comme les dieux – ou Dieu – pour échapper au vertige de la toute-puissance de son imagination.

Philippe Derchain

The specific landscape of the valley of the river Nile played an important role in the development of the civilisation of ancient Egypt. At first sight describing this habitat seems to be a simple matter. Using topographical criteria we can say that the inner structure of the Lower Nile Valley is clear. Also there is an abundance of ancient sources we may analyse to obtain information dealing with the landscape occupied by the Ancient Egyptians over more than three millennia from around 3200 BC to AD 400. However, when considered in more depth the situation turns out to be more complicated. There is no proof that an Egyptian in the time of the pyramid-builders of Giza (2579–2486 BC) used the same epistemic structures or applied the same terminologies as his later historical companion of the Ptolemaic period (332–330 BC). Quite to the contrary it would be a rather unexpected result if concepts, technologies, and economic practices related to the environment remained unchanged over a period of 3000 years. Between 3200 BC and AD 400 Egyptian society and culture changed, as did its habitat both in a physical as well as in a conceptual sense. There was a multiplicity of intellectual reactions to the environment they lived in, and so it is more appropriate to speak about the 'landscapes' of Ancient Egypt. In contrast to other studies in this volume this

M. Bollig, O. Bubenzer (eds.), *African Landscapes*,
doi: 10.1007/978-0-387-78682-7_7, © Springer Science+Business Media, LLC 2009

chapter deals with a wealth of sources stretching over millennia of intense produc-tion of physical environmental features, meaning(s), and symbols.

It is the aim of this contribution to offer a first impression of the manifold dimensions of the theme. First we describe the country from a geographi-cal point of view trying to produce an objective point of reference. Then we introduce some very common Egyptian words which have been used over a long period of time designating the land in order to obtain ideas concer-ning a few basic landscape conceptualisations. In a third step we have a look at an epigraphic programme of a tomb of the Old Kingdom and analyse the way in which it mirrors the environment. Finally we speak about a religious theory of the structure of the world which in the opinion of the Ancient Egyptians resembles the environment of the river oasis of the Lower Nile Valley, that is, the landscape of Ancient Egypt.

7.1. THE GEOGRAPHICAL POINT OF VIEW

The land of Ancient Egypt differs profoundly from other regions of northern Africa. A long time before the rise of the ancient culture around 3200 BC it became part of the huge arid zone impressing its constraints on northern Africa (see the contribution by Riemer, this volume). In a narrow sense Ancient Egypt can be defined as the last part, that is, the last 1000 km, of the river Nile streaming over more than 6500 km from its headwaters in Central Africa and the Ethiopian Highlands to its delta at the Mediterranean Sea. This country is not identical with the present state of the Arab Republic of Egypt which includes not only the river valley but also the Sinai peninsula, the desert regions east and west of the Nile, and a region south of the First Cataract up to 22°N (Baines & Málek, 1980, pp. 12ff.; Ibrahim, 1996; Bard, 1999, pp. 1ff.). Ancient Egypt was limited to the oasis of the Lower Nile Valley. The profound contrast between the riverine landscape abounding with water and the adjacent hyperarid environments charac-terised this basic situation. The annual inundation of the Nile was the base for the rise of the Egyptian civilisation. Rain did not play a role in ancient Egypt culture or economy (Said, 1993, 82ff.; Ibrahim, 1996, pp. 27ff., 50). The long stretch of the Nile from its sources to the Mediterranean Sea shows different riverine land-scapes (Said, 1993, pp. 28ff.). The width of the Nile valley from Khartoum and the 6th cataract (16° 20′N) to the Gebel el-Silsila (24° 39′N) is on average less than 6 km. Between the Gebel el-Silsila and the end of the characteristic Nile-turn at Nag Hammadi (26° 00′N) the width is approximately 6 km. Then it extends to 20 km up to Beni Suëf (29° 00′N) near the Fayum. In the Delta (30° 00′N–31° 20′N) the situation differs extremely. Here the extension of the areas flooded by the Nile reached more than 100 km in width.

Starting again at the First Cataract near Aswan (24° 00′N) then going north passing Thebes (25° 44′N), Asyut (27° 11′N), and Memphis (29° 51′N), the number of watercourses, channels, and rivers forming what we call 'Nile' increased and the system branched out more and more. From a historical point of view it is difficult to describe the development of this riverine environment and its changes. The numerous watercourses and even the main river itself often shifted their courses. The high-dams having been built since the nineteenth century make it even more difficult to map the exact historical situation. Finally in the Delta north of Cairo there have been two, five, and seven different rivers over time transporting the waters to the northern marshes and the Mediterranean Sea (Bietak, 1975, pp. 67ff.; Said, 1993, pp. 57ff.; Ibrahim, 1996, pp. 52ff.).

This kind of 'river multiplication' and the expansion of the riverine system at the same time led to a significant reduction of velocity of waterflow. This is one of the reasons why the development of the riverine environment from Khartoum (16° 10'N) to the coast of the Mediterranean Sea (31° 20'N) in the north influenced both the quantities of arable land and the qualities of their soils. In Nubia and in the southern parts of Egypt up to the Gebel el-Silsila (24° 39'N) there were only relatively small areas within the Nile Valley which could be used for settlements, agriculture, cattle-keeping, or other cultural activities. Going down to Thebes (25° 44'N) in Upper Egypt then going northwards via Dendera (26° 08'N), Abydos (26° 11'N), Asyut (27° 11'N), and Beni Suëf (29° 27'N) and at last reaching the Delta (30° 00'N–31° 20'N) the areas being flooded seasonally in ancient times increased more and more. So the speed of the water was reduced and consequently sedimentation increased. This is the main reason why the qualities of the soils used by the Egyptians were much better in the northern than in the southern regions. This may be one reason why the culture of Ancient Egypt was developed in the northernmost part of the long-stretched riverine environment of the Nile.

Principally speaking there were five factors constituting the habitat of Ancient Egypt (Abu al-ʿIzz, 1971; Herb, 2001 pp. 379ff.): (i) the river or the system of watercourses; (ii) the natural high-dams and sandy islands (Arabic *gazîras*); (iii) the alluvial grounds; (iv) the marshes and swamps; and – not really belonging to Ancient Egypt – (v) the desert regions beyond the valley rendering strict borders to the culture (Figure 7.1). The main feature of the environment was the Nile and its flooding. The land's water supply was completely based upon the river and its inundation mechanism. In antiquity the annual flood began in July. It marked the beginning of the Egyptian calendar year which comprised the seasons *ḥt* 'inundation', *Prt* 'emergence (of the alluvial grounds from the flood)' and *šmw* 'deficiency (of water)'. Regularly the inundation reached its peak in September. The 'low-phase' was between the middle of May and June (Stricker, 1956; Bonneau, 1993; Said, 1993, pp. 96ff., 127ff.; Seidlmayer, 2001).

The Nile had three basic functions: (i) the river 'presented' the water and made the valley so extraordinarily attractive, (ii) the mechanism of flooding determined agriculture, the most important sector of economy of Ancient Egypt, and (iii) the branches and channels of the Nile were the streets for inner-Egyptian merchandise traffic. The Nile is best understood as a natural net of waterways. Natural dams were frequently located directly at the banks of watercourses. Dams and *gazîras* were on higher grounds. Normally they were not reached by the inundations of the Nile. Being nonflooded, these areas were dry and safe places

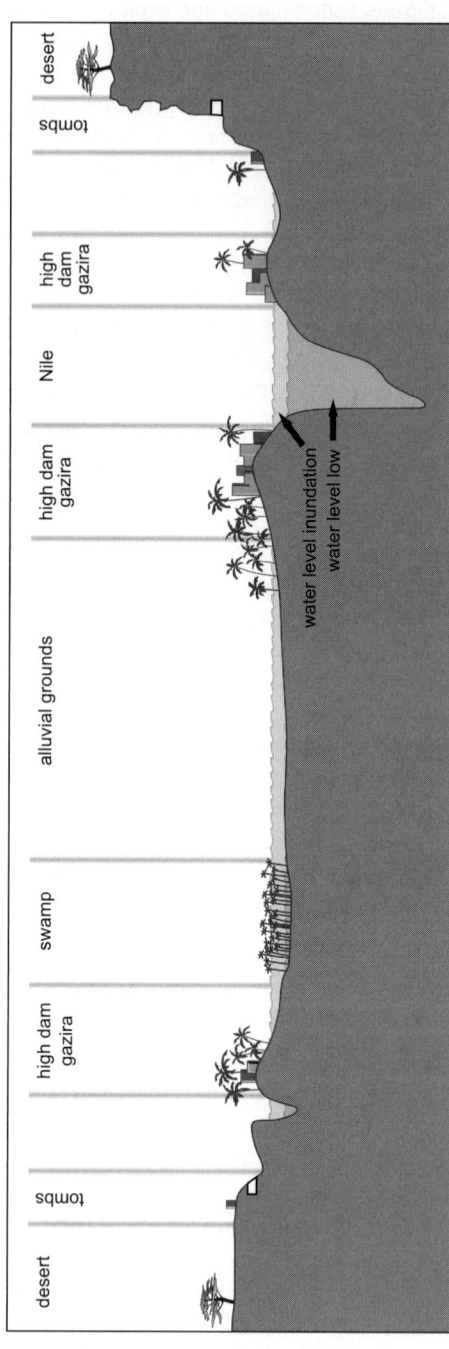

Figure 7.1. Scheme of the landscape of the Nile Valley giving the topographical factors constituting the habitat of Ancient Egypt (Drawing by the author)

and most attractive for settlements (Bietak, 1975, pp. 27ff., 49ff; Butzer, 1976, pp. 22ff.; Herb, 2001, pp. 381ff.). Large areas of the Egyptian Nile Valley consisted of alluvial grounds. These were flooded seasonally for three to four months (Said, 1993, pp. 61ff., 128ff., 188; Herb, 2001, pp. 382ff.). These areas were used mainly for agriculture which – as the inundation itself – determined the economic system of Ancient Egypt from the beginning of its history round about 3200 BC until its end in AD 400. Sowing took place after the decline of the inundation when the brackish alluvial grounds 'emerged' (*pr*) from the water. The harvest followed five months later before the beginning of the dry period in May. The seasonally flooded grounds had two characteristics which were important for the economic development since 3200 BC: Both the settlements and the alluvia were nearby. The ways between the fields and the villages with their threshing floors, storages, and so on, were short and thus transport was easy to handle. The alluvial grounds offered a gigantic agricultural potential, and the development in pharaonic times was marked by a growing efficiency to exploit this potential.

At the beginning of the Old Kingdom 2700 BC there were four extensive marsh and swamp regions in Egypt (Herb, 2001, pp. 383ff.). Here in the areas at the edges of the river-oasis groundwater came to the surface, the depth of water was low, and the velocity of waterflow was very low, so that the temperature of water became relatively high, important conditions for the growth of reed, papyrus, and lotos. These were used as important economic resources, especially in the Old and the Middle Kingdoms (2600–1800 BC). The region flanking the east side of the Birqet Qarun (29° 27′N), the lake in the centre of the Fayum depression, was the most famous swampy region of Ancient Egypt (Said 1993: 78ff.). Two other ancient swamps in Upper Egypt extended along the western border of the valley. Large inundated areas can be assumed between el-Qusija (27° 27′N) and the southern border of the Fayum near Beni Suëf (29° 00′N). Another marsh can be reconstructed between Akhmim (26° 34′N) and Asyut (27° 11′N). Finally the huge marshes (31° 30′N) in northern Lower Egypt separated the whole country from the Mediterranean Sea. Especially in the Old Kingdom (2700–2200 BC) these *mḥjjt* 'northern swamps' were centres of fishing, bird-catching, cattle-keeping, and different kinds of using plant materials, especially papyrus.

7.2. TERMINOLOGICAL ASPECTS

In the time of the Middle Kingdom between 2000 and 1800 BC a poet named Neferti looks back on the history of his land. He complains about the adverse situation in a time we call First Intermediate Period (2150–2025 BC). Neferti does not use the name 'Egypt'. The term was created much later by the Greeks. It was derived from the Egyptian *ḥwt-k3-ptḥ* 'house of the Ka of (the god) Ptah', one of the names of Memphis. Neferti describes what had happened in *t3* 'the land' written with the hieroglyphic sign ⊂▭⊃. The reader immediately understands this is the designation of *his* land, the native country of the poet extending from the first cataract-region in the South to the Northern regions of the Delta (Lichtheim,

1973, pp. 139ff.). Characterising the homeland of a person as 'the land' in the sense of 'my land' (i.e., landscape of home), is not a specific Egyptian way of thinking. It probably exists in every culture. So at first sight there is no direct help for our problem in the words of Neferti but they lead us to ☰☰☰ *t3wj* 'the two lands', a *terminus technicus* widespread in Ancient Egypt referring to the original divide of Ancient Egypt.

Since the beginning of their history in 3200 BC the Egyptians divided their native country into two parts drawing the line near Memphis (29° 51′N), the later capital of the Old Kingdom. Some 900 years later during the reigns of kings Unas and Teti (2342–2312 BC) we get for the first time the information how the Egyptians named these 'two lands': 𓏃𓃀𓈖𓈘 *t3-mḥ*, shortly *mḥw* 'land of the northern-plant' (Erman & Grapow, 1957b, pp. 123.12–14, 124; 1957e, pp. 224.10–13), that is, the Delta, and 𓇗𓈘 *šmʿw*, literally perhaps *t3-šmʿw* '(land of the flowering) southern plant' (Erman & Grapow, 1957d, pp. 472–476; 1957e, pp. 227.4–14), that is, Upper Egypt. Obviously the Egyptians felt specific plants to be significant elements of the landscape of both Lower and Upper Egypt. In Lower Egypt there are extensive papyrus swamps and marshes. A gigantic marshy area extends in the northwestern part of the Delta (Herb, 2001, pp. 384ff.). So it is understandable why papyrus 𓇉 was seen as a kind of 'landscape-marker', the plant being the symbol of the Delta. But the word *mḥjjt* 'papyrus' (Erman & Grapow, 1957b, p. 124.8) is a *terminus technicus* only found in medical texts and definitely was not the root of -*mḥw*– in *t3-mḥw*.

The situation with 𓇗𓈘 *šmʿw* 'Upper Egypt' is more complicated. The exact species 𓇗 'the southern-plant' refers to is unknown, although it was often mentioned in texts and pictured in decorations (Smith, 1981, pp. 93ff.). The most plausible theory connects the designations *t3-mḥw* 'Lower Egypt' and *šmʿw* 'Upper Egypt' with the old roots **mḥ*– 'norther' and **šmʿ*– 'southern'. This construction reflects a strong bipartite conception explaining the homeland of the Ancient Egyptians in relation to the characteristic south-north extension of the river-oasis.

A good example of this kind of basic thinking is found on a stela from the second year of King Thutmosis' I reign (1503 BC). He gave a short but significant description of his power (Sethe, 1914, pp. 85.13–14). He determined the complete extension of his territory mentioning only the southern and northern borders: *t3š=frsjj r-ḫntj t3 pn mḥtj = fr mw qdw ḫddjj m-ḫntj* 'its (i.e., the kings might) southern border is up to this land (i.e., Nubia) and (its) northern one (is) up to that circling water going downstream and southwards (i.e., the river Euphrates)'. The attempt to fix the geographical situation of the river Euphrates within Egyptian terminology is astonishing and leads to the question of why the ancient writer avoided the well-known term *jtrw* 'river' using *mw* 'water' for the Euphrates.

The Egyptians themselves seemed to have problems in explaining the terminological relations of *t3-mḥw* 'Lower Egypt' and *šmʿw* 'Upper Egypt'. The two names were created at the beginning of the historical epoch around 3200 BC. Some 1000 years later the Egyptians restricted the use of the complicated phrases to official texts. More often they used the well-known abbreviation ☰☰☰ *t3wj* 'the two lands' (Erman & Grapow, 1957e, pp. 217.1–219.3) stating correctly the known facts and avoiding at the same time the difficulties of etymology. Surprisingly they did not construct any name of its own for the very complicated system of rivers, watercourses, and water channels but used neutral terms as *jtrw* 'river'(Erman & Grapow, 1957a, pp. 146.12–17).

The terminological development gives a good impression of the ancient perception of the environment. The Greek word is an abbreviation of the telling description given by the Egyptians to their country: 'Nile' means *nȝjtrw(.w)ʿȝw* 'the great rivers' (Smith, 1979, pp. 163ff.). So through the eyes of the first millennium BC Egypt is '(the land of) the rivers'. In the eighth to sixth centuries BC the Greeks colonised the Mediterranean world. Arriving in Egypt they had to find a specific name for the river with which they were confronted. Using the common phrase 'Nile' they made a clear distinction between this and all the other rivers in their world. In the eyes of an Ancient Egyptian living 2000 years earlier these terminological efforts were not necessary because there was only one river in his world. From the point of view of the Old and Middle Kingdoms phrasing the idea of the importance of the Nile with words such as 'the land of the river' or 'the land of the great rivers' would not be wrong. But it would be trivial, or even obvious in a way people would not understand the message.

7.3. RIVERINE LANDSCAPE AND DESERT

The basic riverine situation of the Nile was the source of other phrases denoting typical environment situations. A very significant example is *jdbwj* 'the two banks' as a description of the riverine landscape (Erman & Grapow, 1957a, pp. 153.5, 7). Again the intellectual construction is very clear. Egypt was felt to be everywhere where there were two banks of a river. In other words Egypt was 'the land of the riverbanks'. Although the terminological relations seem to be very simple they give us a good idea of why the Fayum and the great oases in the Western Desert were never felt as part of the Egyptian homeland. As with riverine Egypt these huge fertile regions are embedded in an hyperarid environment. In the case of the Fayum there is even a large lake providing people with water. But there are no rivers and because of that a 'two banks' situation does not exist, a condition which is characteristic of the Egyptian Nile valley. Furthermore in the oases there is no inundation mechanism, and so the interrelations between the cycle of seasons and economy (agriculture, gardening, cattle-keeping) differed profoundly from the Nile valley.

In ancient times one of the most popular terms characterising the environment of the riparian landscape is *kmt* 'the Black'. The term came into use at the end of the Old Kingdom 2200 BC (Erman & Grapow, 1957e, pp. 126–127; Shaw, 1993). Literally the word *km* '(to be) black' designated the very dark colour of any material element, for example, of stone as black granite, of animals as dark cattle, and even of the dirt under the fingernails of men working hard outside (Erman & Grapow, 1957e, pp. 123.4, 12, 124.11, 125.4–9). In this line the term was applied to the large alluvial areas of the Nile Valley and by extension to the land itself.

After the inundation the alluvial grounds were covered with fertile layers of dark brackish mud which was the basic natural element of agriculture, the most important economic sector of Ancient Egypt. The area designated by the word *kmt* bore its characteristic colour only during the months of the *prt*-season, that is, from November to February. So the term did not refer to the valley as a whole, but only to one of its landscape factors. To designate Egypt as *kmt* 'the Black' is less a description of the Egyptian Nile Valley or a part of it but the expression of a deep process of conceptualisation of the riverine habitat relating to its economic value. It is the supraregional importance of agriculture which permits the transfer of meaning from the alluvia to Egypt as a whole. 'Egypt (is) the Black (land)' and '. . . (is) the land of agriculture' are two sentences describing the same idea in different words, and the modern spectator may have the feeling that the respective word of the English language reflects the interrelations between culture and environment at this point: agri-*culture*.

Likewise *kmt* offers a key for the understanding of the famous phrase summarising Egypt as 'gift of the Nile' attributed to Herodot, a Greek historian of the fifth century BC The real gift presented by the Nile to the Egyptians was the inundation and the mud brought by it and the water as such. Significantly the Egyptians did not construct 'the White' as the opposite to *kmt* 'the Black' although there are many relating phenomena and of course such words as *ḥd* '(to be) white' (Erman & Grapow, 1957c, pp. 206–208). The homeland of the Egyptians is surrounded by huge deserts. So it seems logical that these regions are perceived as the main contrast to the 'black' river landscape and the 'red' (*dšr*) colour as their distinctive feature. Speaking about the 'black' Nile valley as contrasted with any 'white' lands definitely was not the Egyptian way of thinking even though the white colour concept was used in describing one of the most typical Egyptian desert animals: *m3-ḥd* 'the white desert-animal', the generic term for different species of antelopes terminologically not being separated by the Egyptians (Osborn, 1998, pp. 160ff.).

The term characterising the deserts outside the Nile Valley and being conceptualised as the opposite of *kmt* 'the Black' was *dšrt* 'the Red' (Erman & Grapow, 1957e, pp. 494.5–13; Shaw, 1993). It goes back to a word *dšr* '(to be) red' (Erman & Grapow, 1957e, pp. 488–490). As a designation of deserts it was used from around 2400 BC onwards. One explanation connects the red colour with the sun turning red the sky and the land during his daily rising and setting, respectively. Another one suggests that it comes from the reddish-brown colour of many grounds in the deserts and puts it – as mentioned – in contrast with the black colour of the alluvial grounds of the Nile Valley. The most important term designating the deserts was ⌇ *ḥ3st* which literally means '(region of the) hills' and is the opposite of ⌇ *t3* '(region of flat) land' (Erman & Grapow, 1957c, pp. 234.7, 234.15–16, 235.1). Accordingly there are two very frequently used ideograms representing the areas mentioned and the relating words, respectively. The *ḥ3st*-sign ⌇ shows three usually red-coloured hills. In hieroglyphic writing the number 'three' designates plural. So with ⌇ we get a simple but a very clear picture of the '(lands of many) hills', that is, the deserts on both sides of the valley (Gardiner, 1957, p. 488). On the other hand the ideogramm ⟪⟫ *t3* consists of a long black stroke often completed by three small dots. The sign gives an impression of the flat alluvial grounds inside the Nile valley often mixed with grains of sand (Gardiner, 1957, p. 487). Again terminology and writing demonstrate the strict 'nilocentric conceptualisation' of the Egyptians. Ideogramms mirror the conceptualisation of the world. Now we can better understand the words spoken by the poet Neferti some 1000 years later.

At the beginning of history ⟪⟫ *t3* '(region of flat) land' indeed is a designation of Egypt. The flat lands pictured in the sign ⟪⟫ are the alluvial grounds of the Nile valley. From a local perspective the mountains and hills ⟪⟫ are the characteristics for the world beyond the valley. Through the centuries there was a tendency of generalising the meaning of signs and words. From the sixth dynasty on (2300 BC and later) ⟪⟫ *ḫ3st* mostly meant 'not being Egypt', in the sense of kinds of foreign countries (Erman & Grapow, 1957c, pp. 234.8–10, 234.16, 235.2–17). Again ⟪⟫ *t3* is not any longer a word marking flat alluvial grounds but grounds and lands generally, and this is the meaning Neferti makes use of when speaking about 'his' land.

7.4. SWAMPS

Describing the environment they lived in the Egyptians created a highly complicated vocabulary often overtaxing the possibilities of a modern translation, and in some cases even our understanding. At this point we restrict our statements to a small but characteristic part of the landscape of the Egyptians: the marshes and swamps (Herb, 2001, pp. 378ff.). Developing an economic system which integrates all potential natural resources was one of the important achievements of the economy and culture (Herb, 2001, pp. 388ff.). Especially the marshes near the Mediterranean Sea were intensively used in the Old Kingdom (2600–1800 BC) for cattle-keeping, fishing, bird-catching, and plant-collecting. These economic activities were described in pictures and texts from the time of King Sneferu on (2614 BC and later). The relating processes of conceptualisation date from some 400–500 years earlier.

The principal way people used to go from their settlements to the swamps far away and vice versa was described as *h3tr* 'going down to (the swampy regions)' and *prtr* (*hrj-tp*) 'going up (to the settlements)'. Looking at the valley structure given in Figure 7.1 we get an idea why this terminology operated in this way. The ground levels of settlements and cemeteries were relatively high. Swamps and marshes were low, the most part of them being flooded the entire year. Also we begin to understand why the Egyptians characterised the festival they celebrated after having worked in the swamps for some months as *prt m mḥt* 'coming out of the northern swamps and going up (to their villages)', or in the short version which is typical for the Egyptian terminology: *prt* 'coming out and going up' (Herb, 2001, pp. 412ff.).

𓈅𓌙 *mḥt*, the extensive 'northern swamps' (Erman & Grapow, 1957b, p. 125.4) were located in 𓈅𓇋𓃀𓌙𓊖 *t3-mḥw* 'the land of the northern plant', both names being derived from one and the same root **mḥ-* 'northern' but designating areas of completely different extensions. There are two further words connected with the root **mḥ-* 'northern'. The designation *mḥtj* 'northerner' was directly derived from *mḥt* 'northern swamps'. Literally it meant 'the one who belongs to the northern swamps'. So *mḥtjw* was the designation of the people working there (cf. Erman & Grapow, 1957b, p. 126.4). Taking the picture-cycles of the tombs of the Old Kingdom as main sources for the relevant activities *mḥtjw* 'the northerners' were the men who were specialised to work in the swamps. In this period the designation had definitely not any national or racial character, and there were never any women working in the swamps and accordingly there is no word such as **mḥtjt* 'northern woman'.

The collective noun 𓆛𓆜𓆝 *mḥjjt* (*mḥyt*) 'the northern (fishes)' summarizes different species of fish living in the waters of the swamps and being caught not only by the men working there but also by sporting nobles (Erman & Grapow, 1957b, pp. 127.10–11; Herb, 2001, pp. 404ff.). Surprisingly the ancient sources give information that 𓆛𓆜𓆝 *mḥjjt* (*mḥyt*) 'the northern (fishes)' were caught in Upper Egyptian waters too. So the translation seems unsuitable. We may speculate that in early dynastic periods (3200–2800 BC) the terminology was created in the north. Some 400–500 years later in the fourth and fifth dynasties it moved upstream together with the technical know-how. In later times it was no contradiction to catch fishes in Upper Egyptian waters on the one side and to call them *mḥjjt* (*mḥyt*) 'the northern (fishes)' on the other.

Another designation of the Egyptian swamps was 𓂺𓇋𓇋𓂋𓌙 *dwjt* (*djjt☐dt*) 'papyrus' (Erman & Grapow, 1957e, pp. 511.6–9). First it denoted the particular plant with which the *mḥtjw* 'northerners' were working. Papyrus was uprooted and carried by the workers to their camp. There papyrus was split and men made ropes, baskets, and rafts using the plant. In a broad sense the word designated the area in which the plant was significant (Herb, 2001, pp. 403f.). Working 'in the papyrus' covered the same meaning as working 'in the papyrus-fields'. More often than *dwjt* (*djjt☐dt*) 'papyrus' the enigmatic term 𓈖𓏏𓏭 *pḥww* (Erman & Grapow, 1957a, pp. 538.8–10) was used. Perhaps it is connected with a root **pḥ-* 'at the end, backward' (cf. Erman & Grapow, 1957a, pp. 535ff.).

The tentative translations 'backwaters' or 'distant (marsh)lands' show the problems the modern translator has. Indeed there is no equivalent in any modern language. The common use of 𓈖𓏏𓏭 *pḥww* corresponds to 𓈅𓌙 *mḥt* 'northern swamps'. In some inscriptions the former seems to take the place of the latter. On the other side there are indications that the 𓈖𓏏𓏭 *pḥww* were not identical to, but parts of 𓈅𓌙 *mḥt* 'the northern swamps'. Analysing the scenes containing the word we can conclude that 𓈖𓏏𓏭 *pḥww* marked specific areas near or inside the swamps which were important for working activities. For example, the 'backwaters' were the places where the men installed the nets for bird-catching or grazed their cattle. The 'backwaters' seemed to be a region consisting of small pools and waterways, swampy areas, and solid grounds, a mixture which was very suitable for the different activities of the *mḥtjw*-workers. So the ⌂⌂≡ *pḥww* seemed to refer to the places where the water channels and rivers ended and the actual swamps began (Herb, 2001, pp. 409ff.).

All terms stress the spatial aspects of the Egyptian swamps. In contrast to them the words *šȝw* and *sšw* accentuate some seasonal aspects of the marshes. Again there are no modern equivalents and we must paraphrase the underlying situation. ⌂⌂⌂⌂⌂⌂ *šȝw* were the grounds where the *mḥtjw*-workers collected lotos. This plant needs warm and shallow waters. Buds and blossoms come out of the waters only for a few weeks in the *prt*-season. The term designates a field possessing the important quality of being equipped with lotos. So '(fields of the) lotos-buds and -blossoms' seems to be the best paraphrase of ⌂⌂⌂⌂⌂⌂ *šˁw*, and it is this that the sign shows (Erman & Grapow, 1957d, pp. 399.7–11; Gardiner, 1957, p. 480). The circumstances are similar to ⌂⌂⌂⌂⌂⌂ *sšw*. Again in the *prt*-season the extensive papyrus thickets were suitable places for the birds building their nests and brooding there.

But only for a few weeks between November and February some areas were filled with masses of birds and the loud noise of their calls. Only for a few weeks in the *prt*-season the *mḥtjw*-workers could exhaust these important swamp resources. So '(grounds of the) nests (filled with young birds)' best covers the meaning of the sign ⌂⌂⌂⌂⌂⌂ and the relating word *sšw* (Erman & Grapow, 1957c, p. 483.12, 484.1–11; Gardiner, 1957, p. 473). The seasonal changes of the areas were deeply influenced by outer circumstances. The water level of the swamps depended upon the quantities of the inundations. So ⌂⌂⌂⌂⌂⌂ *šȝw* 'the (fields of the) lotos-buds and -blossoms' were located in different years at different places. The *mḥtjw*-workers had to look for them anew every season. Similarly the migrant birds did not reach Egypt at the same days or weeks every year. Sometimes they came at the end of October, sometimes their arrival took place in December. Also they occupied alternating breeding places every year. Again the *mḥtjw*-workers had to observe these changes of the ⌂⌂⌂⌂⌂⌂ *sšw* 'the (grounds of the) nests (filled with young birds)' (Herb, 2001, pp. 407ff.).

7.5. PICTURES AND DECORATIONS

Funerary architecture delivers a lot of important informations about the culture of Ancient Egypt. From the time of the pyramid builders in (2614 BC and later) up to the Ptolemaic and even Roman epochs (332 BC–AD 395) the walls of temples and tombs belonging to kings and nobles were covered regularly with pictorial scenes and inscriptions. Often the modern spectator feels them to be one of the significant features of the culture. But we should not forget that these sources reflect only the cultural tradition of the uppermost peak of the social pyramid. Although not directly depicted, the landscape of the river valley played an important role in the compositions of the pictorial programmes. Often the figures of animals and plants were pictured in such a naturalistic way that from a present point of view it is no problem to make botanical or zoological identifications. The Egyptological literature is full of entries such as 'Gazella dorcas Linné' or 'Nymphaea caerulea Savigny', producing the impression of scientific exactness. One wonders how easy it is to categorise fauna and flora described in sources from 5000 years ago within the scope of the terminology of the twenty-first century AD. Indeed modern interpretation sometimes produces the impression that the ancient pictures work like photographs and can be seen and 'read' accordingly.

In order to obtain a deeper understanding of these concepts it is interesting to hear an ancient voice speaking on the subject. In the reign of king Pepi II (2254–2194 BC) there was a man named *Snj* living in the town of Akhmim and working nearby in the necropolis of El-Hawawish. One of his jobs was to cover the walls of the tombs with pictures and inscriptions. We would say *Snj* 'decorated' the walls. In the tombs of *K3=j-h3p:Ttj-jqr* and his son *Špsjpw-mnw:Hnj*, two local rulers of Akhmim, *Snj* obtained the right to picture himself and to give a short description of his activities (Kanawati, 1980, p. 19, fig. 8, pl. 5): '. . . I wrote the tomb of the count *Hnj*. I wrote this tomb (i.e., the tomb of *K3=j-h3p:Ttj-jqr*) too. Me alone'. The terminology *Snj* used is different from ours. Distinctly he formulated he 'wrote' the walls. He did not look at himself as a 'painter' or an 'artist' but as a scribe. Accordingly his title was *zš–qdwt* 'writer of the walls'. In the eyes of the Egyptians the immense collections of figures, motifs, and scenes covering the walls were inscriptions and writings (*zšw*), and the men having made them were scribes (Herb, 2006, pp. 169ff.). In other words: 'decoration' and writing are one and the same, and the problem of differentiating them is a modern one (Gardiner, 1957, pp. 438ff., 1989; Kahl, 1994).

Today we know more than 800 tombs dating to the Old and Middle Kingdoms (2600–1800 BC) and containing walls covered with epigraphic or pictorial programmes (Harpur, 1987). The basic ritual inside the tomb was the worship of the owner starting after his burial. Accordingly the main theme of every programme was to describe the ritual processes connected to the tomb. The funerary meal condensed this purpose, and the unit of the programme of the tomb which localised its position inside was the false-door having both epigraphic and architectural characteristics (Wiebach, 1981). The relating scene showed the owner sitting before a table and looking at the offerings presented to him by his priests. Typically the construction of the scene was accompanied by a second 'table': a list giving the components and the chronological course of the meal. The offering-list consisted of the most important provisions of the country, that is, different sorts of bread, cake, and meat, of fruits and vegetables, and so on. Water, wine, and beer were mentioned and thus the most important beverages of ancient Egypt. The chronological order of the meal was reflected by the rowing of the entries of the list. Again being highly standardised the funerary meal was more a schedule of the products of the country than the fixation of a real meal (Barta, 1963).

'Environment' was not a theme of any pictorial programme. Accordingly 'landscapes' were not depicted directly. But they played an indirect role in the themes of which every programme was composed. The main content was the activity of humans. Whatever was pictured, people were doing something, and the target and source of all their activities was the riverine landscape of Egypt. From an ancient point of view environment and landscape were basic and unchangeable. The knowledge concerning it seemed to be a kind of truism and speaking about it is superfluous.

In Ancient Egypt the material aspect of the cult of a deity or a person was mainly based on food and goods. Producing them was a basic requirement and considered as part of the cult too. The goal of the ritual was the celebration of the meal near the door inside the tomb, and the activities of producing the meal preceded the ritual, chronologically and topographically. Scenes and inscriptions of the epigraphic programme reported the essential data of the biography of the owner, described the making of food of the funerary meal and informed the persons who were concerned with the cult and would celebrate it in the future, for example, the farmers, scribes, priests, and the members of the owner's family. Corresponding to the *Preceding* character of the activities introducing the meal the relating units describing the producing of its components covered the walls leading to the false-door and escorted the priests on their way to the funerary meal (Herb, 2006).

The places where the preparatory activities were conducted can be easily located in the structure of the riverine landscape. The tombs and cemeteries were the places where the priests worked. The relating archaeological sites lie in the desert zone near the valley. Having called them *zmjt* (Erman & Grapow, 1957c pp. 444.8–445.3) the Egyptians developed a term of its own for these regions directly adjoining the riverine landscape. The tombs were the final points of ritual movement in antiquity and are the starting points of scientific investigation today.

The settlements the Ancient Egyptians lived in were close to the cemeteries. They were founded both at the edges of the oasis and often on *jww* 'islands,' that is, high-dams and gazîras inside the valley. In towns people lived together. Here were the marketplaces and the houses of the craftsmen manufacturing stones, vessels, furniture, and so on. Here were bakeries and breweries, and the enclosures to which people took cattle, goats, asses, and captured desert animals too. Here in the settlements were the threshing floors and granaries where the corn was stored. The alluvial grounds covering the main part of the valley were the places where the farmers sowed and harvested. From here they transported the corn to the villages and the granaries. In the extensive swamps the *mḥtjw*-workers caught birds and fish. Here they fed their cattle, and here in the swamps they harvested the papyrus, the material base of the culture of writing of Egypt. In the deserts far away from the Nile the Egyptian hunters installed extensive corrals. From year to year they came back to these hunting grounds, looked for hares and hedgehogs, collected newborn gazelles and antelopes, and captured the adult mammals using throwsticks and lassos. Finally they transported the animals to the settlements in the Nile valley (Herb, 2001, pp. 377ff.).

Principally we can recognise two modes of landscape visualisation in the epigraphic programmes. First the country of Ancient Egypt is visible in the localities where the people worked. Second, the landscape is the economic basis of the culture. We obtain information on different kinds of activities, that is, agriculture, gardening, working in swamps, or hunting in deserts. Using the landscape for ritual purposes is the key of understanding its role in the programmes of temples and tombs of the Old and Middle Kingdoms (2600–1800 BC), and the tomb of *K3 = j-m-nfrt* is a good example explaining this idea.

In the reigns of kings Isesi and Unas (2380–2322 BC) a man named *K3 = j-m-nfrt* obtained the right to be worshipped in a tomb of his own in Saqqara, the necropolis of the capital of the Old Kingdom (Simpson, 1992; Fitzenreiter & Herb, 2006). In this time about 250 years after their introduction under Sneferu (2614–2579 BC) the principles of 'programming' a tomb were established. In his chapel *K3 = j-m-nfrt* was not represented with wife, children, or any other person of his family. The construction of the programme was not disturbed by a plurality of addressees. 'The focus is entirely on the self-thematisation of the tomb owner' *K3 = j-m-nfrt* (Simpson, 1992, p. 1). So the programme is relatively simple. Having passed a long corridor the priest reached an antechamber opening the way to the chapel (Figure 7.2). He read the name of the owner on the lintel of the doorway. Thus he made sure he was at the right place. Then the priest entered the chapel, the room in which he was to celebrate the funerary meal (Simpson, 1992, pp. 1ff., fig. 2). Vis-à-vis he saw the false-door as marking the place where he 'met' *K3 = j-m-nfrt* and thus had to work (Figure 7.6).

Inside the chapel all walls were covered with scenes and inscriptions (Figures 7.3–7.6); in other words, doing his work the priest was surrounded by the programme. Only the sections near the floor were not decorated. In comparison with other tombs the programme of *K3 = j-m-nfrt* seems to be small and simple. The false-door on the western wall of the chapel marks the place of the funerary meal and is the starting or final point of our reading (cf. Figures 7.2 and 7.6). Having made a 'cut' at the edge between the western and northern walls the *K3 = j-m-nfrt*-programme becomes a kind of folder. How the scenes are arranged reflects the ordering of texts in a papyrus roll. Opening this 'roll' we recognize that the themes

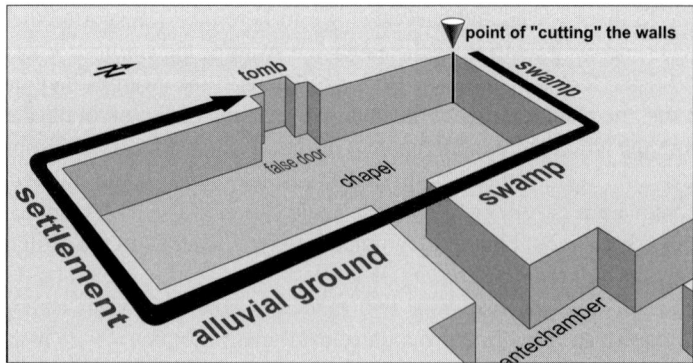

Figure 7.2. Sketch of the chapel of *KA=j-m-nfrt* with the line of reading the decoration
(Drawing by the author)

of the programme are arranged according to the localisation of the activities of the people; that means it reflects the landscapes in which they worked.

The reading begins with the activities in marshes pictured at the northern wall of the chapel and the northern section of the eastern wall, and it ends with the funerary meal at the western wall containing the false-door, offerings, and the offering list. The northern wall gives a schematic papyrus thicket which localises the activities pictured nearby in the swamps at the edges of the valley (Figure 7.3)[1]. The economic activities consist of a sequence of three steps. The first step covers the lower part of the northern section of the eastern wall (Figure 7.3, right) and shows fishing, bird catching, and the main features of cattle breeding. Continuing on the northern wall the representation of *K3 = j-m-nfrt* himself follows (Figure 7.3 left). He is represented visiting the swamps and hunting birds with a throwing stick. But the main intention of his journey is to control *mḥtjw* 'the northerners' working for him and being connected with the cult of his tomb (Harpur, 1987, pp. 176ff.; Herb, 2001, pp. 358ff.; Herb, 2006, pp. 315ff.).

Now the reading turns in a way *boustrophedon*. The motifs directly located at the thicket show men collecting papyrus and working with papyrus fibres in a

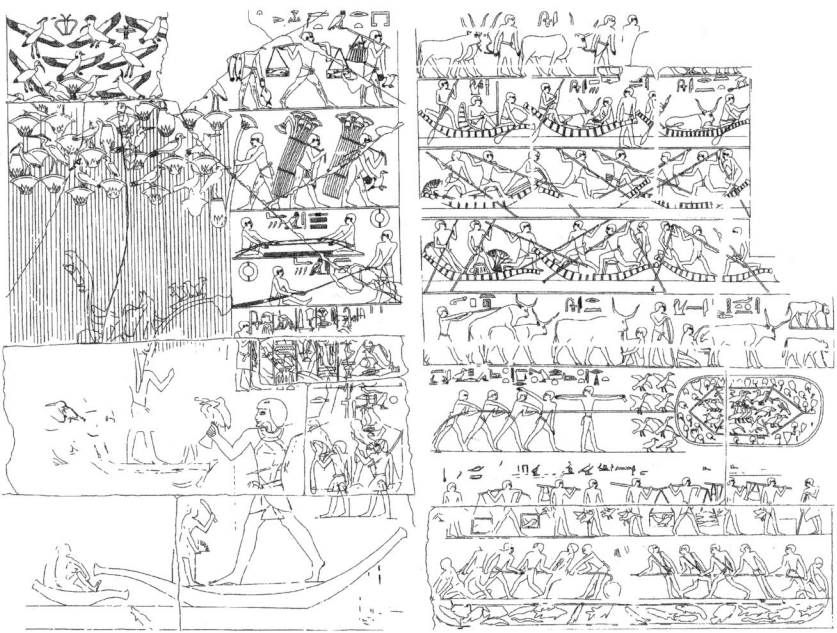

Figure 7.3. Chapel of *KA = j-m-nfrt*, decorations of the north wall and the northern section of the east wall (Simpson, 1992: Plates A, G)

---

[1] The palimpsest- situation does not affect our considerations (Dunham, 1935; Simpson, 1992, pp. 4ff).

camp. A large ensemble of figures and scenes follows representing the coming home of the *mḥtjw*-men after having worked for months inside the swamps. They carry the fish and birds they caught and the bundles of papyrus they harvested. Also they drive their cattle to the villages. Partly the men use the rafts which they have constructed in the camps near the swamps for transporting their products. These scenes cover the upper parts both of the northern wall and the following northern section of the eastern wall (Figure 7.3–7.4, left).

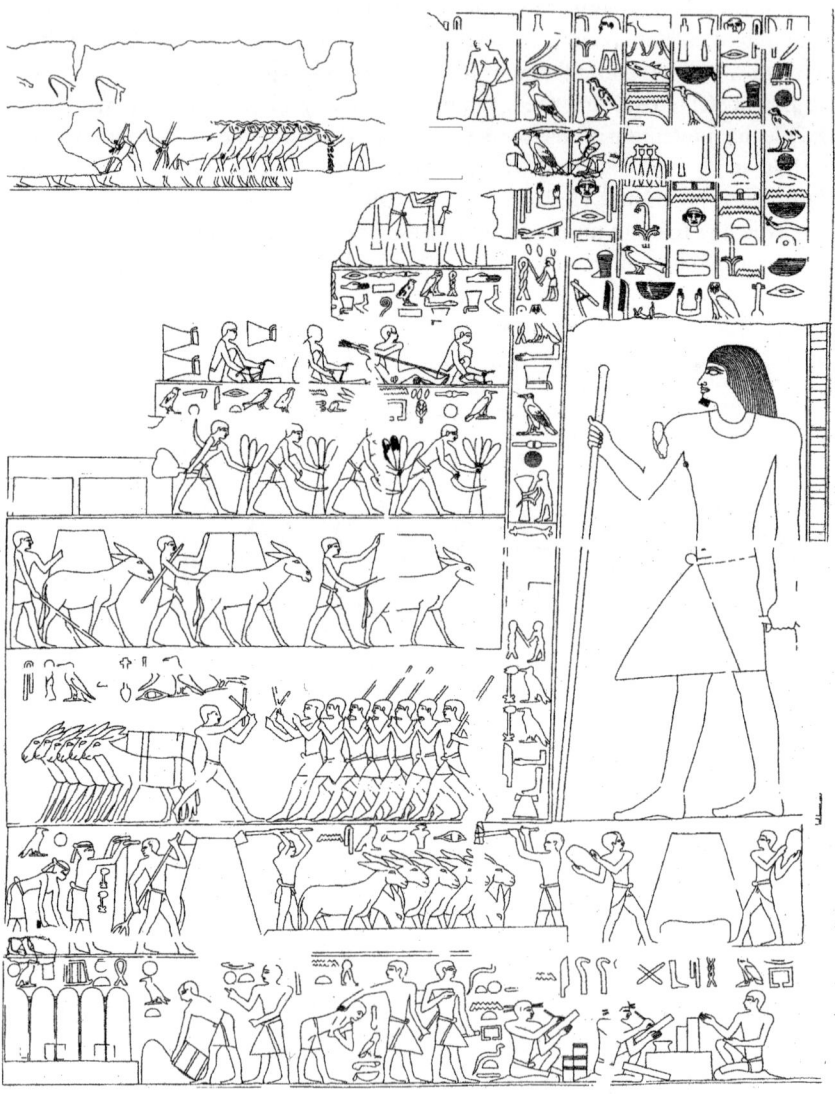

Figure 7.4. Chapel of *KA = j-m-nfrt*, decorations of the southern section of the east wall (Simpson 1992: Plate F)

Leaving the swamps one reaches the alluvia and the agricultural activities. The relating scenes cover the southern section of the eastern wall (Figure 7.4). Agriculture begins with a scene above the doorway showing the sowing of grain in the *prt*-season (Figure 7.4, inset). Harvesting follows combined with the motifs of rope making in the fields (Figure 7.4, above). The lower part of the wall deals with the ways the farmers go to their villages and with their activities there. With the help of donkeys the men transport large bundles of grain to the threshing floors inside the villages (Harpur, 1987, pp. 204ff.). After having prepared the grain the results are registered and stored in granaries (Figure 7.4, below). Again a large figure of the tomb owner accompanies the sequences of small figured scenes. *K3=j-m-nfrt* watches the agricultural activities during the different seasons of the year. A short inscription tells of his duties: *m33 sk3 jt hj mhᶜw 3sh šdj*[...] *hjh3h3jᶜb* 'Viewing (i.e., controlling) the cultivating of grain, (of) pulling flax, reaping, loading (donkeys), striking, winnowing, and heaping' (Simpson, 1992, p. 16).

The activities described on the southern wall (Figure 7.5) are connected with the settlements which have signed on working for *K3=j-m-nfrt* and his cult (Simpson, 1992, pp. 15ff., fig. E, pls. 16ff.). There are two alternative interpretations. It is possible that the workers come from their towns to the residence of the tomb owner and present the results of their foregoing activities. Alternatively *K3=j-m-nfrt* himself makes the journey travelling from village to village and controlling the different economic activities there. The inscription reads *m33 sšnndt-hr jnnt m njwwt=f nt pr-dt t3-mhw šmᶜw ᶜ3-wrt* 'Viewing (i.e., controlling) the accounting of the presentation brought from his towns of the funerary estate in the northland and southland very abundantly' (Simpson, 1992, p. 15). According to the general theme the reading begins with the journey of the owner to the registration places. This trip is pictured in the lower part of the southern wall (Figure 7.5, below). Then the direction changes again . Moving to the upper registers we see the scribes writing down the quantities of the animals *K3=j-m-nfrt* possesses. We recognize cattle, antelopes, gazelles, and ibexes standing being pegged in enclosures or led by herdsmen (Figure 7.5, above).

The programme of the western wall (Figure 7.6) deals with the funerary meal the priests have to celebrate for perpetuating the memory of *K3=j-m-nfrt* (Simpson, 1992, pp. 8ff., fig. Bff., pls. 7ff.). Accordingly the activities are localised near or inside the chapel. In a kind of introductory sequence seven priests are pictured at the bottom of the left part of the wall bringing the offerings into the chapel. What they are doing is explained as *shpt stpt nw hmwk3 n pr-dt* 'delivering the offerings of the ka-priests of the funerary estate' (Simpson, 1992, p. 12, fig. 12). With the word we meet a meaning 'all kinds of products for the (funerary) meal' (Lapp, 1986, pp. 235ff.). The registers of the upper part show the offerings the priests have brought and have carefully arranged in rows inside the chapel and the rooms in front: The meat of cattle and birds from the swamps; the cakes and bread made of the grain of the alluvial grounds; the meat of the animals of the deserts; the wine, vegetables, and fruits from the gardens and plantations; and even the cool and pure waters of the Nile. The offering-list covering the right part

Figure 7.5. Chapel of *KA* = *j-m-nfrt*, decorations of the south wall (Simpson, 1992: Plate E)

of the wall reports the chronology of the funerary meal and determines the order of the products. The false-door between the offerings and the offering-list in the centre of the wall symbolises the addressee of celebration. The door represents the house of *K3*=*j-m-nfrt* localising his place in the vastness of the netherworld, and the scene of *K3*=*j-m-nfrt* sitting at the table and eating bread, meat, and so on reveals the information about his physical prosperity.

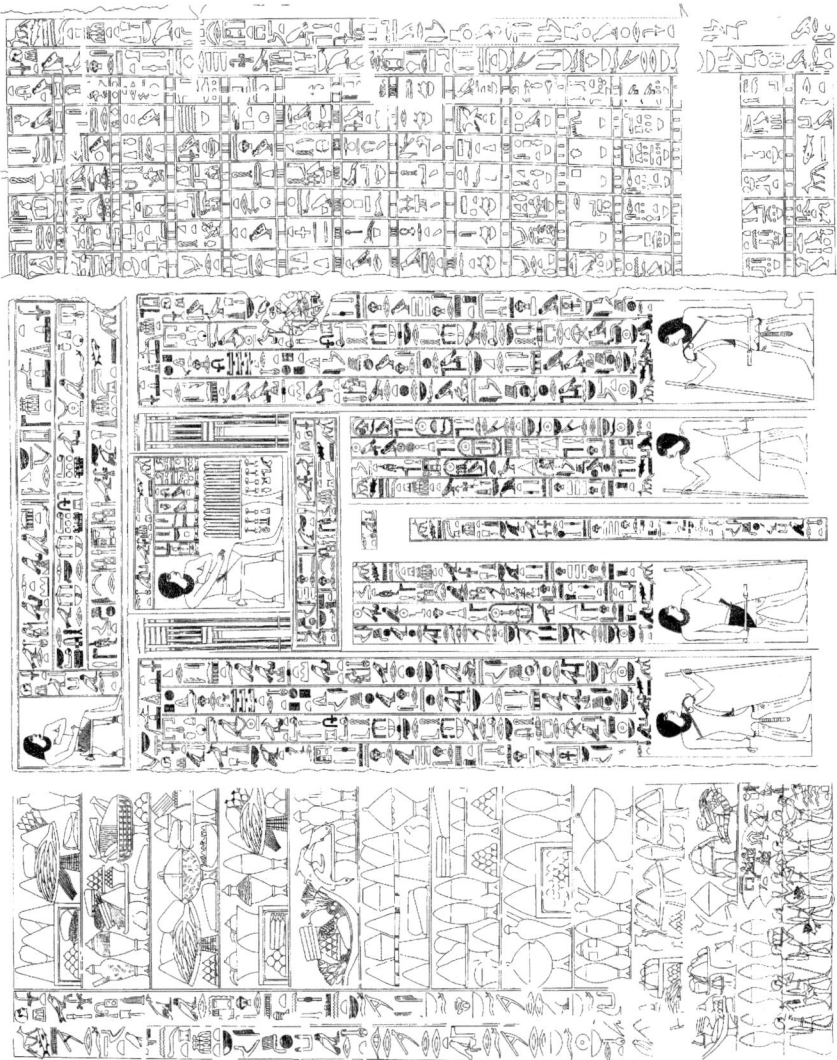

Figure 7.6. Chapel of *KA = j-m-nfrt*, decorations of the west wall (Simpson, 1992: Plates C, B, D)

The environment of the Nile Valley was never a direct theme of the pictorial constructions and epigraphic programmes of the tombs. But using the natural resources of their habitat the Ancient Egyptians described and pictured the relating activities in detail,and the deep interconnection between economy and religious practice opens the way to obtain information, not only about the Ancient

Egyptians, but also about their perception of 'landscapes', that is, the environment seen through the eyes of the people in ancient times.

7.6. LANDSCAPE, MAN, AND GODS

Temples were the houses where the Ancient Egyptians visited their gods, and indeed the Egyptian temple was not only the house of a deity but also and even a symbolic condensation of the landscape of the Nile. It was a model of the world in which the gods and deities animating it were represented. Often the lower parts of the walls were decorated with figures and motifs of plants striving upwards to the sun. Sometimes the floor was also covered with thin layers of silver creating the image of the black colour of mud. The columns of the halls represented other kinds of plants. They supported the roof of the temple which symbolised the sky covered with stars. As in the famous 'room of the seasons' in the sun-temple *Šsp-jb-r^c* of Niuserre the epigraphic programmes and scenes dealing with fauna, flora, and the natural life of Egypt suggested the stage of the appearances of the gods or their performances, respectively.

Since the beginning of the fourth dynasty and the reign of King Sneferu (2614–2579 BC) allusions to environment are often part of the epigraphic programmes of the Egyptian temples. But in the first millennium BC a new system was developed. Now the lower part of the walls of the temples were covered with long rows or 'processions' of fecundity figures and those bringing offerings (Baines, 1985). These human figures symbolised the nomes of Egypt and the different countries of the world having produced these offerings. The scenes went back to the idea of the so-called 'domaines', often pictured in the temples and tombs of the Old Kingdom. In a similar way these figures localised the regions of production of the offerings and represented the complicated processes of transporting them to the temples and tombs (Jacquet-Gordon, 1962). In the first millennium BC the large processions consisting of fecundity figures translated the simple decoration of plants into an anthropomorphous system explaining the relating regions with the help of short inscriptions, that is, the names of the human figures. But basically there was no alteration of meaning from plants to fecundity figures or those bringing offerings.

It is the Ptolemaic façade of the temple of Esna built in the second century AD which presents the most extraordinary abstraction of 'landscape' we know from Ancient Egypt. An unknown artist – or shall we say 'philosopher'? – surpassed the basic procedures of the temple-programmes of the Old, Middle, and New Kingdoms. He created a vision of the world in three dimensions combining the very stereotyped pictorial constructions often described in Egyptology as 'offering-scenes' (cf. Figure 7.8). The main idea underlying these picture-cycles was to fix the meeting of the pharaoh and the gods. The acts of offering on the one side and the ritual scenes on the other describe the event of the meeting in a way that the succession on the wall finally constitutes a coherent message. The pictorial and textual programme of the Ptolemaic façade of the temple of Esna presents

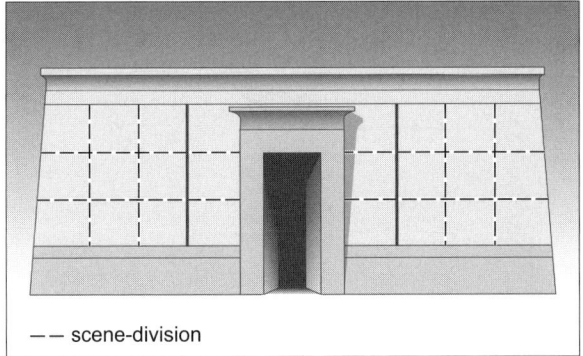

—— scene-division

Figure 7.7. Sketch of the inner façade of the temple of Esna showing the scene-division (Drawing by the author)

the world looking forward to the gods who still are concealed inside the rooms of the sanctuary (Sauneron, 1963, pl. 1). In total 24 scenes in three registers are arranged on the two walls of the façade. So one counts 12 scenes on both sides of the central gate (cf. Figure 7.7). Both parts show a corresponding scheme of scene distribution: they are divided into a rectangle containing nine scenes followed by a column with three scenes near the gate.

The registers and scenes on the left side ascending from bottom to top present the aquatic sphere, the atmosphere, and at last the firmament, that is, a vertical projection of the environmental space labelled by the relating essential components. The corresponding scheme on the right side shows the offerings which consisted of victuals, mineral resources, and clothes. Accordingly they are arranged from bottom to top but reflect a view of the horizontal layout of space from which the offerings being indispensable for life originate. Therefore the programme of the Ptolemaic façade of the temple analysed in this way describes the landscape in which the gods will appear. In this model the landscape is defined by the fundamental elements 'water', 'air', and 'light' on which 'nature' or 'world' is brought into being. On the other side there are the components constituting life and culture. This 'landscape' gets its dynamism from the two figures in the centre of the composition of each wall. On the left side Khnum-Re is depicted, solar might and local god. On the other side Osiris is shown in the relating scene, mediator of the resurrection of life and of all cosmic phenomena corresponding to each other and complementing one another.

In the Egyptian temple decoration Man was symbolised by pharaoh appearing in every ritual scene. Because of that Man was omnipresent in the epigraphic programmes of the temples describing world, nature, and landscape although there was never any 'realistic' motif or scene. It was the action of the pharaoh which determined the activities of the gods in the world. This kind of interrelation finds expression in the composition of the dialogues existing between them. The king spoke to the god using a present-perfect formula: 'I have come to you . . .', and

Figure 7.8. Inner façade of the temple of Esna, offering scene (Photograph by Dagmar Budde, Mainz) (*See also Color Plates*)

the god answered using a future formula: 'I will give you in exchange for . . .' (cf. Figure 7.8). It was this kind of interaction between the king and the gods which guaranteed a harmonious world, that is, the landscape of Ancient Egypt.

But the initial creation began on divine initiative. This idea substantiated the exceptional position of one picture being positioned in the centre of the whole decoration of the Esna façade. It is directly located over the gate the god will appear in and thus in the middle between the two scene-cycles on the left and the right side of the gate (Figure 7.9). The nocturnal sun personified by the ram-headed god Khnum is accompanied by eight persons symbolising the forces escorting him out of the depth of the night and ensuring his stability. The textual commentary announces the act of the first division of the Creator into male and female for making gods and mankind, and the lunar cycle at its starting point. Beginning with the first time of procreation and appearance of the sun and the moon it is Man who totally has under control the repetition which ensures the existence and the working of the world. The methods of expression the Egyptians used in decorating their temples made it possible to picture the environment of the Nile Valley as a specific Egyptian kind of landscape: the stage where Man

Figure 7.9. Decoration over the gate of the inner façade of the temple of Esna, showing the nocturnal sun personified by the god Khnum (Photograph by Dagmar Budde, Mainz) (*See also Color Plates*)

meets God: a deep declaration of the intimate union of nature and culture, and the predominance of the last.

REFERENCES

Abu al-'Izz, M.S. (1971). *Landforms of Egypt*. Kairo: American University of Cairo.

Baines, J. (1985). *Fecundity Figures. Egyptian Personification and the Iconology of a Genre*. Warminster: Aris & Phillips.

Baines, J. (1989). Communication and display: The integrating of early Egyptian art and writing. *Antiquity*, 63, 471–482.

Baines, J. & Málek, J. (1980). *Atlas of Ancient Egypt*. Oxford: Phaidon.

Bard, K.A. (Ed.) (1999), *Encyclopedia of the Archaelogy of Ancient Egypt*. London/New York: Routledge.

Barta, W. (1963). *Die altägyptische Opferliste von der Frühzeit bis zur griechischrömischen Epoche*. Münchner Ägyptologische Studien, 3. Berlin: Bruno Hessling.

Bietak, M. (1975). *Tell el-Dabᶜa II. Der Fundort im Rahmen einer archäologisch-geographischen Untersuchung über das ägyptische Ostdelta*. Denkschriften der Gesamtakademie, IV. Wien: Österreichische Akademie der Wissenschaften.

Bonneau, D. (1993). *La régime administratif de l' ean du Nil dans l' Égypte greeque romaine et byzantine*. Leiden: Brill Academic.

Butzer, K.W. (1976). *Early Hydraulic Civilization in Egypt. A Study in Cultural Ecology*. Chicago: University of Chicago Press.

Dunham, D. (1935). A 'palimpsest' on an Egyptian Mastaba Wall. *American Journal of Archaelogy*, 39, 300–309.

Erman, A. & Grapow, H. (1957a–e). *Wörterbuch der aegyptischen Sprache I-V*. Berlin: Akademie-Verlag.

Fitzenreiter, M. & Herb, M. (Eds.) (2006). *Dekorierte Grabanlagen im Alten Reich Methodik und Interpretation*. Internet-Beiträge zur Ägyptologie und Sudanarchäologie, 6, London: Golden House.

Gardiner, A.H. (1957). *Egyptian Grammar*. Oxford: Oxford University Press for the Griffith Institute.

Harpur, Y.M. (1987). *Decoration in Egyptian Tombs of the Old Kingdom. Studies in Orientation and Sciene Content. Studies in Egyptology*. London, New York: Kegan Paul.

Herb, M. (2001). Der Wettkampf in den Marschen. Nikephoros. Beiträge zu Sport und Kultur im Altertum. Beihefte, 5. Hildesheim: Olms-Weidmann.

Herb, M. (2006). Ikonographie – Schreiben mit Bildern. Ein Essay zur Historizität der Grabdekorationen des Alten Reiches. In M. Fitzenreiter & M. Herb (Eds.), *Dekorierte Grabanlagen im Alten Reich. Methodik und Interpretation* (pp. 111–124). Internet-Beiträge zur Ägyptologie und Sudanarchäologie, 6, London: Golden House.

Ibrahim, F.N. (1996). *Ägypten. Eine geographische Landeskunde*. Wissenschaftliche Länderkunden, 42. Darmstadt: Wissenschaftliche Buchgesellschaft.

Jacquet-Gordon, H.K. (1962). *Les noms des domaines funéraires sous l' Ancien Empire*. Bibliotheque d'Étude, 34. Kairo: IFAO.

Kahl, J. (1994). *Das System der ägyptischen Hieroglyphenschrift in der 0.-3. Dynasite*. Göttinger Orientforschungen, IV. Reihe: Ägypten, Band 29. Wiesbaden: Harrassowitz Verlag.

Kanawati, N. (1980). *The Rock Tombs of El-Hawawish. The Cemetery of Akhmin I*. Warminster: Aris & Phillips.

Lapp, G. (1986). *Die Opferformel des Alten Reiches unter Berücksichtigung einiger späterer Formen.*. Deutsches Archäologisches Institut Abteilung Kairo, Sonderschrift, 21. Mainz: Philipp von Zabern.

Lichtheim, M. (1973). *Ancient Egyptian literature. ABook or Readings I: The Old and Middle Kingsdoms*. Berkeley/Los Angeles/London: University of California Press.

Osborn, D.J. (1998). *The Mammals of Ancient Egypt*. The Natural History of Egypt, 4. Warminster: Aris & Phillips.

Said, R. (1993). *The River Nile Geology, Hydrology and Utilization*. Oxford/New York: Pergamon Press.

Sauneron, S. (1963). *Le temple d' Esna II. Publications de l' Institut Français d' Archéologie Orientale*. Kairo: IFAO.

Seidlmayer, S. (2001). *Historische und moderne Nilstände. Untersuchungen zu den Pegelablesungen zu den Pegelablesugen des Nils von der Frühzeit bis in die Gegenwart*. Achet: Schriften zur Ägyptologie, A.1. Berlin: Verlag.

Sethe, K. (1914). *Urkunden der 18. Dynastie*. Urkunden des aegyptischen Altertums, IV. Leipzig: Verlag.

Shaw, I. (1993). The black land – The red land. In J. Málek (Ed.), (pp. 12–27). Norman: University of Oklahoma Press.

Simpson, W.K. (1992). *The Offering Chapel of Kayemnofret in the Museum of Fine Arts, Boston*. Boston: William Kelly Simpson.

Smith, H.S. (1979). Varia Ptolemaica. In J. Ruffle, G.A. Gaballa & K.A. Kitchen (Eds.), *Glimpses of Ancient Egypt. Studies in Honor of H.W. Fairman* (pp. 161–166). Warminster: Aris & Phillips.

Smith, W.S. (1981). *The Art and Architecture of Ancient Egypt. Revised with Additions by W.K. Simpson*. London/New York: Yale University Press.

Stricker, B.H. (1956). *De overstroming van den Nijl*. Leiden: Brill Academic.

Wiebach, S. (1981). *Die ägyptische Scheintür. Morphologische Studien Zur Entwicklung und Bedeutung der Hauptkultstelle inden Privat-Gräbern des Alten Reiches*. Hamburger Ägyptologische Studien, 1. PhD thesis. Hamburg: Verlag.

Chapter 8

A Land of Goshen: Landscape and Kingdom in Nineteenth Century Eastern Owambo (Namibia)

PATRICIA HAYES

The general appearance was that of the most abundant fertility. It was a land of Goshen to us.

Galton, 1858, p. 195[1]

The term 'landscape' is used in this chapter to denote land that is marked by historical and cultural layers of meaning which have accumulated over time. It explores the precolonial histories of eastern Ovambo kingdoms in the second half of the nineteenth century, and the interrelationship between historical accounts of royal succession and power with the local floodplain ecologies. The region attracted geographical explorers from 1850, notably Galton and Andersson, who brought radically different methodologies of viewing the land. The chapter examines travel narrative and oral tradition, cartography and oral history, each as a separate medium, but whose selective juxtaposition helps to expose their codes and practices. Whereas Ovambo marked trees and graves as powerful sites of hypomnesia, Galton and Andersson sought to erase autochtonous readings of landscape through an ominous discourse of blankness, propped up by empirical techniques, remote from both the human senses and local memory.

---

[1] Goshen suggests a place of light or plenty. See Genesis xlv, 10 etc.; Exodus viii, 22, ix, 26.

M. Bollig, O. Bubenzer (eds.), *African Landscapes*,
doi: 10.1007/978-0-387-78682-7_8, © Springer Science + Business Media, LLC 2009

8.1. READING THE EDGES

This chapter is concerned with different apprehensions of the landscape in eastern Owambo in the mid- to late nineteenth century. The term 'landscape' has gone through its own shifts in conceptual emphasis. The visual theorist W. J. T. Mitchell initially proposed landscape as *inter alia* 'a natural scene mediated by culture' (1994, p. 5), pertaining to all societies. More recently Mitchell presents landscape as a 'cognitive encounter with a place', and he notes that 'an appreciation of landscape may well include a reading – or an inability to read – its narrative tracks or symbolic features' (2002, p. x). This more nuanced approach avoids the reductiveness of an unproblematised nature/culture divide, and allows for the indeterminacy and sometimes 'passive force' of a setting. Broadly speaking, however, the objective here is to track both African and European views from a variety of historical media that are imprinted with a particular local environment.

We begin with a crossing of paths. When the traveller Charles John Andersson reached the kingdom of Ondonga[2] on a journey in search of the Kunene river in 1867, one of his encounters with the king, Shikongo, took place at his camp on the edge of the kingdom. Shikongo was on his way to go hunting, leaving the 'realm of men' to enter the 'kingdom of the animals', in transition between settled and open space, quitting the civilised area of his kingdom (*oshilongo*) to enter the wilderness (*ofuka*) for the hunt.[3] When Andersson moved further north to Mweshipandeka's kingdom in Oukwanyama, a similar pattern emerged, where the king stopped to see Andersson while crossing a threshold.

These seemingly innocuous references to the spaces of interaction between kings and travellers point to many things. The kings controlled such meetings, the occasion for various kinds of exchange in which gifts, food, guns, and ammunition traded hands. In a zone that was hundreds of miles from any colonial foothold (and which was only formally occupied in 1915), Andersson was kept strictly on the edges of society together with his group of companions from further south, although on one occasion he was allowed inside Mweshipandeka's palisaded residence.

For the kingdoms north of the thirstbelt beyond 'Damaraland', contact with outsiders such as Andersson or the Angolan Nogueira had only begun in the 1850s. In Nick Thomas's phrase, these cross-cultural contacts carried with them certain incommensurabilities: that is, actions, purposes, or understandings which could not be decoded by the other (Thomas, 1999, p. 1). For instance, Andersson complained about the way the hunting parties of the Ndonga king frightened off the elephants and other animals, which made his own attempts to obtain ivory more

---

[2] The kingdom now falls within the present-day Omusati region, and its main town is Ondangwa. This was not Andersson's first visit to the northern kingdoms of the Ovambo (see below).

[3] The distinction between the space of humans and animals in eastern Owambo was first made in Salokoski et al. (1991); Kreike (1996) uses the oshiKwanyama terms *oshilongo* and *ofuka* to make a similar point.

taxing. This did not simply signal a different order of extraction from that of the professional European hunter with precision weapons. For Ovambo kings, hunting was a form of reiterating power because the slaying of animals under their aegis was not just the overpowering of nature, it was also about their fitness to rule. It was a means of rehearsing and displaying authority and power. There was a kind of homology between the control over nature and those native to the place. Such features of power in eastern Owambo were seemingly underpinned by an ontological distinction made between human and natural/animal worlds. At one level this separation enabled the hierarchical management and domination of the boundaries between the two.

The edges of settlement were understood to have certain sensitivities. In Owambo, fairly wide woodland belts surrounded the settled spaces and formed buffer zones between the different polities on the floodplain (Siiskonen, 1990, p. 41). Across different African imaginaries in fact, such in-between spaces were thick with meaning. McCaskie argues, for example, that in nineteenth century Asante, people believed that intruders – whether human or supernatural – should be halted at such places. For Asante, the crossroads between settlement and paths out of it into the forest were 'nodes of great spiritual power'. At such fringes, 'both the welcome and the unwelcome were prepared to manifest themselves' (McCaskie, 1995, p. 164).

In Owambo a similar ambivalence operated. These belts represented spaces both of danger and of refuge. Numerous dynastic assassinations are recounted as happening in the interstitial forests between kingdoms, some very close to the time we speak of here. Mweshipandeka's older brother Nghishiimonima was, for instance, murdered on the border between Ondonga and Oukwanyama on the orders of the incumbent Kwanyama king Haimbili. This was a journey that, prophetically, the young Mweshipandeka had refused to join (Kaulinge, 1997, p. 30). But these edges were equally places of refuge, beyond civility and the law. Such ambiguities were recognised and formalised in the *okakulukadhi*, places in the forests between Ondonga, Oukwanyama, and Uukwambi where 'blood-peace' became ritually marked between representatives of the respective kings. Wayfarers entering the *okakulukadhi* were obliged to leave offerings of grain for ancestral spirits believed to reside there, but they also offered sanctuary to the criminal and the oppressed.[4]

---

[4] The first such blood-peace was reached by the kings of Ondonga and Uukwambi in 1868, marked by the sacrifice of a cow at Omagonzati. This was one year after Andersson's last visit through Ondonga. Even later, in the forest area between Oukwanyama and Ondonga at Ondugulugu in 1891, a black cow was sacrificed to mark the Ndonga–Kwanyama 'blood-peace'. According to Petrus Amutenya, 'it symbolised peace and the sweeping away of blood'. Petrus Amutenya, 'Okakulukadhi' translated by Escher Luanda (unpublished manuscript, Okahau, ca. 1990). See also AVEM No 1.477, *Sckär,Historisches, Ethnographisches, Animismus*, ca. 1901--1913. There is disagreement as to dates of this blood-peace, but Finnish sources quoted by Siiskonen give 1891 as the date. See Siiskonen (1990, p. 209).

Andersson's frustration at the checks, cautions, and put-offs as he waited at the edge of kingdoms – at times calling them 'nonsense' – showed an inability to read such codes. There were other codes that passed him by. The landscape was 'speechless' in the sense that the traveller could not take in local meanings derived from the 'sensory experience afforded by biological reality, weather conditions, and actions that take place in the environment' (Puranen, 1999, p. 11). He had a different history of the senses. His apprehension of local idioms was almost nil. Indeed, far from being receptive, Andersson was preoccupied and distracted by his scientific absorptions, and his senses soon became completely clouded over by the illness (black water fever) that claimed his life a few weeks later.[5]

It must be emphasised that the crossing of paths by the European explorer and Ovambo kings was not simply literal, but deeply representational as well. Andersson's travel notes constitute one of the few texts that register interactions and encounters with the Kwanyama and Ndonga kings Mweshipandeka and Shikongo, respectively. Read in the present, his writing intersects with at least one major insider account of Kwanyama history, narrated by the local historian Vilho Kaulinge. Born in 1900, the late Reverend Kaulinge was regarded as the greatest living authority on Kwanyama history until his death in 1992; he was in fact the grandson of Mweshipandeka. The account used here is based on transcripts of long recordings made in 1989 and 1990.[6] This and other texts based on oral transmission and artistry are read next to selected works of Andersson and his erstwhile travel companion, Francis Galton, both of whom produced journals, letters, books, maps, and sketches. Such texts as I discuss here are very select, and by no means comprehensive, but the effect is to place travel narrative uneasily beside 'oral tradition' as two modes of historical recording. My main interest lies in the way the latter interrupts, and is interrupted by, the local forms of 'documenting' (Figure 8.1). This enables us to question the particular assumptions of each form of historical representation.

The notion that real time and real history only begin with the arrival of Europeans and their documentary methods has a long critique in African historiography. It is equally germane to ask how time and history figure in the residues of precolonial African discourse. One reason for juxtaposing these historical 'sources' is that each one operates within a different epistemological spectrum (McCaskie, 1995, p. 144). Whereas travel narrative contains a set of references to metropolitan knowledge systems, these are located outside, and the information concerning Owambo appears no more than a surface index of signs being

---

[5] Ryan's emphasis on cartography in his study of exploration overlooks many of the problematics of exploration. In view of Andersson's difficulties in this terrain, which led to his decline into delirium and death (from blackwater fever), Johannes Fabian's recent work entitled *Out of Our Minds* comes much closer to representing the contingent, vulnerable, at times inchoate, nature of such expeditions (see Fabian, 2000).

[6] The longest statement by Kaulinge was transcribed, translated, and published (see Kaulinge, 1997), with an introduction by this author.

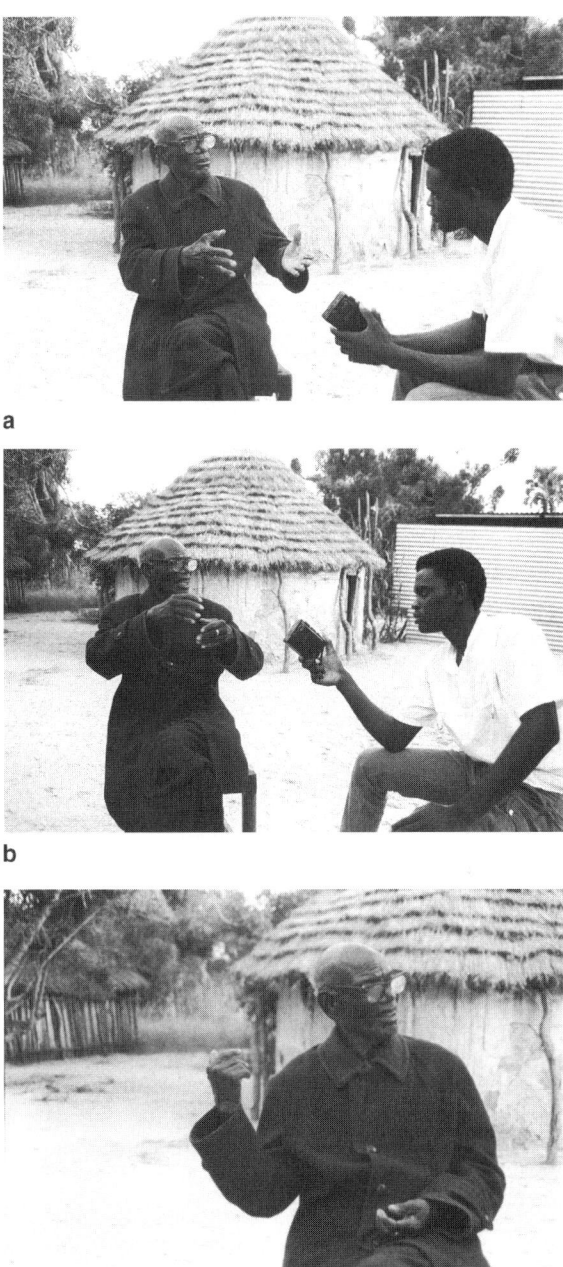

a

b

c

Figure 8.1. Narrating the past. The Reverend Vilho Kaulinge interviewed by Patricia
Hayes and Natangwe Laban Shapange, April 1990.

d

Figure 8.1. (continued)

interpreted and grafted on to an emerging knowledge base in the discipline of geography. By contrast the Kwanyama oral tradition transmitted by Kaulinge seems to offer an internal and regional reference system going back hundreds of years within this part of the African continent.

Placing travel narrative and oral tradition side by side reveals vastly different genealogies and archives. This most obviously refers to their content, but the question must extend to form as well. I read the lack of fit, how the different preoccupations in narrative structure and mode of communication of the two media at times interrupt each other. Such interruptions produce fissures that go behind the surface of things, pointing to each medium's narrative constraints and evidentiary paradigms. Charles John Andersson's journal from his travels in 1867, for instance, makes no reference to any governmental measures taken by Mweshipandeka, or to the Kwanyama past. Andersson's account represents a moment of intersection where two trajectories in time passed each other. In Virilio's spatial metaphor, he was looking at Kwanyama history as if through a slit (2000, p. 38). This is not simply because his own research mission dwelt on other issues. To some extent it was the product of the journal medium and its temporal rhythm, in which travellers wrote day-by-day accounts in the immediate past tense. The space covered in a day, or the delay in movement, were the moments for writing, and the period covered by each entry was necessarily short. The journal was diurnal and descriptive, but cumulatively acted to establish clear sequential chronology. Galton's *Travel Narrative* (1853) closely followed his journal entries, a structure that added to its authority.

By contrast, in Kaulinge's account, time cannot be conceived of in the same way. The specific local historical discourses that are conveyed by Kaulinge concerning landscape and kingdom, productivity and power, indicate that the Kwanyama (like the Akan in Asante) could be located within a centuries-long process of colonising the land. In this 'oral tradition' time is not linear. The temporalities are marked out in a specific way, and as I attempt to explain, part of this has to do with the nature of the medium. Pasts are not so much layered as compacted by the weight of new histories, through ongoing transmission and interpretation across generations, losing their proximity. In a process of historio-graphical maintenance these are fitted into a paradigm – an ancient and meaning-ful schema – that works and which, superficially, appears cyclical. But the earliest tranches become misshapen, increasingly unrecognizable as verifiable 'fact'. They are riddled with old codes; these remain, whereas the contents are emptied. Such features are shared by praise songs and poems (see Kavari, this volume), and sometimes become mnemonic devices. Of these earlier layers in oral tradition Wrigley writes: '[W]e are dealing with a past that is not remembered but taught' (1996, p. 9).

My concern here is not with how valid these traditions which purport to go back centuries are as historical record, and to go over the relative merits of oral tradition. I am more interested in evaluating the currency of the images and codes in Kaulinge's and similar accounts, and assessing how close they might be to those circulating more generally in the second half of the nineteenth century amongst Kwanyama hearers. Historians such as Wrigley and Vansina have argued that first- and second-generation oral traditions usually maintain a great deal of the accuracy of the prior tellers. In particular Wrigley points to an intermediate category that he calls first-stage tradition, 'It consists of the reported experiences of those who were old when today's elderly informants were young' (1996, p. 8). If this is accurate, then Kaulinge (who was born in 1900) was hearing a great deal about Mweshipandeka's rule, and his account is arguably still very close to what many people were hearing in Mweshipandeka's time about events, genealo-gies, and origins, together with their explicit and implicit values.[7] This is deeply suggestive of the possible values inscribed and read off the landscape here in the 1860s when this chapter begins.

8.2. HUNTING AND TRADITIONS OF ORIGIN

Andersson's narration in 1867 of meeting kings as they go for the hunt consti-tutes one of the most productive 'interruptions' of medium and genre. As the record of an encounter, it crosses the path of a profound body of Kwanyama history, the very tradition of their origin as a *hunting* people. This tradition of

---

[7] Although Kaulinge's account seems to be separate from oral traditions recorded by missionaries in Oukwanyama, the latter were solicited between 1890 and 1915, a period very close to Kaulinge's boyhood and a time he emphasizes emphasises as one of learn-ing about the royal histories. The most prominent missionary-ethnographer was Sckär of the Rhenish mission; his material was later extensively used by the American ethnologist Edwin Loeb who published *In Feudal Africa* (1962). These ethnographers and Kaulinge are frequently covering the same ground, but in very different ways.

origins contains the fundaments of the landscape codes which travellers could not read, but which inflected Kwanyama daily life. These codes contained the elements around which the beginning of life was narrativised and carried through into Mweshipandeka's time and beyond. I do not say that each person in Oukwanyama shared a common cognisance of each and every semiotic root of the values with which they imbued their ecological surroundings. Indeed, one of the key aspects of landscape was to function as a social hieroglyph, which concealed the basis of value and naturalised the conventions in which it was viewed or experienced. But the land retained its capacity to act as a bank of signs. These could be drawn upon and imbued with renewed meanings: possible to find in a multiplicity of places, made concrete by markers, re-enacted through many ritual gestures, and conveyed across time by stories through generations. Thus the hunting activity witnessed by Andersson was not only an implicit state-ment about the plenitude of the environment, but a reminder that the latter was a cultural and historical space. That the Swede cannot allude to it in such terms signals one of the incommensurabilities of this precolonial encounter.

Kwanyama origins allegedly lay in the very act of hunting. The traditions of origin invariably state that the progenitors of the Kwanyama were an offshoot of a prior group. Indeed, there are shared ancestors and shared myths about the begin-nings of all those who came to be known generically as Ovambo. These include narratives concerning how the group later known as Ndonga occupied the first place of settlement, and the other Ovambo progenitors branched off from there. In some accounts they are even individualised as Mukwambi, Mungandjera, and so on. Across the floodplain, the genesis of Ovambo kingdoms is therefore presented like a starburst: an explosion of local migrations from one point which came to be known as Ondonga. Remote ancestors allegedly arrived at this loca-tion after a long series of migrations, beginning in the Great Lakes region of central Africa. There are also larger traditions concerning joint Herero–Ovambo origins, but the starburst narrative is in keeping with the striking linguistic, cul-tural, political, and other affinities shared by discrete Ovambo polities on the floodplain. Their oral traditions and histories share common themes, with similar apocryphal stories about cruelty and transgression by their rulers, although not all have the same outcomes (see McKittrick, 2002). They interacted with largely the same environment, apart from small variations. Archaeological research and linguistic histories that link the Ovambo with other African clusters and move-ments have been negligible thus far. But a vast body of transcribed oral tradition and history exists, whose poetics and symbolics warrant close attention. It is the imagery of this history that concerns me here.

Kaulinge's account states that a large body of people said to have migrated 'in many groups' from 'the lake region in East Africa' settled in a place called Ombwenge in 'what is now called Ondonga'. From here the progenitor of the Kwanyama split off with a group for hunting purposes:

> In olden days people lived mainly on hunting, and because of this, the Kwanyama left the rest of the group at Ondonga and came to settle here. They went hunting, found a good place for both hunting and cattle-rearing

and decided to settle there indefinitely. There was also enough water. At the beginning when they started to go for hunting, they used to return to Ondonga, but one time they never came back. We were called Kwanyama because the people who came here first were after meat. Nyama is meat, hence the name Kwanyama (1997, p. 24).

Their migration to this place was assumed on the basis of hunting animals, and of movement and mobility through the wilderness. Implicit too in this set of hunting signifiers is the vestigial empowerment ingrained in it, through the overcoming of nature. Cattle-rearing is mentioned in the account of Kwanyama origins, but it is not as important to their foundation as hunting.

The Kwanyama and their alleged offshoot the Mbandja were hardly alone in representing their origins as stemming from a hunter-founder figure.[8] In East African traditions of origin there are Kisolo of Buganda and Munyama of Ila legend, who gave out meat to their followers after being pushed to the outside of hypothetical prior society and coming to inhabit the world of animals. In this way they attached followers, and from this point some kind of transition was effected towards colonisation and settlement (Wrigley, 1996, p. 100). The beginning of lineages is also traced in studies of Rwandan traditions of origin to the ancestors who cleared such new lands, whose dense history congealed in the term *ubukonde* from the root *kunde* (Mulinda, 2002, p. 75).[9] There is a tempting linguistic resemblance with the floodplain, for in oshiKwanyama the term for a local district of settlement is *omukunda*. Perhaps this word also derives from a process of clearing new lands, the initiation of settlement, and the beginnings of sedentarisation. How the etymologies might have become lost – in migration, separation, and distance from affinities, or simply the way the dictionary work in Owambo has shrunk down the meanings and reduced them to bare functional bones – remains very unclear.

Given that the story shares so much resonance with other African traditions of origin, this lifts it out of mere 'local history' to a more continental ('universalist') discourse. Wrigley warns unnecessarily that stories of progenitors should not be conferred with the 'weight of historical evidence'. But the analysis of East African traditions that refer back through centuries nonetheless opens up important methodological possibilities. Without taking a directly comparative approach, the interpretative methods developed from other African traditions of origin, with their depictions of parallel processes elsewhere, suggest a very rich symbolic bank of cultural recall available to groups who have thought and spoken about their history with reference to accounts transferred orally through generations.

The reality and vividness of the story for Kwanyama audiences of the 1860s is taken further if we consider Wrigley's argument that the narrative device where rulers migrate into a place 'serves to give primary myths a spatial dimension'. He

---

[8] The name Mbandja refers to the breast, the highest status body part of the animal slain by the hunter.

[9] According to Vansina the root *kundu* in central Africa also suggests pregnancy, matrilineage, and familial domain (2001, p. 270).

argues in the case of Buganda that 'As the past is divided into Sasa, real time, and Zamani, the time of myth, so was space divided into here and elsewhere; the crossing of the frontier was the beginning of proper human life' (1996, p. 146). He also observes a pattern where 'very often the formation of the [ruling] dynasty is merged with the transition from nature to culture'. This last point speaks to the act of hunting in eastern Owambo in 1867. For the Kwanyama specifically, it was in a sense a re-enactment: power congealed for the king each time he crossed the boundary.

Although oral traditions are legitimising and normalising discourses for a ruling dynasty, they succeed in marking out temporal and spatial platforms. This undermines the assumption that real time only begins with the arrival of docu-mentarists such as Galton and Andersson, for we are presented with a precise sense of when time began meaningfully for these inhabitants of the floodplain. In Kwanyama oral tradition as relayed by Kaulinge, the transition from one relationship with the environment to another comes with an explicit passage into sedentarisation. The narrative clearly pegs the shift to cultivation (from mere hunting) in the reigns of Kapuleko and Hautolonde: 'As a tradition these people also had their ruler. The first one I understand was king Kapuleko who later on was succeeded by king Hautolonde. Both men were kings from the royal family. People started to cultivate their pieces of land. They started with small fields and later extended them' (1997, p. 24).

But there was trouble in paradise. With the first bad Kwanyama king, the beginnings of a problematic of centralisation appear:

> King Hautolonde died and was succeeded by King Shimbilinga. King Shimbilinga did not have good control over his people. His people suffered from hunger and malnutrition because of the lack of sufficient food supply. There were no rules or orders which compelled people to go for work. He is the one who is reputed to have cut people's fingers. He was upset by the fact that people were not doing anything and that there was no food in the country. In order to assert his authority, he thought of a plan to discipline his top leadership.
>
> One day he summoned all senior and junior headmen in the country to his palace and posed the following questions to them: 'Why is it the people are not working in the country? Why are they not working at all?' At that time there was hunger in the country. He told his headmen that because you don't let people work I will give you work to do so that from here you will learn a lesson of making people work. There was a big tree in his field known as *omukwe* with big and long roots. He ordered them to chop it down with their bare hands. There was nothing they could do but start scratching immedi-ately. If they refused they would have been executed one by one. They spent many days scratching that tree with their fingers. Sometimes they went home and came back in the early morning hours, and sometimes they spent the night at the king's palace. In reality, one cannot cut down a tree with his bare hands, and that is exactly why he ordered them in order to discipline them. Snuff-takers were lucky because they used their tobacco containers to scratch the tree, but in secrecy. Because Shimbilinga's watchmen were around and could report them if they were found using any other tool than a person's own

hands. After a long and hard struggle to cut the tree down with their hands,
the king pardoned them and let them go home to start working hard. But that
tree remained there intact. Myself and many others of my own generation
saw it. (1997, p. 24–25)

Here, cultivation is not only civilisation but tied up with the centralisation
of power. Although this extract falls into a clear genre of stories about cruel
kings in the floodplain, it also addresses themes that run like a backbone through
Kwanyama and other Ovambo oral traditions. Lack of 'good control' leads to
hunger and disorder. Here the king reasserts authority by forcing senior men
to perform a futile activity. They are coerced into attempting the impossible,
without tools. The futility is highlighted by the tree remaining 'there intact'.
The use of manmade tools has become so naturalised that the king's order is
seen as going against nature: the scarcely appropriate snuff boxes are seen
as an advantage. This denaturing of work increases the likelihood of remembrance, both in time and through narrative. The story concerning hunger, the
importance of work, and the arbitrariness of kings is seared into memory, fus-
ing economic necessity and autocratic royal power in one ambiguous moral
universe.

The story of Shimbilinga also instates the tree as a permanent inviolable
marker in the landscape. In this floodplain environment devoid of gradients, hills,
and rocks, the physical and cultural significance of trees in such territory cannot
be overstated. More than anything else they symbolise the peculiar degree of
plenitude, for a complex water system had to nourish the marulas, figs, palms,
sandalwood, thorn, and many other trees, whose scale and magnificence over-
whelmed later visitors to the region.

8.3. THE ORDERING OF THE LANDSCAPE

Kaulinge's narrative continues with the precedent of cruelty, this time emanating
from the king Heita. His grandson and successor Hamangulu was killed after two
years in power, in the forest on his way northwest to Onkhumbi. The renowned
Haimbili then became king, and his rival Mulonda caused the kingdom to experience fission as he set up his own palace at Onanbambi. The latter, however,
was negligible in his historical import, and the kingdom was reunited after his
demise. The narrator Kaulinge places himself firmly in the camp of Haimbili as
the 'real' king. One of the most crucial initiatives of Haimbili was to inaugurate
a set of laws during the sitting of an institution called *epena*. The royal prerogative of control over the timing of cultivation was introduced here. 'Epena was
a very important feast for the Kwanyama people. It was during this feast that
new laws and regulations were established and given to the headmen for execution throughout the country. Laws on issues such as the cutting of the grass in
the fields were given . . . in the same way no-one was allowed to harvest before the
crops ripened' (1997, pp. 26–27).

A close reading of the Kwanyama royal cycle so far shows the enunciation of specific platforms of time, which mark off a prior temporality from the next phase. These could be summarised as follows: the foundation of the Kwanyama kingdom by the first group of hunter-progenitors, the start of crop cultivation, and the official regulation of the timing when the population could plant and harvest. The last of these platforms, with the establishment of the *epena*, is very important. It institutionalised the capacity to intervene in production, as opposed to the earlier platform that dwelt on the use of coercion by the cruel Shimbilinga. Jan Vansina argues that it is at this point, where 'these institutions of governance' come into being, that 'societies are born' (2004, p. 101). This is distinct from Wrigley's proposition, that in oral tradition the earlier point of dynastic foundation and the move 'from nature to culture' marks the birth of a society (1996, p. 146). What is so fascinating from the Kwanyama material is to see how the latter is embedded in the former, and both are understood to be crucial.

In Kaulinge's account, the difficulty facing the new king Haimbili in the first half of the nineteenth century was that the number of royals born during the time of Heita now posed a threat to him. Generally speaking this was frequently a feature of matrilineal systems, at times resulting in fission and purging. 'King Haimbili had tough laws, and members of the royal family both men and women disliked him. So many of them were born in the reign of King Heita. These very people posed a threat to King Haimbili because he believed that they would kill him as he himself had killed King Hamungulu' (1997, p. 26). Thus there is the setting of the scene at the *epena*, where momentous laws are passed and dangers to the polity identified and eliminated or expelled. Social good was simultaneous with dynastic purging. There is a theatrical chill about the narrative:

> In that feast which was organised by Haimbili, two prominent members of the royal family were executed. While the people were busy assembling, Haimbili had already instructed his men to carry out the executions. They were told to kill specific people during the feast. He told them that they too will be executed if they fail to fulfil their tasks. In fact four men died during the feast while others fled and took refuge in neighbouring countries (p. 27).

Haimbili used the public assembly to restore order; social legitimacy was pushed and possible challenges to royal legitimacy expunged. It instilled awesome fear and expelled certain traits of blood. Ensuring the nation's fertility was also about trimming that of royalty. Power and genealogy had to run within certain lines, therefore much was driven underground. All of this in Kaulinge's oral account was enacted upon a public stage. It was a stroke of theatre as much as dynastic politics. But the real political skill was in preparing and plotting secretly off-stage; Haimbili could then choreograph the expansion of royal prerogative at a massive public occasion. Besides the scattering of the family, it had enormous consequences.

It was a new demarcation of royal power: a precedent that became a tradition through repetition, through the serial practice by successive kings. Like most traditions it was of course double-edged, subject to 'interpretation'. *Epena* had to be followed but it could be adapted. It was a form whose content could shift; it carried its past even as later kings transcended it.

8.4. LINEAGES AND LANDSCAPE

In Wrigley's analysis of the forms of historical transmission in central African kingdoms, he distinguishes between 'the remembering of deeds' as opposed to the 'rehearsal of ritual' (1996, p. 12). Certain ritual rehearsals apparently underpinned the proliferated kinship system that Marxist scholarship used to call the lineage mode of production. In their technical language, this was often 'articulated' with a more centralised system, whose capacity to extract was signalled by its description as the tributary mode of production. The rehearsal of ritual as opposed to the remembering of deeds, those are the discursive terrains of those lineages that have come under the dominance of the one that has augmented and sacralised itself. The latter group engages in a dynastic economy of history. It *is* the beginning of history.

But this alignment of memory with ruling group and ritual with lineage overlooks a number of things. First of all, the coining of the new institution of *epena* and its place in the inauguration of a new king, together with channelled violence, took on certain features of rehearsal and ritual. But beyond this, Wrigley's neat distinction tends to obscure the contest between kings and clans, and the way ritual and memory interact, quite ferociously at times. In the historiographical edifice that Kaulinge maintained and transmitted, there are many instances of the way important historical gestures of kings become ritualised, and indications of the way ritual is historicised. Probably the most important of the latter in Oukwanyama is the case of male circumcision.

Within royal history is also a lineage or clan history. Lineaments of the clan histories appear very clearly at certain stages of Kaulinge's narrative; but in addition, kings themselves are part of a history of their own lineages. This might be a ruling clan, but certain practices of all clans apply equally to ruled and ruler. All were, for example, subject to circumcision, which was controlled by older circumcised men. This issue later became a radical platform for political initiative and centralisation in the Kwanyama kingdom: but more of this later.

There is a certain point where Kaulinge's genealogical narrative folds back on itself, and reverts to the earliest period of Kwanyama foundations. In this round, it specifically incorporates a history of prominent clans and resonates as a more collective account. It also reinserts the elements of production, and the elements of landscape, in the new space of 'Kwanyama' migration and settlement.

> When they arrived here they set up a temporary base. At the same time they gathered together to choose a person who would lead them. When the people came from the east to what is now called Ondonga, they were in clans. The revision of clans also took place here. There was a certain clan which possessed a lot of cattle. This clan was known as Ovakwanangombe [those with a cow]. This particular clan was chosen to be the ruler of the nation. That is, anyone who became the king of the country had to come from this clan.

Even at that time they chose their leader from that clan. And from there, the
Ovakwanangombe ruled the country before it expanded (1997, p. 64).

Water, cattle, and wild animals allowed the birth of a new polity. Under the first
king, Hautolonde, 'The people started to build their settlements with the permission of the king. This process continued for a long time till the country finally
expanded. People were provided with axes to go and start their settlements.
No-one was allowed to take up a position on his own initiative' (p. 64). A group
or council of advisors, we learn, monitored this process: 'When the people were
building their settlements the council used to advise them on how to build the
settlements, and worked out in which direction and which place' (p. 65).

Between the two versions of the founding of the Kwanyama kingdom, a
thickening and layering has taken place. The first account is genealogical. But
having set out a rudimentary framework of royal succession and the accretion
of powers, the second cycle of the same phase is administrative and explains the
articulation between central and lineage power. It discusses the appointment and
legitimacies of the ruling clan, and how this shifted. But we should also note how
lineages were replenished by the addition of notables discarded at the succession
of a new king. These were circumcised men who remained at the previous king's
embala and at royal graves, and conducted important rituals. These men drew
power from an important dimension of the African landscape that we have not
yet explored, and which lay underground. For graves in general went beyond the
surface of things, representing a spiritual world that in a sense mirrored the visible world of human beings. To sum up, lineages were historically fluid and could
draw on diverse ritual prerogatives and resources. We have already seen how the
royal clan was dynamic. These were not static layers, and they also shaped and
reshaped each other. The category of *elenga*, a new kind of headman, suggested
the greatest degree of social mobility from the second half of the nineteenth century (see Hayes, 1992).

In this round of Kaulinge's narration, as opposed to simply remembering
the deeds of great men, we hear how the Kwanyama brought with them in their
migration the means of changing the landscape from wilderness to civilisation:
'They came with their seed, especially for the *mahangu*' (1997, p. 68). We have
noted how people were given axes and told where to go, signifying technological
and spatial arbitration. If the axe was the tool of colonisation, the hoe implied
cultivation and the bedrock of civilisation, requiring far more intensive labour
than hunting or herding, and thus the need to command work, and thus centralisation of authority. Providentially, the Kwanyama site was relatively close
to Oshimanya, near Cassinga, with a significant amount of iron ore. But this
was only annexed in Haimbili's time, when his brother Hamulungu 'ordered his
people to open the mine which provided crude iron for making axes, traditional
knives as well as hoes' (p. 68).

The complicated transition from the dominance of one clan to another
(Ovakwanangombe to Ovakwananhali) occurred within a few generations of the
first king, Hautolonde. Significantly, it was the tricky figure of Shimbilinga who

came in as the victorious candidate. His succession was associated with larger ruptures, tensions, and struggles amongst various constituencies who were close to power or sought to usurp it. Kaulinge's second cycle about the cruel king Shimbilinga is given a different emphasis from the first. Instead of detailing the labour on the tree, the account highlights Kwanyama subjects' desire to control the king. 'King Shimbilinga was a very bad king, according to reports. He made people cut a tree with their bare hands because they were lazy, he claimed. From there the people decided that such deeds could not be tolerated any longer. They kept an open eye on him' (pp. 67–68).

We have to go back to the first cycle in order to trace a decisive moment in the contest between kingship and lineage. After Haimbili, Haikukutu, and Sheefeni, we come to the reign of Mweshipandeka, who managed (unlike his brother) to avert assassination in the forests between Ondonga and Oukwanyama when summoned by his predecessor Haimbili. Earlier in the genealogical narration, Kaulinge explains that 'There was another rule which stipulated that anyone taking up the position of king must have undergone a traditional process called *etanda* (circumcision). . . . This process also applied to males of any clan who wanted to take a wife' (p. 28). In an extraordinary move, 'Mweshipandeka had not undergone *etanda* and when he was asked to take over his brother's palace, he refused, saying that he was going to build his own palace. And so he established his kingdom at Ondjiva' (p. 31). From this point onwards in the Kwanyama kingdom, male circumcision fell totally away. Its demise gradually spread through most of the floodplain, although it was a complex and varied process (see Salokoski, 1992).

This was not the only sleight-of-hand effected by the king whom Charles John Andersson met in 1867. Mweshipandeka brought all headmen to central court at Ondjiva when legal cases were heard, housing them for a month at a time and attracting big local audiences for the hearings. 'The people's court was at the palace' (p. 33).

Mweshipandeka also set a crucial precedent for the expansion of Kwanyama power regionally. The proximity of his reign to Kaulinge's own lifetime has a heightened impact on the story-telling:

> When we were brought up we were told many stories about King Mweshipandeka. We were told that he liked war very much. The main reason for waging these wars was to go and capture prisoners, and to bring as many cattle as possible back home. . . . When I grew up in my father's home, he used to tell me that it had not taken him much to get the cattle he owned. All of them came from Ombwenge where they used to go for war (p. 32).

This is by no means all. Under Mweshipandeka the expansion of power was accompanied by an expansion of productivity, and the intensified cultivation of the immediate environs. As Kaulinge concludes, 'In short, Mweshipandeka's rule was accepted amongst his people. He urged them to cultivate their crops and punished those who were lazy. It was during Mweshipandeka's rule that people began to cultivate bigger lands. People had enough food to eat' (p. 32).

In the case of the Kwanyama kingdom, huge changes over time with their long processes of environmental interaction and political consolidation led to what Galton and Andersson saw in 1851: the 'diffused opulence' of an Ovambo kingdom.

8.5. A LAND OF GOSHEN

We now turn to this earlier episode in the accounts by travellers. When the expeditionary party finally reached the end of the wide thirstbelt from Hereroland in 1851, the fecund terrain of Ondonga burst upon Francis Galton's sight. It is worth quoting in full.

> It is difficult for me to express the delight that we all felt when in the evening of the next day we suddenly emerged out of the dense and thorny coppice in which we had so long been journeying, and the charming corn country of Ondonga lay stretched like a sea before us. The agricultural wealth of the land, so far exceeding our most sanguine expectations, – the beautifully grouped groves of palms, – the dense, magnificent, park-like trees, – the broad, level fields of corn interspersed with pasturage, and the orderly villages on every side, gave an appearance of diffused opulence and content, with which I know no other country that I could refer to for a parallel (1852, p. 151).

This verbal description, unlike numerous nineteenth century travel narratives, does not elevate the scene to the realm of the sublime. The latter was usually triggered by scenes that were rugged, mountainous, and wild, causing an elevation of sentiment. The foremost quality recognised here is the orderliness that accompanies (and gives rise to) fertility, abundance, and variety, and this seemingly ignites Galton's lyrical response. The sculpturing of the landscape through cultivation and settlement presented an unexpected scene to the travellers, who were the first Europeans to arrive in Owambo from a southerly direction.

Galton's published account of the Ovambo landscape in the early 1850s struck many readers in turn, and was commented upon in the *Cape Monthly Magazine*: 'One can scarcely imagine the pleasing picture sketched by Galton to have its locale among the wildernesses of interior Africa'. The particular passage that most affected readers went:

> Fine dense timber trees, and innumerable palms of all sizes, were scattered over it; part was bare for pasturage, part was thickly covered with high corn stubble: palisadings, each of which enclosed a homestead, were scattered everywhere over the country. The general appearance was that of the most abundant fertility. It was a land of Goshen to us; and even my phlegmatic wagon-driver burst out into exclamations of delight (1858, p. 195).

Galton's admission that he could not find a comparable place signals the peculiar sentiment that European viewers often remarked with regard to Owambo. It was different but familiar, unique but recognizable. Its fecundity was unexpected; its trees prompting Galton to speak of parks that aristocratic estates or metropolitan publics enjoyed in Europe. In such terms, it was a sight that

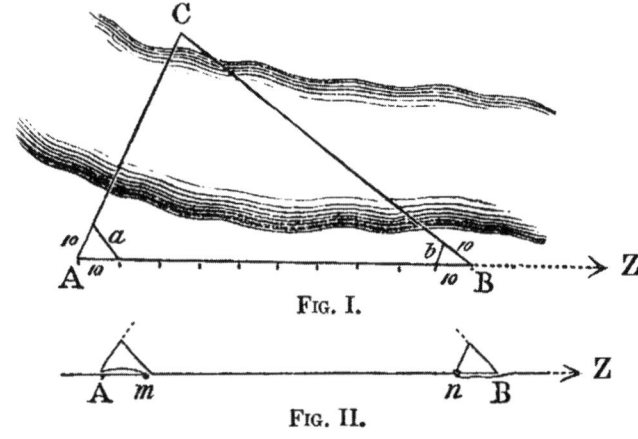

FIG. I.

FIG. II.

To find B C and angle B :—Enter with *b* at the side and *a* at the top.

Figure 8.2. Galton's 'Hints to Travellers': How travellers may ascertain the breadth of a river, valley, or any distance of an object by taking an additional 60 paces and making reference to the Table for Rough Triangulation (Galton, 1864, p. 282)

invoked recognition rather than a sense of alterity. Galton's depiction suggests how this view subtended by the eye evoked a Europeanised landscape with biblical overtones, as opposed to the troublesome African landscape (the Thirstbelt) through which the travellers had just passed.

The purpose of Galton and Andersson's expedition was to reach the Kunene River to the north. Galton had won the backing of the Royal Geographical Society in London for a South African expedition in the wake of Livingstone's 'discovery' of Lake Ngami. Galton's privileged class position and prominent family connections had gained him initial support in the Royal Geographical Society (RGS), and later facilitated his rise to leadership of the organisation. He was the wealthy gentleman-explorer, which explains some of his aristocratic landscape references. Andersson was of a different class, always having to pursue his own ambitions by the seat of his pants, and with a different aesthetic range. His predominant interests lay in science and natural history, and in 1851 his role was more that of an apprentice to Galton the 'explorer'.

While preparing for the sea voyage to Africa in 1850, Galton had picked up the impecunious young Andersson who had just landed in England from Sweden. He came with a veritable menagerie of animals that he tried to sell to any interested party. Andersson's father was the Englishman Llewellyn Lloyd, a man who 'was for the wilds and the life of the wilds, and became a hunter off the beaten track' in Scandinavia (Wallis, 1936, p. 29).[10] Andersson took his Swedish mother's name

---

[10] The book is a biography of Andersson, with a foreword by General Smuts in recognition of his exploratory work and status as Galton's companion.

but thoroughly embraced his father's love of the wilds. Andersson's passions included hunting, natural history, botany, ornithology, taxidermy, exploration, astronomy, cartography, and many other matters. It was very telling that while the amateur geographer Francis Galton was busy learning how to read a sextant for the first time on the ship out to Cape Town, Andersson harpooned his own hand and had an old musket explode on him, the first indications that he was adventurous and 'seemingly accident-prone' (Gillham, 2001, p. 63).

Water as the intimate clue to the continent's geography, a means of passage as well as survival, was consistently the goal of African exploration, before and after Livingstone. Galton had intended to reach Lake Ngami but was deterred by Governor Harry Smith in the Cape, who warned him of the potentially troublesome Trekboer presence in the region. Galton's ambitions turned towards the Kunene River, which had never been reached from the south. This was how Galton's party redirected itself towards Owambo, a trope of isolation for explorers, hunters, and cartographers. After journeying through Damaraland and sojourning some time in the kingdom of Ondonga, all the subject of his famous 1853 travelogue *Narrative of an Explorer in Tropical South Africa*, Galton was refused permission to proceed through Oukwanyama to the river by Nangolo, then king of Ondonga. Galton acquiesced and turned his wagons back. But his publications concerning the expedition were an incitement to further exploration.

The next travellers to attempt this route were the Rhenish missionaries Hugo Hahn and Johannes Rath in 1857, who sought new mission fields outside Herero territory, then referred to as Damaraland. They were accompanied by Frederick Green, the elephant hunter. The trip to Ondonga ended in misunderstanding and Nangolo's son perished in the ensuing skirmish. Publication of the incident caused some revision of the initial favourable opinions of the floodplain. But after Green, Hahn, and Rath's trip, Andersson wrote to Galton in London that he intended to pursue a new expedition in search of the Kunene. 'I don't think I could rest in my grave while the position and character of this much talked of stream is more fully ascertained and established'.[11] He attempted this in 1858, which ended perversely in the 'discovery' of the Okavango River instead. Green in fact reached the Kunene in 1865 (see Baines, 1866, p. 248), and Andersson's final attempt to reach the river was made in 1867 when we meet him at beginning of this chapter.

Andersson was for a long time engaged in writing and cartographic work, both of which merit some remarks here. If local knowledges were banked in the landscape, what were the implications of travellers' writings concerning the landscape here? One of the most telling remarks about writing came from Frederick Green, the professional hunter who travelled with Hahn and Rath on the ill-fated visit to Ondonga in 1857. Green had gone in search of Andersson in 1858 in the area near the Kunene, rather in the manner of Stanley seeking out Livingstone with attendant publicity. The hunter sought to allay public anxieties in the Cape

---

[11] University of London Archives, Galton Papers, Andersson – Galton, Otjimbingue, 21 September 1857.

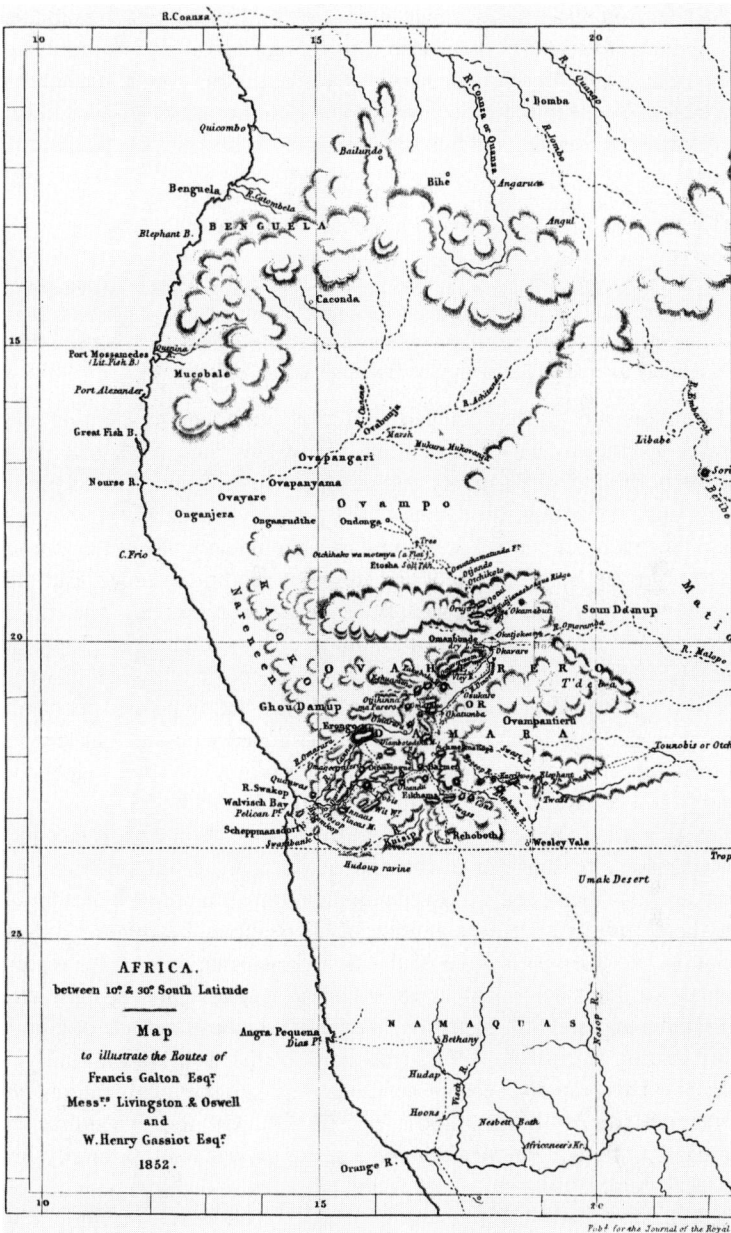

Figure 8.3. Detail of the map to illustrate the routes of Francis Galton, 1852

where Andersson was a well-known figure, as Andersson's path would seem to have led him into territory now hostile to foreign travellers. He published two chapters in the *Cape Monthly Magazine* of 1860. He transmitted the substance of a letter Andersson wrote him as a reassurance to the public of his safety, interspersed with fairly sensational hunting anecdotes of his own. He justified himself as follows.

> The two anecdotes I have related happened to be entwined more or less with the general thread of my narrative, and I ventured to obtrude them, at the risk of being thought boastful of my own exploits. Adverse criticism will not, however, reach me; for, pursuing the tactics I occasionally practice in hunting, after firing my shot, I have rushed into the recesses of the interior, from whence I am not likely to emerge for the next two years (Green, 1860, p. 362).

'Firing his shot' is thus equated with inscription, and the hunting metaphor acknowledges the taking of something out of this landscape, and the power implicit in both modern weaponry and writing. This was an extremely extractive time of elephant hunting, be it noted, and possibly travel writing as well. The two activities were, of course, happening in tandem every day of the journey, and were the norm for many travelers and missionaries on the move. But although Green sought ivory and wrote but little, Andersson's dream was taxidermy, cartography, and authorship. After the failure of his commercial enterprises, his ambition was to publish his travel journals and scientific observations in order to support his wife and children living in Cape Town, and to pay for his natural history research and specimen collections that he hoped would garner him international recognition. For years after their joint expedition he vainly sought adequate backing from Galton and through him influential figures in the RGS, once going so far as to accuse the latter of cold indifference. Galton had in fact for years used Andersson's presence in southern Africa as a means of continuing his own research agendas there, and the Swede did his utmost to oblige by return of post, despite the at times disastrous condition of his health and fortunes.[12]

Galton had attempted some of the first sketch-mapping of the region. The biographer Gillham notes that in 1850 Galton had assembled 'some charts to aid him in his exploration', including 'a detailed map of the Cape of Good Hope and surrounding areas from John Arrowsmith of the well-known family of cartographers'. This 'rapidly became devoid of any geographical features north of the Orange River' (Gillham, 2001, pp. 62–63). On the expedition through Damaraland as far as Ondonga, Galton's geographical work primarily involved taking measurements with a sextant, fixing latitudes, and sketchmapping, but also included sketching and painting watercolours.[13] The sketch-mapping was later

---

[12] See University of London Special Collections, Galton Papers, correspondence.

[13] This included a facetious episode in which he used his sextant to measure the dimensions of a Khoe woman from a distance.

professionally transcribed into a map by Livingstone, Oswell, and Gassiot of London, and published in 1852 (Galton, 1852, pp. 140–141) (Figure 8.3).

The presence of information from further north, as well as central and southern Africa, attests to Galton's access to and derivative use of numerous other African maps. This bears out Ryan's contention that critically we should see mapping as 'temporally embedded and transformative of previous maps, rather than as an innocent inscription started afresh on blank paper' (Ryan, 1996, p. 103). However, it is the very discourse of blankness in the cartographic exercise that so powerfully defines the emplacement of African spaces within the grids of universal measurement. To wit, in his 'Notes for a Projected Book' Section 24 under the heading 'Observing – Astronomical Instruments. Mapping and Mapping Implements', Andersson urged the cartographic project in these terms: 'Every traveler who succeeds in penetrating into parts hitherto unknown or unrecorded would greatly enhance the value of any information that he might be able to render by accompanying it with a sketch map of his route . . . for there are as yet such vast blanks in the African charts that any such light thrown on this immense continent would be acceptable.' [14]

'Parts hitherto unknown or unrecorded' was the stock-in-trade of European cartographic discourse in the mid- to late nineteenth century, which banked on a putative emptiness. As Bollig and Heinemann note, however, earlier maps were not necessarily empty but filled out on hearsay (2002, pp. 271–272). Travellers such as Galton and later Andersson then 'emptied' sections of older maps and attempted to chart the spaces in a manner that qualified as more scientific.

Part of Andersson's construction of blankness lies in the disparagement of what indigenous informants could offer. This is no surprise, given his own blank or critical response in the face of autochtonous codes noted earlier. He warns the traveller that 'In laying down on [sic] a map from native information, there is but little to guide the judgement . . .' but offers a few hints:

> The natives of course only count time and distance by the number of days they consume on a journey, and if less than a day it is by the height of the sun in the sky at the time of starting and arrival. But, then the question arises, what is a day's journey in their estimation? The only way to arrive at anything like the truth is to enquire what proportion a certain distance that you yourself have traveled and that your acquaintances may be acquainted with, near [est] to their day.[15]

Andersson's 'Notes for a Projected Book' are couched as advice to the would-be traveller, a participatory project with implications for the future. Andersson recommended practical, amateur methods. 'It is quite remarkable to what accuracy a person may bring his observations by such simple dead reckoning i.e. courses and distances traveled'. As textual genre, this provided as much incitement as the medium of the map itself. The Notes were never published, although Andersson's

---

[14] National Archives of Namibia (NAN), A83, Charles John Andersson papers, 'Notes for a Projected Book'.

[15] NAN A83, Charles John Andersson papers, 'Notes for a Projected Book'.

final and lachrymose travel journal appeared posthumously as *Travel in South Africa*. Correspondence over the years with Galton reveals the way the eminent man in London demanded and extracted heavily from Andersson's years of practical African experience for his book, *The Art of Travel*, which became a runaway success in London and elsewhere.

The correspondence between Galton and Andersson points to the way the metropolitan scientist could pose the foremost questions of the day to his southern African field researcher, as it were. In turn, this meant that although Andersson was largely based in Damaraland or the Cape, he was kept abreast with the latest turns in geographical debate on the continent that were happening in Europe. Access to libraries and periodicals in Cape Town further assisted Andersson, who also corresponded with a variety of scientific persons and institutions in Sweden. But it was mainly with Galton that Andersson discussed Green's findings in the Kunene region, as this all had a bearing on the map to which Andersson was devoting much effort. Brigitte Lau suggests that Andersson was 'courted by British officials and scientists for his then highly interesting and precious map of the interior' (1989, p. vi), but Andersson needed to reach the Kunene for the map to really provide geographical answers, or so he thought. The enigma of the Kunene, as Bollig and Heinemann have shown, puzzled cartographers for nearly two centuries, with successive maps showing the river going underground, or directed towards the Kavango or elsewhere (2002, pp. 271–272). For a time it represented a 'riddle' rather like that of the Nile, its resolution quite a prize.

Andersson's health, however, had been poor ever since he had been shot in the leg at Otjimbingwe in 1864. Here his trading ventures had prompted him to 'assist' local Herero fighting against Oorlam cattle-raiders led by Jan Jonker Afrikaner's sons. Impoverished and ailing, Andersson could not complete the map to his satisfaction, and the artist-traveller Thomas Baines wrote the accompanying notes when it was finally published by the RGS in 1866. Andersson managed to obtain the necessary credit to stock his final expedition, and returned to the floodplain in search of the river in 1867 when we meet him at the opening of this chapter.

8.6. PLENITUDE AND SCARCITY

It is intriguing to notice how in following the trajectories of Andersson and Galton, we seem to have strayed far from the original question of landscape. We have entered much more the conceptual domain of space, with its 'connotations of abstraction and geometry' (Mitchell 2002, p. ix). Mitchell cites Lefebvre's distinction among perceived, conceived, and lived space in his recent revisionist approach to landscape, and suggests that these correspond broadly to space, place, and landscape, which need to be considered in relation to (rather than isolated from) one another. What precisely are these distinctions, and how useful are they for the nineteenth-century African case under consideration here? After considering oral tradition and travellers' accounts (forms of knowledge that are constituted very differently) alongside each other in this chapter, how do the two kinds of account invoke space, place, and landscape, if it is appropriate to ask? More especially, does 'landscape' still have the implicit explanatory and suggestive power ascribed to it at the outset of this chapter?

Mitchell points to the daily spatial practices that mark out a perceived space, the 'intellectually worked out' quality of conceived space (the 'planned, administered and consciously constructed terrain'), and lived space that offers signs that appeal to the imagination (2002, p. x). The potential for exploring the ways that Kaulinge might refer to all three in his account, and likewise Galton and Andersson, is very rich. It is impossible to do justice to it here. But one issue emerges very forcefully from setting up these conceptually distinct approaches to viewing a place, and that is the way the travellers' accounts give Africa a geography rather than a history, conceptualising space in a highly presentist way. The oral tradition gives layers of time, after an initial setting out of the space of people's settlement. Perhaps it boils down to geography being a space-based discipline, as opposed to time-based history. But in these different agendas, the importance of landscape is retained most strikingly for the African accounts of the place: for certain signifiers continue to exist in the place that refer back to the meaningful pasts of generations before, and continued to stir the imagination in Kaulinge's time. I explain this below by returning to the accounts in question, not forgetting that Andersson ultimately failed in his mission in 1867.

Water drew the first travellers to the floodplain. This is true even while commercial and evangelical interests were tied to exploration. But oral traditions say very little about water, they say much more about forests and fields. Apart from the reference to originary lakes in east-central Africa where the ancestors emerged and from where they migrated, finally, to the floodplain, water is remarkable for its near absence. By contrast Herero praise poems from groups elsewhere in Namibia feature the naming of waterholes as strong nodes, drawing the lines of genealogy and landscape together (Henrichsen, 1999). These, however, were the markers of pastoralists who moved their herds over huge territories. The sedentarisation of the Ovambo on the Cuvelai floodplain has perhaps been of such long date, that water has receded from oral narrative as a marker and become completely naturalised and invisible within it. Perhaps too the seasonal nature of water on the floodplain has made it recede from larger historical accounts, just as the *efundja* (floodwater) itself sinks from view.[16]

Water that was permanent, and thus visible, lay on either side of the floodplain, at some distance from the population clusters generically termed Ovambo.

---

[16] Waterholes do emerge in certain kinds of oral narrative, for instance, the Mbandja oral accounts of fighting the Portuguese in 1904 and 1907. See Hayes (1992, vol 2: 6).

These were the Kunene and Okavango rivers to the northwest and northeast, respectively. These two rivers had nothing to do with the annual flooding of the plain and the filling of the Etosha pan to the south. This was at the root of the confusion amongst early travellers. Galton, Andersson, Hahn, Green, and later the French missionary-explorer Duparquet, all believed the source of the *efundja* to be the Kunene River (Hahn, 1867, pp. 295–296).[17] The name translates locally as the Great River, and it eventually debouches into the Atlantic Ocean. The entire plain was in fact fed by the *oshanas*, the network of water channels like veins and arteries through which water seeped when the rains fell to the north. The flooding had its origin in the Cuvelai river system in the Angolan highlands further to the north and consequently seeped most heavily into the northern areas of the floodplain (Wellington, 1967, pp. 19–20; Williams, 1991, p. 37; Clarence-Smith & Moorsom, 1977, p. 96; Siiskonen, 1990, p. 39). If rains affecting the Angolan headwaters were plentiful, the *efundja* travelled as far as the Etosha pan (Wellington, 1967, pp. 19–20), and floodwaters remained in *oshanas* (channels) and reservoirs from January until June. Southern *oshanas* tended to dry out first. In poor rainfall seasons in the Angolan headwater region, the *efundja* in Owambo was brief and transient, or did not reach the main floodplain at all (Tönjes, 1911, p. 18; Nitsche, 1913, p. 47; Serton, 1954, pp. 104–105).

This water system, characterised by seepage and transience, was like a secret map – the enigma of the landscape – that eluded scientific reading through to the late nineteenth century. The exploratory work of Galton, Andersson, and Green, for example, stumbled through a variety of false and evanescent clues, which did little to reveal the lie of the land. Nor could their routes encompass the wider terrain that would have supplied the answers they were seeking. When Duparquet travelled up through the floodplain in 1880 and thence Onkhumbi across the river, making extensive observations concerning watercourses and societies on the floodplain, he still categorised Ovambo groups as related to the Kunene river system.

It was the obsession of European exploration to prove this hypothesis. Groups on the floodplain itself had other preoccupations that revolved around periodic scarcity. This is amply shown in the kind of oral tradition we have examined here, with narration moving in and out of scarcity and royal control. The *efundja* was only one of three sources of water that sustained the dense population and crop cultivation in the floodplain region by the nineteenth century. In scientific terms, average rainfall was approximately 550 mm, but with high variability (Salokoski et al., 1991, p. 220). Partial droughts were frequent and widespread drought occasional.[18] As rains dropped off and the water standing in

---

[17] See also *Archives Générales de la Congrégation du Saint-Esprit*, 465-III, Duparquet, *Notes sur les Omarambas*, ca. 1880.

[18] See Clarence-Smith for a chronological treatment of regional droughts. P. D. Tyson (1986, p. 68) counters arguments that progressive dessication has occurred in the last two hundred 200 years.

oshanas from the *efundja* dried up, the Ovambo depended increasingly on their third source, groundwater. Other environmental elements interconnected with the presence or absence of water. For instance, the silt carried down from the more fertile headwater region was deposited in the *oshana* floors once the *efundja* retreated. Thus soil fertility, and by implication vegetation, crop cultivation, the herding of cattle and small stock, hunting, and gathering, were all affected by the complex hydrology of the floodplain and the variations in geographical position.

These interlocking features of the environment in their turn influenced demographic and political factors and went some way to explain the emergence of densely populated kingdoms in the region, and possibly why their features of state control developed in a certain way.[19] It also had a bearing on the cultural map of water that was meaningful throughout Owambo. All three water sources tended to favour the more northern areas in terms of plenitude, duration, and salinity. The northernmost kingdom of Evale, in particular, usually continued to receive plentiful rain in April, whereas southern areas such as Ondonga saw scarce rainfall at this time (Siiskonen, 1990, p. 3–4). This doubtless shaped the particular belief that the Vale kings had the greatest rainmaking powers on the floodplain: it coincided with Evale's favoured climatic position over southern polities (Shityuwete, 1990, p. 1; Williams, 1991, p. 46). This mutual intelligibility between climate and magical credibility points very directly to local understandings of landscape. Ritual grew out of this water map. The various kingdoms sent black bulls as offerings to the Vale king who belonged to the rainmaking clan, usually in the middle of the season when crops had begun to grow and rain was short.[20]

The Kwanyama were favourably situated regarding water, located towards the north of the floodplain and centered on the *oshana* system. Although we need to look outside oral tradition in order to grasp the singular environment in which the Kwanyama lived and cultivated, the medium is fundamentally concerned with production. The evolving relationship between people and the land is elaborated and systematised over time in Kaulinge's account. It profiles an expanding social order. Power was derived increasingly in Mweshipandeka's time through a growing population supported by a mixed economy of cultivation and herding which made use of a natural environment over which kings presided, but with which people had a very complex material and symbolic relationship. Thus it is necessary to consider landscape and kingdom together (Figure 8.4).

---

[19] Duparquet in 1880 noted the construction of a reservoir in Ombandja, which Clarence-Smith and Moorsom (1977) have argued would have required centralizsed kingship.

[20] Interview with Aromas Ashipala, Jason Ambole, Petrus Eelu, Vilho Tshilongo, and Jason Amakutuwa, Elim, 26.9.1989.

Figure 8.4. Postcolonial landscape. Visit to the graves of the Kwanyama kings near Namakunde in southern Angola, officially designated as national patrimony, 1999 (Photographs Patricia Hayes) (*See also Color Plates*)

d

Figure 8.4. (continued)

8.7. CONCLUSION

One of the purposes of this chapter has been to explore the different perceptual systems underlying Kwanyama and Ndonga interpretations of the landscape, and those of travellers such as Galton and Andersson. It might also be helpful to take account of histories of 'the senses' more broadly, especially of those who recorded the nineteenth century past in Owambo. This conclusion focusses on two striking facets of the accounts rendered by Kaulinge and Andersson.

We return to Kaulinge's story of the reign of Shimbilinga, who put his headmen to work in cutting down the *omukwa* tree. Significantly, it is related that the headmen left their mark on the tree. 'But that tree remained there intact. Myself and many others of my own generation saw it,' said Kaulinge (1997, p. 25). This landscape marker then becomes the 'evidence', the form of witness to the story, even though Kaulinge acknowledges that Shimbilinga was only 'reputed' to have cut people's fingers. The evidentiary paradigm is tactile and rooted in the landscape, naturalised in more ways than one. The tree carries the physicality of history that young people of Kaulinge's generation saw and perhaps themselves touched, serving a cross-generational function. The story is mapped into the landscape and thence into the mind in a way that makes forgetting difficult.

Because of this story, the tree acts as an archival entry point into Kwanyama history, one that is not removed from the senses because it has a somatic impact through both the visual and tactile apprehension of new generations. It affects their cognitive processing of the story concerning hunger, the importance of work, and the arbitrariness of kings. The point is also that it is a story *in* the landscape. As Mitchell says, landscape is both a space and a represented space (1994, p. 5).

   As opposed to this clustering and proximity of signs and senses, Andersson's writings locate him in an intellectual history where the senses had been increasingly stratified over several centuries. Vision had assumed primacy in postenlightenment European cultures. Cartography in particular represented its radical abstraction: 'The founding assumption of cartography is that it presents the user with a view of the land from above' (Ryan, 1996, p. 102). To achieve this perspective, conventions for shading are used as well as 'colouring and iconographical codes for objects seen from above'. This is a specific kind of visual abstraction allied with text and coded symbols, creating a unique specular position. 'Separate from "the seen", the viewer (as textually constructed) is at the same time at its center'; this is either literally or in the sense of possessing a controlling and privileged viewpoint (1996, p. 101). The ability of the map to convince the viewer that it is accurate stems in part from the representational devices used. The conventionality of the signs used becomes forgotten in most map reading and, Ryan concludes, 'in the forgetting arises the peculiar access to the real granted to maps' (1996, p. 103).

   Accuracy, realism: these are the discursive terms of the map, facilitated by a variety of technical and prosthetic devices. Plus, forgetting. By contrast, Kaulinge's body of oral historiography is hypomnesic, especially around nodes such as Shimbilinga and the tree. Displacement also characterised the documentary activity of Galton and Andersson, as opposed to the didactics of Kaulinge, for unlike local residents they did not bank knowledge in the environment itself but far away in metropolitan institutions.

   This came with an alienation from – and abstraction of (Harley, 1992) – the landscape, as the floodplain was constituted as one of 'the vast blanks in the African charts'. In Andersson's travel notes, the entire floodplain was a vast open space in which European science and extractive methods could be brought into play. Ryan suggests in relation to such assertions: '[T]he cartographic practice of representing the unknown as a blank does not simply or innocently reflect gaps in European knowledge but actively erases (and legitimizes the erasure of) existing social and geo-cultural formations in preparation for the subsequent emplacement of a new order' (1996, p. 104).

   This is why, far from being a neutral representation, maps are 'an incitement' in the way they conceive space. All of this points to the fact that cartography and oral tradition are diametrically opposed in their effects. Scientific cartography tries 'to convert culture into nature' (Harley, 1992, p. 242). Oral tradition, however, seems constantly to seek the opposite.

REFERENCES

Baines, T. (1866). Notes to accompany Mr C.J. Andersson's Map of Damaraland. *Journal of the Royal Geographical Society*, 36, 247–248.
Bollig, M.&Heinemann H. (2002). Nomadic savages, ochre people and heroic herders: Visual presentations of the Himba of Namibia's Kaokoland. *Visual Anthropology*, 15, 267–312.

Clarence-Smith, G. (1974). Drought in Southern Angola and Northern Namibia, 1837–1945. Collected Seminor Papers, Southern Africa Seminor (ed. by Shula Marks), Institute of Common Wealth Studies, University of London.

Clarence-Smith, G.&Moorsom, R. (1977). Underdevelopment and class formation in Ovamboland, 1844–1917. In R. Palmer and N. Parsons (Eds.), *The Roots of Rural Poverty in Central and Southern Africa*. London: Heinemann.

Fabian, J. (2000). *Out of Our Minds. Reason and Madness in the Exploration of Central Africa*. Berkeley: University of California Press.

Galton, F. (1852). Recent expedition into the interior of South-Western Africa. *Journal of the Royal Geographical Society*, 22, 140–160.

Galton, F. (1853). *Narrative of an Explorer in Tropical South Africa*. London: John Murray.

Galton, F. (1858). *Quoted in Cape Monthly Magazine*, 3, 193–197.

Galton, F. (1864). Hints to travellers. *Journal of the Royal Geographical Society of London*, 34, 272–316.

Gillham, N.W. (2001). *A Life of Sir Francis Galton*. Oxford: Oxford University Press.

Green, F. (1860). Narrative of a journey to Ovampoland, in two chapters, *Cape Monthly Magazine*, 7, 303–307 and 353–362.

Hahn, H. (1867). Neueste deutsche Forschungen in Süd-Afrika, *Petermann's Mitteilungen*, 12, 8–12 and 281–311.

Harley, J.B. (1992). Deconstructing the map. In T.J. Barnes&J.S. Duncan (Eds.), *Writing Worlds: Discourse, Text and Metaphor in the Representation of Landscape* (pp. 231–247). London: Routledge.

Hayes, P. (1992). A history of the Ovambo of Namibia, 1880-ca 1930. Unpublished PhD thesis, University of Cambridge, Cambridge.

Henrichsen, D. (1999). Claiming space and power in pre-colonial central Namibia: The relevance of Herero praise songs. Basel: BAB Working Paper No 1: 1999; presented at the *African Studies Association Meeting*, Chicago, 1998.

Kaulinge, V. (1997). *Healing the Land. Kaulinge's History of Kwanyama*. In P. Hayes and D. Haipinge (Eds.), Köln: Rüdiger Köppe Verlag.

Kreike, E. (1996). Recreating Eden: Agro-ecological change, food security and environmental diversity in southern Angola and northern Namibia, *1890–1960*. Unpublished PhD dissertation, Yale University, New Haven, CT.

Lau, B. (Ed.) (1989). *Trade and Politics in Central Namibia 1860–1864* (Charles John Andersson Papers Vol. 2). Windhoek: National Archives of Namibia.

Loeb, E. (1962). In feudal Africa. Published as an annex to *International Journal of American Linguistics*, 28.

McCaskie, T.C. (1995). *State and Society in Precolonial Asante*. Cambridge: Cambridge University Press.

McKittrick, M. (2002). *To Dwell Secure. Generation, Christianity, and Colonialism in Ovamboland*. Portsmouth, NH: Heinemann.

Mitchell, W.J.T.(Ed.) (1994) (First Edition); 2002 (Second Edition). *Landscape and Power*. Chicago: Chicago University Press.

Mulinda, C.K. (2002). The dynamic aspect of some traditional institutions in precolonial Rwanda. Unpublished MA thesis, University of the Western Cape.

Nitsche, G. (1913). *Ovamboland*. Kiel: University of Kiel.

Puranen, J. (1999). *Imaginary Homecoming*. Oulu: Pohjoinen.

Ryan, S. (1996). *The Cartographic Eye. How Explorers Saw Australia*. Cambridge: Cambridge University Press.

Salokoski, M. (1992). Symbolic power of kings in pre-colonial Ovambo societies. Licensiate thesis in Sociology/Social Anthropology, University of Helsinki.

Salokoski, M. Eirola, M.&Siiskonen H. (1991). The Ovambo kingdoms on the eve of European domination. In M. Mörner&T. Svensson (Eds.), *The Transformation of Rural Society in the Third World*. London: Routledge.

Serton, P. (Ed.) (1954). *The Narrative and Journal of Gerald McKiernan in South West Africa*. Cape Town: Van Riebeeck Society.

Shityuwete, H. (1990). *Never Follow the Wolf. The Autobiography of a Namibian Freedom Fighter*. London: Kliptown Books.

Siiskonen, H. (1990). *Trade and Socioeconomic Change in Ovamboland 1850–1906*. Helsinki: SHS.

Thomas, N. (1999). Introduction. In N. Thomas&D. Losche (Eds.), *Double Vision* (pp. 1–26). Cambridge: Cambridge University Press.

Tönjes, H. (1911). *Ovamboland. Land, Leute, Mission. Mit besondere Berücksichtigung seines grossten Stammes Oukuanjama*. Berlin: Martin Warneck.

Tyson, P.D. (1986). *Climatic Change and Variability in Southern Africa*. Oxford: Oxford University Press.

Vansina, J. (2001). *Le Rwanda Ancien. Le Royaume Nyiginya*. Paris: Karthala.

Vansina, J. (2004). *How Societies Are Born. Governance in West Central Africa before 1600*. Charlottesville: University of Virginia Press.

Virilio, P. (2000). *A Landscape of Events*. Cambridge, MA: MIT Press.

Wallis, J.P.R. (1936). *Fortune My Foe*. London: Jonathan Cape.

Wellington, J.H. (1967). *South West Africa and Its Human Issues*. London: Oxford University Press.

Williams, F.-N. (1991). *Precolonial Communities of Southern Africa. A History of Ovambo Kingdoms 1600–1920*. Windhoek: National Archives of Namibia.

Wrigley, C. (1996). *Kingship and State. The Buganda dynasty*. Cambridge: Cambridge University Press.

Chapter 9

From the Old Location to Bishops Hill: The Politics of Urban Planning and Landscape History in Windhoek, Namibia

Jan-Bart Gewald

Standing in the Old Location graveyard one can look across the river at Hochland Park where the Old Location used to stand. But we don't know how it was laid out. Where exactly did the 1959 Shootings take place?

Unam History Society: Windhoek's Monuments[1]

Who controls the past controls the future:
who controls the present controls the past.

George Orwell[2]

---

[1] The full text can be accessed at: http://www.iafrica.com.na/nht/windmon.html.
[2] Orwell, George (1950, p. 248).

M. Bollig, O. Bubenzer (eds.), *African Landscapes*,
doi: 10.1007/978-0-387-78682-7_9, © Springer Science+Business Media, LLC 2009

What lies beyond the windowpane of our apprehension . . . needs a design
before we can properly discern form, let alone derive pleasure from its perception.
And it is culture, convention, and cognition that makes that design; that invests a
retinal impression with the quality we experience as beauty.

Simon Schama[3]

9.1. INTRODUCTION

In this contribution the concept of landscape is used to demonstrate the manner in
which within an urban setting nation-states deal with the historical past. Within
the physical urban environment of Windhoek (the capital city of the Republic
of Namibia which gained political independence in 1990) the current Namibian
government and municipality of Windhoek have sought to inscribe upon the landscape a specific understanding and interpretation of the historical past. In doing so
the current Namibian administration follows on in the tradition of the many and
varied administrations that preceded it. In this contribution the urban landscape of
Windhoek is followed from its emergence in the 1880s, its segregationist phase,
its apartheid phase, and culminating in its nationalist phase in the years after 1990.
Throughout this period changes have been wrought upon the urban landscape of
Windhoek in the interests of a wide and varied combination of factors that relate
to specific understandings of the past, profit, racism, tourism, heartfelt ideology,
and much more. Each of these historically specific phases within the history of
Windhoek has left its traces in the urban landscape of the city. In this contribution
a case study is presented of the manner in which city planning, being the manipu-
lation of urban landscape, can be used to obliterate history.

In dealing with contemporary Namibian society, the historian Jeremy Silvester
has been wont to note that the majority of tourists who visit Namibia tend to
travel and interpret the country with a colonial reading and understanding of
the landscape (Silvester, 2001, 1999b). That is, the country through which they
travel is interpreted and understood from within a very particular understanding
of Namibia's past, an understanding which is reinforced by both the rural and
urban landscapes of Namibia south of the 'red-line'. Indeed more often than not
foreign visitors to the country will remark on the 'European' or even 'German'
atmosphere that appears to characterise some of the settlements in the country.

Using and drawing on his youth in Sussex, Simon Schama, in his magisterial
Landscape and Memory, has sought to put his finger on the fickle feelings, emotions, memories, and understandings that can be and often are brought to the fore
by landscape. In the same manner that the smell of disinfectant may engender
memories and visions of alarming visits to hospitals as a child, so too, the image
of sculpted landscapes can bring to the fore unbidden emotions and understandings
on the part of the observer. Landscapes and their impacts are culturally determined

---

[3] Schama; (1995, p. 12).

products of human agency. More often than not attempts are undertaken by human societies to mould and develop the environment in such a manner so as to create landscapes in keeping with specific sociocultural and historical understandings as to what is correct and proper; what Schama refers to as 'our shaping perception'. Fittingly Schama (1995, p. 9) notes: 'Even the landscapes that we suppose to be most free of our culture may turn out, on closer inspection, to be its product.'

Referring to Yosemite National Park in the United States of America, Schama (1995, p. 9) notes: 'The brilliant meadow-floor which suggested to its first [settler] eulogists a pristine Eden was in fact the result of regular fire-clearances by its Ahwahneechee Indian occupants.' Thus far, in dealing with landscape, there has, almost by definition, been an emphasis on rural landscapes. However, the development of urban landscapes is in no way different to the process described above by Schama. Indeed, much like the 'brilliant meadow-floor' of Yosemite, that which appears to be obvious at first hand can often obscure a hitherto unforeseen past.

9.2. INDEPENDENCE

In the run-up to independence in 1990, the all-white municipal council of Windhoek, Namibia's capital city, had land surveyed for the establishment of a new residential suburb to be known as Hochland Park. Situated less than 2 km from the central business district, the newly surveyed stands of freehold land were made available at extremely attractive prices. Members of the interim government and the administration of Namibia were provided with soft loans and were the main beneficiaries of this bargain sale.[4] In exchange for the soft loans and extremely competitive prices, the new owners were legally obliged to construct permanent housing, in keeping with specific minimum conditions, within a year of the purchasing date. The project was a great success. By independence, in March 1990, all the stands had been sold, and by March 1991, the residential suburb of Hochland Park was a reality.

In establishing Hochland Park, the outgoing South African regime, which had been in illegal occupation of Namibia, ensured that all physical traces of a crime committed against Windhoek's African inhabitants were obliterated from the city's urban landscape. The crime committed and carried out by the South African administration in accordance with apartheid legislation, entailed the forced removal of the city's African inhabitants from the Old Location, and the

---

[4]Those entitled to soft loans also included people employed by semi-official government institutions, such as the employees of the *Republikein* newspapers and the *Democratic Turnhalle Alliance*. The loan conditions were so favourable that there were cases of couples, already in possession of property and land elsewhere in the city, who bought plots in their own name.

subsequent razing of all the buildings that had stood there.[5] By building Hochland Park the outgoing South African regime had ensured that the *Old Location* would forever be no more than an image existent only in ever-failing memory and without any form of binding to the physical world.

The extreme cynicism displayed by colonial regimes is not something new; troubling though, is the patent lack of interest displayed by the current government of Namibia's political elite towards Windhoek's urban history. This is all the more startling when one considers that the current government of Namibia prides itself on a history of struggle, the 'heroes' of which it urges one and all to revere and honour. In keeping with this an enormous monument and graveyard was inaugurated on the outskirts of Windhoek in August 2002.

9.3. WINDHOEK HISTORY

In terms of sub-Saharan African cities, Windhoek is a settlement with a comparatively long history of permanent human occupation.[6] Situated in the Auas Mountains in central Namibia Windhoek has long been an important location. The city's Otjiherero (*Otjomuise*) and Nama-Dama (*Ai-Gams*) names both refer to the perennial hot water springs that first drew people to actively settle in the area of present day Windhoek. The name of the town itself is derived from the Dutch words *Winter* (winter) and *Hoek* (literally corner, but used here to mean place), which also refer directly to the abundant water and grazing that were to be found in the area in the dry winter months of central Namibia. Archaeological excavations in and around the city centre indicate that the Windhoek area has been almost continuously occupied for the past 10,000 years. In the 1840s, when the first European missionaries arrived in the settlement, Windhoek was a trading centre under the command of the Oorlam leader Jonker Afrikaner (Lau, 1987). The urban settlement of Windhoek appears not to have been very extensive, although it did include a number of permanent stone buildings, the most notable of which was the church built and established by Jonker Afrikaner. Pastoralists made seasonal use of the settlement's water and grazing resources; artisans, such as carpenters and smiths, lived in the settlement, and a number of horticulturalists made a living by being specialised in the production of food and cash crops.

---

[5] In terms of international law South Africa was in illegal occupation of Namibia from 27 October 1966 when the United Nations ended South Africa's 1920 League of Nations Mandate for the administration of the territory. In terms of the United Nations 1973 International Convention on the Suppression and Punishment of the Crime of Apartheid, which has been ratified by 101 states, apartheid is defined as a crime against humanity. In terms of the Geneva Conventions, Additional Protocol 1, apartheid has been labelled a war crime in international conflicts. The Rome Statute of the International Criminal Court (ICC) adopted 17 July 1998 lists apartheid as a crime against humanity.

[6] There have been two doctoral thesies written on the history of Windhoek. Kotze, Carol E. (1990), A Social History of Windhoek, 1915 – 1939, unpublished PhD thesis, University of South Africa. and Wallace, Marion (2002) *Health, Power and Politics in Windhoek, Namibia, 1915 – 1945*,. Basel, 2002.

Indeed, wheat, maize, sorghum, finger millet, beans, tobacco, marihuana, pumpkins, calabashes, and other crops were grown (Gewald, 2002a, pp. 68–69).

In 1890 German troops, under the command of Curt von Francois, occupied a hilltop overlooking Windhoek and began supervising the construction of the military fort currently known as *Die Alte Feste*. Interestingly the fort, which was built in such a manner as to ensure a commanding position over the surrounding area, was built around a water well, which still stands today within the confines of the walls. This could be taken as the first serious colonial attempt to implant political hierarchy and power upon the urban landscape in Namibia, and settler historiography credits von Francois with having established the city of Windhoek.

Previously churches, homesteads, and warehouses had been established in urban settings in Namibia. Although some of these buildings definitely doubled as military fortresses, they were in no way comparable to the imperial plan that was being unfolded by von Francois and his colleagues. Responding to the demand created by the German garrison, traders, homesteaders, barkeepers, and the like established themselves in the immediate vicinity of the fort. Wagon drivers, labourers, horticulturalists, *Bambusen*,[7] prostitutes, and general taggers-on, drawn from the indigenous population, moved to Windhoek, serviced the needs of the garrison and the settlers, and established themselves in the rapidly growing town. Although there are reports of a separate African location in the 1890s, enforced segregation on the basis of race only came about during the course of the Herero–German war. In the war a number of forced labour camps came to be established in the settlement (Figures 9.1 and 9.2). These camps, which housed Herero and later Nama prisoners of war, were established at various locations.

Figure 9.1. Alte Feste (Old Fort) in Windhoek ±1906, with prisoner of war encampment in the foreground

---

[7] Servants, generally boys or young men, attached to the German army and settlers.

Figure 9.2. Original caption: 'Windhoek – View towards native quarter' – View towards west showing Rhenish Mission Church with site of old location (*See also Color Plates*)

In 1908 the camps were finally abolished and locations developed at the sites of some of the camps (Wallace, 2002, pp. 40–41). Thus, a location developed in the immediate vicinity of the railway station, whilst another, which supplied German shopkeepers and traders with labour, developed on the site of what is today the suburb Hochland Park, and another developed in the vicinity of the residential area, Klein Windhoek. In addition there were a number of smaller locations that developed, wherever labour was needed, on the fringes of the settlement (Gewald, 2002a, pp. 68–71).

In 1915 South African forces invaded German South West Africa, and swiftly defeated the much vaunted German colonial army. Windhoek came to be the residence of a South African military administrator and the territory adminis-tered by military magistrates. The incoming administration sought to maintain the urban segregation that had been enforced by the German authorities. To this end Captain Octavus Bowker was placed in charge of the Native Affairs Department in Windhoek. In 1922 Bowker took up employment with the municipality of Windhoek as superintendent of the locations, a position which he held until his formal retirement in 1946 (Gewald, 2002b).

9.4. OLD LOCATION

In 1919 the South African military administration estimated that there were approxi-mately 5500 Africans living in Windhoek. The majority of these lived in what would later become known as the Old Location, the site of what is today Hochland Park.

In 1932 the Old Location was reorganised. Straight streets were laid out, plots of land marked and numbered, housing torn down, and so forth. Only those who could prove legal employment were permitted to rebuild houses and remain in the location. Effectively this meant that a large number of people, particularly those living in female-headed households, were left without a home and liable for deportation (Gewald, 2002a, pp. 80–81). People living in the Old Location were not permitted to claim title to the plots of land on which they had built their houses. But the actual houses themselves were considered to be private property and could be sold, inherited, and rented out. Indeed there were cases where complete houses were taken down and re-erected elsewhere in the location, much to the frustration and irritation of Bowker (Gewald, 2002a, p. 78).

When the Old Location was reorganised, an attempt was made to estab-lish formal ethnic group sections. Thus sections within the location came to be referred to as Herero One, or Damara Two, and so forth. However, the strict ethnic segregation of people in the Old Location did not take place, and in principle, an inhabitant of the location could live within any of the ethnic sections.

Within the Old Location the 'sanitation syndrome' was used to justify systematic racial segregation of African inhabitants from whites in the city. That is, in a circular argument, adequate sanitation was not made available to the African inhabitants of Windhoek on the grounds that they were considered to be without need of sanitation. Throughout the years that the location existed the issue of public sanitation was used to legitimate and justify racist attitudes and treatment of the city's African inhabitants.[8] In effect, this was an attempt to bind ethnicity to a particular place, and corresponds directly to the developing reserve policy.

9.5. URBAN AREAS ACT FORCED REMOVALS

Writing in the early 1960s, long before apartheid in its vicious forms was abol-ished, Brian Bunting noted the ideological and personal links that existed between the National Socialist party in Germany and the Nationalist Party in South Africa. Introducing a chapter entitled 'South Africa's Nuremberg Laws', in which he listed the legalised and institutionalised racist legislation that characterised the apartheid state, Bunting noted: 'Nationalist [party] legislation has been aimed . . . at preventing all forms of integration which might lead to the establishment of a united South African nation with a common citizenship and loyalty irrespective of race' (Bunting, 1969, p. 142).[9]

---

[8] On the issue of the 'sanitation syndrome' see Deacon, On the issue of sanitation as a socio-political weapon see Gewald (2002a), Chapter 5.

[9] The complete text of Bunting's work is accessible at http://www.anc.org.za/books/reich. html.

Coming into power in 1948, the South African Nationalist Party set about transforming South Africa into the apartheid state. Central to this policy was the drawing up of laws that regulated residence on the basis of race throughout South Africa. The cities of South Africa were transformed from cities in which segregation had been varyingly applied, into the Apartheid City where each of the racial groups had to reside within designated areas according to ascribed race (Smith, 1992). In the 1950 session of the South African parliament, the Group Areas Act, in terms of which ownership and occupation of land were to be restricted to specific population groups, was passed. This act was coupled to a welter of other acts as well as local government and municipal legislation, which ensured that all urban Africans, classified as 'nonwhites' by the South African government, were not permitted to enter, let alone live, within 5 miles (8 km) of a designated 'European' residential area.[10] Only Africans who were in registered employment in 'European' areas were permitted to enter. Tracts of empty land acted as buffer strips separating the residential areas of the various races. Within the logic of apartheid, also known by the euphemism 'separate development', each racial group would have access to its own public facilities, schools, health facilities, and so forth within its own designated area, this to ensure maximum racial segregation. People who happened to be living in areas not in keeping with their ascribed racial status were forced to move (Smith, 1992).

Some of the spectacularly bloody atrocities committed in the name of apartheid are well known to the world, yet the forced removal of over three million Africans from their urban as well as rural homes is a lesser known fact.[11] In South Africa government bulldozers razed the multiracial suburb of Sophiatown in Johannesburg to the ground. In its place it established the 'White' suburb of Triomf (Van Niekerk, 1995). In Cape Town the multiracial suburb of District Six was razed to the ground.[12] In keeping with the hypocritical nature of apartheid legislation all that remained were graveyards, churches, and mosques. This process was repeated in all of South Africa's 'European' towns and cities. De jura, in terms of international law, Namibia was not an integral part of South Africa, yet de facto Namibia was governed as the fifth province of South Africa, and apartheid legislation promulgated in South Africa was applied equally rigorously in Namibia (Figure 9.3).

---

[10] Population Registration Act (1950), the Natives (abolition of Passes and Coordination of Documents) Act (1952), the Native Laws Amendment Act (1952), the Natives Resettlement Act (1954), and the Natives (Urban Areas) Amendment Act (1955).

[11] See the materials published by the Surplus Peoples Project (1983) and Platzkyland Walker 1985.

[12] The case of District Six, as with Sophiatown, is well known, and over the years a large body of literature has appeared (Bohlin, 1998; McEachern, C., 1998; Rassool, & Prosalendis, 2001).

Figure 9.3. View of Old Location 1930s; note the trees (Steinhoff, 1938)

In 1951 the German anthropologist Gunther Wagner, employed by the Native Affairs Department to provide an ethnographic survey of the Windhoek district, submitted an enormous manuscript that provided findings that ran counter to the ideological wishes of the apartheid government.[13] Wagner's findings indicated that within the urban setting of Windhoek, ethnic groups did not necessarily live independently of one another and that in numerous instances it was simply impossible to make ethnic distinctions or characterisations that had any objective reality in society. Referring to football teams, burial clubs, dance societies, and the like, Wagner noted that membership of these organisations did not inevitably correspond with ethnic divisions. Instead, the African inhabitants of Windhoek, although well aware of their ethnicity, did not organise their lives around ethnic divisions. This ran totally counter to the wishes and propaganda of the Nationalist party and its programme of apartheid.

Oswin Köhler, who would later found the Institute for African studies in Cologne, was appointed as government ethnologist in Namibia in 1955, following the untimely death of Gunther Wagner in 1952. In the first years following Wagner's death Köhler edited and prepared the manuscripts of his late colleague for publication. After which, in 1957 he began work, in keeping with the tenets of apartheid ideology, on designing a model location for the city of Windhoek.

---

[13] For a short overview of Gunther Wagner and his work see Gewald, (2002a).

In August of 1957 the plans for the newly designed and ethnically categorised location were submitted to the municipality.[14] Köhler had neatly divided the new location into ethnic sections. In a horseshoe shape, the ethnic sections were grouped according to ethnic affinity as perceived by Oswin Köhler. Thus, the Ovambo were placed next to the Herero, and next to them the Damara, the Nama, and so forth.

Even though significant improvements of the infrastructure had been undertaken in the Old Location, a decision was taken by the municipality in 1957 to have the Old Location destroyed, and work on the new location begun. With the passing of apartheid legislation, and the white settler community's ardently expressed support for the policies of the National Party, the residents of the Old Location were aware of the intentions of the municipality. In 1956 the community's spokesmen informed the municipality of their refusal to leave the Old Location and stated that they had named the proposed township 'Katutura', a name to be interpreted as 'a place where we will never feel settled' (Lau, 1995, p. 6). The majority of the inhabitants of the Old Location were opposed to the closing down of the location, and did not wish to move to the ethnically segregated location that was being constructed by the South African administration. The Old Location inhabitants dismissed the administration's claims that housing, sanitation, health care, and the like, would be better provided for in the new township. Residents pointed out that apart from transport costs, rents to the township board in the new township would be 14 times as high as in the Old Location, and they would not be entitled to freehold ownership of either land or property. In the Old Location, although inhabitants had no freehold to the land on which their buildings were built, they did have ownership rights to the buildings themselves, and these could be sold.

Old Location residents suggested that instead of establishing a new township 8 km away, improvements could be carried out in the already existing location. The mere suggestion of this was deemed unacceptable by the administration which, operating within the context of apartheid legislation, could seek to create a 'pure white' city, unblemished by the presence of Africans, other than those in the employ of the city's white inhabitants. That is, apartheid legislation provided the white residents of the city with the opportunity to remove the presence of Africans from their sight. By September 1959 location residents had taken to protesting against the move. Noting that Namibia was a mandated territory and that the land belonged to Namibians, Old Location resident N. Mbaeva stated:

> This apartheid that you are coming here to impose, you are trying to impose on a place that does not belong to you. Do you not know that this place belongs to us and to us alone? We are people that are in our own land, and it is not necessary for us to go to another place. We will not condone apartheid. If we move to Katutura, we are condoning apartheid. (Lau, 1995, pp. 27–28)

---

[14] In the event Köhler returned to Germany in November 1957 to establish the Institute for African Studies at the University of Cologne.

In early December 1959 things came to a head. Whilst residents boycotted municipal services, the electricity supply was curtailed and police cordoned off the Old Location. Youths attempting to get out of the township were turned back by armed police. On the tenth of December women, who played a central part in the protests, led a protest march to the Municipal Offices (Angula, 1998). The protest ended in bloodshed. The police, holed up in the Municipal Offices, opened fire on the crowd of approximately 2000 people (out of an Old Location total population of about 16,000). It is believed that 13 people were killed and at least 44 wounded on account of the shooting. The shooting in the Old Location was a turning point in Namibian history. Within days of the shooting, Namibian political leaders had been issued with deportation orders to the 'Native' reserves, or arrest warrants. Most political activists, who had not been detained, escaped to the Bechuanaland Protectorate. In the event, the shootings foreshadowed the events in Sharpeville, South Africa, three months later where 69 people were killed on 21 March 1960 (Silvester, 1999a).[15] The single-member Hall Commission of Enquiry, appointed by the South African administration to investigate the 'direct causes' of the Old Location shooting, concluded that: '[T]he opposition to the removals from the existing location to the new one, was organised by the Hereros in Windhoek at the instigation of their protagonists in New York' (Hall cited in Silvester, 1999a). The opposition of local people to the higher rents, higher bus prices, and the principles of apartheid, were dismissed in favour of alleged foreign instigators.

Immediately after the shootings, people started moving to Katutura. Services were cut off to the Old Location, and pressure was placed on employers not to employ people from the Old Location. In early 1968 the last building still standing in the Old Location was bulldozed to the ground (Figure 9.4).

Figure 9.4. Destruction of the last structures in the Old Location 1967

---

[15] For further information regarding the Sharpeville massacre, see, Sibeko., David M. (1976) *"The Sharpeville Massacre: Its Historic Significance in the Struggle Against Apartheid"*, http://www.anc.org.za/ancdocs/history/misc/sharplle.html

Due to the strange, not to say hypocritical, nature of apartheid legislation graveyards, churches, and mosques, were not torn down. Yet, on account of the pass laws, places of worship and the graveyards, left standing at the sites of razed townships, could no longer be visited, let alone tended. As a consequence, all across the territories under South African administration, the graveyards, churches, and mosques of destroyed communities fell into disrepair.[16]

When the Old Location was removed from the vicinity of Windhoek's city centre, the location associated with the suburb of Klein Windhoek was similarly moved and razed to the ground. All that remained of the Klein Windhoek location was a Roman Catholic mission church, perched on a hill overlooking an abandoned graveyard, and the wrecked remains of a community. From the late 1960s onwards, the church, the only standing building of the Klein Windhoek location, was used as a Scout hall by the Scouts of '3rd Windhoek'. On specific days of the week Boy Scouts would swarm around the Scout hall oblivious of its history, whilst in the valley below boys from St. Paul's College played football on their school's B fields. The remains of headstones, the carcasses of abandoned motorcars, and scraps of barbed wire and corrugated iron, betrayed the prior existence of a human community. But in all the time that I attended football practice or was a Scout at 3 Windhoek, I never realised and was never told, what it was that I was playing amongst.

9.6. RUN-UP TO INDEPENDENCE

In early 1988 a South African military offensive, launched to capture the strategic city of Cuito Cuanavale in southern Angola, ended in defeat. The human, material, and capital cost of maintaining apartheid and the illegal occupation of Namibia, had finally become too high for the South African administration. In the second half of 1988 negotiations regarding the independence of Namibia and the withdrawal of South African forces from the territory were completed, and Namibia's independence became inevitable. In early 1989 the first UN peacekeepers arrived; a year later Namibia became an independent state.

In the second half of 1988, when it became clear to one and all that Namibia would be granted independence, the Windhoek municipal council decided to establish a new suburb on the site of the Old Location. For this purpose the site was resurveyed and divided into residential plots (Figure 9.5). The newly surveyed plots and roads bore no relation whatsoever to the street-plan of the Old Location as it had existed in the past. Indeed, the newly surveyed roads lay at an angle to the roads previously existing in the Old Location. Thus the plans for the total obliteration of the Old Location from the landscape were laid. In interviews

---

[16] As child I grew up in Pioneerspark, a whites- only suburb that came to be built next to where the Old Location had existed. Cycling through the scrubland next to the suburb I was well aware of the grave yard, which was now overgrown by bush and scrub, although I had no idea as to who could possibly be buried there.

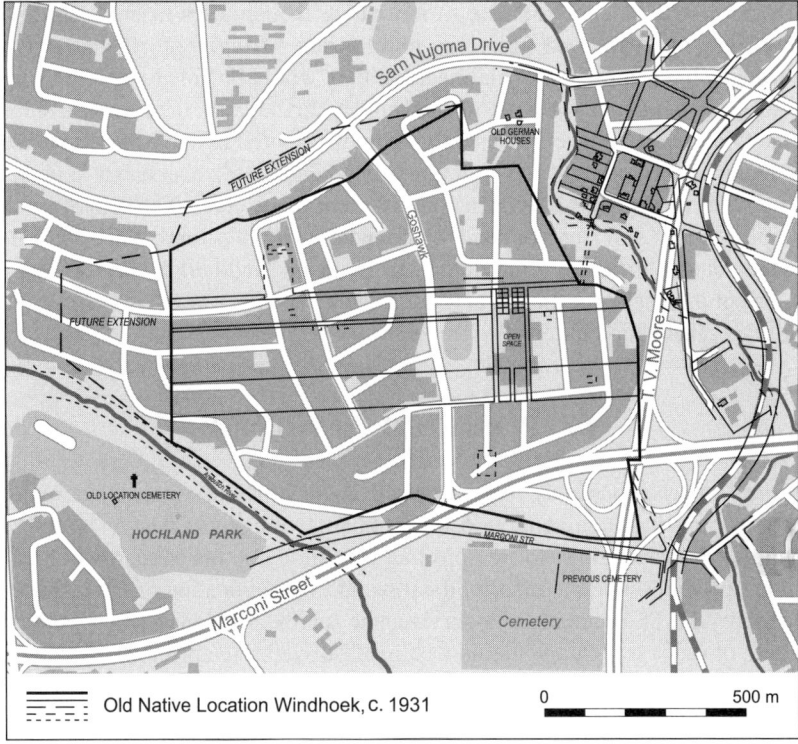

Figure 9.5. Map illustrating the site of the Old Native Location in Windhoek circa 1931

conducted with one of the surveyors involved in the survey, it was suggested by the surveyor that the original stands were ill-suited to the survey as it was to be carried out for the establishment of plots for sale. The surveyor suggested that the survey was conducted with an eye for the terrain, 'so that people can enjoy the view'.[17] This may be so, but this does not diminish the fact that in so doing all traces of the past were consciously obliterated. The plots were put up for sale, and, as noted earlier, were bought primarily by people in the employ of the South African-sponsored Namibian interim government. Granted easy access to loans, and cheap residential plots, the new owners were under obligation to have permanent residential structures erected on the plots within 12 months of purchase. On Namibia's independence in 1990 very few plots had been built upon, by the first anniversary of Namibian independence, Hochland Park, as a residential suburb was in existence.

The resurvey and conditions of sale initiated by the Windhoek Municipal Council in the area of what had been the Old Location, ensured that the past was obliterated forever. As a historian I have often taken people to Hochland

---

[17] Interview conducted property surveyor Glenn T. Windhoek, August 2002.

Park and asked them to indicate to me where their houses existed. Orienting themselves on roads and buildings on the outer fringes of Hochland Park, informants have consistently claimed that they would be able to find the site of their former residences. Yet, upon entering Hochland Park, the former residents of the Old Location have consistently been disoriented by the new road system and residential buildings. The structure of Hochland Park, as it exists at present, is such that it is impossible for people to orient themselves. The graveyard and Catholic church still stand beyond the boundaries of the suburb, yet upon entering the suburb, the terrain provides them with no landmarks with which they can orient themselves with any certainty. The past, as it existed in traces on the landscape, has been obliterated forever.

9.7. POST-INDEPENDENCE

Contrary to what one may have expected, in post-independence Namibia, the new government is continuing to equally obliterate the physical traces of the past from the landscape, albeit for reasons that bear little relation to apartheid legislation but far more with the museumification and Disneyfication of the landscape. The graveyards and churches, that survived the apartheid era due to the hypocritical niceties of racist legislation, are now being destroyed. Freed from the constraints of apartheid legislation, which for all its faults did prohibitthe destruction of places of worship and graveyards, urban planners in post-independence Namibia have seen no harm in destroying graveyards and churches, and obliterating the past in the interests of modern urban planning and, more cynically, substantial profit.

Prior to Namibian independence the elite of Windhoek lived in a cluster of mansions scattered upon 'Luxury Hill'. Following independence a new luxury suburb came to be established on the fringes of Klein Windhoek. This new suburb, which bears the name Ludwigsdorf, is partially built on the remains of the Klein Windhoek location, graveyard, and destroyed Roman Catholic mission church building.[18] Following independence, and the boom in new luxury housing associated with the arrival of foreign diplomatic missions and NGOs, Ludwigsdorf, farthest away from the African high-density suburb of Katutura, expanded dramatically. Parts of the suburb were resurveyed and even more plots of land were made available by the municipality for freehold purchase. In this rush for cheap land in a favourable location the remains of the Klein Windhoek location graveyard were obliterated. Houses currently occupied by the diplomats of foreign countries and Namibia's rich and famous, overlie the graves of hundreds of people, graves which have become permanently lost to the relatives and descendants of the deceased.

---

[18] Ludwigsdorf was named after a German settler who established a beer garden, swimming pool, and bowling alley in the Klein Windhoek river valley. It bears no direct relationship with its namesake Ludwigsdorf, which was one of about 60 sub-camps associated with the Gross Rosen Nazi concentration camp located in lower Silesia.

In the rampant expansion of Ludwigsdorf, the sole remaining standing building of the Klein Windhoek Old Location, the Roman Catholic church that had become a Scout hall, was destroyed in 1994. In 1993 plans were submitted to the Windhoek municipality for the construction of a series of semidetached houses and apartments in Ludwigsdorf. The plans were passed by the municipality and permission was granted for construction. The imminent destruction of the abandoned Roman Catholic church was not considered to be an obstacle. Indeed, in the papers submitted to the municipality, the building was not referred to as a former church, and it was noted that it did not fall within the definition of the city council for 'heritage buildings'.

Not included in a list of buildings drawn up by the Namibia Institute of Architects' Heritage Committee in 1989 (in the year prior to independence) the church building was deemed to have no 'historical or architectural merit'. When news of the imminent destruction of the church became publicly known, concerned former residents of the Klein Windhoek location contacted the National Monuments Council of Namibia (NMC), who in turn approached the construction company involved, SWABOU. In discussions with Mr. Potgieter, the representative of SWABOU, a gentleman's agreement was reached whereby the NMC would provide SWABOU with proposals that would retain the building. Four proposals were submitted to SWABOU before 15 February 1994, the chosen cut-off date. SWABOU rejected the proposals.[19] Before an appeal could be launched and just as protest came to be heard in the popular media, the building was torn down. In its stead the upmarket 'Bishops Hill' housing complex came to be built. The tongue-in-cheek reference to the Roman Catholic church in the choice of the name 'Bishops Hill' for the new housing complex, indicates that the building company was well aware of the historical antecedents of the site upon which they were planning to build.

The wanton destruction of historical sites is not confined to the capital city of Namibia. The municipal council of the beach resort Swakopmund has recently allowed building to begin on the site of mass graves that date back to the Herero–German war of 1904–1908. There is legislation, both at a municipal level as well as at a national level, that was promulgated in such a manner so as to ensure that the obliteration of buildings and sites of historical value would not take place.

9.8. URBAN MONUMENTS IN A NEW STATE

A thread linking new nation-states across time and continents is the seeming need of newly established states to emplace their identity upon the landscape in the form of monuments. The Arc de Triomphe in France, the Brandenburger Gate in

---

[19] Personal communication conducted on 22 April 2002 with Dr. Andreas Vogt, formerly of the NMC.

Berlin, The Voortrekker Memorial in Pretoria, and the Heroes Acre in Windhoek all express the apparent need felt by incipient nation-states to focus their history on a small and specific group of people, who within themselves, appeared to embody the aims and aspirations of the incipient nation-state, and to transfer this abstract thought into a physical embellishment of the landscape. The monuments established to the memory of the select few dominate the landscape and seek to impress in physical form the revised history, hopes, and aspirations of the new state. In Namibia, the SWAPO government, through initiating the building of a Heroes Acre, has consciously sought to change the landscape of Windhoek in such a manner that it has come to be dominated by a new and currently accept-able reading of the past.

At the moment, in terms of memorials and landscape, Windhoek is a city of contradictions. The memorial to the Ovambo campaign, which was erected to commemorate South African soldiers killed in 1917, has been surrounded by corrugated iron sheeting, ostensibly to keep out hoboes, yet at the same time the Rider Memorial, erected to commemorate German soldiers who died between 1904–1908, is kept in a perfectly maintained condition. The current government of Namibia wishes to ensure that its version of a heroic historical past is presented to the world. In this, commemoration of the defeat of Ovambo forces by South African forces in 1917 is considered to be inappropriate. Whereas commemora-tion of German soldiers who exacted and participated in the first genocide of the twentieth century, is not considered to be inappropriate or offensive.

The current Namibian government has a strong interest in history, albeit a specific account of history. It is an officially sanctioned history of heroes and struggle, which includes those killed in the Old Location, but not the Old Location itself. It is a monumental history of struggle in the interests of which the govern-ment funded the establishment of an officially sanctioned Heroes Acre to the tune of N$71 million (€10 million). Directly inspired by the monumental Heroes Acre built by North Korean contractors in Zimbabwe (Namibia's closest political ally), the Namibian government commissioned a series of contractors to begin construction in 2000.[20] For the Heroes Acre project the Windhoek City Council set aside 135 ha of land to the south of the city. City Mayor Immanuel Ngatjizeko (*The Namibian*, 28. January 2000) stated that: 'The central point will be a rocky hill, which stands 1878 metres above sea level and commands an attractive view of the Windhoek Valley and the city'. Originally budgeted at N$35.8 million and destined to ready by March 2002, Heroes Acre had doubled in cost within a year of having been commissioned (Menges, 2000; Amuphadi, 2001; CoD, 2001). In late August 2002 Heroes Acre was officially opened to the public.[21]

---

[20] See for instance, http://www.stewartscott.com/news_views_namibia_heroes_acre.htm,a South African road construction company contracted to lay out the access roads, water supply, and parking lot.

[21] http://www.grnnet.gov.na/News/Archive/2002/August/Week5/tourist.htm& and http://www.namibian.com.na/2002/august/news/027EBA1F48.html.

The current Namibian government is prepared to sink millions of dollars into transforming the landscape of Windhoek in the name of history, whilst at the same time downplaying, neglecting, forgetting, and obliterating traces of crimes against humanity from the urban landscape of Windhoek. A generous view would argue that there is a lack of historical understanding on the part of the government. Indeed, the current Namibian government is dominated by people who cannot trace their history to events that took place in central and southern Namibia. German colonial rule, and the genocides committed upon the Nama and Herero of central and southern Namibia, did not affect the inhabitants of Ovamboland. Not surprisingly the sites of memory, the sites of the former prisoner of war camps that dot the urban landscape of Windhoek, are of no significance to those whose ancestors were not incarcerated there.

For the majority of the current government of Namibia, the past that was inscribed in the urban landscape is simply not their past. Although formally colonised, Ovamboland was never colonised on the ground by German administration, neither did the inhabitants of Ovamboland lose any of their land, stock, or people at the hands of German colonisers. Thus, for the inhabitants of Ovamboland, the German colonial era has none of the negative historical connotations of genocide, land and stock loss, impoverishment, and colonial subjugation that it has for the inhabitants of central and southern Namibia. For the Namibian government, Germany is seen as the first colonising state, yet it is the South African occupation of Ovamboland that rankles. For the current government of Namibia the sites of the former South African army bases in Ovamboland form the sites of anguished memory. It is in seeking to draw attention to either the heroism or the horrors of the war fought primarily in northern Namibia against the inequities of South African rule, that the current government of Namibia has designed and built a monument that literally dominates the landscape. Within the monument the government has sought to include all the aspects of a past that is considered to be of paramount importance to the new nation-state of Namibia, a monument which, by virtue of its size and setting, dominates the capital city of Namibia.

Heroes' Acre Expected to Become Tourist Attraction

The newly inaugurated Heroes' Acre, situated about 15 km south of the Namibian capital, Windhoek is expected to become one of the country's greatest tourist attractions.

The national monument, which was built in remembrance of the country's fallen heroes and heroines, is a magnificent sight, with granite finishing and impressive statues.

One of the greatest features of the Heroes' Acre is the statue of the 'Unknown Soldier' dressed in full combat gear. The statue, standing on granite, symbolises all heroes and heroines who sacrificed their lives for the liberation struggle of the country. In front of the statue is a golden inscription in the

(continued)

(continued)

Namibian President His Excellency Dr Sam Nujoma's handwriting saying: 'Glory to the heroes and heroines of Namibia'.

Soil which was taken from mass graves at Angola and Zambia were placed at the foot of the statue. The graves have been set up in a pyramid form with the first nine graves at the top and the rest of them coming below. They are covered with plastic membrane. Another great feature found at the Heroes' Acre is the eternal flame, which is below the enlarged bronze heroes' medal that will burn for eternity to show appreciation to the fallen heroes.

President Nujoma lit the flame and wreaths were thereafter laid at its foot in remembrance of the fallen heroes and heroines.

The other interesting feature is the concrete tower in white marble and granite called the obelisk, symbolising a sword for the integrity and dignity of the nation and this can be seen all the way from the city centre. The obelisk is situated in front of a curved wall in bronze, which depicts the story of the struggle of Namibia from the time of foreign occupation until the independence celebrations in 1990.

Among the carvings found on this wall is the depiction of the mobilisation of the people, the input of the United Nations (UN), the armed struggle, women fighting alongside their male counterparts, the massacre at the Old Location, the Ongulumbashe scene, as well as a caption showing village life that signifies the fact that soldiers were dependent on village people to help them with shelter, food, and information.

Another feature at the Heroes' Acre site is the restaurant, which was built to serve as an income-generator and which can accommodate 150 people. All the stones used in the construction of the Acre were purchased locally.

9.9. CONCLUSION

In the run-up to Namibian independence, the outgoing South African administration, through building Hochland Park on the site of the Old Location, successfully obliterated all traces of a crime against humanity. In independent Namibia the current administration, through allowing the building of new suburbs and monuments continues to obliterate the past, whilst emphasising a very specific vision of that past. Through building an enormous monument, and thereby transforming the landscape of Windhoek itself, the Namibian government has sought to draw attention to, and demand respect for, what it sees as its respectable and honourable struggle for independence and the establishment of the Namibian nation-state. In control of the apparatus of the newly established state, the government has sought to enforce its vision of the past upon its population, and it has done this by physically transforming the landscape in such a manner that each and every visitor to the city of Windhoek blessed by sight cannot help but see and take notice of the new history of Namibia.

During the era of apartheid legislation, legal niceties ensured that graveyards and places of worship were not destroyed in the forced removal and resurvey of city suburbs. Following Namibian independence the government has not been constrained by such niceties. As a result, in one of the ironic twists of history, a SWAPO government, which was diametrically opposed to South African apartheid rule, has accelerated and, in some cases, completed the task of historical destruction. In Windhoek no traces of African urban settlement prior to the establishment of the apartheid township of Katutura in 1958 exist on the landscape.

Acknowledgements With thanks to Monika Feinen for her excellent cartographic skills, and to Michael Bollig, Casper Erichsen, Nancy Jacobs, and Jeremy Silvester for comments and suggestions.

REFERENCES

Amuphadi, T. (2001). Heroes Acre on way to becoming reality. *The Namibian*, May 11, 2001.

Angula, N. (1998). The 'gendering' of the anti-colonial struggle. *The Namibian*, December 4, 1998.

Bohlin, A. (1998). The politics of locality: Memories of District Six in Cape Town. In N. Lovell (Ed.), *Locality and Belonging* (pp. 168–188). London: Routledge.

Bunting, B. (1969). *The Rise of the South African Reich*. London: Penguin.

Christ of Maletsky, *Namibian, The* (2000). Green light for Heroes' Acre. January 28, 2000.

Congress of Democrats (CoD) (2001). CoD rejects a "Heroes" Acre" for politicians. Press Release, April 19, 2001.

Deacon, H. (1996). Racial segregation and medical discourse in nineteenth-century Cape Town. *Journal of Southern African Studies*, 22(2), 287–308.

Gewald, J.-B. (2002a). *"We Thought We Would Be Free…": Socio Cultural Aspects of Herero History in Namibia, 1915 – 1940*. Cologne: Köppe.

Gewald, J.B. (2002b). A Teutonic ethnologist in the Windhoek district. In R. Gordon. & D. Lebeau (Eds.), *Challenges for Anthropology in the 'African Renaissance'. A Southern African Contribution* (pp. 19–30). Windhoek: Gamsberg Macmillan.

Hall, C. (1961). Report of the commission of enquiry into the occurrences in the Windhoek Location on the night of the 10th and 11th December, 1959, and into the Direct Causes which Led to those Occurrences. Windhoek.

Kotze, C.E. (1990). A social history of Windhoek, 1915 – 1939. Unpublished PhD thesis, University of South Africa.

Lau, B. (Ed.) (1995). *An Investigation of the Shooting at the Old Location on 10 December 1959*. Windhoek: DISCOURSE/MSORP.

Lau, B. (1987). *Namibia in Jonker Afrikaner's Time*. Windhoek: National Archives.

McEachern, C. (1998). Working with memory: The District Six museum in the New South Africa. *Social analysis*, 42(2), 48–72.

Menges, W. (2000). Costs of new State House & Heroes' Acre Outlined. *The Namibian*, April 6, 2000.

Orwell, G. (1990 [1950]). *1984*. New York: Penguin.

Platzky, L. & Walker, Ch. (1985). *The Surplus People: Forced Removals in South Africa*. Johannesburg: Ravan Press.

Surplus People Project (1983). *Forced Removals in South Africa*. The Surplus People Project report, Vol. 1–5. Cape Town: Surplus People Project.

Rassool, C. & Prosalendis, S. (Eds.) (2001). *Recalling Community in Cape Town: Creating and Curating the District Six Museum*. Cape Town: District Six Museum.

Schama, S. (1995). *Landscape and Memory*. New York: HarperCollins.

Sibeko, D.M. (1976) The Sharpeville Massacre: Its historic significance in the struggle against apartheid, http://www.anc.org.za/ancdocs/history/misc/sharplle.html.

Silvester, J. (2001). Picture the past: Signs of the times? *The Namibian*, April 26, 2001.

Silvester, J. (1999a). A turning point in Namibian history. *The Namibian*, July 16, 1999.

Silvester, J. (1999b). Public history – Forgotten history. *The Namibian*, August 27, 1999.

Smith, D.M. (1992). *The Apartheid City and Beyond: Urbanization and Social Change in South Africa*. London: Routledge.

Steinhoff, I. (1938). *Deutsche Heimat in Afrika: Ein Bildbuch aus unseren Kolonien*. Berlin: Reichskolonialbund.

Van Niekerk, M. (1995). *Triomf*. Cape Town: Queillerie.

Wallace, M. (2002). *Health, Power and Politics in Windhoek, Namibia, 1915 – 1945*. Basel: P. Schlettwein.

Windhoek City Council (1993). *Conservation of heritage buildings*. Windhoek.

Chapter 10

Landscape and Nostalgia: Angolan Refugees in Namibia Remembering Home and Forced Removals

INGE BRINKMAN

In the 1970s the land along the Kavango River in Northern Namibia was entirely cleared of its inhabitants. The people of Mangarangandja and Sarasungu, east of the small town of Rundu, were all taken to Kaisosi and Kehemu, two locations that later grew into sprawling communities. Quite a few of those removed were immigrants from southeast Angola in the region collectively known as 'Nyemba'. This chapter deals with the ways in which Angolan immigrants remember their home country, their former Namibian home near the river, and their assessment of the landscape in their current abode. It is argued that it is impossible to assume landscape as being outside of politics, culture, and history. Instead, it is proposed to view landscape as a feature firmly embedded in the context in which it is spoken about.

Although the forced removals were greatly resented at the time, Angolan immigrants hardly referred to their former abode in Namibia. In the memories of their former homes, they focus on their homeland Angola rather than on Mangarangandja or Sarasungu. Their homeland is recalled as an area of agricultural bounty, a land with fertile soil and many rivers. Invariably they complain about Namibia's infertile dry soils and compare them negatively to their well-watered fields in Angola. In this nostalgic framework, their former Namibian abode, although situated near the Kavango River hardly features: only Angolan rivers are remembered well.

When speaking about the history of Kaisosi and Kehemu, many immigrants stress that they were the ones who had actually built the locations. It is

M. Bollig, O. Bubenzer (eds.), *African Landscapes*,
doi: 10.1007/978-0-387-78682-7_10, © Springer Science+Business Media, LLC 2009

emphasised that the area was formerly bush and that all the houses were built by Angolan immigrants. Such statements establish the Angolan immigrants as agents in the transformation of a 'landscape of suffering' to human settlement and challenge the opposition between immigrants and owners, which the Angolans feel is used against them by Namibian inhabitants of the region.

10.1. INTRODUCTION

During fieldwork periods in 1996, 1997, and 1999 in Kaisosi and Kehemu, two informal settlements east of Rundu in Northern Namibia, I spoke with Angolan immigrants, some of whom had lived in Namibia for a very long time. Although they explicitly referred to their country of origin, Angola, it was only by accident that I learnt that a number of them had also lived in another area in Namibia, closer to the Kavango River. My interest in this matter started after one of the interviewees referred to the school in the middle of Kaisosi as 'Sarasungu school'. I knew Sarasungu to be the name of another area, on the other side of the old road between Rundu and Bagani. The interviewee explained that the school had been moved. At the time of these removals the new area did not yet have a name and so before the new settlement started to be called Kaisosi, people continued to refer to the school as 'Sarasungu school'. A former pupil of the school remembered a song in Kwangali, reflecting this ambivalence: 'Our school Kaisosi, known as Sarasungu. Our school Sarasungu, known as Kaisosi' (Interview 1, sung by the younger woman).

I was struck by the lack of attention for their former villages in the immigrants' accounts. Why, when telling about their personal history, did these people 'skip over' their memories of Mangarangandja and restrict their accounts to their life in Angola and their present home in Kaisosi? This question leads to us to the relationship between landscape, violence, and memory. In this chapter I address the issue of flight and forced removal by focussing on the memory of Angolan immigrants. The ways in which immigrants remember their former homes, after an involuntary migration, has received limited attention in scholarly research. In an introduction to a *Paideuma* issue on landscape in Africa, Ute Luig and Achim von Oppen discuss a book written by David Coplan on the songs of Basotho migrants in South Africa. In the 'social wilderness of South Africa' (Coplan, 1994, p. 119), these migrants imagine landscapes in which historical praxis and moral meanings are combined (Luig & von Oppen, 1997, pp. 38–41). Such attempts to 'make sense of a landscape of suffering' (ibid., p. 40) can be compared with the nostalgic framework that the Angola refugees in Rundu employ to refer to their home country.

As the concept landscape has come to be defined more and more as 'a way of seeing' (ibid., p. 10, referring to Cosgrove, 1984, p. 21), distance is regarded as a precondition for any aesthetic perception of landscape (ibid., p. 21). The memories of migrants of their former homelands, physically and historically distanced from the landscape they refer to, may hence be particularly interesting.

As Terence Ranger points out in a paper, involuntary migrants may remember their home sharply and view their new environment with extremely critical eyes (Ranger, 1997, p. 17). Julia Powles argues in an article that such sharp memories of home may be linked to a wish to return to the home country. She presents the example of a woman, who initially expresses vivid memories of 'home', but when she no longer wishes to repatriate, 'her sense of home gradually fades and disappears' (Powles, 2002a, pp. 81–101). This example can help us to understand the memories of Angolans in Kehemu and Kaisosi, both with regard to their country of origin and of Mangarangandja.

In another article, Julia Powles explains that the memories of home and the nostalgia with which refugees relate to Angola are not sentimental, but 'emotionally and bodily, acutely painful' (Powles, 2002b, referring to Serematakis, 1994, p. 4). In her interviews with Angolan refugees who fled to Zambia, it became clear that people especially recall the smell and size of fish. They express such memories not only in speech, but also in dance and song. Although this sensory social memory of fish looms large in the refugees' accounts, standardised narratives of violence hardly featured in the interviews held by Julia Powles.

This was different for Angolan refugees in Namibia. In their memories war and violence were constant themes, told about in a typical manner. Their memories invariably took the form of speech, except where it concerned the history of staying with the MPLA guerrillas in the bush during the independence war; in these cases people also used narratives and songs to express themselves. When I once asked women to come and perform dances, they stated that this made them happy. They complained about the lack of possibility for Angolan refugees to culturally express themselves in Namibia and referred with jealousy to the Zambian context, where, they said, Angolans even had a radio programme. Whereas for Angolans in Zambia reliving of their past experiences was possible, Angolans in Namibia felt very much restricted where it concerned reconstructing their past and their future (ibid.; Brinkman, 2000, pp. 1–24).

Instead of viewing landscape as a given, Luig and von Oppen have stressed that landscape is a process of interaction between perception, nature, history, religion, and aesthetics (Luig & von Oppen, 1997, pp. 7–45). Terence Ranger pointed out that no single model of African landscape exists: colonial as well as African constructions of landscape are contested and subject to change (Ranger, 1997). In this chapter the focus is on precisely such dynamics.

10.2. STILL IN MANGARANGANDJA

The Kavango River is central to an understanding of the history of the Kavango peoples. As fishing, canoeing, and other water-related activities play an important role, most people settled near the riverbanks. In formal politics the river was turned into an international boundary, but for the people of the region, the Kavango River formed a centre of economic, political, social, and cultural activities. In the course of colonial history, several relocations took place that

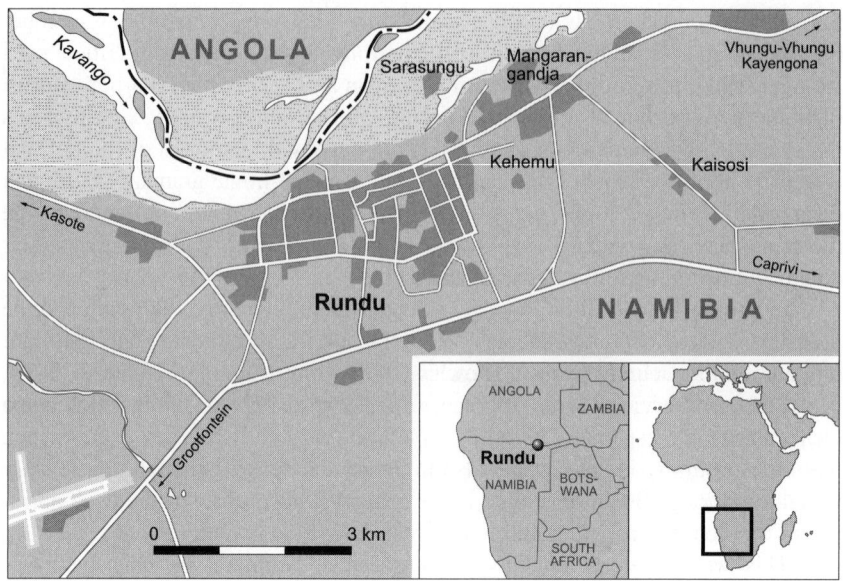

Figure 10.1. Map drawn by Monika Feinen, University of Cologne

drastically altered the place of the river in society. A first large-scale resettlement was carried out by the South African government in 1954–1955. To the east of the settlement of Rundu, a group of people, mainly Angolan immigrants, lived on a small strip of land between the river and the Ncua Lake. This area, known as Nkondo, regularly flooded and when in 1954 the entire settlement was inundated, people were asked to move across the lake and resettle in Mangarangandja and Sarasungu. By the Angolan immigrants I spoke with, this first move was remembered without anger: it was a voluntary change (Interview 2), although in Likuwa's thesis it is suggested that force was used (Likuwa, 2005, pp. 20–22).

Nkondo community life was nearly completely discontinued as the former residents dispersed over Sarasungu and Mangarangandja (Likuwa, 2005, p. 24). Sarasungu had started as a settlement for African policemen working in Rundu in the 1930s, whereas many Mangarangandja residents focused on small-scale trading along the roadside and farming. Both areas had been sparsely populated, but the coming of the Nkondo immigrants meant a population increase, and especially after 1960 the settlements became densely populated. Likuwa suggests that the two villages were different in character, arguing that Sarasungu was of a much more harmonious character than Mangarangandja (Likuwa, 2005, pp. 74–75, 91–93). Angolan immigrants who had lived in the area, however, held that with the population increase the boundary between the two areas became fluid and many people referred to the entire area, including

Sarasungu, as 'Mangarangandja', whereas at the same time people in Rundu often spoke of 'Ncua village'.

It became a mixed community of Kwangali, Shambiu, Nyemba, and Cokwe families, and Mbukushu single men looking for wage labour. There were people who only stayed temporarily: after or before their contract in the South African mines. Others were looking for wage labour in Rundu and yet others were engaged in local trade or farming to provide for themselves and their families. Some people were born in Angola, others on the Namibian side; as the border could be crossed without any permit this was not so important. This relative flexibility is also reflected in the name explanations: it was said that Mangarangandja may be translated as 'no planning' (Interview 3; although Likuwa, 2005, pp. 89–90, offers other explanations).

Because of the growing number of inhabitants, initiatives were taken to provide community facilities and services. Women started brewing and selling beer, a large meeting-place was built, and three churches opened outposts in the settlement: the Full Gospel church, a Lutheran ELOK church, and a bit farther on, in the Sarasungu area itself, a Roman Catholic church. The latter two also functioned as schools during weekdays. In particular the Roman Catholic church was well-established: they had the only stone building in the area and the amount of school pupils quickly rose from 85 in 1965, to 183 in 1970 (Interview 4, Interview 5; 'Die skooljoernaal, 1965–1996', unpublished).

By the end of the 1960s, the South African government had developed plans to remove people living near the river in the entire Kavangoland. The idea was to create a no-go zone, just as the Portuguese had done between Angola and Zambia in an attempt to cut contact between incoming guerrillas and the local population of southeast Angola. The overt political aims converged with general colonial fantasies of clearing the land and reorganising settlements in standardised manners for the sake of state bureaucracy (Scott, 1998). Such colonial fantasies hardly ever succeeded completely, yet had very real consequences for the people involved. Also in this case, the plans were never materialised to the full, but around the regional capital Rundu several removals were carried out. By the end of 1971 people from various locations along the river were called to a meeting in which it was announced that they had to move from the riverside of the Kavango to new numbered plots across the road. Removals took place to the west as well as to the east of Rundu, which must have involved a considerable amount of people, although no statistics are available as to how many people were actually moved. To the west people were moved to Kairaira, and to and from Nkarapamwe. To the east of Rundu, people ended up in Kaisosi or Kehemu.

When the trucks and the armed men arrived, the people from Sarasungu and Mangarangandja were ordered to pack their things. Their hut poles were loaded onto trucks and depending on what truck people happened to mount, they were taken to either Kaisosi or Kehemu. Some people said that the South Africans took the Mbukushu to Kehemu and the Nyemba to Kaisosi, but others maintained there had been no planning behind this (Interview 4, Interview 6, respectively; also Likuwa, 2005, p. 101). The trucks drove to the new locations and the families were dropped, each on one plot, where they lived in the open until they had rebuilt their huts. The community re-established itself with remarkable resilience: the huts were built, beer was brewed, services in church were attended, and pupils went to school. The settlements quickly expanded, especially with refugees from Angola who fled the civil war in their country. By the year 1999 the uninhabited stretch of land between the settlements of Kaisosi and Kehemu had become only very small: Kaisosi had at that time some 5000 inhabitants, Kehemu even more (Census, carried out in 1995, consulted during interview 7; Rundu Town Council, 1995).

10.3. REMOVALS AND ANGER

When recalling the events of the removals most people stress that they were very angry at the time. Although the new locations were only a few kilometres farther on, people strongly resented being moved. The women and children reportedly sang sad songs, and the men quizzed the local leadership about the possibilities to turn the tide. Their anger did not so much concern the disruption of community structures or the destruction of the houses; these were quickly rebuilt. There were three reasons for the people's anger.

Firstly, the plans and implementation of the removals did not in any way include the inhabitants' point of view. The South Africans brutally bypassed any local say in the matter, carried out the removals by means of force, and during the meeting came up with false arguments to justify their actions. Usually these false arguments were retold in direct speech during the interviews: 'They [the whites] said: "Here at the river the war will come. It is near Angola. And here in Namibia you have to move away from the river". We asked: "Where will you take us?" "To that place over there"' (Interview 1 (elderly woman); Interview 8). Other arguments were also mentioned, such as bad health conditions near the river (Interview 3).

Yet people felt that these were not the genuine reasons for the removals. In their view, the South Africans saw it as a means to prevent SWAPO, the Namibian liberation movement, from contacting the population (Interview 6 (elderly man); also Likuwa, 2005, p. 76, p. 94), and, this was especially stressed as a way to get access to the fertile lands next to the river (Interview 6 (both men); Interview 3). Although presented as a measure to protect the people, people held that self-interest was the prime motivation for the removals. No discussion was possible; even visitors were ordered onto the trucks, and if the owner of a house happened to be away the hut was nevertheless demolished; the poles and personal belongings were dropped on a plot in the new settlement. Again the soldiers' comments were added in direct speech: 'They will sleep where their properties are' (Interview 6, younger man).

The Kavango region had just become a Homeland, and the removal plans that were developed by the South African administrator had to be endorsed by the new so-called independent government. According to some people, the headman and the local chiefs were also implicated. Rumours had it that bribes and pieces of land were given in return for consent (Interview 6, both men; Interview 1). A song was created about the headman that implied his role in the removals: 'The place of Kaisosi, ee. Haimbili, Kaisosi, ee. The headman, Kaisosi, ee.' The headman reportedly reacted by fining the singer, stating that he was wrong to blame the headman for the crime and misery that ruled in Kaisosi (Interview 1; Interview 4, Haimbili being the name of the headman at the time). According to the present headman, the South Africans had not consulted any local leaders at all and his predecessor had stood powerless against the force used by the South Africans (Interview 7).

This perspective of powerlessness and naiveté was also underlined by a man who had been involved in politics at the time. He maintained that they had truly believed in the good intentions of the South Africans and only at a later stage discovered the divisive strategy behind the Homeland policy. These accusations and defences stand in contrast with current scholarly debate, in which calls are made to abandon the notions of resistance and collaboration, as these terms suggest a too clear-cut dichotomy (Gengenbach, 2002, pp. 19–47). For many Africans, however, collaboration and resistance remain a source of intense debate. Members of the elite may try to deny the importance of collaboration as a valuable category. At the grass-roots level resistance and collaboration still inspire much discussion. Although I do acknowledge the need for a refined perspective on these concepts, it may be worthwhile, when considering such a reinterpretation, to take into account current debates in African societies on these issues.

A second reason for resenting the removals was the lack of water in the new locations. As Kaisosi and Kehemu are farther from the river, people's instant access to water was stopped. There was only one tap in Kaisosi. Many women had to go to the Kavango River every day to fetch water as the supply in Kaisosi and Kehemu did not suffice. At the one tap there were frequent fights: people tried to fill as many containers as possible, waiting times were endless, arguments about the queue would ensue, and people would start fighting. In some cases the arguments even led to witchcraft accusations (Interview 1; Likuwa, 2005, pp. 87–88). The water was different; one man commented: 'You were used to staying near the river, and there was water everywhere. Coming here they had to dig holes to get water. And it was salty water. "It is salty water!" And they just said: "What are you complaining about? Here is your water!" "No, it is salty." The people suffered' (Interview 6, elderly man). Although it was mentioned that in Mangarangandja there had been a lot of water, in most of the accounts the stress was on the lack of water in the new abode.

The same holds for other references to the environment. Resentment against the removals was not so much expressed in terms of the beauty, fertility, or goodness of Mangarangandja, but rather in a negative evaluation of conditions in the new settlement. The South Africans, it was said, had dropped people in 'the bush'. The new place was considered unsuitable for human settlement. This image of Kaisosi as 'bush' was mostly expressed by referring to animals. Thus it was stated that people had to chase elephants and rhinos from their yards during the night (Interview 6, elderly man; Interview 9). The presence of rhinos would be exceptional, but that is not the issue here. This place, it was held, was unfit for human settlement. Images both of wild and domestic animals were invoked to show that the place was not well prepared by the South Africans:

> All the adults present asked: 'What can we do?' And they [local leaders: the headman and other prominent men in the community] said: 'Our hands are tied by the white people. There is nothing we can do. We will just have to stay like that. We are suffering. There is nowhere else to stay. We sleep in the bush, like pigs and like goats. We cannot do anything, because of these people with the red skin. If we say anything against it, they will kill us'. (Interview 6, elderly man).

If elephants and rhinos were still roaming around and people had to live like 'goats and pigs', it was still 'bush' and not a human settlement. This formed the third reason of resentment: the South Africans were accused of not having taken steps to ensure that human activities – agriculture, cooking, sleep, and so on – could be continued as they should.

SWAPO called upon people to resist the removals:

> Do you want to live in the camps that the Boers are making for you? Do you want to lose your properties such as cattle? They are making camps for your cattle too. How do we call it when they put you in a place that you do not want and he who gives it to you is not his? If you look at the BANTUSTAN, what do you think? They are camps and this is slavery [exile], brothers, men and women, come, the voice is calling you.[1]

As is clear from the interview quotes, however, people were too afraid to openly resist the removals and, despite the resentment, people could not do but obey the authorities' orders.

10.4. MANGARANGANDJA REMEMBERED

The memory of the removals still informs present local political discourse and action. Thus when in the 1990s chief Matumbo expressed plans to move the Angolan inhabitants of Kaisosi to an area farther inland, the anger resurfaced again: people feared that they would once more end up in the bush without any

---

[1] 'Will julle in kampte bly wat die Boere vir julle maak? Wil julle julle bessittings verloor soos julle beeste, hulle maak ook kampe vir julle beeste. Hoe noem ons dit as hulle jou in 'n plek sit waar jy nie wil he nie en hy wat dit vir jou gee is nie syne nie. As julle kyk na die BANTOESTAN wat dink julle? Dit is die kampe en dit is Ballingskap, broerders, mans en vrouens kom, die stem roep julle' [all errors in Afrikaans and in the translation into English are either in the original letter or in Likuwa] Namibia National Archive, Windhoek, Native Affairs Reports 1/1/55, Native Affairs Reports 9, Vol: 13, Swapo letter dated 1--1--1970 quoted in: Likuwa, 2005, p. 83.

water. The sour memory of the removals in the early 1970s was used to argue against the plans. The plans were stalled (Interview 6).

Mangarangandja has become the rubbish heap of Kehemu, and the town's sewerage dam is in Sarasungu. By now the area has become nearly uninhabited except for some tourist lodges, the Ekongoro Youth Camp (that in the 1970s and 1980s was used by the South African authorities to give biblical teachings and anti-SWAPO propaganda; Likuwa, 2005, p. 95), and a detention camp a bit farther on. The old Catholic school/church came to lie on the premises of a golf club, mainly for the South African army, where it was used as a bar. Protest of former inhabitants against the consumption of alcohol in the former church building made the golf club move in the end (see Figure 10.2). This could, however, not prevent the complete obliteration of the graveyard that surrounded the church. Former inhabitants speak of the rubbish and the ruins as a great shame (Interview 1; Interview 5; Interview 9). These examples show that concern over the past of Mangarangandja and Sarasungu still exist: these former homes have not been forgotten. In specific contexts, such as the launching of new plans to move people, the memories of the removals are politically important.

Despite these examples, it is clear that Mangarangandja does not feature to a large extent in the memories of early Angolan immigrants. None of the informants took the initiative to speak about life in Mangarangandja. When asked about it, the evaluation was one without much negative or positive comment. References were made to the community facilities (Interview 5), excessive drinking, labour migration (Interview 1), and to interethnic quarrels (Interview 4).

Figure 10.2. Picture of Mangarangandja area with remains of the golf course
(Likuwa, 2005, p. 106)

In other words, aspects of community life were given that could be interpreted positively, negatively, or in a neutral way.

Although the memory of the way in which the people first arrived in the new settlements is strongly negative, a transformation has taken place. As explained above, anger was expressed at the way in which the removals were carried out, and there were complaints about the lack of water and the wilderness in which people found themselves. Despite this, people said that they would not return to Mangarangandja if it were possible. 'We have got used now', a woman remarked, although she immediately added that by now there were more taps in the settlement (Interview 1; Interview 6). The past in Mangarangandja is not longed for. This is reminiscent of the conclusions of the article on Susanna Mwana-Uta's memories of home by Julia Powles: the fading notions of Mangarangandja may be connected with the fact that people would not wish to return if it were possible. Yet, as shown, in some contexts, the memories of the forced removals from Mangarangandja can be revived and used in political debates.

10.5. CONSTRUCTING KAISOSI AND KEHEMU

In the period when the removals had just taken place, the women and children sang: 'Njelele, Kaisosi the place which displeases me'. It did not take long, however, before other songs started and the women and children sung in the Full Gospel church: 'Kaisosi, Kaisosi, we have found joy. Our place, which fills us with pride' (Interview 1, both women). The assessment of the place Kaisosi changed, because the place changed. Or rather, the place was changed: the changes were due to the intervention of the inhabitants. People took up the challenge: 'Evil landscapes exist to be converted and redeemed' (Ranger, 1997, p. 22). They turned the bush into a place fit for humans to live.

A first sign of this change came in the process of naming. At the time of the removals, the locations had not yet been named. Soon the people started calling the places *Kaisosi*, meaning 'provocation' and *Kehemu*, that can be translated as 'carelessness' (Interview 1– 4, 6). Often the explanations for these names were given not in nouns, but through examples: 'When somebody hits or insults you without reason and you ask: 'Why do you hit me?' that person can say: 'It is *kaisosi*' (Interview 4). The association was with violence and delinquency, but it was the name that people had chosen themselves: 'The South Africans did not want this name, but they could not change it. The people said: 'This is the name' (Interview 3). Although the removals had been forced and the place was not good, the inhabitants now took the initiative from the government and started building up the community in their own way. The naming of the new settlements was seen as a first step in this process towards a new community life. Houses were built, fields were cleared, the pupils went to school, on Sundays there were church services, and women started selling beer, food, and vegetables in the shade of the trees.

Precisely because the place had not been well-prepared and the people felt as if they had been left behind in the 'bush', the contrast was strongly emphasised.

'There was nothing here. We had to sleep in the open until we had built our own houses' (Interview 6). The construction of houses, the act of naming the place, and the building of such 'modern' institutes like schools were at once framed as a process of appropriation and a process of civilization. The 'landscape of suffering' was transformed into 'our place, which fills us with pride'.

Angolan immigrants were eager to tell about this metamorphosis and especially keen to stress that they had been important agents in this change. Both in Kaisosi and Kehemu, the immigrants stressed that they were the ones who had actually built the locations. It was asserted that it had been a bush and that all the houses were built by Angolan immigrants (especially: Interview 8; Interview 10). With such statements the immigrants were clearly referring to residential and ownership rights. In the Kavango region there have been tensions between autochthones and people born in Angola. Angolans feel that the Namibians only see them as a problem, that they deliberately create an opposition between refugees and owners of the land.

The immigrants argue against this opposition with several arguments. Firstly, they stress that not all of the Angolans are refugees: many came to live in Namibia long before the war started. Furthermore, many of the so-called Namibians were also born in Angola, and in any case, all the local ethnic groups migrated from Angola to Namibia in the course of the twentieth century. They also bolster their claims by stating that SWAPO was hosted by Angola for many years during its struggle, so Namibia could do something in return for the Angolans. Finally, emphasis is placed on the fact that they constructed their homesteads in Kaisosi and Kehemu themselves; they changed it from bush into human settlement. This assertion is made with the implication that no government could ever expel them from the place they had created, because it is 'their settlement'. With such statements the immigrant/owner opposition is challenged and Angolan immigrants assert their rights to stay in Namibia (cf. Brinkman, 1999, pp. 417–439).

10.6. REMEMBERING ANGOLA

To summarise, we can conclude that the removals were resented and the remembrance of the landscape in Kaisosi and Kehemu as people found it when they were moved is not positive. There is, however, also pride in the creation and construction of the new settlements. The past in Mangarangandja is not idealised: Mangarangandja is not spoken of in any remarkably positive way; it does not feature in any nostalgic framework.

This is very different for the memories of Angola. This is already shown in the amount of attention Angola receives in the accounts of the immigrants. When speaking with Angolan immigrants, who constitute a majority of the population of Kaisosi and Kehemu,[2] there is a strong tendency to talk about the Angolan

---

[2] For Kaisosi the percentage of people who saw themselves as belonging to one of the Nyemba sub-groups was well over 50, for Kehemu the percentage appears to be lower. Sources: Census, carried out in 1995, consulted during interview 7; Rundu Town Council, 1995.

past, Angolan culture, and the way things were in Angola. It is true that many of the informants were not present during the removals, but also those who had been were more likely to start talking about the Angolan part of their past than about Mangarangandja or Sarasungu. It was only by accident that I learnt about the removals when it was briefly mentioned in one of the interviews, and I asked about the school's name. The stress on Angola was the informants' own initiative: Mangarangandja was only discussed when I asked about it.

The assessments of Mangarangandja and Angola are very different. The immigrants describe Angola as a paradise: a fertile well-watered land. There are extensive references in the accounts to the fruitful combination of rivers, agriculture, and trade. In southeast Angola, so it appears in the accounts, the villages and the small, regional 'towns' stood in ideal interaction. The loss of this combination is strongly regretted. This image of southeast Angola as a land of plenty is all the more striking, considering the assessment of Western scholars studying the region. Southeast Angola has been called 'passive' or even 'dead' and classified as a semiarid, infertile zone. One of the few geographers who visited this area stated: 'Nothing marks the landscape, if only the uniformity, the invariability' (Borchert, 1963, p. 4, p. 67 (my translation); see also Kuder, 1971; Delachaux & Thiébaud, 1936, p. 41). In the Portuguese colonial discourse, southeast Angola was known as 'the land at the end of the earth' and only few whites chose to live in the region (Galvão, 1929, p. 242; Kuder, 1971, p. 225).

Such evaluations sharply contrast with the central place that rivers take up in the refugees' memories of the Angolan landscape. A woman explained that had it not been for the rivers nobody would have survived the war in Angola (Interview 11). The English question: 'Where are you from?' may be translated with: 'Ndonga?' the Nyemba word for river. The answer refers to the name of a river (see also Kubik, 1992/1993, pp. 182–183). In this manner, notions of origin, home, and identity are linked to the landscape. Instead of focusing on one fixed abode, people relate to home as a river: fluid, moving, and connecting many different places. Home and travel, often assumed as opposites, thus become intrinsically related. For the people of southeast Angola, the concept 'dwelling-in-travel' may be especially apt (Clifford, 1997, pp. 81–101; Malkki, 1995).

The attention given to rivers in the peoples' memories does not stand alone: as already mentioned people relate rivers, villages, and 'towns'. People in the dispersed homesteads, that frequently changed location and encompassed a changing group of inhabitants, tried to ensure that through their agricultural activities they had enough food and even some surplus to be traded in the small administrative centres that the Portuguese colonial state had created in the course of colonialism. Products such as soap, salt, sugar, blankets, and clothes could be obtained in this manner. In practice many people may have encountered many problems to make ends meet, but in the nostalgic framework fertility, fields, and trade together ensured that people had all they needed. Instead of an opposition between town and country, so prevalent in Western landscape (Lemaire, 1996, pp. 116–123), rivers, village, and town were felt to form a symbiotic relationship.

The war in Angola changed this. Not only were the people taken from their villages, trading contact between town and the area around was stopped. Town and bush came to stand in opposition and visits were no longer possible. Furthermore, the fighting parties regularly accused each other of poisoning the drinking water and one young man held that UNITA's long-time leader Jonas Savimbi had changed the Lomba River into blood, so that the MPLA could not drink the water.[3] According to the people who had lived through the war, the entire network of relations between rivers, villages, and towns was destroyed.

It is only the Angolan rivers that are remembered so well. Within the nostalgic framework Mangarangandja is not mentioned despite its proximity to the Kavango River and to the market of Rundu. The immigrants create an opposition between a well-watered, fruitful motherland Angola and aridity and drought in Namibia. Even Calai, just across the Kavango River is said to have 'nice farming, nice land' (Interview 13). The former experience of Mangarangandja does not fit into this opposition, and it is not mentioned when people relate to the opposition between drought and abundance. When referring specifically to the history of the removals, the abundance of water in the former Namibian abode is sometimes brought up, but this is stated in a matter-of-fact way and not much regret is expressed. The emphasis is usually on the lack of water in the new settlements of Kaisosi and Kehemu.

It is only when discussing the Angola/Namibia contrast that nostalgia comes in: people express their regret over the loss of the way of life in Angola, and negatively assess life in Namibia. In the following quote these aspects are combined.

> That is why we say that we will go back if there is peace in our country. It is such a nice place. We can go and plant crops, eat good food and possess our own farms. Here we are suffering from hunger. In the bush the owner of the land plants, but when we want to start farming, we are told: 'No, this is our place! Don't plant here (Interview 14).

The Angolan refugees in Rundu are willing to contribute to human culture; they want to convert bush into farm, but they are not allowed. Neither, so the immigrants say, are they allowed to practice their cultural, ritual, and social ceremonies. In Namibia there are no masked dancers, no male or female initiation rites, and no meeting places as there used to be in Angola.

---

[3] Arquivos Nacionais, Torre Do Tombo, Pide/Dgs, Delegação de Angola, P.Inf. 110.00.30, vol. 21, p. 217; Luanda, 12 May 1972 (*Zambia Daily Mail*, 25 April 1972); idem, vol. 9, pp. 108–109, 111–114: Luanda, 12 March 1967; Interview 12; Jonas Savimbi (1934–2002) was the long-time leader of UNITA, the movement that opposed the MPLA government in Angola since independence.

10.7. PAST AND PRESENT

Despite the negative assessment of Namibia and a longing for Angola, not all the Angolan immigrants wish to return to their country of origin. Some have lived in Namibia for a long time and all their children were born in Namibia. Even so, in their accounts, the contrast between good old Angola and dry infertile Namibia comes to the fore. They explain that life used to be good in Angola and that there is not enough water in Namibia, that the local leaders and the Namibian government are trying to evict them, that they do not have enough land, and so on.

The nostalgia is thus doubled. It is not only Angola – home – and Namibia – exile – that are opposed; also the past is placed in opposition to the present. When the war started, the Angolan countryside became the scene of widespread destruction, theft, and rampage. As there were no strategic war targets, the aim of the fighting parties was to control as many inhabitants of the region as they could. People were taken from their villages and led away: by the guerrillas to the bush, by the Portuguese to wired camps near the military garrisons. This part of the Angolan past functions within the same framework as the dichotomy Namibia/Angola: the good Angolan past before the war is contrasted with the bad present of violence, destruction, and flight in Angola and exile and suffering in Namibia (Werbner, 1991, p. 109; Ranger, 1997). In order to understand the double focus of the nostalgic framework, the discussion is expanded to assess the relationship between landscape and the war in Angola.

The war started in eastern Angola in 1966. MPLA as well as UNITA guerrillas entered Angola through Zambia and the Portuguese started a counterinsurgency campaign. For the local population the most drastic consequence of the war was the destruction of village life. Villagers were captured and led away, either to town or into the bush. The farmers lost their livestock, their huts, their fields, and crops. Usually they were only allowed to carry a small load of clothes and personal belongings. It is striking that in their accounts, the informants did not dwell on the situation in the bush or in the concentrated settlements. Descriptions were minimal and little attention was paid to the living conditions in these new homes. The focus in the accounts of these forced movements is on the movement itself and on the destruction of agriculture. The changes in human settlement were barely discussed, whereas the ways in which the war affected patterns of mobility and livelihood received ample attention.

The spatial layout of the Portuguese concentrated settlements does not figure in the accounts. This is remarkable, because in a contemporary Portuguese report on Moxico district it is stressed that the Portuguese resettlement of people led to considerable changes in the settlement patterns in the region: a change from dispersed, family-based small settlements to much larger concentrations of people, a change from a semicircular shaped homestead replaced by lined rows in a grid pattern, and a change from wood and thatch huts to bricks and zinc constructions (N.N., 1974, p. 32, p. 35). Such changes are at most cursorily mentioned in the accounts; they do not constitute a topic.

The same holds for life in guerrilla camps: in the accounts very little information is given as to what these camps looked like, how they were organised, and in what situation the civilians found themselves. In their writings, outside observers and the MPLA leadership focus on these aspects, but in the civilians' reminiscences they hardly figure at all. The accounts focus on aspects of mobility and immobility: process of the movement and the end of village life were discussed extensively and the restrictions on mobility imposed by the fighting parties were greatly resented. As in the accounts of the removals from Mangarangandja, the conditions of movement and the way in which the removals were carried out are given more attention than the situation in the new residence. The new place to which people were taken was in most cases only briefly described. This is similar to Ranger's discussion of migrants' memories of travel and migration itself, rather than homeland and destination (Ranger, 1997). Such views challenge the relationship between landscape, home, and permanence prevalent in common 'territorializing' conceptions of landscape (Malkki, 1995, pp. 15–16).

10.8. THE WAR AND THE END OF INDEPENDENT FARMING

During the war for independence in southeast Angola, clearing the bush and thus creating new farming land independently was considered to be dangerous and difficult. The Portuguese assumed that anyone in the bush was a guerrilla: in their bombing they did not distinguish between guerrilla camps and civilian homesteads. People who were not in Portuguese towns, whether in guerrilla company or not, had to constantly flee from one place to another and this made agricultural activities next to impossible. On top of this, farmers sometimes faced food theft by hungry guerrillas. The war thus meant the end of village life and independent farming: many people fled abroad, and the remainder were taken to the Portuguese settlements or ended up in the bush with the guerrillas.

In 'town', the term used to name the Portuguese resettlement camps, people were generally not allowed to clear the bush for agricultural purposes. The Portuguese would tell people not to create farms of their own in the bush, not to leave the town, not to go to the 'bandits' (Interview 15, elderly woman). People clearing the bush would be too hard to control for the colonial army. People were forced to live on the produce of small gardens within the settlements and on the food rations distributed by the Portuguese on a weekly basis. People realised why the Portuguese handed out seedlings to plant near their houses; as a woman explained: 'They feared that we would say: 'Let us go into the bush, to the MPLA' (Interview 16).

In the bush with the guerrillas, there were also hardly any opportunities to create a farm. The guerrillas depended on civilian farming and induced the women to produce food for them: during rehearsal sessions, women had to answer the question: 'Who are you?' in choir: 'I am a farmer. I like seeds, I like agriculture' (Interview 17). But, this expectation was very hard to fulfil. Livestock was usually killed upon departure from the village, as it was deemed incompatible with guerrilla life. Frequent Portuguese bombing made harvest impossible; napalm and herbicides were used by the Portuguese to destroy crops (Marcum, 1978, p. 201). Despite attempts to minimise the risks, a widespread complaint was that a full agricultural cycle was impossible in the bush, because people were always fleeing from one place to another, trying to avoid Portuguese helicopter attacks.

So this bush, in contrast to Kaisosi and Kehemu, was never converted into a place fit for human dwelling: it remained bush. This was a source of extreme resentment. Here the houses were not rebuilt: people had to sleep in the open, without blankets 'like animals' (Interview 18). During these removals and flights, social relationships were completely torn apart: the frequent references to incest, due to the fact that relatives did not know each other, testify this concern. Here villages were reconverted into bush, instead of bush being changed into settlement. The war thus resulted in a topsy-turvy process. The civilians stressed that proper movements, such as visits to relatives and the founding of new villages as a means to improve agriculture, were made impossible during the war, whereas improper movement, which led to the destruction of agriculture, increased. They stressed that they were forced to leave their villages on the riverbanks and had to stay in inhospitable places, where violence and hunger reigned (Brinkman, 2005).

This perhaps explains the different ways in which the experience of flight and removal are remembered. When moving from Mangarangandja to Kaisosi and Kehemu, there were problems: force was used and people ended up in a hostile environment with only limited access to water. To a certain extent this memory is still alive: the memory of Mangarangandja at times leads to political concern, as the examples of the old graveyard and the recent removal plans show. Yet in Kaisosi and Kehemu there was a prospect of creating a new human settlement. The bush did not last: houses, schools, churches were built, and social life – visits, beer consumption, church services, funerals, familial judicial discussions, community leadership structures – was continued and rebuilt along the same lines. People cleared land in the bush and made new farms in order to sustain themselves. Construction, agriculture, and social life converted an initially hostile environment into human settlement. Mangarangandja was not forgotten, but a return, initially impossible, was later not wished for by those moved. Angolan immigrants use the founding history of Kaisosi and Kehemu as an argument in their discourse about their right to stay in Namibia.

In contrast, when the war came to southeast Angola, farmers lost their villages to the bush, and in the bush there was no opportunity to convert the wilderness into orderly life. Independent farming was dangerous, impossible, or not allowed. Many people merely fled from place to place and had no opportunity to start agricultural activities. In exile in Namibia, the Angolan refugees feel that they are not given a chance to create a full social and economic life. Land access is hard, they are often not allowed to create fields for agriculture, most of them do not have a work permit, and they feel that they are not allowed to live their cultural and social lives fully. The Angolan immigrants describe the time before the war with nostalgia. In their view, nature, agriculture, and trade had stood in ideal interaction in the past. The war destroyed this and in the Namibian context, the Angolan immigrants feel unable to reconstruct this ideal.

10.9. FINAL REMARKS

This essay on memory and landscape shows that we cannot assume a landscape outside politics, culture, and history. In this sense a term such as 'cultural land-scape' is tautological. Because of the links between landscape and history, there is no one fixed conceptualisation of landscape. The ways in which landscape is imagined, described, and remembered changes over time. In this case, it changed from 'Njejele, Kaisosi the place which displeases me' to 'Kaisosi, Kaisosi, we have found joy: Our place, which fills us with pride'.

This change has, however, not affected the dichotomy that Angolan immi-grants create between Namibia and Angola. When referring to this contrast, the memory of the former home on the Namibian side of the Kavango River is often skipped over, and Namibia is classified as dry and infertile as opposed to a fertile and well-watered Angola. This shows that landscape views are not only subject to variation and change, but that they may also be contradictory and ambivalent. Landscape does not lie out there, waiting to be discovered, but is politically and morally charged. It is a 'concept of high tension' (Inglis, 1977, pp. 489–513, referred to in Bender, 1993, p. 3).

It follows that it is not enough to state that landscape is related to views and perceptions. The 'ways of seeing' (in the plural) depend on the specifics of political debate. The context in which a landscape is described matters when interpreting the evaluation of a landscape. This remark not only refers to differ-ences in discourses that exist between, say, Western geographers and Southeast Angolan displaced and dispossessed farmers. It also argues that the context of a specific discussion on landscape shapes the way in which people present their memories. Discussing the present chief's plans to move people farther inland may lead people to express bitter memories of the move 'from the river to the bush' in the 1970s in Namibia. Discussing the tensions between what are known as immigrants and owners may at once lead to expressions of pride on the construc-tion of the settlements Kaisosi and Kehemu, and to negative statements about the Namibian landscape as opposed to positive evaluations of the Angolan landscape. Landscape perceptions form embedded histories.

Such contextual circles of memory are given hierarchical order. Thus in this case, Angolan refugees concentrate on Angolan landscape and attribute it with positive qualities. Mangarangandja is not forgotten, but in the accounts of Angolan immigrants it is far less prominent than Angolan history. The past in Angola is described in nostalgic terms; Mangarangandja is not. Both when the war started in southeast Angola and when the riverbanks of the Kavango were depopulated, it concerned forced involuntary movements. Yet, although a return to Angola is much discussed and many people during the fieldwork expressed their wish to go back, none of the informants mentioned the possibility of going back to Mangarangandja. When I suggested a return, the answer was invariably negative. Kaisosi and Kehemu had changed 'from forced to chosen communities.' (Schmidt, 1996, pp. 183–204).

The discussions on water and bush are similarly related to context. Initially Kaisosi and Kehemu had no water, but this aridity was overcome by taps and boreholes. At the beginning the new areas were looked upon as 'bush', but the re-establishment of the community transformed the landscape from wilderness into productive land. In Angola, civilians were barred from living near water sources and from changing the bush into villages and farms. In Namibia, the Angolan immigrants feel that they are not allowed to participate fully in community life: politically, culturally, and economically. This very much constitutes the drought and aridity of exile in Namibia and the fertility of the Angolan past.

INTERVIEWS

In 1996, 1997, and 1999 fieldwork was conducted with Angolan immigrants resident in Kaisosi and Kehemu with the aim of interpreting the recent history of southeast Angola. After learning about the forced removals from Mangarangandja and Sarasungu, I asked people who could tell more about these events. I was referred to several persons and with these persons interviews were held that specifically dealt with the removals (p.e. interviews 2, 3, 5, and 9).

Interview 1, with two women (both born in Mangarangandja, the mother some time near 1945 and her daughter, born about 1965), Kaisosi, 5 July 1999.

Interview 2, with Reverend J.C. Sindano, Nkarapamwe, 20 July 1999.

Interview 3, with Archbishop Hausiku, Windhoek, 30 July 1999.

Interview 4, with woman (particulars: see interview 1, elderly woman), Kaisosi, 5 September 1997.

Interview 5, with Pastor Erastus Ntsamba, Kapako, 14 July 1999.

Interview 6, with two men (one born in Cuito Cuanavale in 1911 (perhaps later), the other in 1958 in Lupire), Kaisosi, 22 June 1999.

Interview 7, with Mr. Moses Haingura and Pastor JosÉ Ntsamba, Kaisosi, 9 July 1997.

Interview 8, with a woman (born in Cuito Cuanavale, around 1935), Kehemu, 13 September 1997.

Interview 9, with Mrs. H. Hausiku, Kaisosi, 20 July 1999.

Interview 10, with a man (particulars, see interview 6, elderly man), Kaisosi, 27 August 1996.

Interview 11, conversation with a woman (born by the Sobi river, 1958), Vungu Vungu, August 1996.

Interview 12, with a young man (place and date of birth: unknown), Kehemu, 16 June 1996.

Interview 13, with a woman (particulars, see interview 10), Kehemu, 27 July 1996.

Interview 14, with a woman (born in Mavinga, about 55 years old), Kaisosi, 26 August 1996.

Interview 15, with two women (one born before 1925, the other about 1950, both in Mavinga), Kaisosi, 19 June 1997.

Interview 16, with an old woman (born by the Cuma river, around 1920) and her grandson (near Cuito Cuanavale, 1965), Kaisosi, 29 June 1999.

Interview 17, with two women (one born in Cundumba, born between 1935–1940, the other near Cunjamba, around 1930) Kehemu, 22 June 1999.

Interview 18, with a woman (born in 1921 in Mavinga), Kehemu, 30 June 1997.

Acknowledgements Research for this chapter was carried out within the framework of the SFB project ACACIA of the University of Cologne, Germany, funded by the DFG. I wish to acknowledge the help of all the people interviewed, Rebecca Kastherody and Dominga Antonio for their work as research assistants, and Michael Bollig, Heike Behrend, Patricia Hayes, Jeremy Silvester, and Robert Ross for their welcome suggestions on the work. It goes without saying, however, that only I can be held responsible for its contents.

REFERENCES

Unpublished

Arquivos Nacionais, Torre Do Tombo, Pide/Dgs, Delegação de Angola, P.Inf. 110.00.30, MPLA.

Likuwa, K.M. (2005). Rundu, Kavango: A case study of forced relocations in Namibia, 1954 to 1972. Unpublished MA thesis, University of Western Cape.

Die skooljoernaal, Sarasungu, Lower Primary, 1965–1996. Unpublished manuscript, Kaisosi, by courtesy of Mrs. H. Hausiku.

Powles, J. (2002b). Like baby minnows we came with the current: Social memory amongst Angolan refugees in Meheba Settlement. Paper presented at the *Conference of the Association of Social Anthropologists of the UK and the Commonwealth: 'Perspectives on Time and Society: Experience, Memory, History'*, Arusha, April 2002. By courtesy of Julia Powles.

Ranger, T. (1997). New approaches to African landscape. Paper presented at the conference: *'Africa and Modernity'*. Berlin 1997. By courtesy of Heike Behrend.

Rundu Town Council (1995). Proposal for: Demarcation of Rundu into seven wards. Unpublished typescript, Rundu 1995. By courtesy of Rundu Town Clerk, Mr. Lucas Muhepa.

Published

Bender, B. (1993). Introduction. Landscape – Meaning and action. In B. Bender (Ed.), *Landscape. Politics and Perspectives*. (pp. 1–18). Oxford: Berg.

Borchert, G. (1963). *Südostangola. Landschaft, Landschaftshaushalt und Entwicklungsmöglichkeiten im Vergleich zum zentralen Hochland von Mittel-Angola*. Hamburg: Institut für Geographie.

Brinkman, I. (1999). Violence, exile and ethnicity: Nyemba refugees in Kaisosi and Kehemu (Rundu, Namibia). *Journal of Southern African Studies*, 25(3), 417–439.

Brinkman, I. (2000). Ways of death: Accounts of terror from Angolan refugees in Namibia. *Africa*, 70 (1), 1–24.

Brinkman, I. (2005). *'A War for People.' Civilians, Mobility, and Legitimacy in South-east Angola During the MPLA's War for Independence*. Cologne: Köppe.

Clifford, J. (1997). *Routes. Travel and Translation in the Late Twentieth Century*. Cambridge, London: Harvard University Press.

Coplan, D.B. (1994). *In the Time of Cannibals. The Word Music of the South Africa's Basotho Migrants*. Johannesburg: Witwatersrand University Press.

Cosgrove, D.E. (1984). *Social Formation and Symbolic Landscape*. London: Croom Helm.

Delachaux, T. & ThiÉbaud, C. (1936). *Land und Völker von Angola. Studien, Erinnerungen, Fotos der II. schweizerischen wissenschaftlichen Mission in Angola*. Neuenburg: Attinger.

Galvão, H. (1929). *Huíla (relatório de Govêrno)*. Vila Nova: Tipografia Minerva.

Gengenbach, H. (2002). "What my heart wanted": Gendered stories of early colonial encounters in Southern Mozambique. In J. Allman et al. (Eds.), *Women in African Colonial Histories* (pp. 19–47). Bloomington: Indiana University Press.

Inglis, F. (1977). Nation and community: A landscape and its morality. *The Sociological Review*, 25, 489–513.

Kubik, G. (1992/1993). Das "ethnische" Panorama Ostangolas und der Nachbargebiete. *Bulletin of the International Committee on Urgent Anthropological and Ethnological Research*, 34/35, 161–195.

Kuder, M. (1971). *Angola. Eine geographische, soziale und wirtschaftliche Landeskunde*. Darmstadt: Wissenschaftliche Buchgesellschaft.

Lemaire, T. (1996). *Filosofie van het landschap*. Baarn: Ambo.

Luig, U. & von Oppen, A. (1997). Landscape in Africa: Process and vision. An introductory essay. *Paideuma*, 43, 7–45.

Malkki, L.H. (1995). *Purity and Exile. Violence, Memory, and National Cosmology among Hutu Refugees in Tanzania*. Chicago/ London: University of Chicago Press.

N.N. (1974). *O Distrito do Moxico. Elementos monográficos*. Luanda.

Marcum, J.A. (1978). *The Angolan Revolution: Exile Politics and Guerrilla Warfare (1962–1976)*. Cambridge, MA: MIT Press.

Powles, J. (2002a). Home and homelessness: The life history of Susanna Mwana-Uta, an Angolan refugee. *Journal of Refugee Studies*, 15(1), 81–101.

Schmidt, H. (1996). Love and healing in forced communities: Borderlands in Zimbabwe's war of liberation. In P. Nugent & A.I. Asiwaju (Eds.), *African Boundaries. Barriers, Conduits and Opportunities* (pp. 183–204). London: Pinter.

Scott, J.C. (1998). *Seeing like a State. How Certain Schemes to Improve the Human Condition Have Failed*. New Haven, CT: Yale University Press.

Serematakis, N. (1994). The memory of the senses, part I: Marks of the transitory. In N. Serematakis (Ed.). *The Senses Still: Perception and Memory as Material Culture in Modernity* (pp. 1–18). Boulder, CO: Westview Press.

Werbner, R. (1991). *Tears of the Dead. The Social Biography of an African Family*. Edinburgh: Edinburgh University Press.

Part III

Identity, Memory, and Power in Africa's Arid Landscapes: Perspectives from Social and Cultural Anthropology

Chapter 11

The Anthropological Study of Landscape

MARTIN RÖSSLER

This chapter provides an overview of how anthropology has examined various aspects of landscape not only in Africa but on a global scale. Treating landscape basically as a cultural concept, anthropologists have investigated a broad spectrum of landscape issues in many different cultures. Among these topics, particular attention is here given to approaches focussing on the relation between nature and culture, studies on historical imaginations and notions of identity connected to landscape, as well as to studies of cognition of and orientation in landscapes. Finally, recent research on the role of landscapes within the political economy of globalisation is addressed. The multitude of issues investigated by the 'anthropology of landscape' has over the years been paralleled by shifting theoretical orientations. Whereas semiotics, hermeneutics, and interpretive as well as cognitive theory were prevailing earlier, many recent studies draw on postmodernist deconstruction. The heterogeneity of systematic issues and theoretical orientations in anthropological studies on landscape have so far prevented establishing a theoretical notion of landscape of cross-cultural applicability. Yet this contribution is intended to set a broader systematic framework for the studies on landscapes in northeast and southern Africa that are presented in this volume.

M. Bollig, O. Bubenzer (eds.), *African Landscapes*,
doi: 10.1007/978-0-387-78682-7_11, © Springer Science + Business Media, LLC 2009

11.1. INTRODUCTION

From an anthropological point of view, landscape is primarily a cultural concept. Although this assertion suggests that anthropology has examined the concept of landscape on the same scale as it has studied other cultural concepts, this is not the case. Although specific local concepts of landscape have tangentially been examined in early ethnographies (e.g., Richards, 1939), the systematic and comparative study of landscape has in fact emerged quite recently as an issue of anthropological analysis. Regardless of its short academic tradition, there is now an immense amount of studies on the subject.

The present volume focuses specifically on arid Africa, however, a more general overview of anthropological perspectives on the study of landscape requires extending this regional focus for at least two reasons. The first reason is that African, and in particular arid African landscapes, represent a rather limited section of landscapes. Confining the discussion to this regional context would, for example, exclude Pacific volcano islands as well as landscapes such as those prevailing in Iceland, Amazonia, or the rugged highlands of Papua New Guinea. Studying landscape on a general or theoretical level requires taking a broad spectrum of environments into account. Those just mentioned, among others, have contributed greatly to the anthropological understanding of landscape, including landscapes in various African settings. The second reason is directly connected to the first, in that it is above all the comparative analysis of different landscapes which can reveal their significance for human individuals and societies, and which may eventually result in a theoretical concept of 'landscape'.

In addition, it should be underlined at the outset that anthropologists have, in many regional settings, studied rather heterogeneous aspects of landscapes. Among these are the social and mythological meanings attributed to landscapes, patterns of spatial orientation and cognition, landscapes as symbolic systems, ecological aspects of landscape and landscape use, the history of cultural landscapes, or the relationship between landscape as a cultural concept and territory in the sense of a basic resource.[1] Although this array of topics looks impressive, it bears serious shortcomings. Most important, it is difficult if not altogether impossible to identify any universals with regard to an 'anthropology of landscape', let alone with respect to landscape as a concept which is universally relevant in all human cultures. Correspondingly, selecting studies that may illustrate what anthropological research on landscape has achieved so far is not an easy endeavour.

As a starting point let us assume that from an anthropological perspective the study of landscape is an 'attempt to understand the involvement of people with the land that over time has generated a "sense of place" for them' (Waterson, 1997, p. 63), this 'sense of place' being derived from many different factors in different societies under varying cultural, historical, and environmental circumstances. It

---

[1] Such heterogeneous issues are not only typical for anthropology. In cultural geography, too, various theoretical approaches to the study of landscape have been employed simultaneously (Norton, 1989).

seems reasonable, therefore, to distinguish among several main topics that have so far been investigated, although many of them are closely interrelated. Naturally the lack of space does not allow full coverage of all the issues examined, let alone the consideration of any details.

11.2. BACKGROUND: CHANGING VIEWS OF THE LANDSCAPE

As indicated above, the anthropological study of landscape does not have a long history. By way of contrast and for obvious reasons, the focus on landscape has long been central to geographical analysis. This holds not only for physical geography but also for social and cultural geography. The latter was of particular importance for anthropology because some major theoretical insights of cultural geography provided significant stimuli for most recent anthropological research. Inasmuch as there is often but a slight difference between geographical and anthropological research, the following sections mainly concentrate on what may be labelled 'genuine' anthropological studies, although reference is repeatedly made to geographical studies as well.[2] First of all, however, some general remarks illustrate what 'studying landscape' has been about thus far.

The concept of landscape as Westerners have long understood it was coined in western Europe in the seventeenth century. The English term, which was derived from the German *Landschaft* or Dutch *landschap*, originally designated not the physical landscape itself, but rather a particular experience of seeing a landscape, namely its representation in painting (Bender, 1993; Abraham, 2000). Accordingly, the idea of landscape emerged 'as a dimension of European elite consciousness at an identifiable period in the evolution of European societies' (Cosgrove, 1984, p. 1; cf. Baker, 1992). Against the background of this culturally specific notion it follows that landscapes, in a cross-cultural perspective, have to be analysed within particular historical, social, economic, and political contexts.

Prior to anthropologists, geographers have pointed out that every landscape is not just something to be looked at but rather a socially constructed way of seeing which is embedded in the wider society and political economy. Applying both the term and the concept of landscape to 'landscapes' in a more general sense has subsequently led to various, and often strongly debated, definitions and theoretical discourses. In this context, geographers, following Buttimer (1969) and Gregory (1978), in contrast to earlier approaches during the last decades

---

[2] Lack of space in this chapter also does not allow for the inclusion of the contribution of anthropological archaeology to the study of landscape. In general terms, in a similar vein as cultural anthropologists, archaeologists in recent studies aim at an understanding of past peoples' cultural constructions of localities and space, including such aspects as place naming and the relation between memories and landscape (see Ucko & Layton, 1999; Rubertone, 2000; Tilley et al., 2000; Tilley & Bennett, 2001; Lenssen-Erz & Linstädter, this volume).

have increasingly focused on interpretive, philosophical, and even postmodernist approaches to the study of landscape (Benko & Strohmayer, 1997; Duncan & Ley, 1993; Soja, 1989, 1996; Thrift, 1996).[3] This discourse was in the main based upon the idea that such concepts as landscape, location, place, and space are not 'given facts' in any sense, as particularly physical geography had long suggested. Thus, whereas physical geographers have treated the features of landscapes as measurable, *objective attributes of space*, more recent culturalist approaches have been concerned with *meanings ascribed to spatiality*.

It was furthermore assumed that every landscape is in fact ambiguous in several respects. Firstly, a particular landscape is 'that characteristic portion of the world visible by an observer from a specific position' (Conzen, 1990, p. 2),[4] which implies that a landscape is both object and subject. Secondly, there is a distinction between a landscape's spatial extent and its contents or material features. Thirdly, people who live in a given landscape, and people who do not may give rather divergent meanings to it (cf. Okely, 2001). For obvious reasons this last point is of particular importance for anthropology.

Against this background, recent approaches to the study of landscape employed by cultural geographers have focused on four main topics, namely environmental awareness, symbolic representation, landscape design, and landscape history (Conzen, 1990, p. 4). These approaches, questioning materialistic interpretations without, however, neglecting material needs and constraints, have at the same time led to the discovery of meaning and ideology, of the 'real' and the symbolic, of the interdependence of nature and culture in landscape (Baker & Biger, 1992; cf. MacNaghten & Urry, 1998). For cultural geographers as well as for anthropologists a main argument lies in the fact that both nature and culture have shaped landscapes.[5] From a present point of view, indeed very few landscapes have escaped human impact. The point is that even in those regions where human impact on the landscape has been comparatively slight – for example, in sparsely populated landscapes such as the arctic, deserts, or high mountains – it has always had its roots in human culture.

Put simply, it is this relationship between culture and nature in which both cultural geographers and anthropologists are interested. Although it is often not

---

[3] Duncan & Ley (1993) mention four major modes of representation in Anglo-American cultural geography, namely descriptive fieldwork, positivist science, postmodernism, and the interpretive mode of representation based upon hermeneutics. The order of these 'modes' as presented by the editors is interesting because anthropologists would commonly list interpretive approaches prior to postmodernism.

[4] Meinig (1979) has distinguished ten versions of perception which ten different observers of the same landscape may have. They may perceive it, respectively, as representing nature, habitat, artefact, system, problem, wealth, ideology, history, place, or aesthetic.

[5] It has been pointed out by the philosopher Kate Soper (1995) that the very notion of 'nature' – which is in a sense an intrinsically Western cultural concept and representation – always implies historical and cultural dimensions.

easy to identify any marked difference between both disciplinary approaches (cf. Feld & Basso, 1996, p. 3ff.), we may cautiously draw a distinction that is based upon both disciplines' academic backgrounds, irrespective of how strongly these may have been blurred. Accordingly, in cultural geography, as a branch of geography, the 'natural'/spatial dimension as the basis of human society and culture is given analytical priority, whereas anthropology takes human culture as a starting point, which inter alia is also concerned with space.[6] Needless to say both approaches are closely interrelated.

11.3. LANDSCAPES AS TEXTS

One of the first mainstream theoretical stances in both the 'new' cultural geography and anthropology was to treat landscapes metaphorically as texts, which means as constructs which can be 'read' and which therefore are also open to interpretation. Under the influence of what was labelled symbolic, or later on, interpretive anthropology, landscapes in much the same way as rituals, oral traditions, dress codes, and the like became the subject of semiotic or hermeneutical analysis (Fairhead & Leach, 1996, p. 15; Cosgrove, 1984; Duncan & Ley, 1993). Moore, (1986), for instance, in her work on the Marakwet of Kenya, treats space as a reflection of social categories and classifications systems, in this sense taking up an idea that was originally formulated by Durkheim & Mauss (1963 [1903]). The argument strongly focusses on the relationship between ideas and meanings on the one hand and spatial organisation on the other. The Marakwet stress the significance of space for the ordering of social perception and experience. In their local discourse, people permanently refer to physical positions of humans, events, and objects, including their relative positions and movements in the landscape. The main point is that landscape is always but one of the multiple cultural 'texts' to be read and interpreted, and that landscapes are always embedded in various discursive fields related to cultural institutions and processes.

Duncan (1990), following Cosgrove and other 'new cultural geographers', also treats landscape – in his case an urban landscape – as a text which can not only be read but which may also be altered repeatedly, thus requiring continuously modified readings and interpretations. Drawing on the work of Turner, Geertz, and Foucault, Duncan considers a culture's signifying system as being composed of various discursive fields which stand in specific relation to one another. The relationship which is emphasised in this study is the one between the landscape as a signification system and the discursive field of kingship. Both fields merge in the spatiality of the royal capital of Kandy in Sri Lanka. The main argument behind the meticulous analysis and interpretation is that landscape, as a signifying system, is one of the key elements of a cultural system as a whole.

---

[6] For the relationship between geographical and archaeological approaches, see Wagstaff (1987).

Cities, as highly complex cultural landscapes, provide particularly suitable examples of this kind of theoretical framework. This is also evident from the investigation of Mexican border cities, to which Arreola & Curtis (1993) apply the method of 'landscape anatomy'. They emphasise that cities are not only political or economic systems but above all cultural creations. Again the notion of landscape is dealt with as a text, as a medium that can be read and interpreted. Because landscapes, quite similar to the built environment (Rapoport, 1982), are identified to convey messages about cultural meanings, the reading and interpretation of these texts are believed to provide important clues about human thought and consciousness, about the human impact on the landscape now and in the past, and therefore about an important component of culture in general.

11.4. NATURE AND CULTURE: THE ECOLOGY OF LANDSCAPES

One of the most important fields within the study of landscape has been the analysis of its ecological dimension, or, perhaps more precisely, of the interdependence of human culture and the 'hard facts' of the physical environment. This perspective transcends more conventional anthropological analysis in that it is explicitly related to interdisciplinary approaches such as human ecology, environmental history, or historical ecology which strongly focus on ecosystems (Balée, 1998; Tress et al., 2001), and which have stimulated the emergence of further subdisciplines. *Landscape ecology* studies how humans modify the habits and distribution of nonhuman species, with the consequence that human landscapes are analytically distinguished from natural landscapes. In this view, landscape is defined as the 'material manifestation of the relation between humans and the environment' (Crumley, 1994, p. 6). By way of contrast, *historical ecology* or *landscape history* is defined as 'the study of past ecosystems by charting the change in landscapes over time.' (ibid.; Balée, 1998).[7] In general terms, from the ecological viewpoint which is at the heart of these approaches, landscapes are seen to be the results of the historical interrelatedness of humans and their physical environment. Accordingly, studies from this academic background also allow for identifying changing human attitudes toward the environment.

It should be pointed out, however, that the distinction between a 'natural environment' and a landscape shaped by humans (e.g., an agrarian landscape), which is often at least implicitly drawn, can be quite misleading analytically. Thus the very notion of 'environment' suggests that it is identical to nature, and thereby separated from or even opposed to human culture (cf. MacNaghten & Urry, 1998; Ingold, 2000, pp. 190ff.). However, there are always blurred boundaries between landscapes modelled by humans and an autonomous nature that is

---

[7]The relationship between landscapes and changing environments from an ecological perspective is discussed in Holland et al. (1991).

supposedly uninfluenced by human agency (Agrawal & Sivaramakrishna, 2000). This is corroborated by the fact that many societies do not recognise any opposition between human social life and what is commonly called 'natural' species on the other hand. For the Kamea of Papua New Guinea, for example, resources such as land or species of fauna and flora 'furnish an important venue through which social distinctions are created' (Bamford, 1998, p. 28).

The idea that the social and natural worlds are mirror images of each other was put forward a century ago by Durkheim & Mauss (1963 [1903]; cf. Mauss, 1974 [1904–05]). These scholars argued that the classification of the social world underlies the classification of elements of the natural world, and that the model of social order is used by humans to regulate their relationship with nature. In contrast to such arguments, Bamford sketches a nondualistic world for the Kamea, where human environmental relations are mutually embedded rather than separated. In a related sense, so-called cultural landscapes – defined as the reflection of the interaction between people and their natural environment over space and time – provide a framework within which nature is conceived not as the opposite of, but as the counterpart to, human agency in shaping landscapes (von Droste et al., 1995; Hanssen, 1998).

Contrasting attitudes to one and the same environment have been among the most prominent issues of anthropological research on landscape. Comparing Australian Aboriginal viewpoints with those of white pastoralist settlers, Strang (1997) points out that for each group, the same landscape can entail very different experiences. This reveals how landscapes as well as relationships between humans and their environment are culturally evaluated[8] and constructed, and that these values and constructs may differ profoundly from one culture to another. Although the land in itself is a constant factor, and therefore 'landscape' a common idiom referred to by both groups, Aborigines and white settlers differ with respect to their affective responses to the land and the values attached to it. It follows that although cognition of environment, responses to ecological pressures, and certain beliefs and values are basic and universal elements of the human environmental relationship, it is culturally specific values which determine the practice of interaction with the environment. The latter is in the main articulated through the way of manipulating the landscape and the methods used for resource control.

The studies mentioned thus far indicate that the relationship between 'nature' and 'culture' is at the very heart of ecological approaches to the study of landscape. In this respect, some authors also investigated how inhabitants themselves view landscape in terms of an ecosystem, often within the context of recent ecological transformations. Jorgensen (1998), for instance, examined the ideas that the Telefolmin of Papua New Guinea have about the natural world, and how these ideas were modified over the course of history. Locals draw a contrast between the human spirit and bush spirits which today is especially articulated in

---

[8] A general account of how environments are valued, drawing primarily on European examples, is provided in Gold & Burgess (1982) and in Hanssen (1998).

the context of peoples' struggle against large-scale mining activities in the area. Besides strongly affecting the sociopolitical domain, these activities considerably transform the local landscape. Within this process, bush spirits, according to local beliefs, have become more active and have encroached upon the human world over the last decades.[9] The belief system is therefore closely connected to, as well as modified by, landscape transformation.

The close relationship between ecological and social transformations, the latter including such features as demographic processes and religious conversion, has been widely neglected or even ignored. Yet, as Olson (1997) demonstrates for the Pacific islands, it is only after local social transformations have been understood, that we may also understand ecological transformations. With regard to religious influences, an earlier study contrasting influences of Christianity and Islam on cultural landscapes in the southern Philippines (Hausherr, 1972) reveals how landscape is shaped by religious forces which are interconnected with the dynamics of social and economic processes. Whereas immigrant Christians along the coastlines of Mindanao radically transformed the landscape by utilising large-scale plantation farming (which followed extensive logging), inland Muslim communities have clung to traditional shifting cultivation. In this region, the land is parcelled out to extremes because of the complex inheritance rules within large social units. The emerging contrast between 'conservative-subsistence oriented' Muslim and 'modern-capitalist' Christian cultural landscapes is, according to Hausherr (1972), largely based upon divergent attitudes toward economic innovation which in turn result from contrasting religious orientations.

Ecological transformation is not necessarily conceived of to be a negative process, as in the sense of environmental degradation. Rather, it is sometimes evaluated quite positively by local people. Thus for the Zafimaniry of Madagascar, the 'clarity' of a landscape, which is a positive cultural-ecological concept irreconcilable with Western ecological ideology, is primarily achieved through deforesting land (Bloch, 1995). Cleared forests provide good views, do not catch mist and clouds and provide evidence that human beings have successfully transformed nature into culture. Examples from the West African transition zone also show contrasting views on ecological transformation. Since early colonial occupation forest patches in the savanna have been regarded by Westerners as relics of an original dense humid forest which had largely disappeared due to shifting cultivation practices. However, local people represent this landscape 'as half-filled and filling with forest, not half-emptied and emptying of it' (Fairhead & Leach, 1996, p. 2). This notion is strictly opposed to environmental policy orthodoxy. The cultural significance lies in the fact that forest islands, each typically concealing a

---

[9] The Telefolmin strictly contrast the human world, including domesticated animals and plants, with the natural world. Ecological damage (such as landslides) resulting from mining activities, alongside proselytisation, monetarisation of the local economy, and other cultural transformations have in this setting provoked multiple new patterns of spirit interventions. All of them are ascribed to the non human world.

village at its centre, are conceived by locals not as relics of destruction but rather as culturally formed by themselves or their ancestors. Hence, in the eyes of the locals these forest islands bear witness to positive human agency rather than to environmental degradation.

Other studies examine human environmental relations from still different theoretical perspectives. Among these approaches, two are of particular interest. The first one, from an archaeological perspective, addresses the question of why people choose particular locations for habitation as opposed to others (Tilley, 1994; cf. Slack & Ward, 2002). Conventionally this question has been answered with reference to environmental factors. Certain conditions of climate, soil, or water supply allow for the establishment of settlements and later on result in specific patterns of demography, technology, or territoriality. Yet the issue of cultural meanings that are involved in the choice of site location has mostly been regarded as insignificant, so that further research is greatly needed (cf. Perry, 1999).

A second perspective that examines the relationship between humans and landscapes is derived from both cultural ecology and formalist economic anthropology. Gragson (1993) in his analysis of foraging strategies in lowland South America distinguishes 'patterns' from 'processes' as two interdependent attributes of landscapes. Whereas 'pattern' refers to the distribution of energy, materials, and species through space, 'process' denotes the interaction of these components through time. It is argued that analysing both pattern and process will not only enable the understanding of how foragers make choices between resource alternatives but also the application of results derived from particularistic studies to new local contexts.

11.5. URBAN AND OTHER LANDSCAPES

Although most anthropological (and geographical) studies of landscape have examined 'traditional' rural landscapes – that is, landscapes coming close to the seventeenth century European meaning of 'landscape' – there are in fact many more types of landscape. A simple but important example is a 'national landscape'. The relationship between landscape and society at a national level is of a different nature than the one met at a regional or local level, where landscapes are inhabited by some scattered groups of foragers or horticulturalists. It is particularly at the national scale that it becomes evident how ever-growing human settlements have transformed a natural landscape (Browning, 1971). Taking this one step further, it is also obvious that the present world is, to a large extent, characterised by spatial relations between towns and rural hinterlands, and that with increasing urbanisation, especially in the Third World, urban landscapes become increasingly important for a theoretical examination of landscape on a general level (Bley, 1997; Correa, 1989).

This holds equally true for European and North American landscapes. The development of landscapes on these continents has heavily been influenced by human impact, which pertains to agrarian landscapes as well as to the development of towns and industrial landscapes (Aalen, 1978; Conzen, 1990). In still further studies it is revealed how, in addition to the preindustrial landscapes of hunter and gatherer societies, a broad spectrum of agrarian landscapes as well as industrial landscapes, new types of landscape have emerged in the modern or postmodern era, such as 'postmodern landscapes' (e.g., DisneyWorld, shopping malls), 'landscapes of power and pleasure' (landscapes of war, landscapes of leisure),[10] or homogenised 'globalised landscapes' (Soja, 1996; Atkins et al., 1998). We return to the latter below.

11.6. MAPPING HISTORY ONTO LANDSCAPES

The investigation of landscape reveals that the spatiality of human life cannot analytically be separated from its sociality and historicity (Soja, 1996, p. 2). Accordingly, one of the aspects studied most intensively by anthropologists has been the association of landscapes with local or regional histories: nature is as historicised as history is naturalised (Hastrup, 1998, p. 120). Although this phenomenon is culturally expressed quite heterogeneously, skimming through the literature reveals that it is apparently an almost universal one.

I remember countless walks through the landscape on the Indonesian island of Sulawesi, accompanied by local friends. What the European outsider sees on such a walk is first of all a very beautiful landscape, and one is immediately inclined to comment on this 'picturesque scenery'. By way of contrast, locals see[11] quite different things, and these are what they find worth commenting on. Such issues mainly cover historical processes, stretching from myths of the origin of the earth and mankind to the ancestors' movements, origins of political

---

In South Sulawesi, the I La Galigo myth preserved in manuscripts dating back to the sixteenth century describes an extremely complex cosmology. According to traditional belief among the Makassar, the sacred mountain (in the background) is the centre of the universe, residence of the deities, and the origin of all forms of earthly life (Figure 11.1).

It is regarded as the hub of the world (*pocci' tana*) in both a topographical and cultural sense. A well-known mythical tradition describes how one of the creator god's great-grandsons, Sawerigading, sailed around the world, which at that time largely consisted of water, on a giant ship. Because the spirits of the underworld wanted him to become their ruler, they caused his ship to sink.

(continued)

---

[10] A variety of culturally produced landscapes is presented by Wilson (1992).

[11] Okely (2001) within the context of Western culture distinguishes between looking at a landscape from seeing a landscape. Whereas for instance non labouring spectators 'look' at a landscape, cultivators 'see' the landscape. Seeing in contrast to looking is linked to all senses.

(continued)

Figure 11.1. The highlands in the regency of Gowa, South Sulawesi (Photograph by author)

The steep mountain ridge in the foreground is locally regarded as the ship's hull which, bottom up, has been lying there for ages. History is in this case naturalised in that the ridge is regarded as material evidence for how the world was created by supernatural beings. On the other hand, nature is historicised in that local people, whenever they see the ridge, always associate it with an ancient myth which provides a historically informed explanation for its odd shape.

There are many similar tales to the one just outlined: narrow trails tucked away in the forests still bear witness to where the ancestors went to war against their enemies. Big holes in the ground are identified as footprints of giant buffaloes descended from heaven. Trees or stones mark places where centuries ago people began to erect houses or performed sacrifice to the gods. In such narrations elements of the landscape are evidently regarded to be more social than natural features. The landscape becomes an order of named and significant places, a representation of society and its cultural heritage. Put differently, myths, oral traditions, and memories of past events and historical processes are deeply embedded in the landscape.

and social structures, and the foundation and shifting of settlements. All of these historical events and processes are always directly related to specific elements of the physical landscape (Rössler, 2003).

These and related issues are widespread in various societies. Particular attention has been paid to the cultural heritage of Australian Aborigines in connection with landscape. One characteristic feature is that among Aborigines, creation myths are superimposed upon the land, thereby transforming temporal processes into spatial structures and at the same time localising social structures based upon chains of ancestors within the landscape (Bender, 1993, S. 2ff.; cf. Kolig, 2000). Quite similarly, Morphy (1993, 1995) shows how in Northern Australia landscapes *of* memory such as represented in colonial maps differ from landscapes *as* memory for Aboriginal society. Among the latter, landscape together with place-names first of all refer to ancestral action and to spiritual forces which are periodically used to reproduce the present in the form of the past. What emerges is a distinct space–time structure.[12] This means that landscape is conceived of as the transformation of the ancestral past in the sense that present topographical features bear evidence of ancestors' actions and movements. Hence, the past is connected to the present in peoples' memories through multiple ties between social organisation and physical features of the natural environment. Australian historical thinking is memorialised in the landscape.[13] The landscape therefore becomes 'story' (or 'text', see above) through certain inscriptive and interpretive practices (Rumsey, 1994). In particular, kinship structures are 'mapped'onto the landscape so that individual kinship relations and perception of landscape are closely entwined. A different view has been suggested by Swain (1993), who contrasts two different spiritual principles in Australian Aboriginal thinking. One is site-based and cojoined with specific human beings, whereas the other is not localised but rather spatially transcendent. The central argument is that Aborigines conceptualised their being not primarily in terms of time but rather of place and space, albeit in a historical dimension, for which contacts with outsiders ranging from Melanesians and Indonesians to Europeans were also important.

People in many parts of South America also inscribe history into their landscape. Among the Yanesha of Peru, 'topographic writing' is based upon 'topograms' which are defined as landscape elements imbued with historical meaning through myth and ritual (Santos-Granero, 1998). In a quite similar fashion, myth and history are used to define both physical space and humans' relationship to physical space on the southern Bolivian altiplano (Sikkink & Choque, 1999). The immediate relationship between people and the local topography results from the fact that the community situates itself socially and politically with reference to the natural environment. Specifically, storytelling regarding the landscape serves to fashion this landscape in cultural terms quite similarly to the Western Apache way of 'place-making' through constructing and reinventing history (Basso, 1996). This sort of storytelling also resembles the traditional songs Malaysian *orang asli* use for 'mapping' the local landscape (Roseman, 1998). These people sing their maps, the lyrics referring to place-names 'weighted with memory' and

---

[12]This is paralleled by space–time concepts as examined by Hugh-Jones (1979) in Amazonia or by Munn (1986) among the Massim of Melanesia, where it represents a value parameter within the *kula* exchange. See also Ingold (2000).

[13]A similar phenomenon is Icelandic notions of the 'Beginning' which is also primarily referred to in terms of places rather than of time (Hastrup, 1998).

inscribing certain forms of knowledge on local history and geography in songs (see also Kavari & Bleckmann, this volume).

In one of the earliest theoretically intriguing anthropological studies of space, Hugh-Jones (1979) examines a concept of 'space–time' among Amazonian Indians which is of major importance for social reproduction and for how experience of the present-day world is linked to the ancestral past. The Indians derive the concrete world from the 'imaginary' ancestral world. It is less the landscape as a totality but rather rivers running through dense rain forest which provide important points of orientation as well as the basis for the space–time structure. In this setting, it is in particular the (spatial) relation between 'upstream' and 'downstream' which by linking the ancestral to the present-day order is closely connected to a temporal dimension.

In Africa many landscapes are largely a product of historical migrations, settlements, and practices of land use, as examples from Kenya and Zimbabwe show (Luig & von Oppen, 1997; Ranger, 1997). Typically, these historical processes have involved human agents from both African and European cultures, who in the course of history developed contrasting mental maps of the landscape. Landscape elements which are of particular importance for Africans are places invested with spiritual power, which are usually connected to a mythical background and which occupy important positions for the periodical cultural reproduction through local religious cults (Colson, 1997).

Contrasting images of landscape as either 'nature' or 'culture' against the background of differences in African and European perceptions of landscapes become evident in the case of the Matopos hills in Zimbabwe (Ranger, 1999). The Matopos hills are by no means 'wild nature' as they are presented to visitors in the National Park. That they are in fact representations of local culture and history has only been disguised by removal of both cultivators and pastoralists from the National Park some decades ago. Rather than reflecting wild nature, the hills have, for a long time, been the site of a symbolic (as well as violent) struggle of religion and politics. They also still bear witness to the protest against colonial authority, covering indigenous traditions as well as white missionary and colonial agency.

Landscapes marked by ancestral power, memories investing present landscapes with meanings, landscapes of memory, and landscapes as memory have been widely examined by anthropologists. In this context particular attention has also been given to characteristic topographic features such as mountains. Especially the notion of 'sacred' mountains almost universally plays a significant role in religious life (e.g., Blondeau & Steinkeller, 1996; Gingrich, 1996). In addition to the examples cited above, there are further studies on the topics summarised in this section from mainland Southeast Asia (e.g., Gesick, 1985, 1995), India (e.g., Gold & Gujar, 1997), the Caribbean (e.g., Olwig, 1999), and Melanesia (e.g., Munn, 1986; Küchler, 1993). However, the issue is not only one of traditional and exotic landscapes (Schama, 1995). Thus for post-Civil War Americans landscape also served to create official public memories and contributed to the way in which the past was remembered (Shackel, 2001). Equally, Mount

Rushmore is perhaps one of the most spectacular examples of how a landscape can be used to memorise a glorious national past. This landscape memorial in South Dakota testifies how much time and energy humans may invest in the endeavour to convert a landscape into a symbol of public memory. It took 14 years and hundreds of workers to carve the heads of four U.S. presidents into the granite rock.

In still another perspective, as Horowitz (2001) shows for New Caledonia, landscape invested with historical memories is not only an important aspect of cultural heritage in terms of spiritual matters, but also in terms of present-day reactions against environmental degradation. Accordingly, the relationship between landscape and history is also linked to the ecological domain as addressed above.

11.7. PLACE-NAMES, PLACE-MAKING, AND IDENTITIES

The issue of cultural identity is closely connected to ties between history and landscapes. In fact, both topics are often inseparable. Bamford (1998), whose work has already been cited above, demonstrates how for the Kamea of Papua New Guinea the local landscape is not only a testimony of the ancestors' actions and of past events, but also significant for the ongoing elicitation of social identities (see also Ingold, 2000, pp. 189ff.). Furthermore, it is especially in the context of identity construction that the naming of places and the processes of 'place-making' have been stressed in a number of studies. In West Timor individuals and groups are strongly associated with particular named places in the landscape, each of them referring to remembered ancestral deeds or experiences (McWilliam, 1997). Kin groups especially regard named places as the most significant markers of their history and identity.

In the Faeroe Islands, place-names also form a very significant aspect of the local construction of landscape (Gaffin, 1993). It is place-names, as texts of social processes and as 'embodiments of personality and personhood' (Gaffin, 1993, p. 53), which transform a physical ('natural') landscape into a cultural one. By naming places, such as 'Gormund's Cove' or 'Haldan's Wife's Field', as well as by naming persons such as in the case of 'Hilmar under the Ledge', people construct close ties between the landscape and their social relations as well as personal and cultural identities. Hastrup (1998), in another European insular setting, came across quite similar relationships between naming practices of topographical features and notions of identity.

By way of contrast, Gray (1999) shows how everyday routines, rather than the naming of places, may shape identities. Shepherding practices of sheep farmers in the Scottish borders create an attachment to the landscape, thereby defining the farm, farmers' way of life, and the region they live in as places where they are 'at home'. At the same time, the spatial dimensions of everyday life refer to locality in the sense of place-making and identity-making as a cultural process. Paralleling such accounts, Raffles (1999), also drawing on the concept of place-making, elucidates how locality is narrated and embodied in the Amazonian context. Rather than constituting something like place in the strict sense, locality is here conceived of as a set of relations, 'a density in which places are discursively and imaginatively materialised and enacted through the practices of variously positioned people and political economies' (Raffles, 1999, p. 324).

The ordering of space is also of paramount importance for the creation of identities among Austronesian societies, where in many cases identity is appropriately described as a localised or spatially defined sense of belonging (Fox, 1997a; Bubandt, 1997; cf. Appadurai, 1995). Named places linking the ancestral past to present social practice are again of particular significance in this context. According to local beliefs, it is above all the domestication of landscape by ancestral beings (i.e., the transition from nature to culture) which is articulated by naming settlement places and paths. Fox (1997a), in search of a systematic device, introduces the concept of *topogeny* which is complementary to genealogy. Whereas genealogies represent an ordering and transmission of knowledge with regard to kinship, topogenies represent social knowledge of an ordering of places or, respectively, a succession of place-names ordered in temporal perspective. Often narratives of previous journeys provide 'maps' of landscapes consisting of named places linked through identifiable tracks. Such topogenies may therefore be understood as 'geographic' texts representing a set of social memories and at the same time constituting a significant means of identity construction (see also Fox, 1997b; Pannell, 1997; Hastrup, 1998).

In other instances landscapes may almost completely fuse with houses and settlements not only in an architectural sense but also conceptually. The famous traditional houses of the Indonesian Toraja, which on the societal level constitute focal points of extensive bilateral kin networks, together with genealogies and myths are embedded in the landscape not only in a material sense but also as a cultural imagination (Waterson, 1997). This may again be considered a cultural text which, by conceptually connecting landscape, architecture, and ancestral actions through space and time, is a strong means of constructing local identity.

What the 'becoming' of a place actually means in a theoretical sense has been formulated by Allan Pred: '. . . the "becoming" of any settled area involves the local coexistence of structuring processes which vary in their geographical extent and temporal duration and which concretely interpenetrate with one another through the time-space specific practices of mediating agents, through the lived biographies of actual people' (Pred, 1986, p. 2). Although this approach has been followed at least implicitly by many subsequent studies, it was much later when it was emphasised that the processes of place-making explicitly involve the construction of identity and difference (Gupta & Ferguson, 1997). In all of these cases the polysemic character of landscape must be stressed. A landscape is nothing given but rather continuously constructed and reconstructed, as well as contested. This issue is further elucidated in the following section.

11.8. EXPERIENCING AND CREATING LANDSCAPES

When considering the significance of landscapes in connection with historical memories and identity constructions, it becomes apparent that in most if not all societies landscapes are culturally worked upon, created, and re-created in terms of past experiences and future potentialities (cf. Luig & von Oppen, 1997; Bender, 1993). Keeping in mind the origins of the Western notion of landscape, one must be aware of the fact that landscape as a general concept is not only a particular, elitist Western European way of seeing, but rather a 'way in which people – all people – understand and engage with the material world around them' (Bender, 2001, p. 3) (Figure 11.2). Furthermore, it is important to note that in contrast to contemporary Western culture, where the dominant quality of landscape is that it is individually 'seen' or 'looked upon', landscapes in other societies are not pri-

Figure 11.2. Individual experience of the landscape in German romanticism: C. D. Friedrich: The Wanderer Above the Sea of Fog, 1818 (Courtesy Hamburger Kunsthalle)
(See also Color Plates)

marily ego-centred. Instead, it is often the nonvisual aspects of landscape which are most significant (Bender, 1993).[14]

This issue has been elaborated by Hirsch (1995) who argues that landscapes constitute cultural processes that are oscillating between a foreground conception of everyday experience and a background conception of ideal, imagined existence. In other words there is a conceptual difference (albeit a connection) between a landscape that we see and a landscape that is a product of local practice. The Western mode of representing landscape (e.g., in painting), is nothing but one particular, culturally specific expression of this foreground–background relationship. Earlier approaches by cultural geographers, defining landscape as 'a cultural image, a pictorial way of representing or symbolising surroundings' (Cosgrove & Daniels 1988, p. 1) have largely ignored its processual character. Landscape, however, should always be treated as a cultural process that relates the foreground of everyday activities to the background of potential social existence. Although this process may be fixed in certain situations as, for example, in landscape painting, it is on a general level deeply embedded in the dynamics of social relations.

The distinction Hirsch draws between foreground and background is analytically equivalent to the contrast between place (*lieu*) and space (*espace*) in the work of De Certeau (1984, p. 117). In this sense, place is conceived of as a fixed order of spatial elements, a homogeneous and isotropic spatiality within which any two elements (say, mountains, trees, or rivers) cannot occupy one and the same place. In contrast, space is conceived of as involving notions of directedness, movement, and perception as well as a temporal dimension. Rather than being composed of fixed elements, space is *practiced* place. By way of illustration, we may consider that whereas a map is basically a representation of places, a walk through the landscape represented on that map transforms these places into space (cf. Pannell, 1997, p. 163). Similarly, a mountain as a place is nothing but physical matter occupying a fixed position within the system of coordinates on a map. However, from the perspective of a person walking through the landscape, the mountain may be continuously altering its position in relation to this person. At one point, it may lie in front, then behind, now to his or her left or right. Put briefly, the mountain becomes part of lived-in space. The point is that such aspects of spatial orientation are often culturally significant. Among the Makassar of Sulawesi, for instance, traditionally any person working in the rice fields must face the holy mountain instead of turning his back toward it. Maintaining a specific position in relation to the mountain, which is regarded as the centre of the macrocosmos, is a cultural rule pertaining to spatial behaviour (cf. Rössler, 2003).

In more general terms, such approaches as cited in this section extend the anthropological perspective toward an examination of how particular landscapes are perceived, experienced, and actively sensed by people themselves. The

---

[14]This does not mean that in European cultures, experiences of landscape are altogether superficial. Some concepts have in fact developed into major cultural themes, such as, for example, the forest (*Der Wald*) which over the centuries has developed into an important, almost mythical space of experience in German culture (see Lehmann & Schriewer, 2000).

ethnographic studies presented in the volume edited by Feld and Basso (1996), to a large extent relying on linguistic data, provide excellent examples for a method of investigation which aims at precisely analysing local knowledge of landscapes and the way this knowledge is expressed and transformed into action. Such arguments are closely related to the field of orientation in landscapes.

11.9. ORIENTATION IN LANDSCAPES

In many recent studies, the focus on human beings as actors and agents in the context of landscapes has implicitly suggested that humans only live mentally in the world. All too often the discussion of the human body and of body movements has been left aside altogether in interpretations of space and landscape. The reason for this appears quite obscure, because it is primarily the human body that is the means through which the world comes into being (Eves, 1997, p. 177). On a more general level, this is intimately related to the difference between space as a rationally objectified geometric 'container' in the Cartesian sense and space as something which is lived-in and experienced (see also Hirsch 1995, pp. 16ff.). Understanding landscape in relation to humans and human culture requires taking a notion of spatiality into account which considers the body as a constitutive part of it.

Another aspect of human bodily interaction with landscape is how humans perceive it, describe it, and talk about it. Most approaches examining these phenomena strongly rely on methods developed by linguistics and cognitive anthropology (Pinxten et al., 1983; Senft, 1997; Lucy, 1998). It appears that there is considerable cross-cultural variation in spatial language, particularly spatial reckoning and description, against the background of theories of spatial cognition (Levinson, 1996). Such contributions are primarily concerned with notions and concepts of space in everyday activities.

Once again the discussion cannot avoid considering the contrast between nature and culture. Cognitive anthropology has questioned, for instance, if there is any perspective that may be termed 'natural' in spatial description (Wassmann, 1994; Wassmann & Dasen, 1998). Whereas Westerners, perceive landscape from an anthropocentric perspective, locating landscape elements relationally to each other (A left of B, C in front of D, etc.), speakers of many non-European languages do not have such relative terms but instead describe space in terms of cardinal edges or absolute geocentric angles (A east of B, C uphill versus D downhill, etc.). Additionally, cardinals in many languages are not only used to describe macrospace but microspace as well. In still other languages, both systems of reference for describing landscape are used side by side, sometimes further combining them with place-names or landmarks. These issues have been intensively studied in the Malay archipelago, where in many regions the ordering of and orientation in landscapes is based upon a system of directional axes or coordinates, which are not only abstract reference systems but also an important feature of social practice (Barnes, 1988, 1993; Fox, 1997c; Bubandt, 1997; Wassmann & Dasen, 1998; Rössler, 2003). For many Austronesian societies linguistically as well as with regard to formal criteria, directional axes such as

'Toward the interior' as opposed to 'Toward the sea' or 'Upriver' as opposed to 'Downriver' are characteristic, although in some regions such patterns of orientation have also developed into fixed cardinal points.

In another study on the cognition of landscape, Wassmann (1993a) has examined route knowledge among the Yupno of Papua New Guinea, experimentally comparing behaviour *en route* (how people use paths in the landscape) with its verbal representation (how people describe their movements in the landscape) as well as with its graphic representation (how people draw routes). Another interesting phenomenon is the transformation of landscape perception through cultural change. Wassmann (1993b) examines how the 'mental images' of landscape and settlement patterns in New Guinea, which traditionally were conceived of as 'closed', have been transformed into much more 'open' spaces through increasing contacts with the external world. This demonstrates from another theoretical angle how spatial thinking is highly correlated with certain cultural requirements. It is suggested that specific cultural features, which are found in societies for whom communication about space or orientation in space are daily necessities (as among Australian Aboriginals or Inuit), evidently stimulate spatial thinking in a way that is nonexistent in other societies.

11.10. THE COLONIAL AESTHETICS OF LANDSCAPES

In addressing the noteworthy question, 'Why do we find landscapes beautiful?', Appleton (1975) demonstrates that there is no generally accepted theory for the aesthetics of both natural and culturally shaped landscapes. I would add that there is not even a common denominator of landscape aesthetics on an empirical level. Whereas Westerners produce and consume landscape painting, landscape architecture, or landscape design (Preece, 1991) in order to express or, respectively, enjoy its 'beauty', people in other cultures often do not see any such beauty in landscapes. Instead, as outlined above, they may see quite different things. What we like about landscapes – particularly their aesthetics – is therefore different from what 'they' like about them.[15]

The issue of landscape aesthetics is of particular interest in the context of the colonial encounter, and has been intensively studied with regard to African landscapes (Luig & von Oppen, 1997). Against the background of their own traditional understanding of landscape, Europeans in Southern Africa converted *land* which was as a resource exploited by the colonial economy into *landscape* in the sense of European imagination or painting (Ranger, 1997). The 'African landscape', typically exhibiting wideness, wilderness, and emptiness, was basically a European cultural construct and invention (which still lives on imaginatively for safari tours; see also Dieckmann, this volume). Early European missionaries in Southeast Africa already

---

[15] Related phenomena are those cultural representations of nature for which the authors in Lesch (1996) coined the German term *Naturbilder*. These representations not only refer to things existing but also imply visions, imaginations, and moral evaluations of nature.

perceived and represented the local landscape rather specifically, and this way of seeing the landscape also shaped the missionaries' attitude toward the local population (Harries, 1997). The way landscape was constructed by Europeans was part of the entire process of the colonisation of consciousness, and it contributed largely to the European cognitive authority over the local population.[16]

Quite similar issues, yet in a broader perspective, are addressed by an interdisciplinary study on two white settler societies in South Africa and Australia (Darian-Smith et al., 1996). In both regional settings, naming and possessing the land, representing the landscape as well as defining a sense of place, have been culturally contested within the relationship between indigenous people and white settlers. The connection of space and power in the colonial and postcolonial context of both regions is of particular interest because, for instance, European art and literature in South Africa as well as white Australian literature between the seventeenth and the twentieth centuries did not refer to any scheme of seeing which was based on indigenous concepts. Instead, it furthered the fiction of an 'empty land' and developed into something like a 'poetry of empty space'. The significance lies in the fact that colonial literature in general appears to have widely shaped and structured the experience of colonisation and also of the colony itself (Noyes, 1992).

In Australia, white explorers, administrators, and settlers created an imperial landscape through European naming and mapping of its geographical features, whereas Aboriginal place-names were ignored or subverted in this process. In South Africa, the landscape was divided into areas of forestry, farming, or game reserves in accordance with European models but quite contrary to the traditional flexible patterns of land use (Bunn, 1996). Such studies confirm how

---

Colonial landscape photography produced quite similar effects. As Hayes (2000) shows for the case of former South West Africa, panoramic photographs taken in the interwar period served as a medium to express a specific landscape aesthetic which represented the 'dreamworks' of colonialism (Hayes, 2000, p. 56). Besides constructing an artificial primitiveness of a tribal population untouched by Western civilisation, many of these photographs, like visual poems emphasising emptiness and vastness, represented the landscape as an open frontier, combining 'natives and nature' into an aesthetic of colonial ideology. A fine example is this photograph, (Figure 11.3) taken in 1943.

(continued)

---

[16] A quite similar colonization of consciousness is found in recent history. In the global discourse on indigenous land rights, international environmental and human rights organisations commonly invent or reinvent 'traditional' and 'sacred' cultural landscapes populated by natives living in perfect harmony with nature. Mostly in contrast to peoples' own concepts of the landscape, these Western constructs are largely based upon romanticising and stereotyped views of indigenous peoples (no matter in which part of the world) who are supposed to be related to their land primarily through magico-religious ties (Bollig, 2001).

(continued)

Figure 11.3. Panorama of two Kaoko hunters posed against landscape (Courtesy National Archives of Namibia, A450 Hahn Collection)

An important technological device which photographer Carl Hugo Hahn, who between 1921 and 1946 was Native Commissioner in Ovambo, used to overemphasise the vastness of the landscape in this picture was a revolving panoramic lens encompassing views of 180 degrees. The photograph shows two Kaoko hunters at the margins of a huge empty space, which is further accentuated by the intense interplay of sunlight and shadows. The pointing arm of the right figure, although occupying but a tiny part of the whole picture, is the only element providing a direction for the beholder's eye. The photograph is neither exclusively a landscape nor an ethnographic picture. Rather, it integrates colonial ideals of 'natives' and 'nature' into a single corpus of Western imagination of Africa.[17]

the colonial landscape and its inhabitants assumed an aestheticised role within the colonisers' imagination. At the same time it becomes evident once again how closely landscapes are related to memories,[18] which in South Africa as well as in Australia are constructed differently by indigenous peoples in contrast to the white settlers. That these memories contrast and conflict is based upon the fact that natives and invaders perceive land, landscapes, and places quite differently (Gunner, 1996). It should finally be added that in the context of South Africa and Australia, space and landscape have also been strongly racialised.

---

[17] The figures in the photograph were certainly posed, if only because the revolution of the panoramic lens took considerable time. In a different mode, the new technology of aerial photography served colonial administrative interests. Starting in the 1930s, it was less connected to aesthetics than to an accurate mapping of colonised landscapes in South West Africa (see Hayes, 2000).

[18] We learn from the work of Schama (1995) that myth and memory are also of paramount importance for the notion of landscape in Western culture.

11.11. RESISTANCE, DELOCALISATION, GLOBALISATION: WHAT IS HAPPENING TO LANDSCAPES?

Over the past years, landscapes have increasingly been related to political and economic processes on a global scale. Irrespective of the fact that these processes involve ecological transformation and environmental degradation, they also brought about serious problems of a different order: it appears that much of the discussion on what is happening to the world today has required the rethinking of the concept of landscape once again, this time perhaps more radically than ever before.

To begin with, landscapes have recently been closely related to political resistance. In this respect it has been argued that 'the articulation of resistance is always contingent upon the spatiocultural conditions of its emergence and the character of its participants' (Routledge, 1997, p. 83). Cultural constructs of landscape often become one of the most important motivations for resistance against state authority in the context of land conflicts. This becomes obvious when considering that resistance, which just like the spatiality of landscape is a form of human experience, is always closely connected to notions of identity and difference, and therefore, with place-making (Gupta & Ferguson, 1997, pp. 17ff.). The political processes of deterritorialisation as well as reterritorialisation refer to localities as products of power struggles; and it has been observed that under the condition of asymmetric power relations, which is typical for many postcolonial nations, political authority produces and defines space (Pile, 1997; Streicker, 1997; Moore, 1998; Roseman, 1998; Rössler, 2003). This issue has also been discussed in the so-called 'geographies of resistance' (Pile & Keith, 1997).

Globalisation, so widely referred to in much of postmodern anthropological writing, affects not only political economies worldwide, but also landscapes. One central argument in this context draws again on the close relationship between human culture and landscape.

During the last two decades, perhaps the majority of anthropological studies focused heavily on the relationship between landscape and cultural identity in the sense of 'rootedness' in place. Human cultures were regarded as more or less strictly localised, localisation in this sense being defined as the establishment of cultural *and* spatial boundaries that distinguish an Inside from an Outside. Because this notion seems to have vanished by the end of the twentieth century, the concept of localised cultures has given way to a focus on the contestation of landscapes in the context of local, national, and global power relations (Gupta & Ferguson, 1992, 1997; Appadurai, 1991, 1995, 1996; Soja, 1996; Bender & Winer 2001; Hastrup & Olwig, 1997). Hence, the relationship among landscapes, migration movements, exile, struggles for land, displacement, poverty, constructions of diaspora, and so forth has meanwhile become a major field of anthropological study, and to a large extent has replaced philosophical interpretations of landscape and space (Feld & Basso, 1996).[19]

The main argument underlying all of these studies is that concepts of space, landscape, and locality must be related to the apparently contradictory concerns of *displacement* of communities and identities. This means at the same time that the interrelatedness of culture and landscape, as soon as it was 'discovered' by anthropology, was in practice already disappearing under the conditions of globalisation, inasmuch as more and more 'cultures' are no longer distinct entities rooted in landscapes and encapsulated within fixed spatial boundaries. To characterise this situation, Appadurai (1991, 1996) coined the term *ethnoscapes*, a concept that denotes a somewhat diffuse landscape of deterritorialised and merely imagined cultural identities.

According to Gupta & Ferguson (1992, 1997), the fact that at the turn of the millennium the isomorphism of space and culture has become an illusion, results from mass migration and transnational culture flows across borders, from cultural differences within a locality (including multiculturalism and the emergence of subcultures), as well as from the phenomenon that most nations as products of postcoloniality are made up of mosaics of cultures without any clearly set boundaries. Directly referring to the landscape metaphor, it is stressed that 'the presumption that spaces are autonomous has enabled the power of topography to conceal successfully the topography of power' (Gupta & Ferguson, 1992, p. 8). Accordingly, it is argued at present that any anthropological analysis of culture relating to landscape must increasingly focus on contested landscapes, landscapes of movement and exile, as well as on contemporary movements leading to dislocations between people and landscape (Bender, 2001; see Brinkmann, Gewald, and Dieckmann, this volume).

In addition to the political domain as stressed in these studies, the fact that in so many cases cultures no longer correspond to particular localities also requires modified theoretical perspectives. This pertains especially to the 'role of place in the conceptualisation and practice of culture' (Hastrup & Olwig, 1997, p. 1). However, even if the conventional 'localising strategies' of ethnographies may have been identified as inappropriate in a largely globalised world, it is only by detailed case studies in specific *localities* that we may understand the altered relations between place and culture. By examining what is meant by 'place' in a world which apparently no longer knows 'places', this seemingly paradoxical endeavour may finally lead to a reconceptualisation of the concept of culture as a whole.

Yet one important, final point should be emphasised. Anthropologists at present are eager to *deconstruct* the notion of localised cultures; on the other hand, people in many parts of the world, mostly those experiencing political pressure, are *constructing* notions of such localised cultures, with landscapes constituting significant elements in this process (Hastrup & Olwig, 1997; Rössler,

---

[19] From another perspective it may also be argued that scholars such as Soja (1996) are attempting to integrate the multiple global problems we are facing into postmodern philosophical thought. In Soja's case, 'thirdspace' is a theoretical concept which tries to capture the ever-changing social spatialities in the modern world.

2003). Notwithstanding the fact that globalisation is a phenomenon whose effects should neither be overlooked nor underestimated, an outstanding figure of contemporary anthropology reminded us that 'no one lives in the world in general' (Geertz, 1996, p. 262). He was right, because even exiled or diasporic individuals live in some confined and limited stretch of this planet, and always in some sort of landscape.

11.12. CONCLUSION

The aim of this chapter was to summarise some of the most important contributions anthropology has made to the study of landscape. As I cautioned the reader at the outset, this is quite a tricky endeavour, because one is easily left with the impression that what has been done so far is little more than 'butterfly collecting'. This holds for the manifold systematic fields investigated as well as for various theoretical orientations employed. Furthermore, as was particularly demonstrated in the final section, we have to face the additional problem that landscapes are rapidly changing worldwide, not only in the sense of environmental degradation or increase of built landscapes, but also with respect to the relationship between social structures and political economies on the one hand and landscapes on the other. Taken together these issues have so far hindered the formulation of a theoretical concept of landscape that is applicable cross-culturally. Achieving this end requires firstly, many more case studies grounded in modern theoretical approaches and secondly, increased interdisciplinary cooperation.

Acknowledgement I am grateful to Aruna Dufft for her assistance in compiling and critically assessing the material for this chapter.

REFERENCES

Aalen, F.H.A. (1978). *Man and the Landscape in Ireland*. London: Academic Press.

Abraham, I. (2000). Landscape and postcolonial science: *Contributions to Indian Sociology*, 34, 163–187.

Agrawal, A. & Sivaramakrishnan K. (2000). *Agrarian Environments: Resources, Representations and Rule in India*. Durham, NC: Duke University Press.

Appadurai, A. (1991). Global ethnoscapes: Notes and queries for a transnational anthropology. In R.G. Fox (Ed.), *Recapturing Anthropology. Working in the Present* (pp. 191–210). Santa Fe, NM: School of American Research Press.

Appadurai, A. (1995). The production of locality. In R. Fardon (Ed.), *Counterworks: Managing the Diversity of Knowledge* (pp. 204–225). London: Routledge.

Appadurai, A. (1996). *Modernity at Large: Cultural Dimensions of Globalization*. Minneapolis: University of Minnesota Press.

Appleton, J. (1975). *The Experience of Landscape*. London: Wiley.

Arreola, D.D. & Curtis, J. R. (1993). *The Mexican Border Cities: Landscape Anatomy and Place Personality*. Tucson: University of Arizona Press.

Atkins, P., Simmons, I. & Roberts, B.K. (1998). *People, Land and Time: An Historical Introduction to the Relations Between Landscape, Culture and Environment*. London: Arnold.

Baker, A.R.H. (1992). Introduction: On ideology and landscape. In Baker. A.R.H & Biger. G (Eds.), *Ideology and Landscape in Historical Perspective: Essays on the Meanings of Some Places in the Past* (pp. 1–14). Cambridge: Cambridge University Press.

Baker, A.R.H. & Biger, G. (Eds.) (1992). *Ideology and Landscape in Historical Perspective: Essays on the Meanings of Some Places in the Past.* Cambridge: Cambridge University Press.

Balée, W. (Ed.) (1998). *Advances in Historical Ecology.* New York: Columbia University Press.

Bamford, S. (1998). Humanized landscapes, embodied worlds: Land and the construction of intergenerational continuity among the Kamea of Papua New Guinea. *Social Analysis*, 42, 28–54.

Barnes, R.H. (1988). Moving and staying space in the Malay Archipelago. In H.J.M. Claessen & D.S. Moyer (Eds.), *Time Past, Time Present, Time Future. Perspectives on Indonesian Culture* (pp. 101–116). Dordrecht: Foris.

Barnes, R.H. (1993). Everyday space: Some considerations on the representation and use of space in Indonesia. In J. Wassmann & P.R. Dasei (Eds.), *Vol. 11. Alltagswissen – Der kognitive Ansatz im interdisziplinären Dialog* (pp. 159–178). Freiburg: Universitäts Verlag.

Basso, K.H. (1996). Wisdom sits in places: Notes on a Western Apache landscape. In S. Feld & K.H. Basso (Eds.), *Senses of Place* (pp. 53–90). Santa Fe, NM: School of American Research Press.

Bender, B. (Ed.) (1993). *Landscape: Politics and Perspectives.* Providence, RI: Berg.

Bender, B. (2001). Introduction. In: B. Bender & M. Winer (Eds.), *Contested Landscapes. Movement, Exile and Place* (pp. 1–18). Oxford: Berg.

Bender, B. & Winer, M. (Eds.) (2001). *Contested Landscapes. Movement, Exile and Place.* Oxford: Berg.

Benko, G. & Strohmayer, U. (Eds.) (1997). *Space and Social Theory. Interpreting Modernity and Postmodernity.* Oxford: Blackwell.

Bley, H. (1997). Die Giriama und Mombasa vor der Kolonialzeit. Naturaneignung und Land-Stadt-Beziehungen an der Küste Ostafrikas. *Paideuma*, 43, 93–120.

Bloch, M. (1995). People into places: Zafimaniry concepts of clarity. In E. Hirsch & M. O'Hanlon (Eds.), *The Anthropology of Landscape. Perspectives on Place and Space* (pp. 63–77). Oxford: Oxford University Press.

Blondeau, A.-M. & Steinkeller, E. (Eds.) (1996). *Reflections of the Mountain. Essays on the History and Social Meaning of the Mountain Cult in Tibet and the Himalaya.* Wien: Verlag der Österreichischen Akademie der Wissenschaft.

Bollig, M. (2001). Kaokoland. Zur Konstruktion einer Kultur-Landschaft in einer globalen Debatte. In S. Eisenhofer (Ed.), *Spuren des Regenbogens. Kunst und Leben im südlichen Afrika* (pp. 474–483). Stuttgart: Arnold.

Browning, D. (1971). *El Salvador: Landscape and Society.* Oxford: Clarendon Press.

Bubandt, N. (1997). Speaking of places: Spatial poesis and localized identity in Buli. In J.J. Fox (Ed.), *The Poetic Power of Place. Comparative Perspectives on Austronesian Ideas of Locality* (pp. 132–162). Canberra: Australian National University.

Bunn, D. (1996). Comparative barbarism: Game reserves, sugar plantations, and the modernization of South African landscape. In K. Darian-Smith, E. Gunner & S. Nuttall (Eds.), *Text, Theory, Space. Land, Literature, and History in South Africa and Australia* (pp. 37–52). London: Routledge.

Buttimer, A. (1969). Social space in interdisciplinary perspective. *Geographical Review*, 59, 417–426.

Certeau, M.de. (1984). *The Practice of Everyday Life.* Berkeley: University of California Press.

Colson, E. (1997). Places of power and shrines of the Land. *Paideuma*, 43, 47–57.

Conzen, M.P. (1990). *The Making of the American Landscape.* Boston: Unwin Hyman.

Correa, C. (1989). *The New Landscape: Urbanisation in the Third World.* Sevenoaks: Butterworth Architecture.

Cosgrove, D. & S. Daniels (Eds.) (1988). *The Iconography Of Landscape: Essays on the Symbolic Representation Design and Use of Past Environments.* Cambridge: Cambridge University Press.

Cosgrove, D.E. (1984). *Social Formation and Symbolic Landscape.* London: Croom Helm.

Crumley, C.L. (1994). Historical ecology: A multidimensional ecological orientation. In C. L. Crumley (Ed.), *Historical Ecology. Cultural Knowledge and Changing Landscapes.* 1st ed. (pp. 1–16). Santa Fe, NM: School of American Research Press.

Darian-Smith, K., Gunner, E. & Nuttall, S. (Eds.) (1996). *Text, Theory, Space. Land, Literature, and History in South Africa and Australia*. London: Routledge.

Droste, B. von, Plachter, H. & Rössler, M. (Eds.) (1995). *Cultural Landscapes of Universal Value. Components of a Global Strategy*. Jena: Fischer Verlag.

Duncan, J. (1990). *The City as Text: The Politics of Landscape Interpretation in the Kandyan Kingdom*. Cambridge: Cambridge University Press.

Duncan, J. & Ley, J. (Eds.) (1993). *Place, Culture, Representation*. London: Routledge.

Durkheim, É. & Mauss, M. (1963 [1903]). *Primitive Classification*. Chicago: University of Chicago Press.

Eves, R. (1997). Seating the place: Tropes of body, movement and space for the people of Lelet Plateau, New Ireland (Papua New Guinea). In J.J. FOX (Ed.), *The Poetic Power of Place: Comparitive Perspectives on Austronesian Ideas of Locality* (pp. 174–196). Canberra: Australian National University.

Fairhead, J. & Leach, M. (1996). *Misreading the African Landscape: Society and Ecology in a Forest-Savanna Mosaic*. Cambridge: Cambridge University Press.

Feld, S. & Basso, K.H. (Eds.) (1996). *Senses of Place*. Santa Fe, NM: School of American Research Press.

Fox, J.J. (1997a). Place and landscape in comparative Austronesian perspective. In J.J. Fox (Ed.), *The Poetic Power of Place: Comparative Perspectives on Austronesian Ideas of Locality* (pp. 1–21). Canberra: Australian National University.

Fox, J.J. (1997b). Genealogy and topogeny: Towards an ethnography of Rotinese ritual place names. In J.J. Fox (Ed.), *The Poetic Power of Place: Comparative Perspectives on Austronesian Ideas of Locality* (pp. 91–102). Canberra: Australian National University.

Fox, J.J. (Ed.) (1997c). *The Poetic Power of Place: Comparative Perspectives on Austronesian Ideas of Locality*. Canberra: Australian National University.

Gaffin, D. (1993). Landscape, personhood, and culture: Names of places and people in the Faeroe Islands. *Ethnos*, 58, 53–72.

Geertz, C. (1996). Afterword. In S. Feld & K.H. Basso (Eds.), *Senses of Place* (pp. 259–262). Santa Fe, NM: School of American Research Press.

Gesick, L.M. (1985). Reading the landscape: Reflections on a sacred site in Southern Thailand. *Journal of the Siam Society*, 73, 157–161.

Gesick, L.M. (1995). *In the Land of Lady White Blood: Southern Thailand and the Meaning of History*. Ithaca, NY: Cornell University Press.

Gingrich, A. (1996). Hierarchical merging and horizontal distinction: A comparative perspective on Tibetan mountain cults. In A.-M. Blondeau & E. Steinkeller (Eds.), *Reflections of the Mountain. Essays on the History and Social Meaning of the Mountain Cult in Tibet and the Himalaya* (pp. 233–262). Wien.

Gold, A.G. & Gujar, B.R. (1997). Wild pigs and kings: Remembered landscapes in Rajasthan. *American Anthropologist*, 99, 70–84.

Gold, J.R. & Burgess, J. (Eds.). (1982). *Valued Environments: Essays on the Place and Landscape*. Lexington, KY: George Allen & Unwin.

Gragson, T.L. (1993). Human foraging in lowland South America: Pattern and process of resource procurement. *Research in Economic Anthropology*, 14, 107–138.

Gray, J. (1999). Open spaces and dwelling places: Being at home on hill farms in the Scottish borders. *American Ethnologist*, 26, 440–459.

Gregory, D. (1978). *Ideology, Science and Human Geography*. London: Hutchinson.

Gunner, L. (1996). Names and the land: Poetry of belonging and unbelonging: a comparative approach. In K. Darian-Smith, E. Gunner & S. Nuttall (Eds.), *Text, Theory, Space. Land, Literature, and History in South Africa and Australia* (pp. 115–130). London: Routledge.

Gupta, A. & Ferguson, J. (1992). Beyond 'culture': Space, identity, and the politics of difference. *Cultural Anthropology*, 7, 6–23.

Gupta, A. & Ferguson, J. (1997). Culture, power, place: Ethnography at the end of an era. In A. Gupta & J. Ferguson (Eds.), *Culture, Power, Place: Explorations in Critical Anthropology* (pp. 1–29). Durham, NC: Duke University Press.

Hanssen, B.L. (1998). Values, ideology and power relations in cultural landscape evaluations. Dissertation, University of Bergen. Bergen.

Harries, P. (1997). Under alpine eyes: Constructing landscape and society in late pre-colonial South-East Africa. *Paideuma*, 43, 171–191.

Hastrup, K. (1998). *A Place Apart: An Anthropological Study of the Icelandic World*. Oxford: Clarendon Press.

Hastrup, K. & Olwig, K.F. (1997). Introduction. In K.F. Olwig & K. Hastrup (Eds.), *Siting Culture: The Shifting Anthropological Object*. (pp. 1–14). London: Routledge.

Hausherr, K. (1972). Die Entwicklung der Kulturlandschaft in den Lanao-Provinzen auf Mindanao (Philippinen) unter Berücksichtigung des Kulturkontaktes zwischen Islam und Christentum. Dissertation, University of Bonn. Bonn.

Hayes, P. (2000). Camera Africa: Indirect rule and landscape photographs of Kaoko. In G. Miescher & D. Henrichsen (Eds.), *New Notes on Kaoko, The Northern Kunene Region (Namibia) in Texts and Photographs* (pp. 48–73). Basel: Basler Afrika Bibliographien.

Hirsch, E. (1995). Landscape: Between place and space. In E. Hirsch & M. O'Hanlon (Eds.), *The Anthropology of Landscape: Perspectives on Place and Space* (pp. 1–30). Oxford: Oxford University Press.

Holland, M., Risser, P.G. & Naiman, R.J. (Eds.) (1991). *Ecotones: The Role of Landscape Boundaries in the Management and Restoration of Changing Environments*. New York: Springer.

Horowitz, L.S. (2001). Perceptions of nature and responses to environmental degradation in New Caledonia. *Ethnology*, 40, 237–250.

Hugh-Jones, C. (1979). *From the Milk River: Spatial and Temporal Processes in Northwest Amazonia*. Cambridge: Cambridge University Press.

Ingold, T. (2000). *The Perception of the Environment: Essays in Livelihood, Dwelling and Skill*. London: Routledge.

Jorgensen, D. (1998). Whose nature? Invading bush spirits, travelling ancestors, and mining in Telefolmin. *Social Analysis*, 42, 100–116.

Kolig, E. (2000). Social causality, human agency and mythology: Some thoughts on history-consciousness and mythical sense among Australian Aborigines. *Anthropological Forum*, 10, 9–30.

Küchler, S. (1993). Landscape as memory: The mapping of process and its representation in a Melanesian society. In B. Bender (Ed.), *Landscape: Politics and Perspectives* (pp. 85–106). Providence, RI: Berg.

Lehmann, A. & Schriewer, K. (2000). *Der Wald – Ein deutscher Mythos? Perspektiven eines Kulturthemas*. Berlin: Reimer.

Lesch, W. (Ed.) (1996). *Naturbilder – Ökologische Kommunikation zwischen Ästhetik und Moral*. Basel: Birkhäuser.

Levinson, S.C. (1996). Language and space. *Annual Review of Anthropology*, 25, 353–382.

Lucy, J. (1998). Space in language and thought: Commentary and discussion. *Ethnos*, 26, 105–111.

Luig, U. & von Oppen, A. (1997). Landscape in Africa: Process and vision; an introductory essay. *Paideuma*, 43, 7–46.

MacNaghten, P. & Urry, J. (1998). *Contested Natures*. London: Sage.

Mauss, M. (1974 [1904–05]). Über den jahreszeitlichen Wandel der Eskimogesellschaften. In M. Mauss. *Soziologie und Anthropologie*, Ed. 1 (pp. 183–278). München: Fischer Verlag.

McWilliam, A. (1997). Mapping with metaphor; cultural topographies in West Timor. In J.J. Fox (Ed.), *The Poetic Power of Place. Comparative Perspectives on Austronesian Ideas of Locality* (pp. 103–115). Canberra: Australian National University.

Meinig, D.W. (1979). *The Interpretation of Ordinary Landscapes*. New York: Oxford University Press.

Moore, H. (1986). *Space, Text, and Gender: An Anthropological Study of the Marakwet of Kenya*. New York: Guilford Press.

Moore, D.S. (1998). Subaltern struggle and the politics of place. Remapping resistance in Zimbabwes eastern highlands. *Cultural Anthropology*, 13, 344–381.

Morphy, H. (1993). Colonialism, history and the construction of place: The politics of landscape in Northern Australia. In B. Bender (Ed.), *Landscape: Politics and Perspectives* (pp. 205–244). Providence, RI: Berg.

Morphy, H. (1995). Landscape and the reproduction of the ancestral past. In E. Hirsch & M. O'Hanlon (Eds.), *The Anthropology of Landscape. Perspectives on Place and Space* (pp. 184–209). Oxford: Oxford University Press.

Munn, N. (1986). *The Fame of Gawa: A Symbolic Study of Value Transformation in a Massim (Papua New Guinea) Society*. Cambridge: Cambridge University Press.

Norton, W. (1989). *Explorations in the Understanding of Landscape: A Cultural Geography*. New York: Greenwood.

Noyes, J. (1992). *Colonial space: Spatiality in the discourse of German South West Africa 1884–1915*. Studies in anthropology and history, 4. Chur: Harwood Academic.

Okely, J. (2001). Visualism and landscape: Looking and seeing in Normandy. *Ethnos*, 66, 99–120.

Olson, M.D. (1997). Re-constructing landscapes: The social forest, nature and spirit-world in Samoa. *The Journal of the Polynesian Society*, 106, 7–32.

Olwig, K.F. (1999). The burden of heritage: Claiming a place for a West Indian culture. *American Ethnologist*, 26, 370–388.

Pannell, S. (1997). From the poetics of place to the politics of space: Redefining cultural landscapes on Damer, Maluku Tenggara. In J.J. Fox (Ed.), *The Poetic Power of Place. Comparative Perspectives on Austronesian Ideas of Locality* (pp. 163–173). Canberra: Australian National University.

Perry, W.R. (1999). *Landscape Transformations and the Archaeology of Impact: Social Disruption and State Formation in Southern Africa*. New York: Springer.

Pile, S. (1997). Introduction: Opposition, political identities and spaces of resistance. In S. Pile & M. Keith (Eds.), *Geographies of Resistance* (pp. 1–32). London: Routledge.

Pile, S. & Keith, M. (Eds.). (1997). *Geographies of Resistance*. London: Routledge.

Pinxten, R., van Dooren, I. & Harvey, F. (1983). *Anthropology of Space: Explorations into the Natural Philosophy and Semantics of the Navajo*. Philadelphia: University of Pennsylvania.

Pred, A. (1986). *Place, Practice and Structure: Social and Spatial Transformation in Southern Sweden: 1750–1850*. Cambridge: Cambridge University Press.

Preece, R.A. (1991). *Designs on the Landscape: Everyday Landscapes, Values and Practice*. London: Belhaven.

Raffles, H. (1999). 'Local theory': Nature and the making of an Amazonian place. *Cultural Anthropology*, 14, 323–360.

Ranger, T. (1997). Making Zimbabwean landscapes: Painters, projectors and priests. *Paideuma*, 43, 59–73.

Ranger, T. (1999). *Voices from the Rocks: Nature, Culture & History in the Matopos Hills of Zimbabwe*. Harare: Baobab.

Rapoport, A. (1982). *The Meaning of the Built Environment*. Beverly Hills, CA: Sage.

Richards, A. (1939). *Land, Labour and Diet in Northern Rhodesia*. London: Oxford University Press.

Roseman, M. (1998). Singers of the landscape. Song, history, and property rights in the Malaysian Rain Forest. *American Anthropologist*, 100, 106–121.

Rössler, M. (2003). Landkonflikt und politische Räumlichkeit: Die Lokalisierung von Identität und Widerstand in der nationalen Krise Indonesiens. In B. Hauser-Schäublin & M. Dickhardt (Eds.), *Kulturelle Räume – Räumliche Kultur* (pp. 171–220). Münster: Lit.

Routledge, P. (1997). A spatiality of resistance: Theory and practice in Nepal's revolution of 1990. In S. Pile & M. Keith (Eds.), *Geographies of Resistance* (pp. 68–86). London: Routledge.

Rubertone, P.E. (2000). The historical archaeology of Native Americans. *Annual Review of Anthropology*, 29, 425–446.

Rumsey, A. (1994). The dreaming, human agency and inscriptive practice. *Oceania*, 65, 116–130.

Santos-Granero, F. (1998). Writing history into the landscape: Space, myth, and ritual in contemporary Amazonia. *American Anthropologist*, 25, 128–148.

Schama, S. (1995). *Landscape and Memory*. London: Vintage Books.

Senft, G. (1997). *Referring to Space. Studies in Austronesian and Papuan Languages*. Oxford: Oxford University Press.

Shackel, P.A. (2001). Public memory and the search for power in American historical archaeology. *American Anthropologist*, 103, 655–670.

Sikkink, L. & Choque, B. (1999). Landscape, gender, and community: Andean mountain stories. *Anthropological Quarterly*, 72, 167–182.

Slack, P. & R. Ward (Eds.) (2002). *The Peopling of Britain: the Shaping of a Human Landscape*. The Linacre Lectures. Oxford: Oxford University Press.

Soja, E.W. (1989). *Postmodern Geographies: The Reassertion of Space in Critical Social Theory*. London: Verso.

Soja, E.W. (1996). *Thirdspace: Journeys to Los Angeles and Other Real-and-Imagined Places*. Cambridge: Blackwell.

Soper, K. (1995). *What Is Nature? Culture, Politics and the Non-Human*. Oxford: Blackwell.

Strang, V. (1997). *Uncommon Ground: Cultural Landscapes and Environmental Values*. Explorations in anthropology. Oxford: Berg.

Streicker, J. (1997). Spatial reconfigurations, imagined geographies, and social conflicts in Cartagena, Colombia. *Cultural Anthropology*, 12, 109–128.

Swain, T. (1993). *A Place for Strangers: Towards a History of Australian Aboriginal Being* Cambridge: Cambridge University Press.

Thrift, N. (1996). *Spatial Formations*. London: Sage.

Tilley, C. (1994). *A Phenomenology of Landscape: Places, Paths and Monuments*. Oxford: Berg.

Tilley, C. & Bennett, W. (2001). An archaeology of supernatural places: The case of West Penwith. *The Journal of the Royal Anthropological Institute*, 7, 335–362.

Tilley, C., Hamilton, S. & Bender, B. (2000). Art and the representation of the past. *The Journal of the Royal Anthropological Institute*, 6, 35–62.

Tress, B., Tress, G., Décamps, H. & D'Hauteserre, A.-M. (Eds.) (2001). Special Issue: Bridging human and natural sciences in landscape research. *Landscape and Urban Planning*, 57(3).

Ucko, P.J. & Layton, R. (Eds.) (1999). The archaeology and anthropology of landscape. London: Routledge.

Wagstaff, J.M. (Ed.) (1987). *Landscape and Culture: Geographical and Archaeological Perspectives*. Oxford: Basil Blackwell.

Wassmann, J. (1993a). *Finding the Right Path: The Route Knowledge of the Yupno of Papua New Guinea* (Working Paper No. 19). Nijmegen: Cognitive Anthropology Research Group, MPI for Psycholinguistics.

Wassmann, J. (1993b). Worlds in mind: The experience of an outside world in a community of the Finisterre Range of Papua New Guinea. *Oceania*, 64, 117–145.

Wassmann, J. (1994). The Yupno as Post-Newtonian scientists: The question of what is 'natural' in spatial description. *Man*, 29, 645–666.

Wassmann, J. & Dasen, P.R. (1998). Balinese spatial orientation: Some empirical evidence of moderate linguistic relativity. *Journal of the Royal Anthropological Institute*, 4, 689–711.

Waterson, R. (1997). The contested landscapes of myth and history in Tana Toraja. In J.J. Fox (Ed.), *The Poetic Power of Place: Comparative Perspectives on Austronesian Ideas of Locality* (pp. 63–90). Canberra: Australian National University.

Wilson, A. (1992). *The Culture of Nature. North American Landscape from Disney to the Exxon Valdez*. Cambridge: Blackwell.

Chapter 12

Kinship, Ritual, and Landscape Amongst the Himba of Northwest Namibia

MICHAEL BOLLIG

The anthropology of landscape has dealt mainly with the intricate relationships amongst memory, identities, and power and how these relations are engraved physically and intellectually in space. Many accounts have delineated how hegemonic discourses and counterdiscourses have reflected upon the conceptualisation of landscape in local cultural settings and during specific periods of time. This contribution develops an older anthropological thread in that it seeks to link kinship and landscape. The northern Namibian Himba inscribe genealogies into the landscape. In commemorative rituals they visit the graves of ancestors and ritually link these graves back to their homesteads. By commemorating their ancestors they re-create a network of places mirroring their selective presentation of genealogies. This chapter takes the ethnography of such a commemorative ritual as a starting point and explores how kinship is linked to space through ritual. In conclusion it discusses why actors perform such rituals: the attempt to stabilise local hierarchies and the quest for land-rights in a situation characterised by uncertainty and conflict are identified as major motives for the ritual.

M. Bollig, O. Bubenzer (eds.), *African Landscapes*,
doi: 10.1007/978-0-387-78682-7_12, © Springer Science+Business Media, LLC 2009

12.1. INTRODUCTION

Mutually reinforcing relations between ritual and kinship were postulated already by Durkheim (1912) and subsequent generations of social anthropologists have added case studies and furthered the argument (Evans-Pritchard, 1940; Lienhardt, 1961): ritual was seen as a major device to foster kinship solidarity and descent group identity and on the other hand elements of ritual were symbolic representations of kinship (Turner, 1967). It was also Durkheim (1912) who hypothesised that concepts of landscape representation were shaped according to patterns of social organisation: landscape became an embodied form of the kinship relations of Aborigine communities. Although the nexus between kinship and ritual has been a major topic in anthropology ever since, landscape disappeared for a long time from the agenda of social anthropological research and theorising. Anthropologists discarded holistic approaches to the environment and turned to ecosystem analysis (Moran, 1979) and the physical consequences of man–environment interactions (e.g., Crumley, 1994; for excellent summaries see Scoones, 1999; Balée, 1998). Emic concepts of the environment were gaining some currency when cognitive anthropologists started to research local knowledge of vegetation, soils, and fauna (see, e.g., Conklin, 1955).

The reintroduction of the landscape project into the anthropological research agenda, however, had little to do with considerations of local knowledge and cognition. The major impetus to reconsider the issue was ideas regarding linkages amongst memory, power, identity, and landscape. Landscape research originating from the social sciences and history has consistently emphasised how issues of power and conflict form local considerations on landscape. In recent years anthropologists and historians working in southern Africa have analysed cultural presentations of landscapes in relation to the political history of the wider region. Terrence Ranger (1999) in his book on the Matopos Hills of Zimbabwe describes how an indigenous landscape ordered by a variety of shrines is superseded by another vision of the landscape: an imperialist militaristic version of the Matopos landscape stressed the importance of graves of colonial soldiers and the grave of Cecil Rhodes became a key reference point in this mnemotope.

Other historians and anthropologists, such as Patrick Harries (2002), Patricia Hayes (2000, also this volume), John K. Noyes (1991), and Larissa Foerster (2004) have added other case studies of confrontation between local interpretations of landscape and colonial inscriptions onto the landscape. Taking this line of thought into the present Bollig and Heinemann (2002) pointed out that photography nowadays tends to link ideas of the indigenous with African landscapes: local people and landscape become inextricably linked both in photography and text. Some publications took up the issue of game parks as a specific way to inscribe European ideas of wilderness, the sublime, and purity onto an African landscape (Carruthers, 1999; Dieckmann, 2001; also this volume). In many papers authors stress that actors actively change features in the landscape so as to make it bear witness to their ideas on history, identity, and man–environment relations. Inscription in many instances is not only a mental but also a physical process.

Little attention, however, has been given to how local African communities actively create a landscape as a mnemotope, a landscape bearing the memory of past generations, reflecting the depth of kinship relations and witnessing the legitimacy of occupation of the contemporary generation (but see Colson, 1997).

This contribution tries to combine both lines of thought: the inscription of kinship onto the landscape in a ritual process is explored and political strategies connected with such efforts in inscription are followed up. The core of the data originates from the participant observation of a large ritual. In August and September 2001 we had the great opportunity to witness a major ritual (*okuyambera*) in which the members of a chiefly Himba lineage (Northwest Namibia) commemorated their ancestors. They had built a special compound for the ritual in which around 120 relatives lived together for about six weeks and commemorated their ancestors. They slaughtered oxen for specific ancestors and visited their graves in various local and distant places. The visiting of a great number of ancestral graves spread over a wide area in ritual processions turns an abstract genealogy into a network of memorable places: collective memory is spatialised and relations between kinship group and land are imbued with a morality and intense emotions. By so doing they created a mnemotope – collective memory embedded in specific places and the landscape in general – and presented a profound political statement vis-à-vis their neighbours and – as they thought – towards the government. The landscape was used as memory (Küchler, 1993) and as a medium to present specific political messages to members of the community but also to political actors beyond the realm of the community.

12.2. METHODS

In short I now outline the methods leading to the data on which this chapter is based. The ritual was conducted by a group of people whom I have been working with since 1994.[1] Many men and women present at the ritual have been members of my core sample since this date and I gathered detailed background information on their status, wealth, kinship relations, personal history, and households. However, there were also a number of people who came from afar (many of them from Angola) for whom my prior data could provide little information.

The ritual started in the second week of August 2001. We joined the festivity at the end of the fourth week of August and stayed throughout September. When we left, the ritual had been finalised and only a few people were still living on the ritual compound. I engaged in informal talks with participants and key actors in the ritual, took down notes on the course of the ritual, its geographical ramifications, and the many explanations our hosts gave. My wife, Heike Heinemann, filmed

---

[1] Altogether I spent a period of some 36 months in the field, spread between a two year stay between 1994 and 1996 and several more months of fieldwork in subsequent years. The last visit to the community took place in spring 2006.

sequences of the ritual. The visual material proved to be an extremely valuable source of data. The digital film-bits were later scrutinised together with some informants connecting the camera to the car battery and turning the interior of the car into a cinema in miniature. Key informants commented upon the visual documentation of the ritual and we were able to ask further questions concerning the course and symbolism of the ritual action.

The application of video techniques helped to review moments of great emotional intensity from some distance and at the same time allowed one of us a great deal more of involvement. Whereas it would have been impossible to elicit any comments during the key parts of the ritual, informants were very willing to comment later on. All conversations taped on video were later transcribed and translated by Uhangatenua Kapi who was also present during the ritual as an assistant. Uhangatenua Kapi also assisted when I took genealogical data. This turned out to be the most arduous exercise with Himba informants as there is a taboo of naming dead ancestors. Ancestors have to be alluded to by nicknames or by their kinship relations. Alternatively somebody had to be found amongst the senior men and women present who did not fall under the naming taboo. The Himba themselves sometimes lost the overview of their genealogical accounts and it was through Uhangatenua's great patience and intimate knowledge of the local culture that we succeeded in putting together consistent genealogical information.

12.3. THE HIMBA

The Himba are nowadays presented as one of southern Africa's last indigenous minorities living a traditional lifestyle (Bollig, 1997a; Bollig & Heinemann, 2002; Miescher & Rizzo, 2000; Warnlöf, 2000). The fact that the region has witnessed extreme shifts in social-ecological relations throughout the past one hundred years due to colonial impact (Bollig, 1997a, 1998a, 2006) is rarely mentioned in adverts, travel literature, and development plans. In the second half of the nineteenth century northwestern Namibia (often addressed as Kaokoveld or Kaokoland and nowadays administratively incorporated into the Kunene Region) saw the intrusion of bands of well-armed raiders preying on the livestock of indigenous populations.

Many local people fled across the Kunene and went to southern Angola. There they had close contact with the Portuguese colonial system between the 1880s and 1920s (Bollig, 1998b). Many men worked as labourers on plantations and as mercenaries for the Portuguese army (Almeida de, 1936). Changes in the Portuguese colonial system, around 1910, urged them to cross the Kunene once again back into Namibia. In 1917 the South African government took hold of the region, disarmed local people, and established three tribal reserves. In the 1920s borders were instituted and any trade across international and internal boundaries was supressed (Bollig, 1998b). Local herders were forced back into subsistence pastoralism.

In the late 1960s the Kaokoveld was attributed a homeland status and in the 1970s it became a battleground in the Namibian war for independence. Since 1990 the area has become popular among tourists as a favourite destination for anthropotourism (Bollig & Heinemann, 2002; Rothfuss, 2000). The plans of the Namibian government to build a dam at the Kunene resulted in a prolonged dispute between the local community and the government (Bollig, 1997b, 2006).

Nowadays about 30,000 people are estimated to identify themselves as Himba, some 20,000 living in northern Namibia and another 10,000 living in southern Angola. Most Himba still live off their herds. Milk, meat, and maize are the most important ingredients of their diet. Most cereals are procured from barter-trade with itinerant traders. However, most households also cultivate small gardens in which they grow maize, millet, and occasionally beans and calabashes. There are major differences in wealth and alongside herders owning a few dozen cattle there are herders owning several hundreds. Patron–client relations are established through donations of livestock and are generally embedded into kinship relations.

The Himba follow a complex system of double descent (Malan, 1980; Crandall, 1992). Whereas most of the livestock property is channelled through the matriline, status (e.g., chief status) and ritual possessions (e.g., the ancestral fire) are channelled through the patriline. Due to patrilocal rules of postnuptial residency and high rates of divorce, members of the matrilineage tend to be dispersed whereas members of a patriline frequently live close to each other. Chiefs have a great say in matters pertaining to local politics and resource management. In precolonial times bigmen (*ovahona*) dominated communities to some extent, however, the South African colonial government appointed chiefs: while there were three acknowledged chiefs at the beginning of the South African period in the late 1980s some 36 chiefs had been named. They control smaller tracts of land with some hundreds to thousands of people. However, where their rights and obligations are exactly located is hard to tell and their portfolio is at the moment under negotiation.

A reform of communal landrights has been passed by the parliament in 2003 and is currently being implemented, but during the period of time the ritual took place the rights of traditional authorities in regard to tenure rights were not at all well defined. Bollig (2006, pp. 325ff.) shows how so-called *oveni vehi* (owners of the land) control access to land at the local level and Bollig (2007) discusses the contemporary ambiguities and shifts in the local land tenure system and environmental problems brought about by high stocking rates and failing common property resource management.

Himba religious beliefs are based on the worship of ancestors. Harmonious relations with ancestors (*ovakuru*) ensure good luck and good health to all the living relatives of a descent group. Symbolically ancestral beliefs are condensed in rituals of commemoration at gravesides and at the ancestral fire. Physically the *okuruwo* is a fireplace framed by a half-circle of stones between the main hut and the main entrance of the stock enclosure. In commemoration rituals the symbolic relation between holy fire and ancestral graves is enforced (Bollig, 1997a).

Graves are, next to and in connection with the ancestral fire, the most important places for carrying out ancestral rituals. A grave is established as a ritual site during the funeral. If a senior man dies, numerous oxen have to be slaughtered to provide skulls for the decoration of the grave. Within a few weeks up to 30 oxen may be slaughtered. After the activities directly connected to the burial have ceased, the grave 'is closed' until the first commemoration ritual (*okuyambera*) takes place about one year later.

12.4. THE GREAT *OKUYAMBERA* RITUAL

Ideally ancestors are commemorated annually or every other year. However, not all ancestors are commemorated that frequently: it is only prominent dead male ancestors who are commemorated more often, whereas female ancestors are rarely commemorated. Commemorations of ancestors become less the longer they are dead. Genealogically more distant ancestors are often commemorated together with less distant ones. A household head will decide upon a commemoration ritual once he feels that a specific ancestor has given him a hint to do so: the wishes of an ancestor may become disclosed either through dreams or through ritual experts. A commemorative ritual usually takes place at the household and centres on the ancestral fire. The ritual may involve a visit to the ancestral grave where certain parts of the ritual are performed. Other relatives are invited to take part in the ritual, but the size of the festivity is determined by the number of cattle somebody wants to invest in the ritual: at least one ox will need to be slaughtered for the ancestor; more may be needed to host guests for a longer period of time. If guests stay for a long time it will also be necessary to sell some livestock in order to buy maize.

The *okuyambera* ritual we observed in 2001 was of another magnitude: instead of conducting the ritual at his home the chief together with other leading figures of the descent group decided to build an entirely new ritual homestead at a place where his forefathers had dwelled and where some of his ancestors had been buried. Many more guests than usual came to participate in this festivity and they stayed together much longer than usual. This implied that the ritual involved enormous costs but also offered a platform to commemorate many more ancestors than is usual at such a ceremony.

12.4.1. The Creation of a Ritual Space

The ancestral ritual had been organised over a long period of time. The local chief, Hikuminwe Kapika, had sent young men to close paternal relatives living on both sides of the Kunene river months ago. For the start of the ritual they had to wait for a time when the Kunene was low enough to allow people to cross. Finally in early August a group of some 80 people congregated, the core being a group of some 15 to 20 close paternal relatives (see Figure 12.1). Most of them had come with their family or at least parts of their family. Those coming from

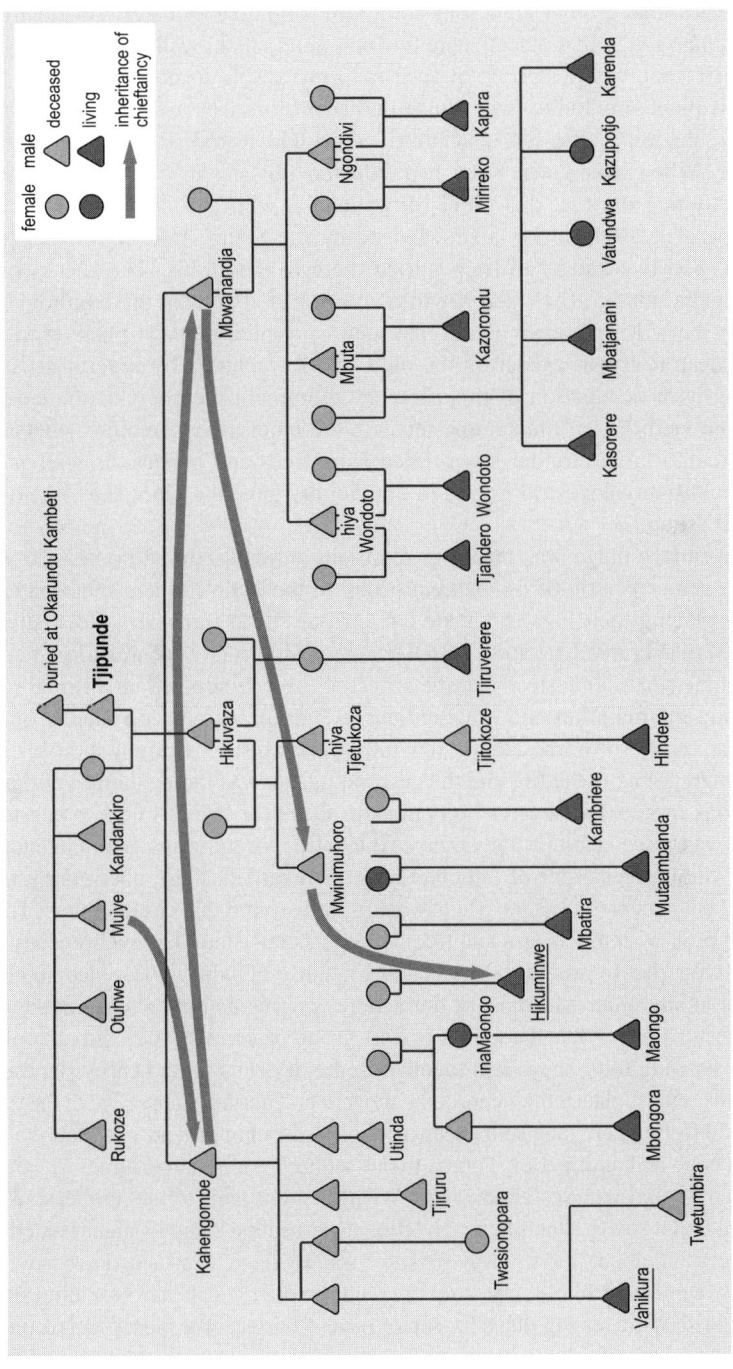

Figure 12.1. Kinship relations of those involved in the ritual

nearby had brought parts of their cattle herd in order to supply meat and milk for all. A few had even brought sacks of maize from their storages. Together they built a ritual village in a place where two past chiefs had dwelled from the 1920s to the 1950s and where a group of ancestral graves were located.

This place, Ombuku, was situated adjacent to a dry riverbed, where trees provided shade and the sands of the riverbed had stored sufficient quantities of water. When asked why they had selected this place, first of all informants mentioned the fact that chief Mbwanandja, who had been the first chief of this descent group to be acknowledged by the South African colonial government, had lived and had been buried there in the 1930s. The successor to Mbwanandja, his brother's son Mwinimuhoro, had also lived there although he had been buried in another place. Interviews revealed that the place even had older ancestral graves predating the days of Mbwanandja by several decades. The long-term occupation of this place by prominent members of the descent group, the visibility of their homesteads in the landscape (remains of huts and enclosures), a large circular space freed from trees and bushes, as well as the presence of many ancestral graves in the vicinity gave the place the fame of an ancestral ground.

The ritual village was meant to represent an ideal Himba homestead with the most senior person of the descent group in the main hut and other paternal relatives placing their huts in a wide circle from left to right with a diameter of about 150 m (Figures 12.2 and 12.3). The main cattle enclosure was placed at the centre of the ritual homestead and the ancestral fire was situated on a straight line connecting the main hut and kraal entrance. The fire was placed exactly on the east–west axis of the homestead so that the ascending sun was right at the back of those congregating at the fire and the descending sun was facing them. For practical reasons there were several exceptions to the ideal layout of the homestead. One hut was placed a bit off the circle next to a tree for more shade and in another instance the correct order of huts had been disregarded; these aberrations were deplored but accepted. Sleeping huts were placed around this circle of huts. These were not built with much care and frequently just consisted of a few branches and twigs put together to provide a bit of shelter against the wind. These sleeping huts were not in any order and most of them were congregated at the eastern edge of the homestead where the shade was best. A group of younger men did not bother to build sleeping huts; they slept together in the dry sands of a nearby riverbed.

Some traders placed their cars at the western and northwestern edge of the ritual homestead from where they sold cheap alcohol. A dancing ground was placed at the northern edge and an itinerant Tjimba trader selling traditional perfumes to women positioned a stand with his goods there. We placed our tents to the northeast of the homestead on a sandy island in the riverbed underneath a stand of giant Faidherbia albida trees and about 200 m away from the ritual village. Most ritual activity took place between the main hut, ancestral fire, and the *otjoto*, a shelter providing shade and designed as the resting place for senior men. At first all the meat from the many slaughtered oxen and also other food was placed in front of the shelter and from there it was distributed by the elders to all participants of the celebration.

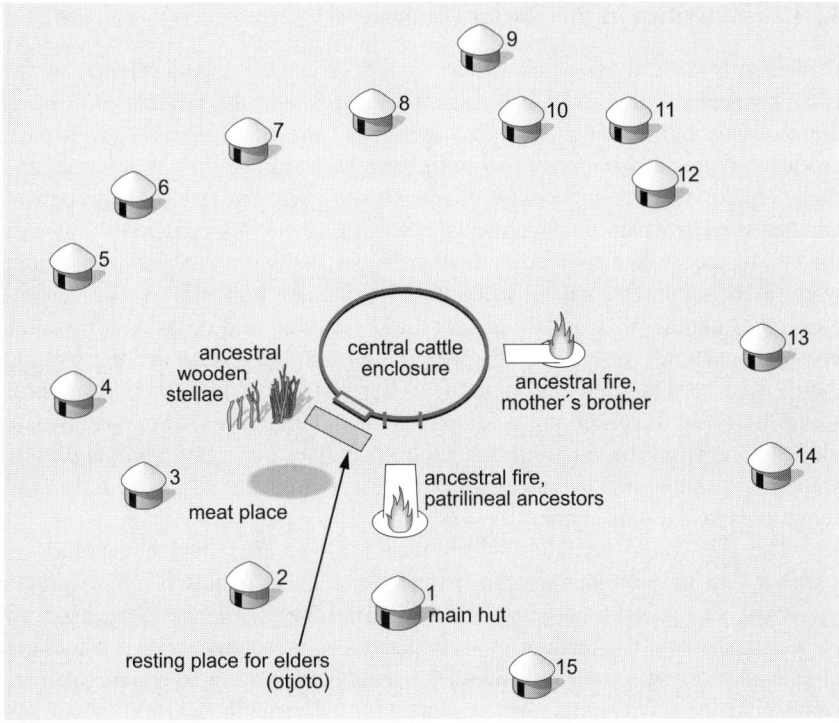

Figure 12.2. Schematic presentation of the ritual homestead; genealogy tied to ritual homestead

Figure 12.3. The ritual ground at Ombuku (ancestral wooden stella, ovipande)

12.4.2. Activities in the Ritual Homestead

Obviously during a six-week ritual a lot of everyday life developed in the ritual homestead. Livestock had to be attended to and the milking of cows in the morning and evening structured the day in the ritual homestead. Women produced butterfat from milk and in fact the seclusion of the ritual homestead even seemed to give them more time to do so. The display of sacks of butterfat was another way in which to document the wealth of the descent group. Towards the end of the ritual I counted 15 leather bags filled with butterfat. These bags were taken home after the celebration. Other women knitted baskets or undertook other handicraft work. Men spent hours talking and occasionally visited nearby households. Several men active in the ritual homestead had not brought their entire herd to the ritual homestead but had left at least their goat herds somewhere else under the supervision of some older children. These men would move to and from the ritual homestead to their livestock camp. Youths present at the ritual exuberated in dances which usually commenced at about 10 pm and sometimes lasted until 3 am.

The core ritual activities which took place in the ritual homestead are connected to the commemoration of ancestors: these include visits to graves (*omavaru*, sing. *evaru*), blessings at the ancestral fire (*okuhuhura*), the slaughter of sacrifical oxen, the reading of their intestines, the demonstrative placement of ever more firewood on the ancestral fire, and the setting of commemorative wooden stellae (*ovipande*). I first comment on the visits to ancestral graves and describe one such event in more detail:

In the morning, after the cattle had been milked (i.e., around 9 am) people congregated near the main hut and the ancestral fire. Men sat around the ancestral fire and women placed themselves by the entrance of the main hut (Figure 12.4). There they started mourning some ancestors whose graves were to be visited later on that day. The women wailed in suffocated voices; the men sat staring at the ground, their eyes wet with tears. Clearly visible and perceivable for others, the group about to set off for the ritual procession was entering the emotional state regarded as appropriate when approaching graves during an *okuyambera* ceremony. For an outsider the scene stood in stark contrast to other episodes of the ritual. As if listening to a command a large group of people went into mourning whereas around them the daily life of the homestead continued: the remaining cattle were milked, women cleaned their milking equipment, and others went about their daily chores. A group of young men isolated a group of cattle – mainly oxen – which they started to drive towards the ancestral graves which were to be visited later that day. The leading herdsboy occasionally blew a local trumpet produced from a Oryx horn underlining the wailing of the women in a very emotional way. After about half an hour of intense mourning a group of people finally left the homestead for the *evaru*, the sacred procession to the graves. Such processions can take a few hours when the graves are nearby, they may also take a full day or even several days and include overnight stays in other

Figure 12.4. Women wailing before setting off to the *evaru* (*See also Color Plates*)

places if the graves are far away. Overnight stays are not rare as there is a rule that people returning from the *evaru* must enter the homestead before sunset. If the graves are far off cars were used to take at least the older people there. In those instances where ancestors were commemorated whose graves were situated in southern Angola the *evaru* would just shortly proceed in the direction of the graves and then return.

The *evaru* itself is described in detail in the following section and the account is taken up here of when the people returned from the procession. They had to enter the homestead at exactly the point they left it. Many men who had taken part in the *evaru* brought some mopane leaves from the graves which they had been visiting during the course of the day. These were put into a finely knit basket (*oruako*) which was filled with water and was placed at the left side of the ancestral fire. In the late afternoon the ritual leader blessed all participants of the ritual from this basket. With the clan of Hikuminwe Kapika, the Ohorongo (Kudu) clan, it is the custom that women receive the blessing first and only then will the men be blessed. The elderly women of the homestead put on their festive gear (those who participated in the *evaru* had already put it on in the morning). 'Out of respect (*ondengere*) for the ancestors' they crawl the last ten or so metres to the ancestral fire (Figure 12.5). The blessing finalised the ritual activities of that day.

The next morning an ox was slaughtered for the ancestors commemorated the previous day. Not for each ancestor is an ox slaughtered. Sometimes just the most prominent one will be selected; sometimes an ox is slaughtered for two ancestors. Usually one of the elders, representing one of the houses of the lineage, acts as the donor of the sacrificial animal. Although ideally all older men should donate some oxen, in reality the majority of oxen for the commemoration were supplied by a few people. The wealthy chief and two wealthy cousins contributed

Figure 12.5. Women crawling to the ancestral fire in festive gear *(See also Color Plates)*

the major number of animals slaughtered. Only the ritual leader, genealogically superior to the chief but in economic terms much less well endowed, also contributed significantly. Out of 28 oxen slaughtered and sold for the ritual (21 slaughtered, 7 sold) 19 were donated by these four people. Figure 12.6 shows which of the participants slaughtered for which ancestor. The figure graphically shows the dominance of these four men. The figure further indicates that the ritual leader Vahikura mainly had his oxen slaughtered for the most senior ancestors whereas the chief, Hikuminwe, had his oxen slaughtered for his chiefly predecessors. In two instances men slaughtered animals for nonpatrilineal relatives. They argued that they had planned to commemorate these ancestors for a long time and that now the opportunity had arisen to do so.

Cattle were suffocated according to Himba fashion. Major chunks of meat were cut and brought to a place laid out with branches. The intestines were 'read' (*okuroora oura*) and the near future was diagnosed for the people and region. The intestines mirror the local landscape with mountains and riverine valleys clearly discernible (see Figure 12.7). Small arteria crossing each other and reddish or whitish spots are read in their relative position to each other. After reading the intestines the meat is put into huge pots and drums in which they are boiled for hours. For each ox slaughtered, a new branch of mopane or an entire young mopane tree is put on the ancestral fire. If several prominent ancestors have been commemorated the preceding day even two or three oxen may be slaughtered. It will take at least two and perhaps even four days until this meat is eaten and only after the meat is finished can the next ancestor be commemorated.

Throughout the ritual and culminating at its end, smaller rituals not directly connected to the commemoration of ancestors were also conducted. Several naming ceremonies, the initiation of two girls, two marriages, and the formal ending of the seclusion-phase of boys who had been circumcised the previous year were

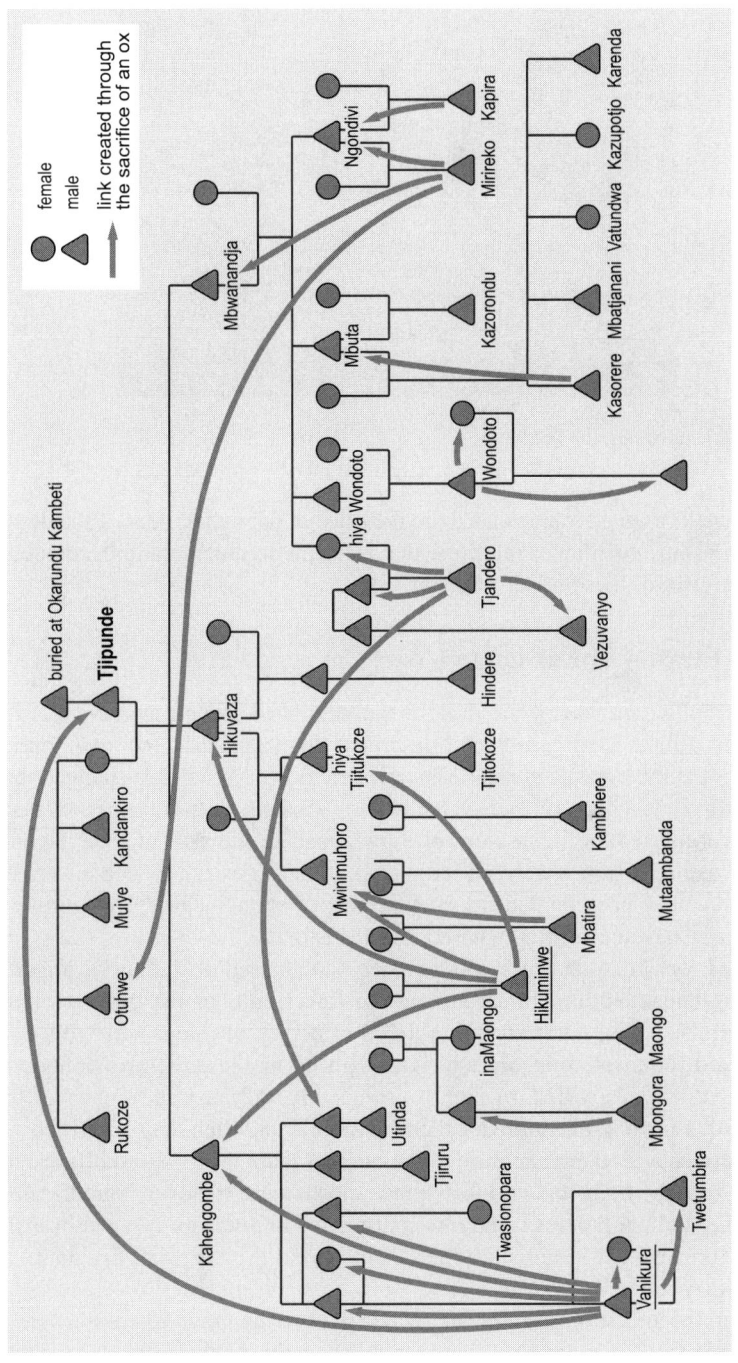

Figure 12.6.   Relationship between actor offering a sacrifical oxen and ancestor commemorated

Figure 12.7. Reading the intestines

carried out. These ceremonies need the presence of the major elder of the descent group and some patrilineal relatives: the ancestral commemoration provided an ideal platform for these lesser rituals.

12.4.3. Evaru – Procession to Grave

The most important part of the ritual is the visit of the ancestral graves (*evaru*). In another article (Bollig, 1997a) I described the physical layout and symbolic meaning of Himba graves in detail. The construction of graves has changed thrice throughout the last century: Before the 1940s/1950s people were generally buried in a crouched position: men were often wrapped in the skin of their favourite oxen. Two or four stones were placed on top of the grave. Later people were buried in a stretched position and stones were heaped upon the grave. More recently many graves have been fitted with costly gravestones.

Poles (*omihi*) with the skulls of oxen slaughtered at the burial are fitted at the top end (and sometimes also at the lower end) of the graves for men (Figure 12.8). These oxen are slaughtered in a very different way from cattle slaughtered for meat consumption. Although normally cattle are strangled in this context they are killed differently: somebody will hack the sinews of their knees with a bush knife to render them immobile and then they are killed with further strikes with the bush knife to the back of their skull. The skulls are then severed from the body and carefully meat and skin are removed from the upper part of the skull. A hole of some few centimetres in diameter is hewn or drilled into the forehead. The skulls are then displayed at the ancestral fire and occasionally they are smeared with ochre.

From the ancestral fire they are then carried to the grave where they are put on poles. Some skulls are also placed in trees nearby (if there are many skulls). For senior women skulls will only be put in nearby trees. Graves are fitted once with skulls and even a light wooden fence may be constructed around them; after

Figure 12.8. Oxen skulls displayed at a gravesite

that, however, nobody will attend to the grave and the skulls and poles are left to decay. Hence, older graves are sometimes hardly discernible (see Figure 12.9). Here I describe the course of one day of visiting the graves.

It had been decided the previous day that the following day would be suitable to visit the ancestral graves of Ongorozu and Omitengundi (Figure 12.11). The graves were placed in four separate locations with between one and four graves each. Only after milking had stopped around 9.30 am did preparations for the *evaru* get off the ground. A herd of about 30 cattle, mainly oxen, were separated from the rest to be driven to the graves. Answers to the question of why cattle were taken to ancestral graves differed slightly: ancestors wanted to see their cattle, the ancestors had to be honoured, some new heads of livestock were to be blessed as ancestral livestock or the herds at the grave were to add to the solemn atmosphere. The first graves to be visited were about 6 km away and the small herd needed about an hour to get there. At the same time the senior women, many of them in their festive gear with their *ekori* headdress, gathered at the right side of the main hut and started wailing. Later I was informed that the right side is the side of mourning in such a ritual. In fact all milking utensils, milk pails, and calabashes were displayed at the left side of the hut. Crandall (1992) has suggested that for the Himba the left is associated with living and matrilineality whereas the right is connected to death, mourning, and patrilineality.

The tone was set by the most senior women present. In low tearful voices they wailed, crying out phrases like 'Oh, child of our eldest mother (i.e., mother's eldest sister)' or 'Oh, beloved one of our fathers'. None of the ancestors to whom these wailing phrases were dedicated was named in person. At the same time some senior men had gathered at the ancestral fire; they were not wailing but looking at the ground with faces full of sorrow and some with tear-filled eyes. All of them had put on the *otjiyahambura*, a type of richly adorned leather cape

Figure 12.9. Ancestral grave at Ombuku (*See also Color Plates*)

usually worn by women. With them was a milking pail filled with some butter-fat which was tied to the travelling stick, *omutenge*.[2] At the fire the men spoke briefly about who would lead the group to the graves. This person (*omuhongore*) should stand in a specific relation to the dead person visited: he should belong to the matriclan of the dead person's father. The person selected remained *omuhongore* for the entire procession although he only stood in this specific kinship relation to some of the ancestors visited.

Meanwhile the cattle herd had left under the blowing of a Himba trumpet and at about 10:30 the group designated to go to the graves also started to move off. The chief had organised three cars, his own, one of a tourist enterprise, and ours. People in this context, however, did not just jump into the cars in the homestead. They left the homestead in the direction of the graves in a long line and in a solemn and dignified way (*oruteto*) and only entered the cars beyond the perimeters of the

---

[2]The *omutenge* is a simple stick of about 120 –150 cm to which things are tied when migrating.

homestead. In later interviews informants emphasised that *evaru* is a kind of migration and that the travelling stick and the walking in a straight line, one person after the other, indicates this. All three cars were filled with people, so that about 10 women and 10 men went on this motorised *evaru*. We drove about 30 min to the first set of graves.

Stopping the cars some few hundred metres away from the graves, people organised themselves into one row of ten males and a second row of about ten women. The men set off before the women, the *omuhongore* leading the group. Both groups now started wailing again and the nearer they got to the graves the louder the wailing became. Some people had to be propped up because their sobbing became so intense. Whereas women just went about wailing, the men mixed loud calls to the ancestors (*okukwa*) with wailing. The ancestors were called upon to acknowledge the arrival of a group of kinsfolk who had come to visit and honour them. Meanwhile the herd of cattle had arrived and was driven near to the grave. Their mooing underpinned the wailing of the visitors to the grave. The graves visited in this instance were all of female ancestors, some directly descending from the lineage, others married to males from the descent group.

Just before reaching the graves the men tore handfuls of mopane leaves and threw them to the ground on their way to the graves. This is said to bring luck (*otjitaarazu*). Then they gathered at the first grave: the men huddled around the grave to form a ring while the women stood aside for a moment underpinning the activities of the men with their continuous wailing thereby creating a highly emotional atmosphere. Amongst the men the wailing stopped and only the loud calls upon the ancestors remained. Then the men crouched around the grave and started clearing it of leaves, twigs, and other things. Stones that had rollen off the grave were placed back upon it. In the process of restoring the site the ancestors were called upon continuously. These calls were fairly uniform; the ancestor was also directly talked to. Requests for good luck (*erao*), fertility, and rain (*ombura ngairoke*) were included in the appeal to the ancestors. The image of an ideal pastoral world was evoked: the bull

---

Hikuminwe:zayambere musuko tupao erao,omukazendu tjakupwa monganda ngakwate omapaha, ngakwate vetatu,omitandu mbiwaa navyo ngarire mbiwetupe tusya navyo, ombura ngairoke tupao otjitaazu, tupao otjiwa, iso otjipyu monganda zaambere, ondwezu ngaipose mozongombe, ngarire putwaisa otjimbwi pove mba katuna omukwao. Iyaa! Iyaa! Iyaa!

They (the cattle) commemorated, girl, give us luck, when the woman is married in the homestead she should give birth to twins, let her give birth to three, the history you went with you should leave it with us. You should give it to us, leave it with us. Let it rain, give greenness to us, give us good things, take away the heat from the homestead, they (the cattle) commemorated, let the bull sound amongst the cattle, let us get all luck from you here. There is no one else. Yes! Yes! Yes!

mooing in the midst of the cattle herd. A small part of such a plea spoken at the grave of an ancestress is presented in the text.

After a short while the men stood up again. They went some metres to the side and tore more mopane leaves from a tree. This time they took the leaves to the grave holding them with both hands folded in front of their abdomen. Once again all the men crouched around the grave and scattered the leaves over the grave. Then the *omuhongore,* the leader of the group, took some butterfat from the milk pail he had carried along and gave a handful of fat to all the men at the grave. On command they started annointing the grave stones and the leaves upon them with the fat loudly evoking the ancestor. Finally they made room for the women who were not sitting around the grave but were crouched in a group to one side of it wailing intensely.

The same ritual activities were also carried out for the two other graves in this small graveyard. There was a slight variation in conduct: at one grave, for example, where the adult daughter of the deceased woman was present, she was called upon by the women and urged to place both hands on one of the grave-stones.

Finally the whole group sat a few metres away from the grave in the shade for a while; two bottles of beer were drunk at the site with the bottle circling from person to person. At this time the wailing had ceased and people talked about everyday things and the procedures for the next grave were discussed. At the next two places single graves were visited. These graves were just a few hundred metres apart. At one grave only key parts of the ritual were conducted. The grave was fairly old, probably dating from the 1940s. At the other grave the wailing was intense. Here the adult son of one of the major persons of the entire ritual was buried. He had committed suicide in the early 1990s. Here the herd was driven near to or along the grave. The first animal to step onto a stone on the grave was to be slaughtered next as it is said that the ancestor has called upon that animal. Then some few kilometres onwards another set of three graves was visited. Again these were very old graves and at least for two of the graves the knowledge of who was buried there was rather faint. A clearer picture only emerged in discussions between the older men and women. In fact, when the group was about to leave the place, somebody more or less stumbled over a set of stones and it was remem-bered, that, yes, in fact this was a grave too. Somebody even remembered who was buried there. Also this grave was cleared and some more stones were placed there in order to improve the grave's visibility. At about 4 pm the last grave had been visited and the group set off to return home, a car ride of about 30 min.

This ritual is carried out at every grave visited with some variations. One important minor ritual which I did not observe the day we participated in the *evaru* was the kindling of a small fire at the foot of a grave. This act is only carried out at the graves of senior men. A very small fire is kindled at the foot of the grave with the help of firesticks, some dry grass, and some dung. The fire only flares up briefly. A part of the ashes of this fire is taken back in a small leather sack to the ancestral fire in the homestead and is used there when lighting the main ancestral fire.

12.5. DISCUSSION: LANDSCAPE, RITUAL, AND KINSHIP

12.5.1. Collapsing Kinship and Landscape

The ritual abounds with metaphorical allusions to the close ties between kin group and deceased ancestors between homestead and graves. The leaves brought back from the graves, where they had been smeared with fat onto the gravestones, are put into the basket next to the ancestral fire. The leaves were in close physical contact with the grave and in fact they were rubbed upon the gravestones in order to increase the contact. From visit to visit to the graves the mass of mopane leaves in the basket accumulates as does the number of mopane branches on the fire. Once all those who went to the graves have returned everyone drinks and is blessed from this basket.

The simple rule is that all those whose fathers are already dead, drink from the basket themselves. For all those whose fathers are alive, the ritual leader takes a sip from the basket and sprays the water onto their foreheads.[3] So they achieve to some extent a physical union with the ancestor via the leaves that have been rubbed onto the grave. At the very end of the ritual when the leaves of all ancestral graves are gathered in the basket the final blessing is administered. The leading ritual elder takes sips from the basket and sprays water into the face, onto the shoulders, and on the breast of those being blessed. At the same time he marks head and chest with several dots of the white ash from the ancestral fire.

For many ancestors commemorated oxen were slaughtered and for each ox one branch of mopane was added to the fire, so that also the ash – just like the leaves – indicates the entity of all ancestors of the descent group. The ashes taken back from the graveside symbolically work in a similar way. The ashes from the grave are mixed with those of other ancestors in the ancestral fire. Other symbolism emphasises the strong ties between the living kinsmen. When at the end of the festivity the lineage-elders convene to plant the poles of the leadwood tree in the ritual homestead close connections and intense cooperation are emphasised. The elders go together to the forest and return together in a long row. When putting the poles into the hole[4] they try to do it at the exact same time and it is important that all senior kinsmen present put their hands to the poles. They erect this commemorative wooden stellae (*ovipande*) which authenticises the ritual and inscribes its meaning onto the landscape for many years (Figure 12.10) together. At the end of the day all the skulls and horns of oxen slaughtered during the festivity are heaped onto the poles.

---

[3] This process (*okupunguha*) in a vague and ambiguous manner could be likened to the Catholic communion: here too, the living create a commonness with a supernatural entity by consuming blessed materia.

[4] The *gastro-oesophageal vestibule* of an ox is put into the hole. This indicates good luck with animals.

Figure 12.10. *Ovipande* act as a symbolic device to memorise larger ceremonies (*See also Color Plates*)

The symbolism used during the celebration underlines and continuously emphasises the unity of the descent group linking the living to each other and connecting them with the deceased members of the lineage. The kin-group jointly produces this marker of the mnemotope signifying their ancestral ground. The *ovipande* act as a symbolic device to memorise the event. All the visits to graves, all oxen slaughtered for commemoration and all blessings given are inscribed onto the landscape by these poles which become just like a zipped version of a complex event.

At the same time ancestral relations are projected into space. For six weeks graves were visited all over the region. Some of these graves are well known. The very ancestor buried there is commemorated frequently and the grave is visited every other year. Some graves are nearly forgotten and are only visited during such major celebrations. Knowledge of these graves is rejuvenated during the ritual. A complex genealogy is reinscribed onto the landscape. A mnemotope is reinvigorated as personal memories and collective memories are readjusted. In a highly emotional setting the younger people taking part in the ritual in general and in the *omavaru* (sing. *evaru*) in particular, learn the position of graves and about their kinship ties to the people buried there. They learn to use the landscape as memory. Figure 12.11 shows how genealogy is merged with places in a landscape. Memories of ancestors and in fact relations within the genealogy become spatialised. Places are imbued with a sense of lineage identity and are reloaded with collective memories. The *omavaru* tie the theoretical enterprise of the reconstruction of a genealogy to space through a highly emotive event.

12.5.2. The Politics of the Ritual

Although the shared interest of a kinship group in the ritual has been highlighted up until now, individual political strategy was also motivating the festivity.

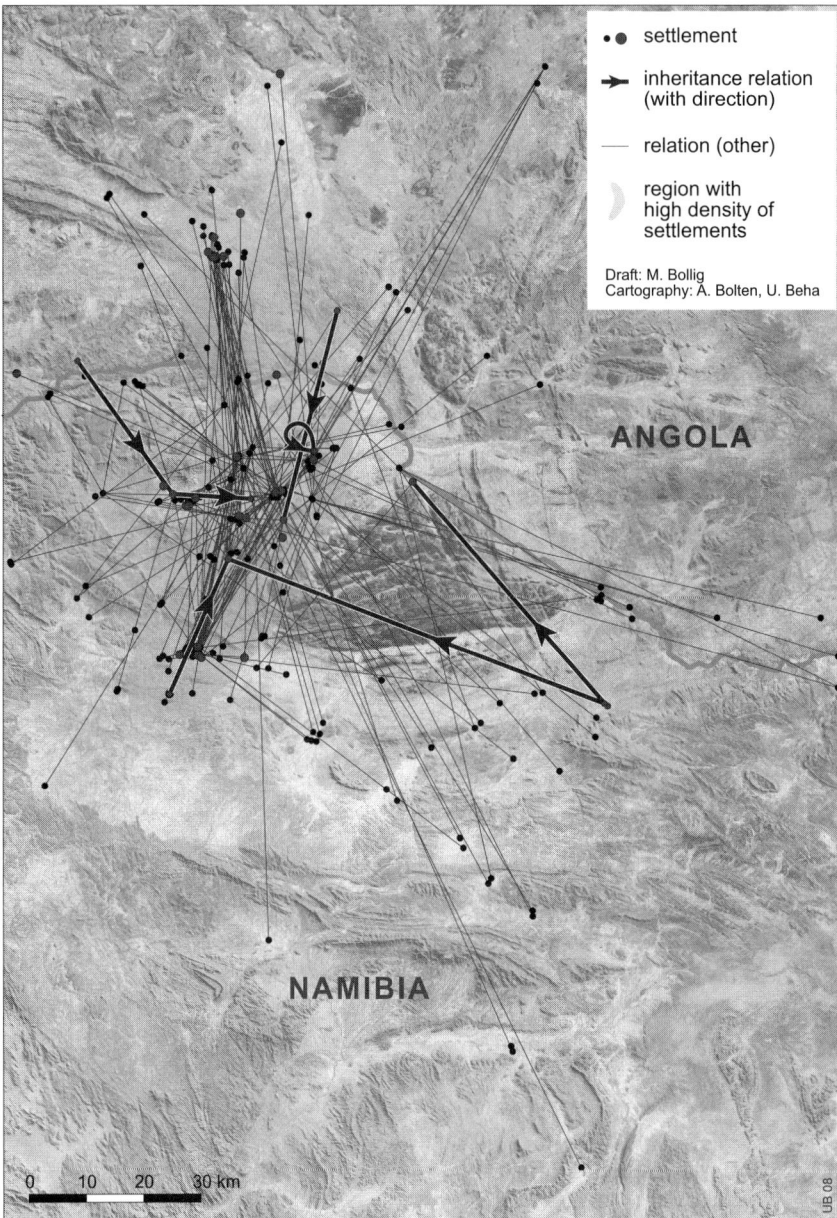

Figure 12.11. Location of graves visited during one day's *evaru* (*See also Color Plates*)

Inscriptions onto the landscape were also adhering to calculated, historically and politically embedded efforts at expressing political statements in relation to the control of land. The local chief used the ritual to underline his power base within the kinship group and in the regional context. On the other hand the local Himba group underlined its claim to places vis-à-vis other, competing groups.

The chief H.K. was the spirit behind the celebration. In fact, when replenishing our supplies in the nearby town everybody spoke of his festivity as if he were the owner of the ritual. Nobody at the festive ground spoke about the ritual in this way. There, amongst equals, the festivity was labelled as the commemoration of a descent group. Much of the ritual activity was conducted by the ritual elder of the lineage, a politically unambitious frail old man. The chief steered the celebration from the background. He contributed his ideas as to which graves should be visited. In the end he owned the car which took the people to the graves farther away. It was also him who dealt with the traders and saw to it that people were well fed and that there was a constant supply of alcohol for the festivity.

When I started my fieldwork with this Himba subgroup the chief's position had been very much contested. Competitors from within his family criticised him comprehensively. When he inherited the position from his father in 1983 the South African administration had allegedly intervened on his behalf. After 1990, with Namibian independence, this obviously did not contribute to his authority. However, during the fight against the planned hydropower scheme at the Kunene he gained reputation and internal recognition inasmuch as he established links to the regional and national Herero leadership. A huge ritual like this one further contributed to his position within the kinship group and also to the larger regional political setting.

It was interesting to hear how people present at the ritual felt about the inheritance of the main ancestral fire of the family. At the time of the ritual this fire was guarded by an old and rather sick senior member of the descent group. Due to the fact that he was the guardian of the ancestral fire, he also had the power to administer blessings at the fire. Throughout the commemoration ritual he led all major ritual activities and administered most of the blessings himself. Although the chief was genealogically not in a position to inherit the ancestral fire and the power to speak blessings at the fire many people thought that an exemption should be made to this rule and that the present chief should inherit this privilege. The ritual ostensibly added to the chief's power and the chief himself spoke in very proud words about the achievements made: in the end the enormous number of 29 oxen had been slaughtered and sold to commemorate a great number of ancestors. When visiting other homesteads during this period I was asked to recount the expenses. Most people uttered words of respect and astonishment at such an achievement. The enormous number of cattle sacrificed signified the wealth of the chief's family and underlined their claim to be the leading descent group in the region. The ceremony reconfirmed Ombuku as the ancestral grove of the Kapika descent group.

Several major patrilines of northern Kaokoland possess ancestral groves like the one the Kapika family has in Ombuku. It is at these places, where many ancestors are buried close to each other, that the ritual homestead should be built and from where the *omavaru* should start off. One key informant claimed that once such rituals are no longer undertaken a descent group's claims to the land would be foresaken. However, the ritual is not only used to underpin the claim to ancestral land but also to emphasise claims to contested lands.

When looking onto the map of graves visited by the Kapika family it is striking to see that graves were also visited which were beyond the boundaries of the chiefdom. To commemorate ancestors in places beyond the boundaries of the chiefdom implies questions to the legitimacy of present boundaries. In several speeches I heard the argument being made, that the old age of graves of one group in an area lent proof to their claim to the land. The chief left little doubt that the ancestral commemoration of August/September 2001 should also be seen in this context. The great number and old age of graves in the area had been reconfirmed and the ritual ties between landscape and genealogy had been restated in an impressive manner. This show of power and legitimacy also aimed at questioning and perhaps even discrediting the claim to land by other competing groups.

SUMMARY

In commemorative rituals Himba link genealogies to the landscape. They imbue the landscape surrounding them with history, with meaning, and with morality. Symbols and dramatic enactments emphasise the corporateness of the descent group. Collectively patrilineally linked people drink water from a basket which is filled with leaves taken back from the ancestral graves. They do not only enact a communion with their ancestors but with the very same act re-create and dramatically enact the unity of their descent group. Symbolic ties between ancestral fire and graves as well as between genealogy and graves are restated. A network of graves is linked through ritual processions of the living descendents of the kinship group. Collectively they clean graves, rub gravestones with fat and 'plant'commemorative poles at the end of the festival. Unity and solidarity are continuously emphasised throughout the ritual. They connect this virtuous behaviour to the ancestors commemorated in ritual: the ancestors are repositories of a worldly morality.

In this sense the ritual served to ensure bonds between kinsmen. The fact that the group performing the ritual was the chief's descent group was definitely of importance in this context: many senior men taking part in the ritual were not only kinsmen of the current chief but also potential rivals. However, taking part in the ritual, which was so strongly connected to and dominated by the present chief, also served to signal respect for his authority and a keen interest to keep up good relations with him. Furthermore, all the wealthy patrons of the descent group were present at the ritual: they are connected through stock loans, through inheritance relationships, and through practical aid in day-to-day herding chores (e.g., using the same shepherds, using the same pastures). The ritual also served to re-emphasise these relationships.

However, beyond the immediate concerns of the kinship group the ritual was also seen to be a statement with regard to landrights. With several reforms currently taking place in Namibia's communal lands the situation of landrights has become fairly dynamic and unpredictable from a local point of view. Whereas in South African times (lasting until 1989) tribal groups were granted certain use rights and their traditional authorities were regarded as champions of local resource management, the liberalisation since 1990 has brought several innovations. People now have the right to settle where they want with no regard to their ethnic affiliation. The entire former administrative region of Kaokoland was reshaped into the Kunene Region with new administrative regulations. The region, which had been a battleground of the Namibian war of independence for almost two decades, became the target for development planners and mineral surveyors. Throughout the 1990s the Himba opposed plans for a huge hydroelectric power scheme at the Kunene. Most important in this context: the legitimacy of many traditional authorities acknowledged by the South African government was called into question by the new government. Local politics soon became entangled with national party politics. These dynamics created a situation of uncertainty regarding the role of local authorities, the relationship between local community and the state, and also (and perhaps most importantly) in relation to landrights. The ritual described in this chapter was a powerful attempt to re-emphasise the old order and to link it with a moral landscape.

REFERENCES

Almeida de, J. (1936). *Sul de Angola. Relatório de um governo de Distrito (1908–1910)*. Lisboa: Typographia do Annuario Commercial.

Balée, W. (1998). Historical ecology: Premises and postulates. In W. Balée (Ed.), *Advances in Historical Ecology*. New York: Columbia University Press.

Bollig, M. (1997a). *"When War Came the Cattle Slept …" Himba Oral Traditions*. Köln: Köppe.

Bollig, M. (1997b). Contested places: Graves and graveyards in Himba culture. *Anthropos*, 92, 35–50.

Bollig, M. (1998a). The colonial encapsulation of the North-Western Namibian pastoral economy. *Africa*, 68(4), 506–536.

Bollig, M. (1998b). Zur Konstruktion ethnischer Grenzen im Nordwesten Namibias: Ethnohistorische Dekonstruktion im Spanungsfeld zwischen indigenen Ethnographien und kolonialen Texten. In H. Behrend & T. Geider (Eds.), *Lokale Ethnographien in Afrika* (pp. 245–274). Köln: Köppe.

Bollig, M. (2006). *Risk Management in a Hazardous Environment: A Comparative Study of Two Pastoral Societies. Studies in Human Ecology and Adaptation*. Berlin: Springer.

Bollig, M. (2007). Success and failure of CPR management in an arid environment: Access to pasture, environment and political economy in Northwestern Namibia. In H. Leser (Ed.),*The Changing Culture and Nature of Namibia: Case Studies. The Sixth Namibia Workshop Basel 2005. In Honour of Dr. H.C. Carl Schlettwein (1925–2005)* (pp. 17–35). Basel: Basler Afrika Bibliographien.

Bollig, M. & Heinemann, H. (2002). Nomadic savages, ochre people and heroic herders – visual presentations of the Himba of Namibia's Kaokoland. *Visual Anthropology*, 15(3), 267–312.

Carruthers, J.H. (1999). *Intellectual Warfare*. Chicago: Third World Press.

Colson, E. (1997). Places of power and shrines of the land. *Paideuma*, 43(1), 47–59.

Conklin, H.C. (1955). Hanunóo color categories. *Southwestern Journal of Anthropology*, 11(4), 339–344.

Crandall, D.P. (1992). *The OvaHimba of Namibia: A Study of Dual Descent and Values*. Oxford: University of Oxford.

Crumley, C.L. (Ed.) (1994). *Historical Ecology: Cultural Knowledge and Changing Landscapes*. Santa Fe, NM: School of American Research Press.

Dieckmann, U. (2001). 'The vast white place': A history of the Etosha National Park in Namibia and the Hai//om. In L. Lenhart & M.J. Casimir (Eds.), *Nomadic Peoples: Vol. 5,2. Environment, Property Resources and the State* (pp. 125–153). Oxford: Berghahn.

Durkheim, E. (1912). *Les formes élémentaires de la vie réligieuse*. Paris: Alcan.

Evans-Pritchard, E.E. (1940). *The Nuer: A Description of the Modes of the Livelihood and Political Institutions of a Nilotic People*. Oxford: Clarendon Press.

Foerster, L. (2004). Zwischen Waterberg und Okakarara: Namibische Erinnerungslandschaften. In L. Foerster, D. Henrichsen, & M. Bollig (Eds.), *Namibia – Deutschland: Eine geteilte Geschichte. Widerstand Gewalt Erinnerung* (pp. 164–179). Köln: Rautenstrauch-Joest-Museum.

Harries, P. (2002). *Photography and the Rise of Anthropology: Henri-Alexandre Junod and the Thonga of Mozambique and South Africa*. Cape Town: Iziko Museum.

Hayes, P. (2000). Camera Africa. Indirect rule and landscape photographs of Kaoko, 1943. In G. Miescher, D. Henrichsen, & J.T. Friedman (Eds.), *New Notes on Kaoko. The Northern Kunene Region (Namibia) in Texts and Photographs* (pp. 48–76). Basel: Basler Afrika Bibliographien.

Küchler, S. (1993). Landscape as memory: The mapping of process and its representation in a melanesian society. In B. Bender (Ed.), *Explorations in Anthropology. Landscape: Politics and Perspectives* (pp. 85–106). Oxford: Providence.

Lienhardt, G. (1961). *Divinity and Experience. The Religion of the Dinka*. Oxford: Clarenden Press.

Malan, J.S. (1980). *Peoples of South West Africa/Namibia* (1st ed.). Pretoria: Haum.

Miescher, G. & Rizzo, L. (2000). Popular pictorial constructions of Kaoko in the 20th century: Continuities and Discontinuities. In G. Miescher, D. Henrichsen, & J.T. Friedman (Eds.), *New Notes on Kaoko. The Northern Kunene Region (Namibia) in Texts and Photographs* (pp. 10–47). Basel: Basler Afrika Bibliographien.

Moran, E.F. (1979). *Human Adaptability: An Introduction to Ecological Anthropology*. Boulder, CO: Westview Press.

Noyes, J.K. (1991). *Colonial Space. Spatiality in the Discourse of German South West Africa 1884–1915. Studies in Anthropology and History*. Chur: Harwood Academic.

Ranger, T.O. (1999). *Voices from the Rocks: Nature, Culture and History in the Matopos Hills of Zimbabwe*. Oxford: Indiana University Press.

Rothfuss, E. (2000). Ethnic tourism in Kaoko: Expectations, frustrations and trends in a postcolonial business. In G. Miescher, D. Henrichsen, & J.T. Friedman (Eds.), *New Notes on Kaoko. The Northern Kunene Region (Namibia) in Texts and Photographs* (pp. 133–158). Basel: Basler Afrika Bibliographien.

Scoones, I. (1999). New ecology and the social sciences: What prospects for a fruitful engagement? *Annual Review of Anthropology*, 28, 479–507.

Turner, V.W. (1967). *The Forest of Symbols: Aspects of Ndembu Ritual*. Ithaca, NY: Cornell University Press.

Warnlöf, C. (2000). The "discovery" of the Himba: The politics of etnographic film making. *Africa*, 70(2), 175–191.

Chapter 13

The Spectator's and the Dweller's Perspectives: Experience and Representation of the Etosha National Park, Namibia

UTE DIECKMANN

In this contribution I exemplarily analyse two different ways of looking at the same environment, that is, the Etosha National Park in north-central Namibia. I portray the view of the western tourists visiting the area and on the other hand the perspective of the Hai‖om, a San group which up to the 1950s resided within the park area and lived predominantly from hunting and gathering. It is argued that the perspectives – the spectator's view and, following Ingold's terminology (Ingold, 2000, p. 189), the 'dweller's perspective' – are influenced by long-established cultural concepts and by the mode in which space is experienced and engaged. Both factors, the conceptualisation of and the engagement with space, are closely intertwined and have to be contextualised politically and historically in order to arrive at meaningful explanations of landscape visions and comprehension. The tourists' view is shaped by the Western aesthetical perspective of landscapes and a broad idea of how African sceneries should look. The tourists are located outside of the environment and visual features dominate their experience. The angle of the Hai‖om is one from within and is affected by their active engagement with the land. For the Hai‖om the Etosha

M. Bollig, O. Bubenzer (eds.), *African Landscapes*,
doi: 10.1007/978-0-387-78682-7_13, © Springer Science + Business Media, LLC 2009

landscape is not merely scenery, but a network of paths, of social relations, and of places imbued with social identity.

*This morning we left Oukuajeko for a game drive which turned out to be the best game drive I've ever been on. Apart from the abundant Common Zebra, Giraffe, Springbok, Gemsbok, Wildebeest, Red Hartebeest and Greater Kudu we saw 2 young male Lions on their own only a few metres from the truck. We then saw a Black Rhino before going to a waterhole where there was a pride of Lions, including a litter of cubs. (Trip Report, Sep-Oct 1998, David Kelly, Prestonpans, East Lothian, UK,*http://www.camacdonald.com/birding/tripreports/Namibia98.html)

This is now the grave of my grandfather. He takes his rest since 1948, that year, he died. They have buried him here. I did not help by myself to bury him but my parents, my father has shown me where they had buried him. This man is Petrus, Oahetama Suxub. Oahetama is the house name and Petrus is the Christian name. The surname is Suxub. It is the father of my father. He was the leader of !Gobaub, of this area. It was his area. They stayed here [pointing out] . . . different families, Suxub and Haneb and ||Khumub and ... |Aib and ||Gamxabeb. . . . (Kadisen ||Khumub, 6-9-01)

13.1. INTRODUCTION

What do both quotes tell us? The first one comments and focusses on animals whilst the second one relates to people. Remarkably, both of them refer to the same site, the Etosha National Park in Namibia. In this contribution, I explore the two perspectives which they illustrate. After a short sketch of the Etosha National Park and of my fieldwork, I take the reader on two distinct journeys to Etosha in order to portray the different perspectives in more detail. I then explore the underlying concepts of the two perspectives and contextualise them historically and politically before finally suggesting a more specified use of the term landscape in social and cultural anthropology.

13.2. THE SITE

The Etosha National Park (22,270 km^2) as one of the world's largest national parks is the premier tourist attraction in Namibia. It draws more than 100,000 visitors every year. The popularity of the park is based on the abundance of wildlife: most of Namibia's lions, elephants, rhinos, and other large animals live within the boundaries of the park. About 54% of the overnight visitors originate from outside southern Africa (Mendelsohn et al., 2000, p. 30, 34). Those are the people to whom I refer when portraying the spectator's perspective. Today, when tourists travel on the park's comfortable roads they think of themselves as traveling in a virgin natural environment. However, traveling in the region has a long tradition. The area was already traversed by various travelers during the nineteenth century on the way to Ovamboland and the Kunene river (e.g., Andersson, Galton, Schinz, and McKiernan; see Hayes, this volume). Their travel records often resemble those of contemporary tourists and are expressions of a similar perspective on tropical savanna landscapes passed down to the tourists of today (see below).

The area south of Etosha Pan, where most of the tourist roads run, has long been the home of a hunter-gatherer community. In the second half of the nineteenth century these people were labelled 'Nama-Bushmen' (Schinz, 1891, p. 127), so-called Bushmen (who were proposed to be in fact regarded as 'impoverished Namaqua', Hahn & Rath, 1859, p. 298) or 'Saen' (Galton, 1889, p. 42) and became known as 'Haiumga' (v. Francois, 1895, p. 233), 'Heikum' (von Zastrow, 1914, pp. 2ff.), 'Hei-²om' (Fourie, 1959 [1931], pp. 211ff.), in the beginning of the twentieth century (Dieckmann, 2007). Nowadays they are referred to as Hai‖om and are generally categorised as one of the 'Bushmen' or San groups of Namibia. The label 'Bushmen' is no longer popular in the official discourse in Namibia, and the term 'San' is used instead. But in informal conversations, people, especially white farmers, still talk of 'Bushmen'. Some Hai‖om even prefer the label 'Bushmen' and do not regard themselves as San. They assume that San mainly refers to Kung[1] whereas 'Bushmen' is a mere description of the former way of life. Without doubt, for the sake of simplicity, Hai‖om were called Bushmen by different parties in former times. The term did not necessarily imply a negative attitude according to the judgement of the Hai‖om.[2]

Nowadays around 7000–8000 Hai‖om live mostly in the Kunene and Oshikoto region of Namibia according to the census data (1991) of the National Planning Commission of Namibia (see Widlok, 1999, p. 19).[3] In precolonial times and during the onset of the colonial period, they were reported to live in the region stretching from Ovamboland, Etosha, Grootfontein, Tsumeb, Otavi, and Outjo to Otjiwarongo in the south (some authors claim that the southern limits extended to Rehoboth, e.g., Bleek, 1927; Schapera, 1930), and were enmeshed in trade networks and sociopolitical relationships with surrounding groups. Sometimes, neighboring groups shared pieces of land and resources with the Hai‖om (Widlok, 2003).

As a consequence of the ongoing European penetration into Namibia since the mid-nineteenth century, these hunter-gatherers were gradually alienated from the land they lived on and thus lost control over economic resources as well as their political autonomy. The German Colonial Administration created

---

[1] The !Kung are another group of (former) hunter and gatherers, living in the eastern parts of north Namibia. Both – Hai‖om and !Kung (as well as, e.g., !Xoon, Khwe, Naro, etc.) – are officially categorized as 'San'.

[2] Academics, members of the public, and diverse NGO's disagree on the politically and/or scientifically correct term; for a discussion see Gordon (1992, pp. 4f., 17ff.) and Widlok (1999, pp. 6ff.).

[3] This number is a rough estimation, due to difficulties involved in the census method which does not mention ethnic status and the "'problem'" of switching identities (see Widlok, 1999, p. 19).

the park in 1907. However, initially and for a long time afterwards, the Hai‖om were accepted as residents within the game reserve. White settlers increasingly occupied the surrounding area with the result that nearly all the land (outside the park) formerly inhabited by Hai‖om was occupied by the settlers in the 1930s. The game reserve became the last refuge where these people were still allowed to practice a hunting and gathering lifestyle. Up to the 1940s, the Hai‖om were regarded as 'part and parcel' of the game reserve, articulated for instance in a letter by the native commissioner of Ovamboland, who was acting game warden of the game reserve, to the secretary of South West Africa in 1940:

> I do not consider the Bushmen population of the Game Reserve excessive; in fact I thought that room could be found for more wild families and that these could be settled at places other than the main springs and game water-ing places, where big concentrations of various species of game even proved so attractive to visitors. I pointed out too that the Bushmen in the Reserve form part and parcel of it and that they have always been a great attraction to tourists. (NAN, SWAA A50/26, 5-9-1940)

The few adventurous tourists visiting Etosha enjoyed this ethnographic ingredi-ent during their otherwise wildlife-focused safaris. Hai‖om men were tem-porarily employed at the Namutoni and Okaukuejo police stations, or obtained seasonal work on the farms in the vicinity of the reserve. In the 1940s, the official attitudes of the park administration changed remarkably. Following the appoint-ment of the first full-time game warden in 1948, a strict limitation was imposed regarding the species that were allowed to be killed, after a period of 20 years without any amendments to the laws concerning hunting by the Bushmen. They had been allowed to hunt with bow and arrow for their own consumption. The only exemption had been protected game. Additionally to the strict regulations with regard to hunting, instructions were issued that stockowners were no longer allowed to possess more than five head of large stock and ten head of small stock each. In 1949, the Commission for the Preservation of Bushmen was appointed to investigate the 'Bushmen question' in South West Africa. The Commission was asked to issue recommendations primarily on the question if 'Bushmen reserves' were advisable. In the final commission's report, the Hai‖om were not regarded 'Bushmen-like' enough to be preserved:

> Nowhere did your [the Administrator's] commissioners receive the impres-sion that it would be worthwhile to preserve either the Heikum or the Barrakwengwe [Khwe, another group labelled 'Bushmen'] as Bushmen. In both cases the process of assimilation has proceeded too far and these Bushmen are already abandoning their nomadic habits and are settling down amongst the neighbouring tribes to agriculture and stock breeding [. . .]. (NAN, SWAA A627/11/1, 1956)

Thus, it was recommended that the Hai‖om be removed from Etosha to work on farms or to settle in Ovamboland. In the beginning of 1954, the native commis-sioner of Ovamboland convened a series of meetings in Etosha to reveal the decision to expel them to the Hai‖om. All Hai‖om with the exception of 12 families, who were employed in the park, had to leave (Dieckmann, 2001). The increasing interest in tourism (NAN, SWAA A511/10, 1938–1951) was undoubtedly a major factor which influenced this decision. At the same time international conservation organisations had begun lobbying for game parks without people.

In a grand-scale vision of national planning the coexistence of intensively used agricultural landscapes and pristine 'natural' landscapes became the epitome of an African modernity. It was the task of the state to implement such 'modernised' landscapes in order to stimulate development. The administration acknowledged the potential of nature conservation in this context. Although the game reserve had still a way to go in order to become the Etosha National Park,[4] by now, the 'national park ideal' (Neumann, 1998, p. 9) had emerged as the underlying concept for further development. This concept of a 'pristine landscape' excluded people from the area to be preserved, in particular people who were not 'pure' in a racial sense. This entails the ideal of the same pristine condition of the people as of the landscape to be preserved.

Today, the Hai‖om are one of the few groups left without legal title to any land in Namibia (Widlok, 1999, p. 32).[5]

13.3. FIELDWORK IN ETOSHA

I went to Etosha on various field trips between 2000 and 2006 to explore the history of the National Park, and in particular the developments in regard to the former population of the southeastern part of the park (Dieckmann, 2007). In 2001, I became involved in a project which was aimed at the creation of maps that take into account the long human history within the area, the documentation of the 'forgotten past' in order to deconstruct the image of Etosha as an untouched and timeless wilderness. Particular realities of the past are not 'forgotten' by chance but with reason, as Silvester et al. (1998, p. 14) have argued for the southern African historiography: 'Empirical gaps are a symptom rather than a disease: they exist for reasons which have to be theorised'. The maps produced in the project present more of the park than the few waterholes accessible to tourists and more than just the fire patterns, animal distribution, or vegetation zones.[6] In this way, they aim at presenting a 'forgotten' landscape, which is silenced by the official representations of the National Park. Much in the same way as there are reasons for gaps in historiography, there are obviously

---

[4]In 1958, The Game Reserve No. 2 became the Etosha Game Park and finally the Etosha National Park in 1967.

[5]The same holds true for the Topnaar living in the Namib Desert, see Gruntkowski, this volume.

[6]Open Channels, a UK- based NGO with the funding support of Comic Relief (collaborating with WIMSA, a Namibian- based NGO, and STRATA 360, a Canadian organisation responsible for the realization of the maps) has sponsored several field trips to Etosha.

Figure 13.1. Photo Pan (Ute Dieckmann, 2007) (*See also Color Plates*)

also reasons for gaps in what is visually represented on maps: the 'national park ideal' being just one of the reasons.

My own perspective on the landscape of the Etosha National Park changed noticeably during the period of my fieldwork. Upon my arrival my perception was certainly not very different from common Western views of African landscapes in general and the Etosha area in particular. I saw an arid environment populated with an abundance of wild animals in an essentially bare and unpleasant landscape, epitomised by the Etosha Pan, a huge salt pan without any vegetation (Figure 13.1).

But due to my previous historical research in the Namibian National Archives on people categorised as Bushmen formerly living in Etosha, I was already aware that hunter-gatherers must have lived rather comfortably within the Etosha area for – at least – some centuries. Driving through Etosha, it was hard to imagine how people survived in this landscape, which did not appear to be very hospitable. With the ongoing work and the permission obtained to get out of the car whilst in the park,[7] my own perspective of the park changed considerably. By walking through the area together with the people who grew up there, I became more familiar with the different plant resources, the seasonal variation of edible plants and animal distribution, the remains of former settlements, the temporality of the landscape, as well as the stories linked to specific places, and so on. The 'hostile' turned into a more habitable environment.

---

[7] Researchers have to apply for special research permits at the Ministry of Environment and Tourism to work in the Etosha National Park. Dependent on the research focus, certain activities, such as walking around in the park, can be carried out with this permit.

13.4. TWO JOURNEYS IN ETOSHA

The two journeys described below aim to illustrate the different perspectives of the landscape. The first paragraph, called 'the spectator's trip' exemplifies the view of tourists visiting Etosha. The second trip called 'the (former) dwellers' trip' reveals an entirely different perspective. Both are notional journeys. I did not travel with the tourists but read the accounts of their trips publicised weekly on the Internet. Further information was obtained by occasional observation whilst traveling with the Hai‖om and via participating observation and informal conversations at the waterholes and other tourist spots in Okaukuejo, Halali, and Namutoni. The second trip is not the description of one single recorded journey with the Hai‖om, but includes information collected during various journeys which I undertook as a researcher with the Hai‖om and through interviews and informal conversations at the rest camps.

13.5. THE SPECTATOR'S TRIP

Many tourists prepare themselves by reading about Etosha, either by surfing on the Web or by reading some travel guides. Etosha is advertised as 'Africa's untamed wilderness' under the tile of 'The Living Edens':

> Southern Africa's Etosha is a vast and ancient land of seasonal paradox. During the blooming of the wet season, this is an Eden of glorious abundance in which springboks, elephants, lions, leopards, cheetahs, jackals, zebras and giraffe thrive. It is also an Eden that slowly disappears when heat, drought and thirst put all life at risk, except for that of opportunistic vultures. (http://www.pbs.org/edens/etosha/)

About half of the tourists enter Etosha from the south and stop first at the Andersson gate, the entrance to the National Park.[8] The name of the gate evokes the image of adventurous explorers arriving in an untamed wilderness (at least for the tourists acquainted with that part of Namibian history), inasmuch as Charles John Andersson was the first European explorer to reach the Etosha Pan together with Francis Galton in 1851 (Dierks, 1999, p. 14; also Hayes, this volume). At the gate, the visitors are informed about the manifold rules one has to obey when visiting the park. The fact that you are prohibited to get out of your car outside of the rest camps limits the experience to just gazing. Usually the holiday-makers go straight to the nearest rest camp, Okaukuejo, to arrange their booking and find their bungalow or put up their tent. On the 17 km long way to the camp, they can

---

[8] Until recently, the Etosha National Park was only accessible for tourists through two gates, the Anderson Gate in the south, close to Okaukuejo and the Lindequist Gate in the north-east, close to Namutoni. In 2003, a new gate was opened in the north (King Nehale Gate), giving additional access to visitors entering from Ovamboland.

Figure 13.2. Waterhole Okaukuejo by sunset (Ute Dieckmann, 2003) (*See also Color Plates*)

catch a glimpse of the mopane grassland along the road and depending on the season they will most likely see some animals crossing the road.

A map of the park, which is useful for sightseeing tours to the various water-holes as well as postcards, mostly with animal shots, is available in the shop.

The rest camp is fenced in. Visitors have to be inside the fences of the rest camp by sunset and may leave the protected area only after dawn. The gates of the camp, as well as the small shop, close at sunset. In the evenings, there is not much to do except ramble along the waterhole (Figure 13.2), which is of course also fenced off. During the peak season at sunset the number of tourists often exceeds the number of animals, which despite their 'wildness' do not seem particularly impressed by the human visitors. The noise of clicking and whirring cameras sometimes disturbs the enjoyment of the romantic atmosphere created by the elephants, zebras, giraffes, kudus, oryx, springboks, and others coming to quench their thirst.

In a travel report it is remarked that:

> The observation area at the water hole is separated by a large wall, a deep ditch and an electrified fence. We grab a seat and wait for the movie to start it didn't take long. A herd of elephants walked in like they owned the joint, which based on the reaction of other animals they do. It is amazing to watch the elephants play with a gorgeous sunset as the backdrop while I have my feet up enjoying a cocktail. I am sure my photos will not do the scene justice. (http://www.worldwander.com/namibia/textetosha.htm)

Tourists usually start a game drive early in the morning as soon as the gates open to have a good chance of seeing some animals. The rest camp is left behind, the birds have started twittering earlier on, but neither their sounds nor their appearance seem to be worth documenting for most of the Western visitors. The tourists drive to the different waterholes and are looking forward to finding the animals for which Etosha is famous.

Driving, for instance, to Gemsbokvlakte ('oryx pan'), Olifantsbad ('elephants bath'), Aus ('fountain'), Rietfontein ('reed fountain'), and Springbokfontein one can reach Namutoni (the third rest camp) in the early afternoon. The visitors don't know the Hai‖om names for these places, ‡Kharios, ‡Gaseb, |Aus, ‖Nasoneb, ‡Arixas, and |Namob, nor anything of the settlements, history, and former life at some of the waterholes. Furthermore they do not get any information on the fact that some of the waterholes are artificial boreholes whereas others are natural springs (some equipped with a pump nowadays) nor about the difficulties of managing such a park.

A waterhole without animals is not worth stopping at to observe, photo-graph, and document; these places are just passed by. On the other hand a herd of elephants with their offspring, for instance, or several hundreds of zebras and springboks, some kudus, black-faced impalas, or giraffes seem attractive enough for an extended halt in the hot and shadeless landscape. The most important ani-mal to be seen and photographed is without any doubt the lion.

From the main road to Namutoni, two dolomite hills near Halali (the second rest camp) are noticeable from afar. They stand out in Etosha's plains (Figure 13.3).

On the other side of the main road, one sometimes gets a glimpse of the huge white-greenish salt pan, reaching as far as the horizon, which may serve as background for an impressive scenery with a herd of zebras, springbok, or a couple of gnus.

> Etosha Lookout extends one kilometer onto the pan, giving ghostly sightings of the odd hazy silhouetted ostrich or antelope in the distance. The pan is such a unique natural wonder and the lack of vegetation is great for viewing game. Reaching Halali campsite, we felt physically drained from the con-stant 47°C temperature inside the car. (http://www.bootsnall.com/cgi-bin/gt/travelogues/taylor/54.shtml)

Reaching Namutoni in the afternoon, those tourists interested in history have the opportunity to visit the old Fort at Namutoni and the one-room museum inside. A model of the battle at Namutoni of 1904 is on display (with miniature figures rep-resenting Ovambo warriors trying to attack or escape) as well as old weapons and historical photos from the German period, the Fort, the German Schutztruppe, and so on. Additionally, some information is offered on the history, in particular the German history of Namutoni.[9]

The visitors may leave the park the next day, many films filled with shots of animals at waterholes or near the pan. Some travellers publish their reports on the Web and others just put their best shots there:

---

[9]The museum in the Fort was closed in 2007. Another permanent exhibition with more compre-hensive information about the history and the management of the park at another location replaced it.

> We didn't see any lions in Etosha, but we did see a couple of leopards head-
> ing off on their evening hunt. A highlight was the black rhino family (dad,
> mum and junior) drinking at the floodlit waterhole near the Halali camp.
> Not sure how the pics will turn out, as it was fairly dark, but it was magical!
> (http://www.horizonsunlimited.com/johnson/Etsoha.shtml)
> To see most of the 115 mammals you should stay at least for one week at the
> park. Etosha is a "paradise" for photographers. (http://www.wildlifephotog-
> raphy.de/reisebericht_e.htm)

Most of the photographs of private tourists displayed on the Web are animal
shots, at waterholes, at sunset, or on the main roads. Views of the landscape
without any animals are hardly ever published. Certainly, this can be ascribed
to the tourists' perception of the landscape as 'desert-like', 'scorching, dry', and
the pan itself as 'an eerie, blinding white landscape' or even 'an eerie, quiet and
lifeless landscape'.[10] I came across one photo of 'bare' landscape, with the following comment.

> This was the wet season but Etosha Park looked pretty dry to me. Gerhard
> explained that in the dry season most of the park becomes a desert like
> the barren area in the background called the Etosha pan. (http://berclo.net/
> page95/95en-namibia.html)

This 'desert landscape' serves well as a background but it is not worth documenting on its own:

> . . . The size of the park is 22 270 km? and it consists mainly of grass and
> bush savanna. This type of landscape provides excellent opportunities for
> game viewing at the numerous waterholes or in the open countryside. (http://
> www.biztravel.com/TRAVEL/SIT/sit_pages/5908.html)

In a word, for the tourists, the landscape serves as a stage for the daylight and
nightly performances of African animals. This implies also that their concept of
African landscape refers to the natural environment without animals and people;
those are not perceived as an integral part of the landscape itself. They are mentioned additionally, that is, as the main actors in front of the prototypical image
of an African savanna landscape.

13.6. THE (FORMER) DWELLERS' TRIP

The second excursion takes some Hai‖om and me to old locations, no longer
documented on contemporary official maps. We want to find a waterhole, which
was culturally important in former times. ‖Nububes, as it is called, is recorded

---

[10]E.g. For example, http://www.go2africa.com/namibia/etosha/4; http://berclo.net/page95/95
en-namibia.html/4; www.rehlh.com/VacationPackages/Africa/Namibia.htm/4; http://africanadrenalin.
co.za/wildernesssafaris/ongava.htm./

Figure 13.3. Halali Koppies from the main road (Ute Dieckmann, 2002) (*See also Color Plates*)

on old German maps of Etosha, originating from the beginning of the twentieth century, but disappeared from the official maps during the South African colonial period. There is no road passing the old waterhole but the men claim to know its location. They are sure that they will recognise where we have to leave the car and how to find the way (see Figure 13.4).

The trip is undertaken under different circumstances than the spectator's trip: we have official approval to leave the tourist roads and we are allowed to get out of the car and to walk around in the park. The Hai‖om who guide me are four elderly men, Mr. Kadison ‖Khumub (born 1940), Mr. Willem Dauxab (born 1938), Mr. Hans Haneb (born 1929), and Mr. Jacob |Uibeb (born 1935).[11] All of them grew up in the park. Kadisen ‖Khumub has worked most of his life in the park, whereas the others worked on farms outside of the park for quite some time. All of them regard Mr. ‖Khumub as fortunate because of his regular employment in the park for more than 40 years. The life on the farms was more difficult: the treatment of the workers was left to the attitude and character of the farm owner. Some farm owners were reported to treat their workers very badly, to beat them,

---

[11] Sadly, Hans Haneb died in November 2006 and Jacob |Uibeb died in June 2007, Willem Dauxab died in August 2008.

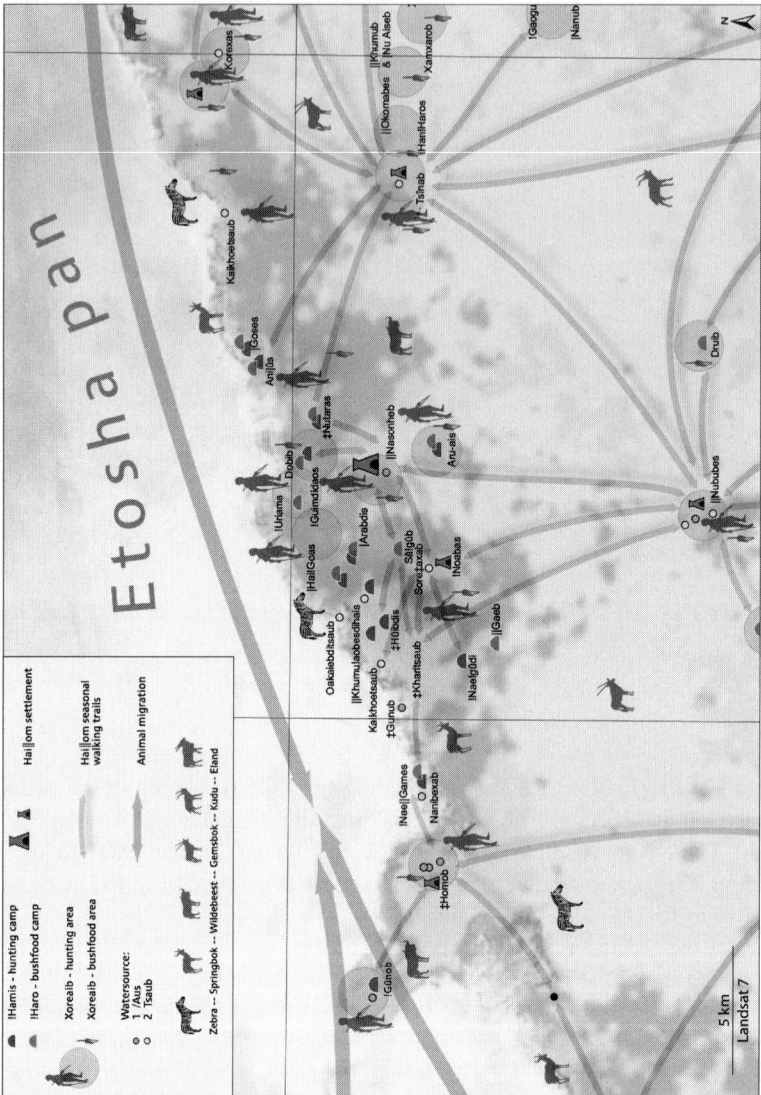

Figure 13.4. Extract of a cultural map, illustrating seasonal mobility, ‖Nububes in the south (*See also Color Plates*)

and not to pay enough money. Hai‖om often reacted to such treatment by moving from one farm to another, working here and there for some months before moving on. Furthermore Mr. ‖Khumub had a better income during his whole life than the others. All of them were retired and receive some income through their pensions as all Namibians older than 60 years receive around N$350 pension. Three of them lived in the *location* of Okaukuejo and Hans Haneb lived at Oshivelo, a rural slum close to the Lindequist gate near Namutoni. He paid regular visits to extended family members living in Okaukuejo.

I leave the rest camp Okaukuejo and drive to the location to pick up the four men. The majority of the ('native') staff and their families live some hundred metres away from the rest camp and out of the tourists` sight. The layout of Okaukuejo with this separated location is an imprint of the apartheid era in Namibia. The game reserve is not the 'natural island' as which it is marketed, but instead provides evidence of the political developments around it. The four men are already waiting at the location. They are familiar with all the Etosha roads; indeed they assisted in constructing most of them. After the eviction of the Hai‖om from the park in 1954, a few Hai‖om were allowed to stay on to work for the Nature Conservation within the park. They had to settle near the stations of Namutoni and Okaukuejo and helped construct the tourist facilities and roads, maintaining vehicles, and the like. Others, such as Mr. ‖Khumub, were encouraged in the late 1950s to start working there again, as the need for labour increased (see Dieckmann, 2001).

On the road, they warn me when approaching a sharp bend; they comment on the animals, mentioning, for example, that the zebras have already moved back from the west, where they spent most of the rainy season. Springboks and gnus have also returned. We leave the car at a sandy road, a firebreak. Firebreaks divide the whole national park into small quarters in order to prevent the spread of bushfires giving evidence of the park management having left visible marks on the landscape.

An old footpath, hardly visible, crosses this newer road.[12] According to Mr. ‖Khumub, this path leads from ‖Nububes to a small hill, called Druib, where the families living at ‖Nububes used to go to collect some bushfood when ripe. We get ready for the walk, taking sufficient water with us. Along the footpath, we find some dry naue (*Termitomyces*); our guides want to collect them on the way back to prepare them for dinner in the evening. We sometimes stop at raisin bushes (‡âun, ‖naraka‖naen, sabiron, different *Grewia* species), as the men are keen on eating the berries; it is the best time of the year now, after the rains. While walking they keep their eyes fixed on the ground most of the time and check for animal tracks. In front of a small elevation we lose the path, moving around in dense thorny bushes. It is not evident how the men distinguish the animal paths from the people's paths.[13]

---

[12] In fact, we went several times to ‖Nububes, and twice, the men had difficulties in telling me exactly where to stop. They orientate themselves on the course of the road, knowing that the path is situated behind several "'normal'" bends followed by a sharp right-hand bend.

[13] Probably different criteria than the sole appearance of the paths are needed, in particular, the course of the sun and their own movement within the landscape.

Figure 13.5. ‖Nububes (Ute Dieckmann, 2003)

They find the right way while walking around within the network of foot-paths and/or animal paths.[14] Finally one of the men finds the path and we continue our march. After one and a half hours of continuous walking we eventually find ‖Nububes (Figure 13.5). Everybody is glad; the Hai‖om haven't been there for more than 50 years. However, in the meantime, the well has dried up.

Mr. ‖Khumub shows us how visitors were greeted (*mainuai*) by the headman (*gaikhoeb*) in former times. He put some ash on the legs of the visitors to welcome them. People of other areas sometimes came to visit, to collect bush-food or to hunt, but first they had to go to the headman to ask for permission. The headman was a respected man, responsible for solving conflicts within the group and between individuals of different groups. He also had a specific role in the management of resources.

We leave the waterhole to look for the former settlements that were usually situated some hundred metres away from the water to avoid irritating the animals coming to drink. At ‖*Nububes*, there were several settlement areas from different periods and various family groups. According to the season they either stayed there or moved to other places to collect bushfood or to hunt game. Mr. ‖Khumub grew up – at this place – here, the father of his mother, ǀNuaiseb, was the head-man of the area. While moving around, Mr. ‖Khumub finds an old cartridge case

---

[14] Once, we tried to find another former historically important waterhole, situated about three kilometres away from the road. Our guide lost the right path at a crossing of several paths. He had been to the waterhole in 1994, but that time, he had the same difficulties in finding the right path. Finally we found the waterhole without returning to the crossing, walking along other paths which ran more or less in circles.

Figure 13.6. Goat kraal at ‖Nububes; some kraals at other locations are easier to recognise (Ute Dieckmann, 2003)

and he tells us that his grandfather must have hidden his rifle around here before the people had to move away. The people traded with the Ovambo; they got the rifles from the Ovambo king (most likely Ondangwa, but the people always refer to Ovambo in general) in exchange for some animal skins. We come across the remains of an old *kraal* (Figure 13.6), either for young goats or for dogs. However, outsiders like myself may not easily recognise the meaning, that is, the former function, of these heaps of stones without the insiders' explanations.

The Hai‖om kept goats up to the 1940s; some also had cattle. The dogs were used for hunting, a fact which was not appreciated by the station commanders of Okaukuejo and Namutoni, but difficult to control. We discover a grave, another heap of stones, which according to Mr. ‖Khumub was the grave of |Nuaiseb's mother. It is very obvious that the Hai‖om did not only have a number of stories attached to the land but also left imprints of their communities in the land in manifold ways.

On the way back to the car, I ask them if they were not afraid of the lions. No, the lions were regarded as 'colleagues', as friends. 'And if they try to attack?' Mr. ‖Khumub explains that there was a saying shouted at approaching lions: '‖Gaisi ai!nakarasa', 'you ugly face, go away!'. Lions were usually not eaten, just two brothers of one family, ‖Oresen, ate lions, and one died later prematurely, a fact related by other people to his predilection for lion meat. A remarkable funeral took place at ‖Nububes, a man with the surname ‖Oreseb had killed a lion, prepared the meat, and ate it at ‖Nububes; the few leftovers of the lion had to be buried and all the people at ‖Nububes were called by ‖Oreseb to attend the burial, crying like the lion roars. Another story is connected to ‖Nububes. There was an elephant that chased the people and roamed around the settlements.

The people said that this elephant was ridden by a ||*gamab* (spiritual agent),[15] creating unrest at ||Nububes. The *!gaiob* (shaman), in this case Mr. ||Khumub's father, was called and he discovered the hidden rider. He took his bow and arrow and shot the ||*gamab* off the elephant; the elephant disappeared and peace returned to ||Nububes. That is to say that numerous stories are connected to the place and that ||Nubube*s* serves as a hub of collective memory.

Later on our way back, one of the men points out a smaller flowering plant, explaining that it is !khores (*Adenium boehmianum*) which is comparatively scarce in the Etosha area. The root of !khores was used to prepare the poison for the arrowheads (!khoreoas). As it grew mostly in the southern area of the park, it was exchanged with people residing closer to the Etosha pan, in return for salt. We continue, most of the time silently, but the men do not forget to collect the naue at the termite hill to take them home.

Back at the car, we decide to make a turn to ‡Homob, where Mr. Dauxab stayed for a couple of years before the Hai||om were evicted from the park. He shows us the settlement areas including the house of his father and the *!hais*. A *!hais*, a specific tree could be found at each settlement and was the ritual place where the men used to bring the prey after a hunting trip, prepare the meat, and divide it up. Depending on the occurrence at the settlement, different species of trees were used as *!hais*. Explaining the meaning of this tree to me, the men jocularly called the tree 'the kitchen of the men'. ‡Homob consists of two waterholes, and Mr. Dauxab explains that one was for drinking and the other was used to hunt. While he shows us the remains of a *!goas*, the shelters used by the hunters while waiting for game, the other two men lament about the condition of the waterhole surroundings nowadays; most of the bushes around the water have been destroyed by elephants by the way, an observation which holds true for quite a few springs. The removal of humans from this habitat and the sole use of resources by game have altered the environment, in some places significantly. Far from being a virgin nonchanging environment, environmental change is of importance and the idea of 'natural equilibrium and stability' a mere fiction.

Mr. Dauxab points out the tree where the Hai||om dwellers waited for the tourists visiting the waterhole, in order to get sweets, oranges, and sometimes clothes from them. During the years immediately before the eviction a game warden regularly patrolled ‡Homob and the men living there were temporarily employed to procure the wood as building material for tourist facilities in Okaukuejo. Mr. Dauxab moves around and points out the footpaths which connected different settlements.

---

[15] Because belief systems were not the focus of my research, I did not analyze the data concerning belief systems including the ||*gamab* systematically. It seems fair to say that several ||*gamab* usually stay in ||Gama||aes ('"||Gama-Nation"', somewhere in heaven) and watch the events on earth from there. The earth and the sky are two separated entities. Furthermore the ||gamab can help people as well as harm people. There are good ||*gamagu* and bad ||*gamagu* (see also Widlok, 1999, pp. 52--56).

In the process of the removal in the 1950s, the Hai‖om were first persuaded to settle close to the police stations at Namutoni and Okaukuejo, which took a while. Only then, the native commissioner of Ovamboland convened a series of gatherings to tell the Hai‖om to leave Etosha completely. Mr. Dauxab and his family left their former place of residence (Tsînab) to move to ‡Homob, closer to Okaukuejo, which was more accessible for the station commanders and later the game wardens. Mr. ‖Khumub and his family left ‖Nububes as well to move to ‖Nasoneb, closer to the main road and easy to access from both the police stations. These memories are revitalised while rambling around at the different sites.

Driving back, everybody is covered in dust, hungry, and scratched by thorns; we experienced the landscape with more than just our eyes.

13.7. LANDSCAPE EXPERIENCE AND REPRESENTATION

What do these journeys tell us about the conceptualisation of space? And what does this suggest for the anthropological approach to a productive use of the term 'landscape'?

13.8. THE SPECTATOR

The spectator is physically separated from the land. He or she can observe the animals in front of an African environment with his or her own eyes but is separated by a fence or surrounded by the bodywork of a vehicle. She looks at the landscape from a distance. This seems to be the crystallisation of the conventional Western landscape concept. 'In the contemporary Western world we "perceive" landscapes, we are the point from which the "seeing" occurs. It is thus an ego-centred landscape. A perspectival landscape, a landscape of views and vistas' (Bender, 1993, p. 1). The spectator has a fixed point (here the parked car) from which he perceives the landscape or she experiences the landscape as a scenic view gliding past the window, as Neumann puts it: One travels 'through the landscape as an observer "taking in" (consuming) the scenery, rather than travelling in the landscape' (Neumann, 1998, p. 20).

Neumann points out that development in transportation technology in the course of the development of the Western landscape concept has strengthened the perception of landscape as a picture (1998, p. 20). The spectator does not move around and experience the landscape with the various senses; he doesn't smell, feel, or taste the environment. Unquestionably this focus on the visual perception can be attributed to the strong linkage of the concept to art (see Rössler, this volume).[16] But this origin

---

[16] In the English language, for instance, the term was introduced as a technical term used by painters in the late 16th sixteenth century (Hirsch, 1995, p. 2; Neumann, 1998, pp. 15ff.). The contemporary popular '"Western landscape concept"' still contains more of the aesthetical connotations which evolved with its artistic origin than characteristics produced by scientific thought later on (see Hirsch, 1995, p. 2; Luig & von Oppen, 1997, p. 10).

is not the only crucial aspect: why did it all of a sudden become interesting to paint the 'landscape'? The genre of landscape painting and the concept of 'landscape' therewith developed in the emergent capitalist world at the same time as the dichotomies between the ruling city and the dominated rural hinterland came into being in the emergent capitalist societies (Bender, 1993, p. 1).

It is significant that the involvement of the people with the land had changed before. The urban dweller was no longer engaged in the cultivation of the land, and the 'countryside' became aesthetically attractive for the urban elite as 'nature'. Nature was now regarded as an entity of its own, detached from the social world. The subject, the perceiver, usually the urban dweller, looked at the landscape as an object of contemplation. By the end of the nineteenth century, landscape and nature had become almost synonymous categories (Neumann, 1998, p. 21). But there were, at the same time and the same place, different ways of understanding and relating to the land and tensions between the elitist aesthetic perspective and the alternative perspective of the peasant dwellers (Bender, 1993, p. 2).

African landscapes have been annexed by the aesthetical view. They became incorporated into the system of Western polarities. Whereas the Western world became more and more associated with urbanity and 'culture', Africa itself was transformed into a symbol of wilderness, nature, and rurality:

> [Eighteen]th and nineteenth century travellers developed an appetite for visions of wilderness, opaqueness, darkness and disaster which coincided with increasing European expansion to other parts of the world. [. . .] With the advancing urbanization and industrialization of Europe, otherness was increasingly identified as 'nature' (for which, since the eighteenth century, 'landscape' was also used as a synonym) or 'wilderness', remote from the world of humans subsequently labelled as 'culture'. (Luig & von Oppen, 1997, p. 12)

European travelers were unquestionably accustomed to the vistas of landscapes back home. Arriving in Africa, the travelers found 'nature', which was in many aspects incomparable to the familiar landscapes they were used to appreciating in Europe. The African landscape (without animals) was experienced and described as 'uniform, monotonous, without end', 'naked', 'monotonous' (see Harris, 1997). Frank Oates noted in his diary in 1873: 'South Africa is sadly dull and monotonous, and I believe the influence is a bad one, and the loss of scenery has a depressing effect on the spirits' (1889, pp. 46ff., cited from Ranger, 1997, p. 60). Harris noted: 'When the landscape did not speak in a language understood by the viewers they described it in inverted and negative terms of what they knew. The land was empty, monotonous, colourless, treeless, silent, naked, devoid of perspective, unmarked by human enterprise and without God' (Harris, 1997, p. 183). However, for the Europeans who settled in Southern Africa (missionaries, settlers, soldiers, etc.), the perception of the landscape changed after a while with the active engagement with the land (for Swiss missionaries see, e.g., Harris, 1997).

The contemporary descriptions of the Etosha landscape as 'desert-like', 'scorching, dry', 'an eerie, quiet and lifeless landscape', 'sparse', or 'parched' by the tourists have a striking similarity to the descriptions of earlier travelers.

The 'wilderness' and 'otherness' of Africa became subsequently epitomised and immortalised in the famous national parks and game reserves throughout the continent: 'The trope of the "wild" has become a particular trade mark of Africa' (Luig & von Oppen, 1997, p. 33). All of them were established by the European colonial authorities, which helped to reinforce and legitimate imperial rule. The idea behind the establishment of these parks was 'the notion that "nature" can be "preserved" from the effects of human agency by legislatively creating a bounded space for nature controlled by a centralised bureaucratic authority' (Neumann, 1998, p. 9; for Kruger National Park see Carruthers, 1995; for Matobo National Park see Ranger, 1989; for Arusha National Park see Neumann, 1998).

Tourists travel there to consume 'nature', which actually means to *look at* the exotic, rich, and primordial. 'Wild' animals are a crucial part of Western fantasies about this 'other' world. They are the symbols of unspoilt nature and untamed wilderness. One German woman in a tourist group made her motivation plain: she visited Etosha because of the animals, whereas the Kaokoveld, northwest of Etosha, was believed to be a suitable destination to watch the 'native' people, the Himba.[17] Thus, nature and culture are separate spheres of interest.

The spectators come equipped with their video cameras, binoculars, and photo cameras to 'capture' the various species of animals. Documenting the scene seems even more important than watching and observing it. The films and movies on Africa as well as photographs in brochures and on postcards have become the framework for perceiving the African landscape. Gordon notes: 'If there is one thing which characterises contemporary tourism it is its visual aspect. Tourism is photography' (Gordon, 1998, p. 111). Indeed, this urge for representation supports Casey's notion that the mere existence of landscape already calls for its representation, that is, that the representation is actually already part of the landscape experience (Casey, 2006, p. 10), although I would limit this notion to the Western landscape concept. In particular for the African context, the representation by Western spectators can also be interpreted as a form of appropriation of the wilderness. At least the parallels between the hunting tradition and the tradition of photographic representation suggest this reading (for Shooting and shooting see, e.g., Landau, 1998). The exotic animals are the key ingredients within the national park context: African scenery becomes worthwhile photographing or filming if and only if there is sufficient game to be seen. The African landscape acts as a stage for the 'wild' animal. A skim through old travel accounts and arti-

---

[17] For the visual presentation and marketing of the Himba see Bollig & Heinemann, (2002); for an historical analysis of the pictorial construction of the Kaoko in the 20th twentieth century see Miescher & Rizzo, (2000).

cles about Etosha proposes that this way of written and pictorial representation of the Etosha landscape has not changed considerably since Etosha was first passed through by early travelers in the nineteenth century.

McKiernan – who undertook several journeys to South West Africa between 1874 and 1879, noted:

> I left Okoquea (Okaukuejo) at sunset; expecting to travel the greater part of the night and get to Okokanna (Okahakana, saltpan west of the Etosha pan) early next day. There was a bright moon and the road was open, but it was cold and frosty, and my naked people seemed to suffer so much that at 9 o'clock I camped by a thicket. . . . We were on the road again at sunrise, and about 10 o'clock came to where there were great numbers of wild animals feeding on the open plan. Gnus, zebras, gemsboks, hartebeest and thousands of springbok were before us, and above the low bush to our left were long necks of six giraffes. It was the Africa I had read of in books of travel. (McKiernan, 1954, p. 96)

When comparing contemporary travel descriptions and this quote from the nineteenth century two aspects become obvious. Firstly, already during that time the actual landscape experience was guided by culturally embedded epistemic structures and secondly, the notion of the landscape serving as the mere background for the African game was also already prominent.

The travel descriptions in the twentieth century demonstrate the same mood. Professor Dr. Lutz Heck – a German tourist visiting Etosha at least twice and publishing a book and articles about his experiences – noted in 1956:

> White and never-ending, the plain of the Etosha Pan lay in front of us. The paths of the game lay criss-cross over it, trodden by the hoofs of thousands of thirsty animals who since the beginning of time visit the watering places. Below us, at the foot of a steep ridge the water of a pool was reflected sparkling in the sunlight. It was midday and the zebra, wildebeest and a few gemsbuck lay peacefully by the water. It was like a picture of the Garden of Eden, and when we looked back on to the steppes the giraffe came into view among the Acacia trees. (Heck, 1956, p. 85)

But photography became an increasingly significant part in the representation of African landscapes. Whilst the written accounts of the 1940s and 1950s, for instance – although seldom – mentioned encounters with people in Etosha (trips with specific game wardens who served as tour guides, or Bushmen dances), the photographs exclusively show the animals (e.g., Heck, 1955, Figure 13.7; Davis & Davis, 1977).

As we have seen, African landscape without animals is frequently perceived as 'desert-like'. By contrast, the Garden of Eden seems to be a prominent metaphor when animals populate the landscape. This Garden of Eden appears as the timeless paradise from which temporality and history are excluded. When looking at it, most visitors do not think about the changes it has undergone over the years or centuries. Instead, they perceive what they see as static scenery, immortalised in the photographs.

Figure 13.7.   Photograph of a traval account to Etosha (Heck 1955)

> Pictoralized nature is fundamental in the history of the national park ideal. 'Framing' nature in painting, whether pastoral or sublime, transformed it into picturesque scenery, where the observer is placed safely outside the landscape. Likewise, surveying, bounding, and legally designating a 'wild' space makes it accessible for the pleasure and appreciation of world-weary urbanities. (Neumann, 1998, p. 17)

The 'framed' or 'fenced' nature is perceived as a relic of the ancient past, but it is in fact a contribution to a particular historical narrative of European imperialism and capitalism and a commodity of the present. Whereas colonialists transformed most of the African 'wilderness' into productive fields and exploitable resources, 'national parks represent remnants of the pre-European landscape, pockets of the remote, unoccupied wildlands preserved as reminders of a "national heritage"' (Neumann, 1998, p. 29). National Parks may *represent* wilderness, although they are in fact, as Luig argues (1999, pp. 24f), often 'inventions of wilderness', an observation which holds true for the Etosha National Park as well. Former human inhabitants as reminders of human impact in the past, firebreaks as indicators of contemporary nature management, and tourist roads as signs of contemporary economic utilisation definitely contradict the wilderness idea of western thought.

For the tourist, the history of the land lies on the other side of the fence, either outside the National Park or in the rest camps, at Fort Namutoni, within the museum, and in some old photographs in the reception of Okaukuejo. The history is encapsulated in the patches of 'civilisation', the rest camps, and separated from the landscape whereas the 'naturalness' of the landscape can be discovered 'out there', within the fenced (sic!) park. This constructed landscape is consumed by the tourists and thus perpetuated as 'the Edenic vision of the landscape' (Neumann, 1998, p. 18),[18] in glaring contrast to the world of 'civilisation'.

---

[18] In fact, most of the African National Parks are advertised as versions of the "Garden of Eden" – a romanticized wilderness (see Neumann, 1998, p. 18, for further references).

13.9. THE DWELLER

The way in which the land is perceived by the (former) dwellers is completely different. Even if the elder Hai‖om men with whom we undertook our second trip are no longer dwellers in the land, their views of it have been shaped and changed over time in the course of both practical and social appropriation of the land first and subsequent dispossession afterwards. They perceive the land from within. It is nothing exotic to watch but a familiar dwelling domain.

The relationship amongst the land, human beings, and nonhuman beings is obviously defined differently than in Western thought. For example, the story of the lion eaters reveals that lions, in particular, were respected by the people as colleagues or friends, as equals. Lions were also sources of divine power. The ‖gamagu (spiritual agents) supervised the land and the people and prevented misbehaviour.[19]

The death of the lion eater ‖Oreseb was interpreted to be caused by the fact that he had eaten lion meat. ‖Gamagu would punish other forms of misbehavior as well, for example, unreasonable hunting. The !gaiob (shaman) could mediate between ‖gamagu and humankind, and ‖gamagu could also be asked for support, for example, for rain.[20]

This perception of the Hai‖om – that the world is not separated into a natural world on the one hand and the world of human society – is shared with other people who live (or have lived in the recent past) primarily from hunting and gathering; see, for example, Turnbull (1976, 1965) for the Mbuti, Endicott (1979) for the Batek, and Bird-David (1990) for the Nayaka:

> Whereas we [Westerners] commonly construct nature in mechanistic terms, for them [some hunter-gatherer groups] nature seems to be a set of agencies, simultaneously natural and human-like. Furthermore they do not inscribe into the nature of things a division between the natural agencies and themselves as we do with our 'nature-culture' dichotomy. They view their world as integrated entity. (Bird-David, 1992, pp. 29–30)

Additionally, for the dweller the land is charged with emotion and personal identities. Specific areas were occupied by specific family groups and personal identity was strongly linked to these family groups and the area in which they lived. Mr. ‖Khumub still feels strong links to his family group and the area occupied by

---

[19] Detailed descriptions of the spiritual world of the Hai‖om are to be found in Wagner-Robbertz (1976,) and Schatz (1993), but different to the information provided there, the Hai‖om in Etosha talk about several //Gamagu, not just a single one. A compilation of religious characteristics in various Bushmen groups is found in Guenther (1999), providing insight into their relation to land and non-human inhabitants as well.

[20] For a discussion of this see Ingold (2000, pp. 40–60). The question, in sofar as this feature can be explained exclusively by the practice of hunting and gathering or which other factors – shared with non hunter-gatherers – might contribute to such a world view, cannot be discussed here.

them. Arriving at ‖Nububes is arriving at 'his' place. He takes over the task of his grandfather to welcome the visitors.

Still today, after more than 50 years, every elder Hai‖om who grew up in Etosha knows the area of his or her family group, the headman, and the seasonal patterns of mobility. This is the space they are most familiar with, whereas the areas of other family groups are not as well known. Social relations are embedded in the environment. The landscape is not something out there, detached from social life as it is for the spectator. Again we notice similarities to other hunter and gatherers, in this case, for example, the Koyukon of Alaska:

> The Koyukon homeland is filled with places . . . invested with significance in personal or family history. Drawing back to view the landscape as a whole, we can see it completely interwoven with these meanings. Each living individual is bound into this pattern of land and people that extends throughout the terrain far back across time. (Nelson, 1983, p. 243, cited in Ingold, 2000, p. 54)[21]

The people and the places are connected with each other via footpaths that form a visible inscription of a network of social relations and of potential movement in the landscape. Individuals and family groups are attached to territories which are connected via paths to each other. Moving from one place to another entails the enactment of family relations.

The relation between people and places is also reflected in some place-names. The two dolomite hills pointed out on our tourist trip as outstanding landscape features are called ‖Khumub and |Nuaiseb by the local people. ‖Khumub and |Nuaiseb, as already mentioned during the journey, are two important family names in the Etosha area. Oral history tells us that |Nuaiseb was the headman of a larger area, including these two hills, ‖Khumub was his nephew. The hills offered a lot of bushfood, in particular berries (sabiron: *Grewia villosa*, ‡âun: *Grewia* Sp., ‡huin: *Berchemia discolor*) but also roots (e.g., !han: *Cyperus* sp.). |Nuaiseb decided to let his nephew ‖Khumub, who was a subheadman of that area, have one hill for collecting the bushfood there.

Other places or areas are referred to in the same way: for example, |Amessihais is the name for an area of shrubs. The meaning of the name is 'the bushes of the woman with the family name |Ames'.

As Adam states:

> . . . [F]or the native dweller the landscape tells – or rather *is* – a story, 'a chronicle of life and dwelling' (Adam, 1998: 54). It enfolds the lives of predecessors who, over the generations, have moved around in it and played their part in its formation. To perceive the landscape is therefore to carry out an act of remembrance. . . . (Ingold, 2000, p. 189)

---

[21] This concept of landscape is not restricted to hunter and gatherer groups;, some pastoral nomads share similar ideas (for the Himba see Bollig, 2001). Probably, the mobility of these groups is the decisive factor in shaping the concept.

Graves are maybe the most obvious landscape markers involving the 'act of remembrance'. They can serve as illustration for another crucial aspect of the dweller's perspective. The process of dwelling is fundamentally temporal and therefore, the dweller recognises the temporality of landscape (Ingold, 2000, p. 208). The dweller is well aware of the changes of the landscape over time and these changes are an integral part of his perception. The animal tracks tell him about their presence and their behavior a couple of hours ago. Seasonality is acknowledged in remarks about the migration of the zebras. The lack of bushes around the waterholes is interpreted as a result of the increase of the elephant population within the park over the last decades. The remains of human settlements and the graves – signs of the human inhabitants living there more than 50 years ago – are remembered. The men who worked in the park still know about the artefacts such as cartridge cases, arrowheads, and broken tins and they can explain their origin.

But the incorporation of the past into the landscape is not limited to the material remnants (or the lack of them); it is also present in the stories connected to particular places. These stories are personal stories about specific individuals. We already heard about it in the place-names for the two dolomite hills and the story of the lion eaters. Travelling through the landscape with the Hai‖om brings up more and more stories, about encounters with lions, conflicts with other groups, about crazy or lazy men. These stories are memorised when passing the places where they happened or when visiting the birthplace, the former place of residence, or the place of death of the specific individuals. The stories are not only remembered and explained to me as a researcher interested in oral history, but the people tell them to each other in Hai‖om, when passing the places. Only after my repeated questions about what they are talking or laughing, did they make the effort to translate the stories into Afrikaans. The stories are inscribed in the landscape itself. New features, new meanings, and new memories are constantly woven into it. The landscape itself is under permanent construction and reconstruction and different parties were and are involved in this process.

The gravestone of Johann E. M. Alberts (1841–1876) near ‖Nasoneb (Rietfontein) testifies to the presence of the Dorsland Trekkers from South Africa. They passed the area several times on their way to Angola where they settled for some years and back to the area of Grootfontein, where they founded the short-lived Republic of Upingtonia during the second half of the nineteenth century (Mouton, 1995, pp. 47–55). Mr. ‖Khumub remembers from the time of his childhood that he and his playmates respected the grave and avoided playing there. During the German colonial period at the beginning of the twentieth century, some settlers tried their luck within the area. The ruins of their houses are proof of their former presence. The Hai‖om are still aware of the German colonial period, even if they were born after the South African Administration took over the territory. The old people tell you that the Germans used to ride on camels on patrols. They identify old shards of glass lying around as German leftovers. During the early South African period, the Hai‖om were temporarily employed at the stations, in road construction, and some were engaged in transporting food relief in government vehicles to Ovamboland at the beginning of the 1930s1998 (see Hayes, , for that period). Remnants of those vehicles were first given to some Hai‖om staying in the park and left at several settlements within the park in the course of the eviction.

The reminiscence of the eviction became attached to the landscape and specific places much later. For decades before the eviction, most of the men were used to finding temporary seasonal employment on the farms around Etosha. Thus they were accustomed to leaving home for a prolonged period of time. Therefore, immediately after the eviction, the former dwellers did not realise that there was no return for them (see Dieckmann, 2003, pp. 71–75).

During the time of the war of liberation, some army camps were put up within the area and patrols were undertaken to chase the liberation fighters. Some Hai‖om, like members of other San groups as well (see Gordon, 1992), actively took part as trackers for the SADF.

All these different threads became woven into the texture of the land-scape. The Hai‖om remember them when moving around in the park. The different fragments are picked up and explained at the very places where the events took place.

During the last decades a new implication was attached to the Etosha Park. Although ancestors of Hai‖om were living all over northern-central Namibia in the nineteenth century and at the beginning of the twentieth century, after independence Hai‖om started to represent and claim the Etosha National park as their 'homeland'. Etosha has become the catchword with respect to the land issue of the Hai‖om. Several factors are crucial: first, Etosha is a demarcated space with clear borders and a fixed name; second, Etosha was the last refuge in the face of increased (white) settlement in the vicinity; and third, Etosha is a catch-phrase well suited to draw national and international public attention. Gupta and Ferguson pointed out that 'homeland' remains 'one of the most powerful unifying symbols for mobile and displaced peoples' which serve as 'symbolic anchors of community for dispersed people' (1992, p. 11).

13.10. CONCLUSION: REFLECTIONS ON TERMINOLOGY

The two views presented here are certainly extreme examples in the whole field of possible perspectives on landscape. There are without doubt some others as well. The maintenance staff or the scientists employed in the park, for example, hold other views, which could be called the 'operator's perspective' or the 'scientist's perspective'. They have to maintain the roads or check the waterholes; they have to count animals, or are preoccupied with fire prevention. They perceive the park in a completely different way than the tourists, but are well aware of the tourist's perspective. The farmers occupying cattle farms on the border of the National Park have still another view of the park. They see it as a nuisance because the lions and other predators cross the border wherever the fence is broken and feed on their cattle. The Oshivambo-speaking people inhabiting the communal area north of Etosha may again have another point of view. It becomes clear that 'landscape perceptions form embedded histories' (Brinkmann, this volume).

The decisive point in the different perspectives is the kind of engagement with the land. The conventional Western concept of the aesthetical landscape developed in a time when the interaction with the environment changed profoundly. This concept still prevails in the tourists' view of the park. Because they are not permitted to change their kind of engagement with the landscape (looking at it from an outsider's viewpoint), the concept itself is perpetuated in a highly specific way.

On the other hand, the conceptualisation of the land is also not the same for the different generations of Hai‖om. The younger Hai‖om regard Etosha as a lost homeland but having lived their whole lives in the location of Okaukuejo or outside the park, they are not or have never been dwellers at all. They are not aware of family groups and social organisation linked to the landscape. They have no or just a vague idea about animal tracks, the ways of hunting, or the bushfood areas. They don't believe that the ‖gamagu have mystical powers to control social-ecological processes in the land. They often do not even know the birthplaces of their parents or the Hai‖om names of specific places. Certainly their consciousness of the temporality of the landscape is very limited. Because the knowledge is no longer of practical use for the younger generation (and there is also a lack of opportunities to be taught in practice), the elder generation does not deem it necessary to pass the knowledge on to the younger people. Furthermore, the young people don't have the possibility to interact with the environment in the same way as their mothers, fathers, and grandparents did. Thus, it is likely that they will start to develop a perspective of the landscape of the park from the 'outside', although certainly with a different meaning/significance than that of the tourist visitor.

In my point of view, the two different perspectives on and understandings of landscape presented in this chapter provoke once again a revision of the concept of landscape. Its inflationary and little reflected use, which is observable both in social anthropology and neighboring disciplines (Luig & von Oppen, 1997, p. 15; see also Rössler, this volume), impede an operational way of analysis of phenomena connected to land, land-morphology, topography, land features, environment, and above all also the varying cultural and individual perceptions of those material features. Depending on the discipline and the scientific paradigm, some authors refer to landscape as a cultural concept or symbolic construct (ibid.), whereas others regard it as the material surface of the land, or a defined section of the earth's surface (with or without human impact).

Ingold suggests the adoption of a 'dwelling perspective' in approaching landscapes in order to move beyond the opposition between a naturalistic view of landscape, which deals with landscape as a neutral external background of human activities and a culturalistic concept that considers every landscape as a cognitive or symbolic ordering of space. According to the 'dwelling perspective', the landscape 'is constituted as an enduring record of – and testimony to – the lives and works of past generations who have dwelt within it, and in so doing, have left there something of themselves' (Ingold, 2000, p. 189). This is what also constitutes the temporality of landscape.

This suggestion alludes to my argument – as illustrated in this chapter – that the term landscape should be used in contexts where the complex physical and mental involvement or interaction of (human) actors with the environment is in the focus of investigation. That is to say that the relationship amongst the actor, the person who experiences (in our case the spectator or the dweller), and the spatial unit is crucial for the term landscape to be used. The term includes thus cultural, social, and cognitive aspects, because it neither refers to something entirely external of the person – the material world – nor to something merely in his mind or internal to the person but is defined by the link between the two.

Furthermore, it is worthwhile emphasising that the term landscape is a heuristic device. One should not fall into the trap of assuming that the actors involved with the environment in varying ways must have a similar concept of 'landscape' when actively engaged with the environment.

REFERENCES

Bender, B. (1993). Landscape – Meaning and action. In B. Bender (Ed.), *Landscape: Politics and Perspectives* (pp. 1–18). Oxford: Berg.

Bird-David, N. (1990). The giving environment: Another perspective in the economic system of hunter-gatherers. *Current Anthropology*, 31(2), 189–196.

Bird-David, N. (1992). Beyond 'The Original Affluent Society': A culturalist reformulation. *Current Anthropology*, 33(1), 25–47.

Bleek, D.F. (1927). The distribution of Bushman languages in Southern Africa. In *Festschrift Meinhof: Sprachwissenschaftliche und andere Studien*. Hamburg: Kommissionsverlag von L. Friedrichsen & Co.

Bollig, M. (2001). Kaokoland. Zur Konstruktion einer Kultur-Landschaft in einer globalen Debatte. In S. Eisenhöfer (Ed.), *Spuren des Regenbogens. Kunst und Leben im südlichen Afrika* (pp. 474–483). Stuttgart: Arnoldsche.

Bollig, M. & Heinemann, H. (2002). Nomadic savages, Ochre People and heroic herders – Visual presentations of the Himba of Namibia's Kaokoland. *Visual Anthropology*, 15(3), 267–312.

Carruthers, J. (1995). *The Kruger National Park: A Social and Political History*. Pietermaritzburg: University of Natal Press.

Casey, E. (2006). *Ortsbeschreibungen: Lanschaftsmalerei und Kartographie*. München: Wilhelm Fink Verlag.

Davis, S. & Davis, S. (1977). Etosha revisited. South West African Annual, 142–144.

Dieckmann, U. (2001). The Vast White Place': A history of the Etosha National Park in Namibia and the Hai‖om. *Nomadic Peoples*, 5(2), 125–153.

Dieckmann, U. (2003). The impact of nature conservation on the San: A case study of Etosha National Park. In T. Hohmann (Ed.), *The San and the State* (pp. 37–86). Cologne: Köppe.

Dieckmann, U. (2007). Hai||om in the Etosha Region: A History of Colonial Settlement, Ethnicity and Nature Conservation. Basel: Basler Afrika Bibliographien.

Dierks, K. (1999). Chronology of Namibian History. Windhoek: Namibian Scientific Society.

Endicott, K. (1979). Batek Negrito Religion: The World View and Rituals of a Hunting and Gathering People of Peninsular Malaysia. Oxford: Clarendon Press.

Fourie, L. (1959 [1931]). Subsistence and society among the Heikom Bushmen. In A.L. Kroeber & T.T. Waterman (Eds.), Source Book in Anthropology (pp. 211–221). New York: Harcourt, Brace & World.

Francois, H. Von (1895). Nama und Damara. Deutsch-Süd-West-Afrika. Magdeburg: Baesch.

Galton, F. (1889). Narrative of an Explorer in Tropical South Africa. London: Ward, Lock and Co.

Gordon, R. (1992). The Bushmen Myth: The Making of a Namibian Underclass. Boulder, CO: Westview Press.

Gordon, R. (1998). Backdrops and Bushmen: An expeditious comment. In W. Hartmann, (Eds.), The Colonising Camera (pp. 111–117). Namibia: Out of Africa.

Gupta, A. & Ferguson, J. (1992). Beyond 'Culture': Space, identity, and the politics of difference. Cultural Anthropology, 7, 6–23.

Hahn, H. & Rath, J. (1859). Reisen der Heren Hahn und Rath im Südwestlichen Afrika, Mai bis September 1857. Mittheilungen aus Justus Perthers' geographischer Anstalt über wichtige neue Erforschungen auf dem Gesamtgebiete der Geographie von Dr. A. Petermann, 295-303.

Harris, P. (1997). Under alpine eyes: Constructing landscape and society in late pre-colonial South-East Africa. In U. Luig & A. von Oppen (Eds.), The Making of African Landscapes (pp. 171–191). Frankfurt: Frobenius Institut.

Hayes, P. (1998). The 'Famine of the Dams': Gender, labour & politics in colonial Ovamboland 1929–1930. In P. Hayes, (Eds.), Namibia Under South African Rule: Mobility & Containment, 1915–1946 (pp. 117–146). Oxford: James Currey.

Heck, L. (1955). Großwild in Etoshaland. Berlin: Ullstein.

Heck, L. (1956). Etoshaland. S.W.A. Annual, 75-85.

Hirsch, E. (1995). Landscape: Between place and space. In E. Hirsch & M. O'Hanlon (Eds.), The Anthropology of Landscape: Perspectives on Place and Space (pp. 1–30). Oxford: Oxford University Press.

Ingold, T. (2000). The Perception of the Environment. London: Routledge.

Landau, P.S. (1998). Hunting with gun and camera: A commentary. In P. Hayes, J. Silvester & W. Hartmann (Eds.), The Colonising Camera: Photographs in the Making of Namibian History (pp. 151–155). Cape Town: University of Cape Town Press.

Luig, U. (1999). Naturschutz im Widerstreit der Interessen im südlichen Afrika. In G. Meyer & A. Timm (Eds.), Naturräume in der Dritten Welt (pp. 11–35). Mainz: Hausdruckerei der Universität Mainz.

Luig, U. & von Oppen, A. (1997). Landscape in Africa: Process and vision. In U. Luig & A. von Oppen (Eds.), The Making of African Landscapes (pp. 7–45). Frankfurt: Frobenius Institut.

McKiernan, G. (1954). The Narrative and Journal of Gerald McKiernan in South West Africa, 1874–1879. Cape Town: The Van Riebeeck Society.

Mendelsohn, J. (2000). A Profile of North-Central Namibia. Windhoek: Gamsberg Macmillan.

Miescher, G. & Rezzo, L. (2000). Popular pictorial constructions of Kaoko in the twentieth century. In G. Miescher & D. Henrichsen (Eds.), New Notes on Kaoko (pp. 10–47). Basel: Baseler Afrika Bibliographien.

Mouton, C.J. (1995). Togte in Trekkersland. Windhoek: Capital Press.

Neumann, R.P. (1998). Imposing Wilderness: Struggles over Livelihood and Nature Conservation in Africa. Berkeley: University of California Press.

Ranger, T. (1989). Whose heritage? The case of the Matobo National Park. Journal of Southern African Studies, 15(2), 217–249.

Ranger, T. (1997). Making Zimbabwean landscapes: Painters, projectors and priests. In U. Luig & A. von Oppen (Eds.), The Making of African Landscapes (pp. 59–73). Frankfurt: Frobenius Institut.

Schapera, L. (1930). The Khoisan People of South Africa. Bushmen and Hottentots. London: Routledge & Kegan Paul.

Schinz, H. (1891). Deutsch-Südwest-Afrika. Leipzig: Schulzesche Verlagsbuchhandlung.

Silvester, J., Wallace, M., & Hayes, P. (1998). 'Trees never meet': Mobility and containment: An overview 1915–1946. In P. Hayes, (Eds.), *Namibia under South African Rule: Mobility and Containment, 1915–1946* (pp. 3–48). Oxford: James Currey.

Von Zastrow, B.G. (1914). Über die Buschleute. *Zeitschrift für Ethnologie*, 46, 1–7.

Wagner-Robbertz, D. (1976). Schamanismus bei den Hain//om in Südwestafrika. *Anthropos*, 71, 533–554.

Widlok, T. (1999). *Living in Mangetti: 'Bushmen' Autonomy and Namibian Independence.* Oxford: Oxford University Press.

Widlok, T. (2003). The needy, the greedy and the state: Dividing Hai‖om land in the Oshikoto region. In T. Hohmann (Ed.), *The San and the State* (pp. 87–119). Köln: Köppe Verlag.

Files, National Archive Namibia

SWAA (South West African Administration)

A50/26: Bushmen Depredations Grootfontein (Part I)

A511/1: Game Reserves General

A511/10: Etosha Pan Game Reserve: Tourist Facilities, 1938–1951

A627/11/1: Native Affairs: Bushmen Reserve

Chapter 14

Is This a Drought or Is This a Drought and What Is Really Beautiful? Different Conceptualisations of the !Khuiseb Catchment (Central Namibia) and Their Consequences

Nina Gruntkowski

This case study focuses on the different concepts of landscapes of different communities making use of the !Khuiseb catchment area (KCA; central Namibia). They conceptualise their environment in different ways due to their different cultural backgrounds and daily experiences with it. It shows how the desert environment of the lower !Khuiseb River is viewed by indigenous Topnaar people and how the more productive highland of the upper !Khuiseb River is valued by farmers of European descent. As we are living in an increasingly globalised world concepts are also imported from foreign environments. These concepts have an impact upon the discourses on the 'new' environment. On the basis of the conceptualisation of 'drought' and 'aesthetics of landscape' by different people this chapter elaborates on how concepts can frame the present use of

M. Bollig, O. Bubenzer (eds.), *African Landscapes*,
doi: 10.1007/978-0-387-78682-7_14, © Springer Science+Business Media, LLC 2009

the environment and the spatial distribution of people within the !Khuiseb catchment area.

14.1. INTRODUCTION

This case study shows the interrelation of social processes and the environmental surrounding as described by Massey (1989):

> The fact that processes take place over space. The fact of distance or close-ness of geographical variation between areas, of the individual character and meaning of specific places and regions – all these are essential the operation of social processes themselves. . . . Nor do any of these processes operate in an environmentally characterless, neutral und undifferentiated world. (Massey in Jackson, 1989, p. 184)

So as not to evoke the old dichotomy of natural versus cultural landscapes the term *social landscapes* is used in the following. The element 'social' indicates that landscape, as an agent, interacts with the social world in order to combine the physical reality and the symbolic attribution through humans as Weichhart (1990) demands. In this case study environmental factors serve as a reference point for the comparison of the conceptualisation of their environment by differ-ent social groups. Therefore I decided to work simultaneously with geographical and ethnographic methods.

During the interviews with the Topnaar[1] and the farmers of European descent it was surprising that drought did not seem to preoccupy the Topnaar living in 12 settlements along the !Khuiseb River in the arid Namib Desert. In contrast to the desert the farms in the highlands receive rain almost annually and remain green for some months. Nevertheless the farmers complained a lot about the dry conditions in the !Khuiseb catchment area (KCA). What was also noticeable was that the farmers constantly stressed the aesthetics of the farm's landscape and the decision to live at one with nature whereas for the Topnaar landscape features seemed less important. Despite the fact that they all live so closely together in the KCA, and that they are all dependent on its water the topics of concern of these two groups differ greatly. Further investigation of the various social landscapes in the KCA gave an insight into different concepts on drought and landscape aesthetics. Soon it became obvious that not even the farmers and the Topnaar are homogenous groups as is set out in the following.

---

[1] The Topnaar speak Nama which belongs to the language family of the Khoisan (see Heine, (1976) for further information). Topnaar is the Dutch-Afrikaans translation of the Nama term = Aonin which Budack (1977) translated as 'people of a marginal area'. The term 'Topnaar' is used equally with = Aonin by the Topnaar themselves.

14.2. METHODS

Data from physical geography provide a framework of the diverse environmental conditions of the KCA. These geographical data serve as a background for the comparison of the different user-regimes in the KCA. Climatic data provide a point of reference for differentiated statements with regard to the environment. Such a reference point is useful for the examination of peoples' different ways of describing arid conditions and precipitation shortfalls. In fact drought is a climatic phenomenon, but for the farming person it becomes apparent in distorted vegetational cycles and detrimental changes to the vegetation. As the study area consists of various subregions with very different climatic conditions the different concepts of drought show the interaction between humans and landscape.

For the investigation of the conceptualisation by the different people living in the KCA I combined formal approaches with qualitative approaches. I started my fieldwork with an investigation of landscape conceptualisations in order to obtain an overview of the diverse situations along the !Khuiseb River. I visited 10 Topnaar settlements along the lower !Khuiseb River and 12 farms in the highlands of the KCA. The farms were chosen according to their diverse use: commercial cattle and sheep farms, weekend farms, tourist lodges, game farms, and holiday farms. The different informants were asked to list elements of their environment and rank them according to the importance for their own livelihood. Furthermore they were asked to build categories out of the listed elements which they perceived as belonging to each other. In the final stage these categories were ranked again according to the importance to their own livelihood.[2]

The analysis of these data gave me an initial idea of the very diverse topics which concern the different people in certain parts of the KCA. They decisively influenced my further qualitative investigation. The results of the farmers showed the inhomogeneity of this group. Subsequently I grouped the farmers into four categories. Largely these groups corresponded to the degree of dependence on the farm's economic output. The cognitive data also provided an insight into the topics which concern the Topnaar along the !Khuiseb. Visiting the informants a second time I presented the results and discussed them with the various informants during the further qualitative interviews.

In this second run of interviews I met 10 Topnaar, one respondent from each larger settlement. The Topnaar informants were of different sex and age. In contrast the 12 farmer informants were mainly men of all ages. In most cases the farmer's wife stayed in the background and just occasionally commented upon the discussion. Only in one case was a woman running the farm by herself. By means of guided qualitative interviews I consolidated the topics of interest which seemed to be of central importance for certain groups. The ethnological approach of participant observation supported the results.

---

[2] With the assistance of my translator Deon Sharuru in the Topnaar settlement I called together all present inhabitants while whereas at the farms I completed the investigation with the person running the farm (farm owner or tenant).

14.3. THE !KHUISEB CATCHMENT AREA–VARIOUS ENVIRONMENTAL CONDITIONS, VARIOUS PEOPLE

The !Khuiseb catchment area (15,500 km²) is situated in central Namibia.[3] Characteristic for the KCA are the very diverse environmental conditions within a relatively small area and the different types of people living adjacent to each other. The easterly part of the Khomas Hochland, with some mountains over 2000 m NN, is relatively productive due to the comparatively high precipitation and is mainly owned by commercial livestock farmers of European origin. The farmlands' productivity decreases from east to west corresponding to the general westward decrease of precipitation. Finally the still relatively productive but fissured, westerly farmland is divided from the desert region by the Great Escarpment which slopes down to approximately 1000 m NN. Adjoining the Great Escarpment is the Namib plain which inclines gently towards the coast. In this desert region of the middle and lower course of the !Khuiseb River the Gravel Namib lies on the northern side and the Dune Namib on the southern side of the !Khuiseb riverbed. The !Khuiseb River oasis incises the plain and prevents the dunes drifting northwards. The Topnaar live along the riverbed of the !Khuiseb although this area is officially a national park. Already in 1903 the German colonial administration established a *Wildschutzgebiet* in the Namib Desert. In 1975, the park was expanded under the South African Mandate and is today known as the Namib-Naukluft Game Park. The Ministry of Wildlife, Conservation and Tourism enacted regulations, some standing in direct conflict with Topnaar traditional practices such as hunting.

14.3.1. Hydrology and Landscape of the KCA

The !Khuiseb River and its tributaries are ephemeral rivers at best flooded for just a few days in the rainy season. The groundwater stored in the riverbeds of the KCA as in many other regions of Namibia is of special importance for the environment, because the rivers themselves only occasionally contain surface water (Jacobson et al., 1995). The distribution of precipitation, which is anyhow very variable, differs greatly within the catchment area. The upper part of the KCA receives up to 450 mm/a precipitation, whereas the lower westerly course near the coast gets less than 50 mm/a (Heyns et al., 1998). Owing to the geomorphological conditions the groundwater of the lower-lying desert region is recharged by the rain in the highlands.

All inhabitants of the KCA are highly dependent on water. Whereas the farmers in the upper parts, with significantly more rainfall, preserve the surface

---

[3] The KCA is also called the !Khuiseb-Gaub drainage system, because the Gaub is the biggest river, which drains into the !Khuiseb.

water of the riverbed through the construction of small dams, the Topnaar, in the desert region, do not in general directly use the surface water of the !Khuiseb River. The construction of dams constrains the Topnaar as they blame the dams constructed by the farmers upriver for the increasingly poor water situation of the !Khuiseb River. The Topnaar seldom mention long-lasting natural water pools; in the past they did not build dams themselves to store extra surface water for their own use. High evaporation rates in the KCA around 3000 mm per annum (Jacobson et al., 1995, p. 16), but especially in the hyperarid Namib Desert, bring the efficiency of dams with open water storage into question.

Farmers and the Topnaar extract groundwater through drilled boreholes. Also the waterworks supplying Walvis Bay and Swakopmund are situated in the !Khuiseb riverbed, extracting groundwater at Rooibank and Swartbank (see Figure 14.1). Water was already the crucial factor at the beginning of the colonisation of the Namibian coast (Moritz, 1999). 1923 the first waterworks

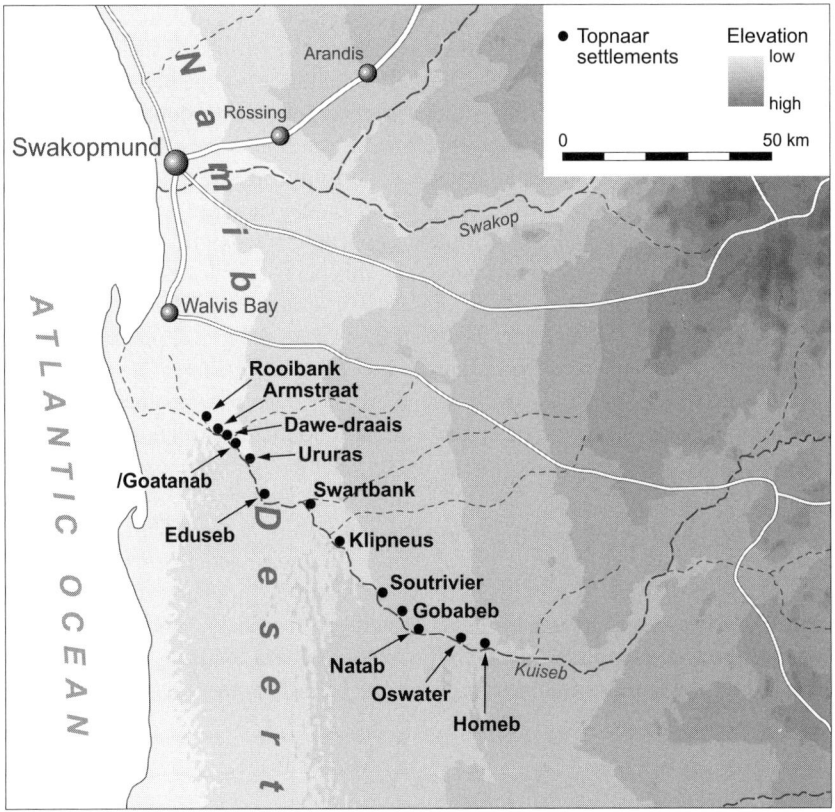

Figure 14.1. Topnaar Settlements (Desert Research Foundation of Namibia)

were installed at Rooibank in the !Khuiseb delta, which is still supplying Walvis Bay. In 1963 Swakopmund was connected to the !Khuiseb water scheme due to the increasingly salinated groundwater of the Swakop River. Consequently, an additional pumping station in the !Khuiseb was erected at Swartbank. With the opening of the Rössing Uranium Mine in 1976 the extraction of water from the !Khuiseb drainage system increased up to 4 million cubic meters per annum (Dausab et al., 1994). The Department of Water Affairs noticed a drastic drop of the groundwater level in the area of Swartbank. Measurements at the Swartbank aquifer from 1982 to 1987 revealed a decreasing ground water level of 2.5 m per annum (Department of Water affairs, 1987). The Topnaar living close by also reported that it is increasingly difficult to reach the groundwater and stated that a natural spring close to Rooibank has dried up.

14.3.2. The Topnaar at the Lower !Khuiseb River

The Topnaar, living in 12 settlements along the !Khuiseb riverbed, mainly farm with goats, which are dependent on the food and shelter of the riverine vegetation. All these settlements are to be found in the middle section of the lower !Khuiseb River. Mainly *Faidherbia albida* and *Nicotiana glauca* grow in the riverbed, which cuts into the gentle valley. After a flood ephemeral species such as *Tribulus zeyheri* germinate and grow. Due to the wetter conditions in the highlands vegetation increases upstream. Here the valley is steep and during floods in the upper part there is no space left between the flowing water and the valley slopes. Due to this fact no permanent settlements can be found in this part of the canyon section. The settlements Homeb and Oswater are located close to the canyon section. As the occurrence of floods increases with the proximity to the highlands and groundwater is recharged much quicker, this part of the lower !Khuiseb River is more productive than the sections closer to the coast. The Topnaar settlements are situated on the slope overlooking the valley or on the gravel plain so as to be out of reach of the seasonal floodwater of the !Khuiseb River.

Towards the coast the vegetation decreases due to a smaller number of floods. In the delta the riverbed spreads out and no distinct valley is observable. The !nara (*Acanthosicyos horridus*) is the dominant vegetation, especially in the southern arm of the !Khuiseb delta. Very little other vegetation occurs in the delta section: a few small trees (mainly *Tamarix usneoides*) and high grasses (mainly *Stipagrostis sabulicola*).[4] In the delta itself there are no settlements as the delta provides no refuge when it is flooded. As the !nara is only harvested seasonally from the end of December until May (Botelle & Kowalski, 1995, p. 40) some Topnaar from all along the !Khuiseb move to the !nara fields and stay there for the harvest in temporary camps or organise the harvest from Armstraat, the settlement closest to the delta.

---

[4] For more information on the !Khuiseb vegetation see Van den Eynden et al. (1992).

The exploitation of the !nara has a long tradition amongst the Topnaar. This also leads to the name !naranin (people of the !nara) often used by themselves to emphasise their relationship with this plant[5]. !Nara plants are owned by families. Still today every Topnaar along the !Khuiseb knows which plant belongs to which family, even if today some harvesters have agreements with other families to pick fruits from their bushes. The family property of the !nara plants is a special characteristic of the Topnaar economy. The object of property relations, however, remains unclear: Widlok points out that it could be the 'land' or 'its resources' the '!nara bushes' or their 'fruit'? (Widlok 2000: Internet Discussion Paper). Most of the !nara plots are situated outside the Namib-Naukluft Game Park and therefore the Topnaars claim to harvest is not in conflict with the strict rules of the national park. Within the park the Topnaar have no legal rights to resources or the land itself. On the contrary many traditional methods, for example, hunting wild animals or the production of charcoal have been forbidden. Furthermore they were forced to stop moving along the !Khuiseb River and settling in some places. As with all game parks in Namibia 'all land and water resources are legally owned by the state. The Topnaar are permitted to live on and use their land in the consent and grace of the Government of Namibia.' (Botelle & Kowalski, 1995, p. 2). With the recent focus on land rights in independent Namibia and the Topnaar claims for land use rights the !nara could gain more political significance.

14.3.3. The Farmers in the KCA

In contrast to the long presence of Khoekhoe-speaking people along the !Khuiseb, settlers of European descent only began to settle and farm in the central part of Namibia during German colonial times (1884–1915). Due to the advantageous climatical conditions the highland around Windhoek became the center of a settler colony. In 1915 South Africa conquered Namibia and finally received the mandate from the United Nations in 1920. Consequently Namibian farmland was increasingly given to white South Africans. Between 1915 and 1920 6,000,000 ha of farmland were distributed among South Africans (Leser, 1982). Today many farms are owned by Afrikaans-speakers originating from South Africa; some have been farming in the KCA for decades and others have just bought their farms recently. Nevertheless still a great number of farms are owned by Germans; some have been the property of the family since German colonial times, whereas others were bought later on. Even today Germans as well as other Europeans buy farms in the KCA, some of which are run by farm administrators.

---

[5] The exploitation of the !nara through Khoekhoe has been recorded since 1677 when Captain Wobma landed on the Namibian coast. Pastoralists made use of the resources of the Namib Desert for a long time. The oldest archaeological site in the Namib Desert, the Mirabib shelter, was dated back 8,400 years by Sandelowsky (1977).

14.4. DIFFERENT ATTITUDES OF THE FARMERS AND THE TOPNAAR AND THE SIGNIFICANCE OF LANDSCAPE

A basic problem for all people living within the KCA is the dependence on the water of the !Khuiseb River which is this area's lifeline. As most of the inhabitants of the KCA are livestock farmers, their attitudes towards landscapes are formed by the experiences gained through livestock farming. Sehoueto (1995) describes the process of appropriation of nature in peasant societies as the result of peasant decisions, which are interpreted as a permanent experiment with nature. This process is restricted by natural and also by socioeconomic facts. In the following I show how different cultural backgrounds and the related concepts of landscape can influence the appropriation of nature.

14.4.1. The Farmers of European Descent

The first investigation into farmers' conceptualisations of landscape led to a typology of four distinct groups of farmers of European descent. This grouping was confirmed by the farmers themselves and further qualitative investigations. The greatest difference amongst the groups is the degree of economic dependence on the farm.

Farmer Group I operate the farm commercially and therefore are highly dependent on the income gained through farming with cattle and in some cases sheep. These farms owned by German- and Afrikaans-speakers have been, in some cases, family property for decades. Without exception all these farmers mentioned a life close to what they conceptualise as nature as an ideal. Thereby the typical European concept of the dichotomy of city and nature plays a central role. Many said that they are attracted by the aesthetics of the landscape surrounding their farm: they often mentioned specific panoramic views from the farmhouse or from viewpoints on their farm. They described the extraordinary panoramic views after rain or at sunrise/sundown in great detail. Additionally the Afrikaans-speakers pointed out that it is a pleasure to work with animals.

Also highly dependent on the farm income is Farmer Group II. Due to environmental or economical changes they struggle for survival. They cannot make their living solely out of the farming business. Therefore they search for additional income through tourism, sale of licences for trophy hunting, sale of milk products, and so on. The farmers of both groups pay a lot of attention to their precarious economic base. Highly variable climatic conditions and economic developments such as the decline of the prices for Karakul sheep fur due to increasing environmental awareness in the 1980s worsened the situation for many farmers. Many Group I and II farmers stressed that the personal affection for their farm and the surrounding nature is a major reason for them to continue farming even when it is no longer profitable. As they love the landscape and the life on the farm, they prefer to search for alternative ways to earn additional income rather than to give up their farm.

Contrasting to these financially dependent farmers, the Farmer Groups III and IV are less independent from the economic output of the farming business. Many of the farms of Farmer Group III have just recently been bought by foreigners, mainly Germans. Animal husbandry is not essential for these farmers whereas the interest in game is distinct. Some of these farmers have built up a guest farm aiming to run the farm as self-financing. The main reason to live on a farm or to own it[6] is the desire for silence and relaxation surrounded by beautiful landscape as a contrast to an urban or European lifestyle. Some of these farmers are dropouts, who gave up their former life to live on their own plot of land; others make enough money in Europe to realise their 'African' dream. For Farmer Group III landscape is extremely important on the aesthetical level as it symbolises the 'other' life in 'wild' nature. The productivity of the land is almost irrelevant.

Farmers of the Farmer Group IV are German- and Afrikaans-speaking weekend farmers. These are mainly businessmen from Walvis Bay or Windhoek who spend their weekends on the farm. In contrast to Farmer Group III they are more interested in the livestock farming business, which is for them a welcome contrast to work in the town. At best the farm makes some extra income or is self-financing. Even if they do not have to live from the land they do not see their livestock farming as purely hobby. Therefore the productivity of the land is just as important as the attractive landscape. Fleeing from their lives in the city Farmer Group IV is in search of a life close to nature at least for the weekends.

14.4.2. The Topnaar

The Topnaar have paid much more attention to natural resources and especially their own history of resource use, whereas for the white farmers the landscape has been historic, either rendering a wilderness or an aesthetically appealing landscape beyond history. The Topnaar often refer to distinct important elements of their surroundings, which besides the use as a natural resource often have nostalgic relevance for the Topnaar. Trees are a good example: The seeds of the acacia trees are essential for goat farming along the lower !Khuiseb. At the same time trees afford shade in this arid environment and it was often said that this same tree had already given shade to the Topnaar forefathers. Therefore a specific tree also has nostalgic value for many Topnaar although it seems less important on an aesthetical level.

It seems that the Topnaars' conceptualisation of their surrounding is much more fine-grained, whereas the farmers' European concept of wilderness seems more coarse-grained. Often it is the little details of their surrounding environment the Topnaar perceive as important inasmuch as they do not refer to striking elements of the !Khuiseb landscape such as the Swartbankberg, a marble mountain close to the settlement Swartbank. It was neither listed in the initial ranking nor

---

[6] Some of the farmers do not live the whole year on the farm but plan to retire there.

mentioned in the qualitative interviews. Only the dunes were occasionally mentioned. The dunes were assigned to the pile with the headline tourism. This indicates that landscape features evaluated highly by Westerners are associated with tourism by the Topnaar. This could be a result of the numerous Namibian tourist brochures which all try to attract foreign tourists with spectacular photographs of the dune landscapes. Recently one campsite was established close to the Topnaar settlement Homeb. This campsite is part of the Namib-Naukluft Game Park run by the Ministry of Environment and Tourism. The Topnaar just gain a small income through the sale of firewood to visitors. Only occasionally do tourists stop at other settlements, but for the future the Topnaar community hopes to build more tourist facilities and to attract more visitors.

Although the Topnaar are tolerated in the national park, they are neither allowed to build their own campsites nor to hunt wild animals as they did in the past. This is often lamented to accentuate the nonexistence of land use rights. Up until today the Topnaar use some wild plants of the !Khuiseb riverbed and annually harvest the !nara fruits. As the !nara fields are situated in the !Khuiseb delta outside the Namib-Naukluft Game Park the harvest is not in conflict with the national park regulations. The dried pips are sold in Walvis Bay. There are few sources of income besides subsistence farming, mainly with goats, and the occasional work of some family members in the harbour town, Walvis Bay. In contrast to the farmers the Topnaar seem to be a more homogeneous group with regard to their views on the environment, which could be attributed to the same cultural background and economic situation. The Topnaar grew up in the same environment, whereas many of the farmers came from South Africa or even overseas.

14.5. CONCEPTUALISATION

Conceptualisation helps to make sense of daily experiences. Two different ways of conceptualising are represented in the following concepts: 'drought' is derived mostly from the bodily experienced environment whereas 'aesthetics of landscape' is more the result of imagination (Lakoff, 1990). The farmer's views of local landscape refer principally to more or less European cultural backgrounds. Existing concepts historically formed in relation to a different environment possibly influence the way these farmers see their 'new' environment and further their process of adaptation.[7]

---

[7] Exemplarily see Harries (1997) on the construction of a south-east African landscape 'under alpine eyes'.

14.6.   WHAT IS A DROUGHT?

The geographic definition of drought concentrates mainly on climatic data. Mensching (1990) defines drought as a time of water shortage due to below-average precipitation during subsequent years. The phenomenon of drought is characteristic for semiarid regions as they are characterised by a high variability in precipitation. (Mensching, 1990, p. 3). However, the perceived drought by humans, especially farmers, is very much related to the land and its productivity and not so much by the availability of water. Therefore the conceptualisation of drought is based on daily farming experiences. However, farmers of European descent and Topnaar farmers responded very differently to the question of what they consider to be characteristic for a drought.

The farmers already predict a drought at the end of the rainy season, which in the KCA lasts from December until April, in some years until May. The prognosis is dependent on the precipitation and how the farm is run. For example, close to the Great Escarpment and the desert region one cattle farmer mentioned 90 mm/a as the minimum for a 'normal' year, whereas the neighbor running a guest farm considers 40 mm/a as sufficient. In the region of the Khomas Hochland (Figure 14.2) with 350 mm/a. average rainfall, however, of 150 mm/a is a cause for concern. More than the water itself, the pasture is of paramount importance for the farming business. Therefore, beyond the amount of precipitation, the temporal distribution of rainfall is decisive for the growth of the grasses. Grasses having sprouted as a result of one heavy rainfall followed by a long dry period will soon stop their growth and dry out. Smaller rainfalls, but well distributed over time, are more useful for the growth of the grasses than one heavy rainfall.

Generally the farmers of all groups complain a lot about the dry conditions of the KCA. For the farmers of Group I and II drought is a threat to their livelihood, but all groups react with dismay when the grass and most of the farms' vegetation vanishes. Even farmers who just run a guest farm without livestock complain a lot when it does not rain sufficiently. It seems that people of European descent can hardly bear to see the land in a dry state. Some experience the dry condition of the farmland as personally depressing; furthermore they describe the land as turning black. In contrast they give enthusiastic detailed descriptions of the beautiful landscape after a good rainfall. In a short time the undulating landscape of the Khomas Hochland and in some years even the region close to the Great Escarpment transform into a green landscape, which some farmers described as resembling European pastures. In these times many farmers take pictures, which they keep in their houses to remember the 'good times'. This memory of the ideal state of the landscape seems to be essential for the enthusiasm with which many continue to farm despite all difficulties. During the rainy season every look to the sky, every observation of the clouds is a hope for rain which will possibly transform the dry and dusty land into green pasture. Many farmers compensate their need for a green surrounding with planted vegetation in the proximity of the farmhouse. Often these plants are not suitable for the arid

Figure 14.2. Khomas Hochland after rain (Photo: N. Gruntkowski) (*See also Color Plates*)

conditions and therefore have to be watered constantly. Green farmstead gardens with palms and flowering bushes at some farms significantly contribute to the landscape and represent neo-European markers in an African landscape.

All farmers constantly measure the rainfall on their farms. The standards set for the productivity of the KCA by the farmers of all Farmer Groups seem not to be completely adapted to the highly variable conditions of this semiarid to arid environment. One unpredictable characteristic of Namibian rainfall is its patchiness. Even in a good rain year the precipitation might not reach all farms, as a consequence these remain without sufficient pasture whereas the neighboring farm is green. The other unpredictable characteristic is the great variation in the amount of rainfall itself. With the normal variation in rainfall the run-off volume of the !Khuiseb River varies greatly. For example at the Schlesien Weir 105.0 mm^3 water passed in 1962–1963 compared to only 0.0065 mm^3 in 1963–1964 (Dausab et al., 1994). At the beginning of the 1980s three years of rain, much below the annual average, were measured at several farms in the KCA. (Department of Water Affairs, 1987) The successive dry years got many farmers into trouble. Due to the decline of the prices for Karakul sheep fur many of the farmers in the less-productive region close to the Great Escarpment and the Namib Desert stopped their business with sheep and began to farm with cattle, which need more pasture and water. These cattle farmers described that consequently the dependence on rain increased and it became more difficult to withstand dry periods.

Despite these difficulties the majority of the farmers seem to remain optimistic. They retain in their memories the 'good times' such as in 1995 when heavy rains fell from February until March in the northwestern catchments of Namibia and the !Khuiseb flowed into the sea for the first time since 1962–1963. Even at Gobabeb in the hyperarid Namib Desert 55 mm rain was measured whereas on average Gobabeb receives less than 25 mm per annum (Jacobson et al., 1995, pp. 118–119). Again and again farmers speak of these good rainfalls. In dry years their hope for better times motivates them to continue with the farming business. This influences the conceptualisation of droughts, because droughts are only noticeable in contrast to wet fertile conditions. Sometimes it seems that the ideas of the land's fertility are still influenced by concepts of European origin. As environmental conditions in Europe are much more stable for the farmers the great variation in the KCA seems difficult to accept. Perhaps the farmers do not set the fertile periods of the KCA as standard but they definitely play a central role in the conceptualisation of the environment. Consequently the idea of the land's fertility and the way of running a farm is sometimes inappropriate.

Despite all these difficulties the farmers remain on their farms. Even after some dry years economic-dependent farmers of Farmer Group I and II continue with their farming business. As the farmers of Farmer Group III and IV they are attracted by the aesthetics of the farm's landscape. However, it is not just the nice panoramic view over the land that attracts the farmers; moreover landscape has a high symbolical significance as it represents the life at one with nature. Some farmers go as far as to mirror the state of the landscape with their own moods. Even when they feel depressed and have to struggle with the harsh conditions in the very dry periods they still remember the 'good' times. Like gold-diggers the farmers are generally optimistic that next time the clouds will bring more rain and transform the farm into green pasture.

Figure 14.3. Namib Desert at the lower !Khuiseb (Photo: N. Gruntkowski) (*See also Color Plates*)

Although the farmers talk a lot about droughts the Topnaar just report an increasing dryness of the region. Most of the Topnaar explain that they do not experience droughts because the trees in the riverbed remain green over the whole year. They seldom report a lack of water, which is generally rare in the Namib Desert throughout the year anyhow (Figure 14.3). Surprisingly some explain droughts in the following terms.

> [B]ut the most threatening time is when the river floods. Than we really understand what drought is, because in other parts when the river floods, the people are happy, because they know we've got water. In our situation we are happy, because the underground water table is recharged, but at the same time the fodder for our livestock is washed away. So we have to find alternative ways to keep our animals through that flood time. (Yesaya Animab)

This informant knew very well how the farmers defined a drought and referred to it, but at the same time confronted the dominant farmer conceptualisation with the different experiences of the Topnaar. Also most other informants explained drought as an isolation from the resources in the riverbed during a flood of the !Khuiseb River. The concept of drought as a lack of water is not generally shared by the Topnaar. The contact with scientists and the Gobabeb research station[8] close to the Topnaar settlements has brought the Topnaar in contact with a scientific concept of drought which is essential for research on arid environments. However the Topnaar still view their environment differently.

At the lower !Khuiseb the Topnaar generally do not experience a direct connection between the rain and the growth of vegetation. Due to the fact that water reaches the lower !Khuiseb as an invisible subsurface flow, it is only possible to observe a remarkable change in the plant formation in the years when the riverbed is flooded for a few days. Suddenly bushes and trees, flowers, and some grasses grow. The rest of the time the vegetation remains more or less stable. As there is almost no rain at the lower !Khuiseb the bushes and trees are more dependent on the groundwater than on floodwater or even direct precipitation. This contrasts with the situation in the upper KCA where the farmers observe the state of the annual and perennial grasses which directly respond to rainfall. When the Topnaar observe the weather in the highland this is mainly to predict floods so that they can start preparations for the flood time such as the collection of seeds from the acacia trees for livestock and the storage of water. As it takes 70 years for the water to reach the sea via underground flow, there is no direct link between rains in the highlands and groundwater recharge observable at the lower !Khuiseb River (Dausab et al., 1994). An immediate improvement of water availability only occurs when a flood reaches this region, which in many years does not happen at all (see Table 14.1). But the floodwater itself is not suitable

---

[8] This research station has been situated at the !Khuiseb since the 1970s and is run by the non-governmental Desert Research Foundation of Namibia (DRFN) with its head office in the capital Windhoek.

Table 14.1. Flooding of the !Khuiseb River at Gobabeb, indicating the number of days of flooding and the highest flood for cases where the river rose overbank (i.e., flowed outside of the central river channel). Data were obtained from and used with permission of the Gobabeb Training & Research Centre (P.O. Box 20232, Windhoek, Namibia)

Year	Flood period (days)	Overbank flooding (m)	Year	Flood period (days)	Overbank flooding (m)
1963	68	*	1982	0	0
1964	–	?	1983	0	0
1965	26	?	1984	0	0
1966	18	?	1985	0	0
1967	22	?	1986	18	3.1
1968	11	?	1987	9	0
1969	18	?	1988	20	0
1970	1	?	1989	9	3.5
1971	34	?	1990	19	2.8
1972	43	*	1991	6	0
1973	15	*	1992	4	0
1974	102	3.0	1993	23	2.9
1975	10	0	1994	6	2.8
1976	61	0	1995	17	0
1977	8	2.5	1996	3	0
1978	7	0	1997	33	4.0
1979	8	0	1998	4	0
1980	0	0	1999	7	0
1981	0	0	2000	23	5.4

*flooding overbank, height not measured, ? no measurement

for human consumption because the water is loaded with sediments, especially tiny silt sediments from the riverbed. The situation improves when the water has seeped away into the riverbed. Then the water is close to the ground and easily accessible for the Topnaar.

Despite all the problems caused by a flood the Topnaar hope for one every year as it improves their water supply. As they are used to the variable climatic conditions in the KCA they do not expect a flood every year. Nevertheless the Topnaar are not content with the harsh situation in the desert and the permanent shortage of water. Some complain that the farmers upriver, as well as the people downriver in the cities on the coast, receive more water through dams on the farms and the water scheme supplying Walvis Bay and Swakopmund.

With the establishment of the national park at the beginning of the twentieth century the former seminomadic Topnaar were forced to settle along the !Khuiseb by the conservation laws. Owing to dry periods in the second part of the twentieth century boreholes with windmills or diesel pumps were installed in the Topnaar settlements along the !Khuiseb River by the government despite resistance from conservationists. Today the permanent settlements are well established, but still the Topnaar accentuate their former nomadic lifestyle. Until today flexibility is observable through their methods of coping with the sometimes difficult environmental conditions. For example, when the groundwater level drops too much and the water extraction is no longer possible the settlements' inhabitants go and fetch water from a neighboring settlement by donkey cart. If this is also impossible then they have even fetched water at the research station where a big water tower is situated. Today the Topnaar do not move towards the water, but carry water to their homesteads over long distances.

On closer examination the concepts of drought by the Topnaar and the farmers are not as divergent as they seem to be at first sight. The reduced access to resources as described by the Topnaar causes a shortage of resources comparable with the decreasing availability of resources mentioned by the farmers. In the case of the Topnaar there is too much water in the riverbed whereas in the case of the farmers there is too little water in the form of rain; in both cases a temporary shortage of natural resources is the result. Although for the farmers drought is a permanent threat, the Topnaar do not make the dry conditions a subject of discussion but consider them as given.

14.7. WHAT IS A BEAUTIFUL LANDSCAPE?

In contrast to the European aesthetic standards[9] of beauty the great expanse of the South African subcontinent was often described by Europeans as desolate and not very picturesque. Gilpin describes the ideal picturesque landscape as a composition of mountains in the distance, a lake in the middle, and a rough foreground in contrast to the smoothness of the middle and far ground (Gilpin in Coetzee, 1988). Many areas of the KCA correspond to these European criteria for a 'picturesque' landscape, apart from the fact that surface water barely exists. But at some farms the dams are situated close to the farmhouse. From there the artificially dammed water, which can last up to the whole year, is visible and gives the impression of a fertile landscape. Another thing often missed by Europeans in Africa is green vegetation as Coetzee refers to the writer Burchell who 'is continually on the lookout for green' (Coetzee, 1988, p. 42). At least the highlands of the KCA are green for part of the year. Therefore the upper part of the !Khuiseb River, especially the Khomas Hochland, distinguishes itself from the surrounding drier parts and the vast expanses of the southern African plateau.

Many farmers, mainly of German descent, are emotionally affected by the KCA landscape. Especially for the economically independent farmers of Farmers Group III and IV the landscape is often decisive for the purchase of the farm. 'When somebody buys a farm . . . you do not go anywhere and say that you want to have a piece of land to farm. If you go to a place, you will have a look at the area, and if you like it you will buy it. . . . For me landscape plays a central role' (Huber).

---

[9] On colonial aesthetics of landscape see Rössler, this volume.

Most of these farmers of Farmer Group III and IV bought their farms after Namibian independence in 1990. As weekend farms they are an opposite pole to the life in the city or as holiday farms to the life in Europe. Some farms are retirement homes or planned as such. Generally the farmers accentuate that the life on a farm is marked by more freedom. This is not just mentioned in view of the few regulations a farm owner has to obey, but in view of the spaciousness of the farm itself. Also many farmers value the wonderful views over the landscape which they often equate with freedom. Many farmers of Group III do not farm with domesticated animals but they often take care of the game by building and running water points for them. As German-speakers put their main emphasis on landscape they view the game as part of the 'natural' landscape. 'The game is part of the new life I'm beginning here. The main reason for this is the joy of life in this beautiful landscape . . . with these animals' (Hannelore Neuffer).

But also economically dependent farmers of Group I and II often accentuate the personal importance of the landscape. Often Farmers Group II run the farm without any profit but nevertheless continue because they like this way of life. In some cases farms have been family property for many decades, up to a century, and therefore the sentimental value increases. 'In former times this kitchen was painted by my mother. I think my mother has copied the colours from the sky, because she liked a certain pink, which nobody managed to mix properly, and also this turquoise-blue. . . . We can have many troubles, but then we watch the sunrise and the sundown. I think I would always miss this if I had to leave this country . . . (Barbara Ahlert).

In this case the identification with the landscape where the informant grew up is very obvious and generally common among the German-speakers. Also Germans who bought their farm during the last few years describe a feeling of home with reference to the aesthetics of a landscape. Often places with a panoramic view over the farm are mentioned; here the typical Namibian 'sundowner' is celebrated by Namibian circles of European descent.

The conceptualisation of an aesthetically appealing landscape has a long tradition in Europe (see also Rössler, this volume). Uninhabited landscape was seen as 'wilderness' in contrast to 'civilisation' and has been valued since romanticism. During colonial times 'wild'[10] landscape was increasingly perceived worth preserving and national parks were established such as in 1907 in the Namib Desert.

The Topnaar seem to have a certain distance to this concept as they have to live with the restriction of the conservation laws. 'We were living in this wild area even before any European came to this area. Normally this area was a wild park. It was wild! Nobody was here, but the first people who came here were the Topnaar. They were brave, because they were not afraid of the lions, they were not afraid of the elephants and what animals were here' (Chief Seth Kooitjie).

---

[10] Luig (1999) shows in a case study of the Matobo national park in Zimbabwe how an African cultural landscape is perceived as 'wild' by Europeans.

In the statement above it becomes obvious that the Topnaar do not share the western concept of wilderness and the demand to preserve it in a national park.[11] More often they discuss their dependence on the resources as well as the difficult economical situation. But here we have to keep in mind that in the interview situation the Topnaar spoke to a Western outsider whom they consider to share the concept of wilderness and conservation. Therefore they probably stressed more the detriment and the underlying demand for land rights as they would do among themselves. As it is prohibited for the Topnaar to hunt game in the national park they just refer to game in their complaints about the strict conservation laws. Vegetation is of paramount importance for the survival at the lower !Khuiseb and trees are often mentioned as very important.

> If trees are dying it hurts me, not because it is a natural resource, but part of me dies with the tree. Why? Because I know how important the tree is. . . . I can feed my animals and things like that, I can sit under the tree. The tree is important, because this tree has grown for many, many years . . . and I link it to my forefathers . . . they were using the tree for shade probably, and for other reasons, too. I've got a very sentimental value towards the !Khuiseb and the environmental resources . . . not only because of the natural resources concept that has been sort of promoted so far, it sound like something that people are bringing to educate us, people living in nature. . . . (Chief Seth Kooitjie)

In view of the desert environment the value of a tree is very obvious. This informant pointed out that it is not just their function as a natural resource, but also a personal relationship, which links the Topnaar with the trees. The Topnaar never talk about their environment in abstract terms as do the farmers who accentuate the aesthetics of the landscape or their life close to nature. But the Topnaar speak of distinct elements of their surroundings which often have a nostalgic or historical relevance. The memory of the ancestors symbolises Topnaar unity with the land they inhabit. This symbolic value has increased recently in the representations to outsiders as a result of the land right discussions in Namibia.[12]

Afrikaans-speakers, like the German-speakers, are attracted by the beauty of the landscape and emphasise the life close to nature. They link this way of life with farming and the work with domestic animals. 'It's an inner feeling, a love you are born with. The attraction towards nature or animals, which some have and other people don't have . . . it's actually something that I can't explain. I think you are born with it and you are attracted with that and at the end of the day you become a farmer . . . to live in nature (Korrei Mensa).

This attitude is common among economically dependent and independent farmers of South African descent. Some have quit their jobs to become farmers and fulfil their dreams; others grew up on a farm (in Namibia or South Africa).

---

[11] On the perception of 'wilderness' by indigenous people see also the case study on the Hai‖// om in the Etosha National Park by Dieckmann, this volume.

[12] See also Widlok (1998).

The work with livestock seems to be of personal importance and is part of what is seen by them as 'natural' life. Only the farming business transforms the land into a farm, therefore the work is part of this life close to nature. In Afrikaans South African literature writing about the farm of one's childhood has a long tradition. 'The farm, rather than nature, however regionally defined, is conceived as the sacral place where the soul can expand in freedom' (Coetzee, 1988, p. 175). In consequence the farm is 'a still point which mediates between the wilderness of lawless nature and the wilderness of the new cities' (Coetzee, 1988, p. 4). This contrasts with German settlers' search for untouched land. Until today the mainly economically independent farmers from Germany lay stress on the 'wild' landscape and game is seen as part of this 'wilderness'. Afrikaans-speakers speak about the game also as an integral part of nature, but view the work with the domestic animals as equally important for their life as a farmer.

14.8. WHO SETTLES WHERE AND WHY?

There is one decisive difference between the farmers of European descent and the Topnaar: the farmers have chosen a life on a farm, whereas the Topnaar do not have many alternative options to choose from due to their general bad economic situation; therefore they have to arrange themselves with the situation at the lower !Khuiseb. Although the spatial distribution of the farms in the upper KCA reflects personal, sometimes economic, and often aesthetic preferences of farmers for certain regions, the settlement of the lower !Khuiseb follows other rules as shown.

Almost all farmers are attracted by the specific landscape of their farm. Many even came from overseas in the search of 'untouched' nature to buy a farm and live there. Consequently they bought a farm which personally pleased them. But also farmers who grew up on a farm spoke about their decision to stay on the farm or to buy another farm. Also the way of life appeals to the farmers and the independence and freedom was mentioned by all farmers. Farm owners in Namibia only have a few restrictions on how to use the farm. Consequently they feel free from the many limitations accompanying a life in a town or even as a farmer in Europe.

This contrasts with the situation of the Topnaar, which is very much restricted by the regulations of the national park. As they were forced to settle in some permanent settlements along the lower !Khuiseb today they can only decide to remain in a settlement or to try to get labor in the coastal towns, Walvis Bay or Swakopmund. As most of the Topnaar lack a good education this greatly reduces the possibilities of making a living somewhere else. Young Topnaar men are constantly on the move between the settlements of the lower !Khuiseb River and these towns. They go to town in search of some temporal labor, but often have to return to the lower !Khuiseb due to lack of labor. They stated that it is always possible to return to the settlements and to make a living with subsistence farming but generally they would prefer life in the town. For the elder people this situation is different. Most elder men and women do not consider a life in town as an option. They are used to the life in the settlements along the lower !Khuiseb River. Even if they wish for an improvement in the infrastructure they are not willing to move to town. But Topnaar of all ages view the park restrictions as an obstacle to improving their life along the lower !Khuiseb River. Consequently the legal situation and the claim for land-use rights are of central importance for the Topnaar.

The farmers have different motives to live on a farm and are more or less in a position to choose where to buy a farm. Therefore an accumulation of farms owned by farmers of a specific Farmer Group is observable in certain parts of the KCA. Farmer Group I farms mainly in the fertile Khomas Hochland which has no spectacular landscape features as it is characterised by a soft undulating countryside. This area is easier to farm and receives higher precipitation, which supports commercial farming. Especially in this productive region the farms have often been family property for a long time and farms are seldom sold to outsiders. Exceptions for Farmer Group I are people who formerly had another business and decided to fulfil a dream. 'I have decided to quit the stress in the city and business, the official business in town, to go out on a farm, which I love very much . . . animals and nature' (Korrei Mensa).

Also businessmen from Europe start running a farm in Namibia with capital from overseas. Mostly they chose a farm in a region which attracted them personally. Even if they intend to run the farm commercially aesthetical aspects are of central importance for the decision of where to buy a farm. Therefore some farms of Farmer Group I are situated in less productive regions, such as close to the Great Escarpment where rainfall varies a lot and makes farming more risky.

Members of Farmer Group II are distributed all over the KCA and no spatial order is discernible as the group is mainly characterised by their economic situation. For various reasons these farmers started farming and got into economic troubles. Some of these farms have been family property for a long time. These farmers are also attracted by the landscape, often due to sentimental reasons as they have gathered a lot of experiences there or even grew up on the farm.

Areas with a spectacular landscape such as the deeply fissured 'Gramadulla' rocks (Rust & Wieneke, 1974; Martin, 1996) close to the Great Escarpment are not suitable for commercial farming with livestock. Mainly farmers of Farmer Group III settle there and the area around the Gamsberg, a dominant table mountain, especially attracts Germans. Some refer to the books of Henno Martin (1996) and Herman Korn (2001), two geologists who lived in the !Khuiseb canyon nearby during the Second World War. As they are economically independent, the farming business is obviously less important than the personal affection for this specific landscape. 'For me nature is the main reason to live here, the game and the nature, the nice nature' (Hannelore Neuffer).

Some of the Farmers Group III run a guesthouse to gain some income. Visiting tourists are a welcome diversion for some of these farmers living more or less isolated on their farms. Often game farming is of personal importance, which further attracts tourists.

Weekend farmers of Farmer Group IV chose their farms due to their personal preferences but also in view of their accessibility from Windhoek or Walvis Bay where their businesses are situated. As the farms in the proximity of Windhoek are more productive, they are expensive and difficult to obtain on the public market. Most weekend farms are situated in the region of the Nausgomab River on the desert margin and easy to get to from the coastal towns. For many of these farmers game is very important as it symbolises the contrast to life in town. 'A farm without game is not a nice farm. I can not see myself on a farm without game' (Huber).

For the Topnaar as for the Farmer Group I and II the agriculturally relevant natural resources seem to have much more relevance than aesthetics: '[W]e care about the nature and that's our pleasure, because we are directly living out of that nature, not in sense of wild animals, but in sense of that we are farming with our goats' (Rudolf Dausab).

As the productivity increases with the distance to the sea, the settlements Oswater and Homeb are relatively productive places. Whereas other settlements are seen as increasingly dry Homeb is perceived as an extraordinary productive place by Chief Seth Kooitjie: 'Homeb is like a village. You come from top and if you come down you see a green village. So it is like a basket and you put food in the basket, than it holds inside. This is the name of Homeb . . . actually it is not Homeb it is !Homeb' (Chief Seth Kooitjie).

In view of its higher productivity one would presume that most Topnaar live in the upper part of the lower !Khuiseb. However, this is not the case as most people live in the four settlements downriver from the Eduseb school outside the Namib-Naukluft Game Park (see Figure 14.1). A water pipeline runs from the water extraction scheme at Swartbank to the waterworks at Rooibank and supplies the four settlements Ururas, Goatanab, Armstraat, and Rooibank. Here the Topnaar have little gardens with vegetables and fruits they have to water. Furthermore, the proximity to the towns, Walvis Bay and Swakopmund, attracts people in view of better access to urban facilities such as hospitals, temporary labor, and commodities. It is possible to get a lift into town as there is some traffic to the waterworks on the gravel road, which is in a good condition compared to the rest of the road along the Topnaar settlements.

The settlements from Swartbank upriver to Homeb are inhabited by farming Topnaar who only occasionally move to the delta to harvest the !naras. Mainly elderly people, children, and some younger Topnaar men who decided against a life in town or did not find a job there live in these settlements far away from the coastal towns. Generally family ties are decisive as to which settlement the Topnaar stay in, even for people returning from town back to the settlements like Chief Seth Kooitjie. In 1975 he rediscovered the former set-

tlement which now is Homeb.[13] Up until the age of seven, Chief Seth Kooitjie
grew up at the Hope copper mine, just a bit upstream from recent Homeb.
When he married he decided in favor of Homeb to build his own house at the
!Khuiseb: 'I talked to my father and brought my wife to show her where I grew
up as a young man and told her "This is the place of my life"' (Chief Seth
Kooitjie). Besides the nostalgic memories he was attracted by the favorable
location, which he described as 'green village'. He is proud of his big garden
close to his house, where he grows vegetables and refers several times to the
relatively high productivity of this place.

The example of Chief Seth Kooitjie shows that also the Topnaar have
preferences for certain parts of the !Khuiseb River. Due to the Topnaars'
generally poor financial means the opportunities to choose, according to personal preferences, where to settle and farm are very limited. Only political
leaders such as the Chief, who is living in Walvis Bay and owns a car, are in
the position to combine life in the town with some farming along the lower
!Khuiseb.

SUMMARY

The inhabitants of the !Khuiseb catchment area (KCA) conceptualise their
environment differently due to their different cultural and historical backgrounds. Exemplarily the concepts of 'drought' and 'beautiful landscape'
show how different the various groups view the landscapes in which they
are living. In the case of the farmers these concepts find expression in the
spatial distribution of the different Farmer Groups in different parts of the
upper KCA. At the lower !Khuiseb River the possibilities for the Topnaar
to move are very restricted due to the conservation laws of the national park
but at the same time by economic constraints. Therefore personal preferences
for a specific landscape are not as easily observable in the spatial order as in
the highlands of the KCA. Despite the different landscapes and the diverse
cultural backgrounds of the people living in the KCA, the different economic
situations influence how people deal with their landscape as the distinction
into four Farmer Groups shows.

Special thanks to the staff of the Desert Research Foundation of Namibia
(DRFN) for their support during my fieldwork and to the anonymous reviewer
for the helpful comments.

---

[13] During his missionary work from 1965 to 1972 at Walvis Bay Moritz noted that Homeb was
not inhabited anymore. In 1959 when Köhler visited this place people still lived there. See in
Moritz (1997).

REFERENCES

Botelle, A. & Kowalski, K. (1995). *Changing Resource Use in Namibia's Lower Khuiseb Valley: Perceptions from the Topnaar Community*. Roma: Institute of Southern African Studies at the University of Lesotho and the Social Sciences Division at the University of Namibia.

Budack, K.F.R. (1977). The //Aonin or Topnaar of the Lower !Kuiseb Valley and the sea. *Khoisan Linguistik Studies*, 3, 1–42.

Coetzee, J.M. (1988). *White Writing. On the Culture of Letters in South Africa*. New Haven, CT: Yale University Press.

Dausab, R. et al. (1994). *Water usage patterns in the Kuiseb catchment area (with emphasis on sustainable use)*. Occasional Paper, 1. Windhoek: Desert Research Foundation of Namibia.

Department of Water Affairs (1987). The Kuiseb environment project: An update of the hydrological, geohydrological and plant ecological aspects. Report No. W87/7. Windhoek: Water Quality Division, Hydrology/Geohydrology Division.

Harries, P. (1997). Under Alpine eyes: Constructing landscape and society in late pre-colonial South-East Africa. *Paideuma*, 43, 171–191.

Heine, B. (1976). *A Typology of African Languages*. Berlin: Dietrich Reimer.

Heyns, P. et al. (Eds.) (1998). *Namibia's Water. A Decision Makers' Guide*. Windhoek: Department of Water Affairs, Desert Research Foundation of Namibia.

Jackson, P. (1989). *Maps of Meaning. An Introduction to Cultural Geography*. London: Routledge.

Jacobson, P. et al. (1995). *Ephemeral Rivers and Their Catchments. Sustaining People and Development in Western Namibia*. Windhoek: Desert Research Foundation of Namibia.

Kinahan, J. (1991). *Pastoral Nomads of the Central Namib Desert*. Windhoek: African Books Collective.

Korn, H. (2001). *Zwiegespräche in der Wüste. Zweitauflage*. Göttingen, Windhoek: Hess.

Lakoff, G. (1990). *Women, Fire, and Dangerous Things. What Categories Reveal About the Mind*. Chicago: University of Chicago Press.

Leser, H. (1982). *Namibia. Geographische daten, strukturen, entwicklungen (Länderprofile)*. Stuttgart: Klett.

Luig, U. (1999). Naturschutz im widerstreit der interessen im südlichen Afrika. In A. Thimm (Ed.), *Studium generale, arbeitskreis dritte welt*. Mainz: University of Mainz.

Marker, M.E. (1977). Aspects of the geomorphology of the Kuiseb River, South West Africa. *Madoqua*, 10(3), 199–206.

Martin, H. (1996). *Wenn es Krieg gibt gehen wir in die Wüste*. Hamburg: Abera.

Mensching, H.G. (1990). *Desertifikation. Ein weltweites problem der ökologischen Verwüstung in den Trockengebieten der Erde*. Darmstadt: Wissenschaftliche Buchgesellschaft.

Moritz, W. (1997). *Verwehte Spuren in der Namibwüste. Alte Ansiedlungen am Kuiseb*. Aus alten Tagen in Südwest, 13. Self-published.

Moritz, W. (Ed.) (1999). *Die 25 frühsten Landreisen, 1760–1842. – Die ältesten Reiseberichte über Namibia, gesammelt und herausgegeben 1915 von Prof. Dr. E. Moritz, Teil 1*. Windhoek: Wissenschaftliche Gesellschaft.

Ranger, T. (1997). Making Zimbabwean landscapes. Painters, projectors and priests. In U. Luig, & v. Oppen U. Luig, (Ed.), *Naturaneignung in Afrika als sozialer und symbolischer Prozeß* (pp. 59–73). Berlin: Das Arabische Buch.

Rust, U. & Wieneke, F. (1974). Studies on gramadulla formation in the middle part of the Kuiseb River, South West Africa. *Madoqua*, 2(3), 69–73.

Sandelowsky, B.H. (1977). Mirabib – An archaeological study in the Namib. *Madoqua*, 10(4), 221–283.

Sehoueto, L.M. (1995). Lokales Wissen und bäuerliche naturaneignung in Benin. – In U. Luig & A.V. Oppen (Eds.), *Naturaneignung in Afrika als sozialer und symbolischer Prozeß* (pp. 83–94). Berlin: Das Arabische Buch.

Van den Eynden et al . (1992). *The Ethnobotany of the Topnaar*. Gent: Universiteit Gent.

Weichhart, P. (1990). *Raumbezogene Identität. Bausteine zu einer Theorie räumlich-sozialer Kognition und Identifikation*. Erdkundliches Wissen, 102. Stuttgart: Steiner.

Widlok, T. (2000). *Dealing with institutional changes in property regimes. An African case study*. Internet Discussion Paper: www.eth.mpg.de.

Chapter 15

Where Settlements and the Landscape Merge: Towards an Integrated Approach to the Spatial Dimension of Social Relations

THOMAS WIDLOK

The separation between 'settlement' and 'landscape' is deeply entrenched in European thought and also in the worldview of many agrarian societies. In anthropology this is reflected in the distinct development of an anthropology of landscape on the one hand and an anthropology of built forms. The comparative use of permeability maps is introduced in this chapter as a promising route towards cross-fertilisation between these two hitherto separate bodies of theory and data. Permeability, the ways in which space allows or prevents humans from passing through places, is particularly relevant for our understanding of the fuzzy zone where settlements and the landscape merge. More generally, permeability maps help us to explore a more dynamic view of the relationship between spatial and social relations because they allow us to consider what one may call the 'social agency of space'. The case material presented in this chapter was collected in the course of field research with ≠ Akhoe Hai//om 'San' or 'Bushmen' and their neighbours in northern Namibia but an explicit comparative perspective is taken that leads beyond this region.

M. Bollig, O. Bubenzer (eds.), *African Landscapes*,
doi: 10.1007/978-0-387-78682-7_15, © Springer Science+Business Media, LLC 2009

15.1. INTRODUCTION

The anthropology of landscape and the anthropology of built forms have largely developed independently of each other. Today there are theoretical challenges and methodological possibilities to create links between these two fields. The aim of this chapter is to explore some of these links. In terms of theory the established divide between settlement patterns and landscape features has to be considered to be the product of a particular cultural tradition which does not necessarily provide the most adequate framework for comparative analysis. An integrated framework that would cover all relationships between people and their spatial environment, irrespective of the old divide, would be able to provide cross-fertilisation between two hitherto separate bodies of theory and data. It may also provide a better understanding of those phenomena that are located at the blurred zone where settlements and the landscape merge.

In terms of methodology the use of permeability maps is suggested as one way of collecting and of systematising data in this integrated anthropology of space. An emphasis on the permeability of landscapes and settlements directs our attention away from isolated spatial features and towards the position of these features in the larger spatial structures that humans inhabit. In comparative research this allows us to detect patterns where there seems to be only a plethora of diverse spatial forms and it prevents us from assuming similarities arising from spatial forms which may only be of a superficial nature. Moreover, this methodology may lead us towards a more dynamic understanding of settlements and landscapes, not only as a static surface onto which humans direct their cultural imagination, but as an important part of the changing social world that is generated and altered in the process of human practice.

Space, in a relational sense, is not only a passive vessel designed by human agents but it may be considered a social agent in itself inasmuch as the humans living within it are also 'patients' in relation to it (see Gell, 1998). This is the case, I maintain, not only for settlements, constituting 'man's largest artefact', but also for the landscape beyond the built environment. In this contribution I use the term 'settlement' to refer to the built environment, 'landscape' to refer to the larger space in which settlements are embedded, and 'space' more generally as covering both, settlements and landscapes.

15.2. RECONCILING THE ANTHROPOLOGY
OF SETTLEMENTS WITH THE ANTHROPOLOGY
OF LANDSCAPE

The anthropological record provides two main bodies of material relating to the position of humans in space, namely the anthropology of settlement patterns and the anthropology of landscape. Research into settlement patterns was particularly strong in the days of structural functionalism and continues to inform

Figure 9.2. Original caption: 'Windhoek – View towards native quarter' – View towards west showing Rhenish Mission Church with site of old location

Figure 11.1. The highlands in the regency of Gowa, South Sulawesi (Photograph by author)

Figure 11.2. Individual experience of the landscape in German romanticism: C. D. Friedrich: The Wanderer Above the Sea of Fog, 1818 (Courtesy National Archives of Namibia)

Figure 12.4. Women wailing before setting off to the *evaru*

Figure 12.5. Women crawling to the ancestral fire in festive gear

Figure 12.9. Ancestral grave at Ombuku

Figure 12.10. *Ovipande* act as a symbolic device to memorise larger ceremonies

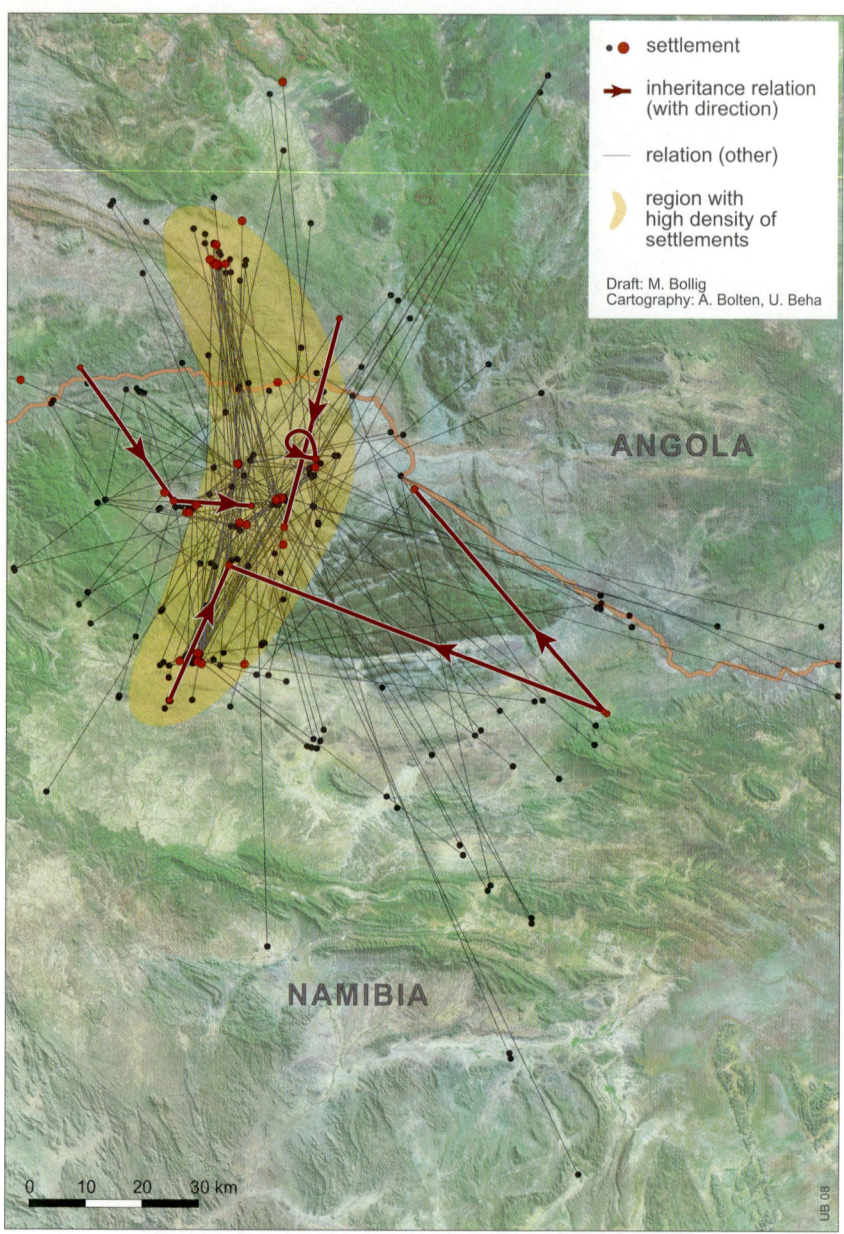

Figure 12.11. Location of graves visited during one day's *evaru*

Figure 13.1. Photo Pan (Ute Dieckmann, 2007)

Figure 13.2. Waterhole Okaukuejo by sunset (Ute Dieckmann, 2003)

Figure 13.3. Halali Koppies from the main road (Ute Dieckmann, 2002)

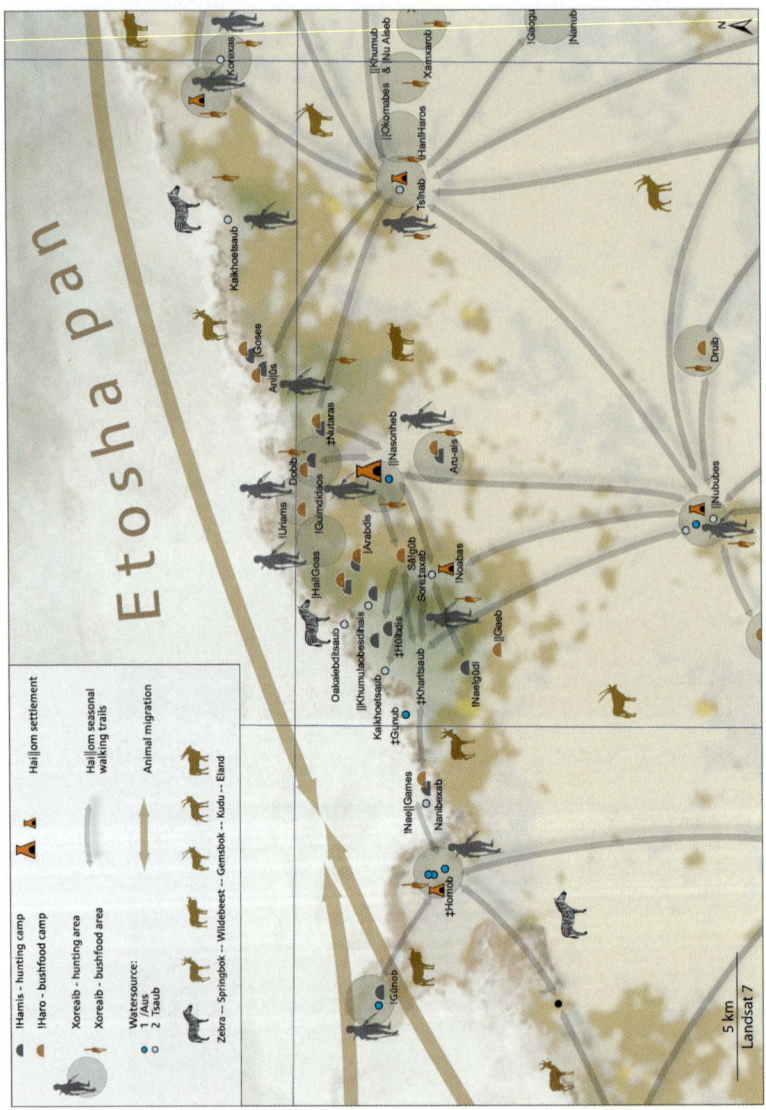

Figure 13.4. Extract of a cultural map, illustrating seasonal mobility. ‖Nububes in the south

Figure 14.2. Khomas Hochland after rain (Photo: N. Gruntkowski)

Figure 14.3. Namib Desert at the lower !Khuiseb (Photo: N. Gruntkowski)

Figure 15.1. ≠Akhoe Hai//om hunter-gatherers in northern Namibia moving camp

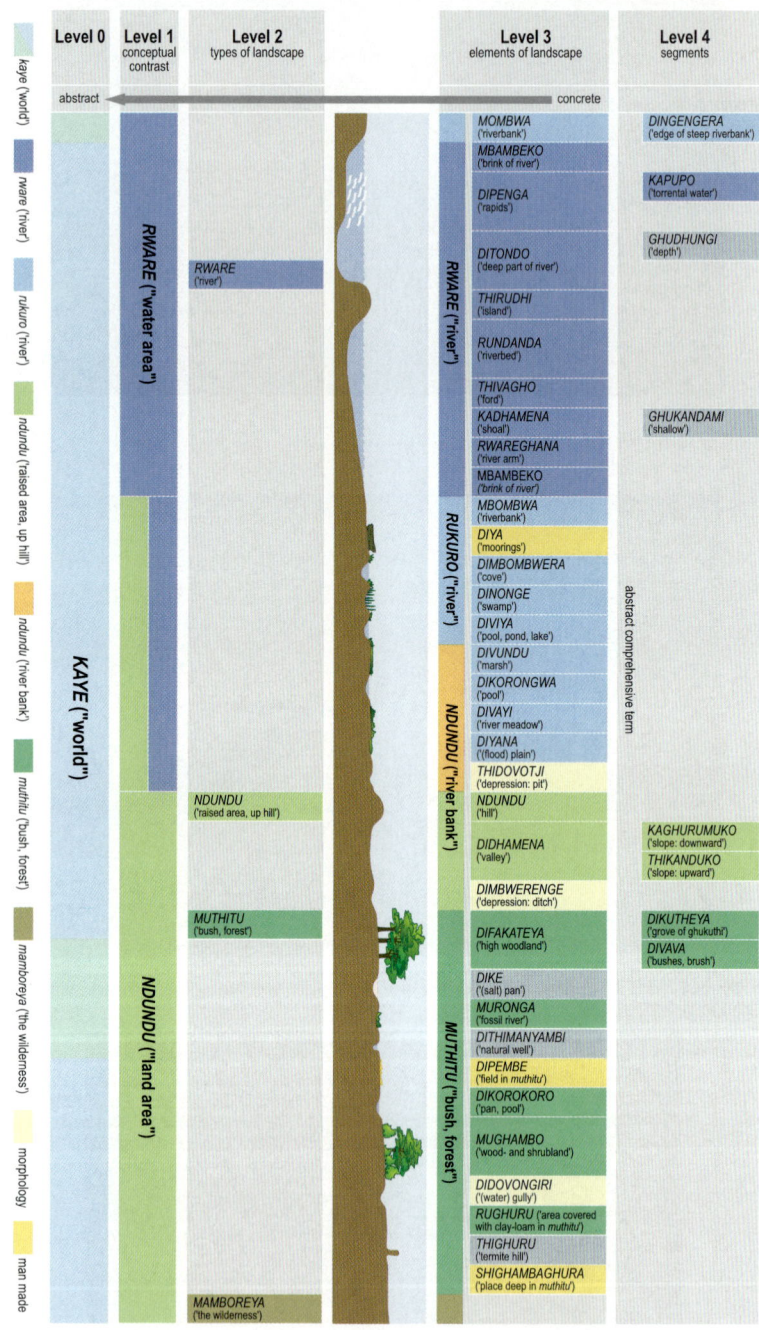

Figure 17.5. Conceptual hierarchy of landscape entities

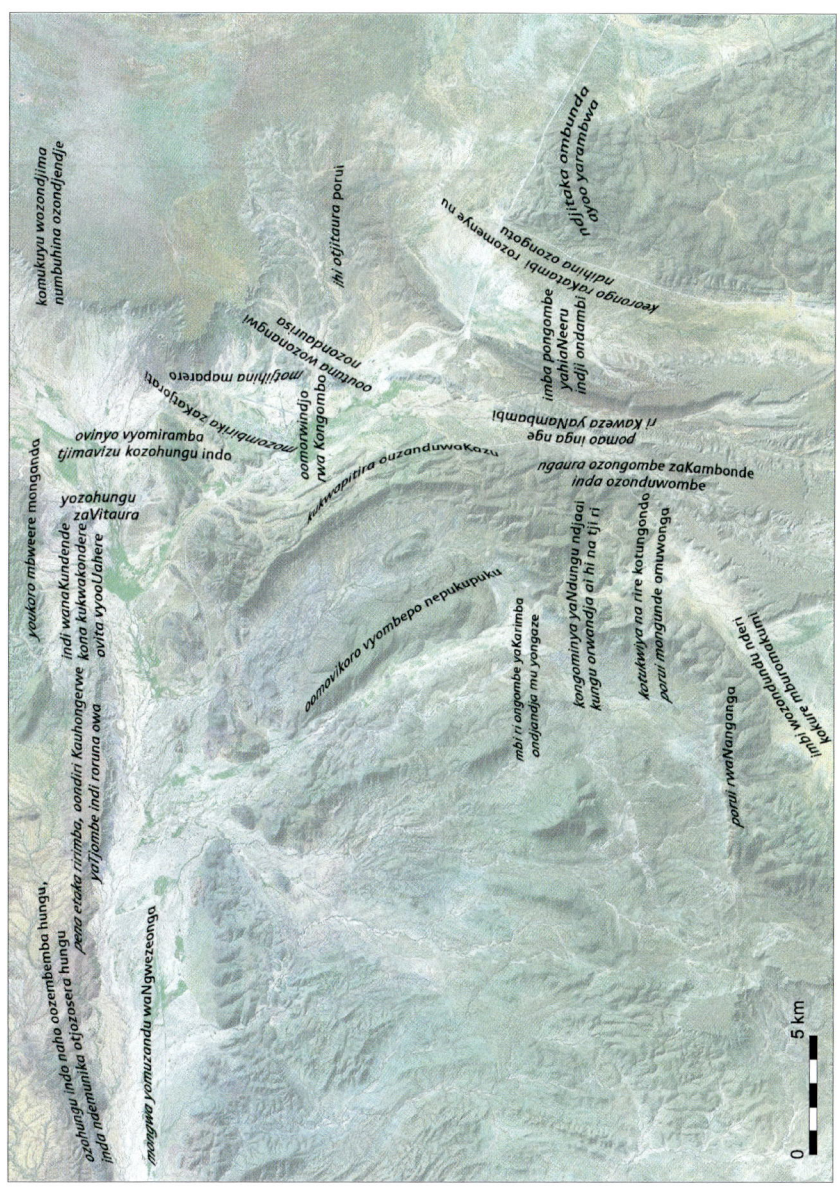

Figure 18.2. Praises of places

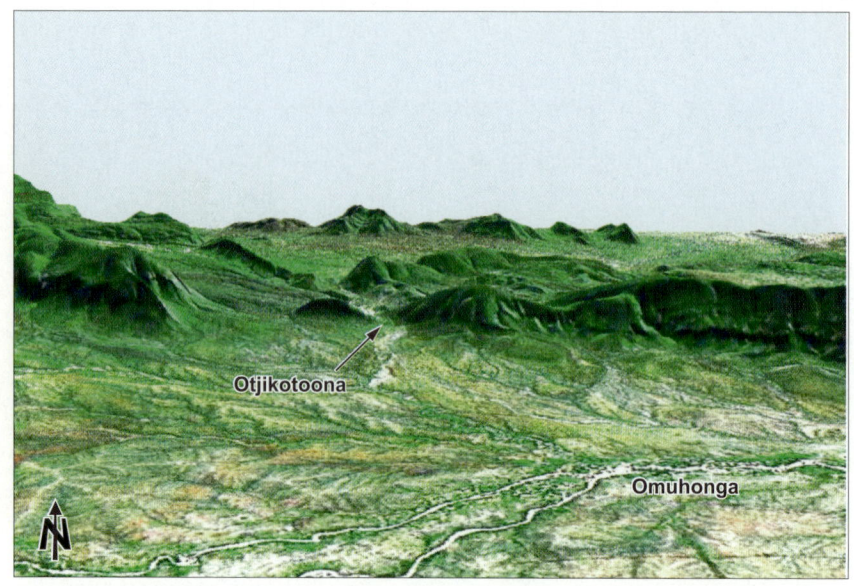

Figure 18.3. Mountain pass of Otjikotoona (3D-satellite image [Landsat, SRTM] subproject E1, CRC 389)

Figure 18.4. Ehomba mountain (3D-satellite image [Landsat, SRTM] subproject E1, CRC 389)

Figure 18.5. Omuhiva (3D-satellite image [Landsat, SRTM] subproject E1, CRC 389)

Figure 18.7. Cattle ombambi

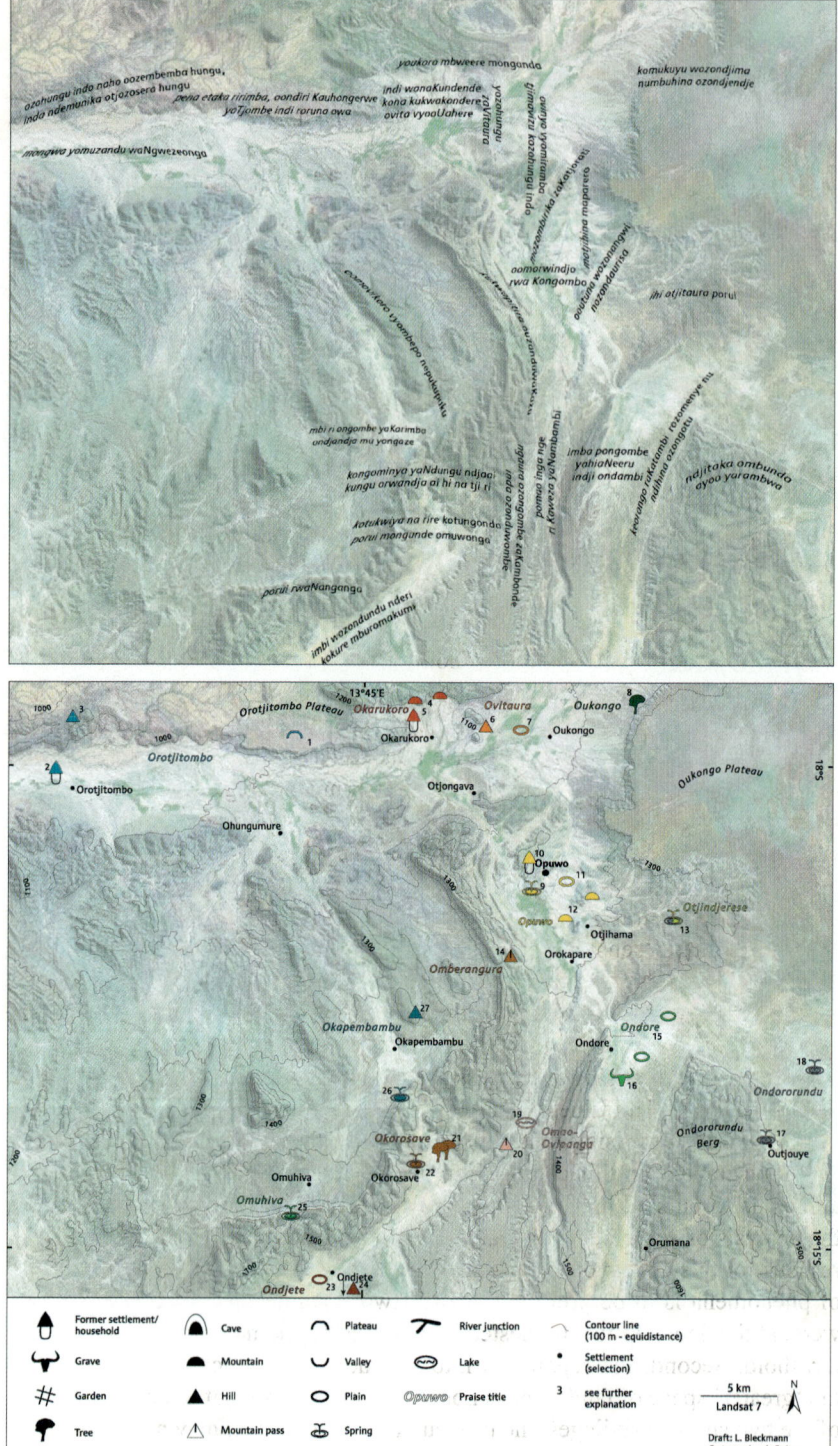

Figure 18.8. Map illustrating the spatialisation of collective memory

science-oriented branches of anthropological research, including archaeology (see, e.g., Kent, 1993). Apart from more specialised research on spatial layouts, drawing ground plans of settlements has become a standard tool of ethnographic descriptions. The anthropology of landscape is a more recent development, at least under this label, and has particularly attracted humanistic and postmodern approaches following an upsurge of studies from history, especially art history (see, e.g., Hirsch & O'Hanlon, 1995). Applied and ecological anthropology have also discovered the importance of landscape as a field of research (see Brody, 2001). It is therefore fair to say that both fields of research are now well established in anthropology.

As in anthropological enquiry more generally, research into settlement patterns and landscape has broadened the known spectrum of how humans conceive and use space. The prime objective has been a better understanding of similarities and differences in settlement layouts and in the perception of the environment. An example for identifying patterns of similarity in diversity is Henrietta Moore's study of Marakwet settlement layouts. She was able to show how the diversity of ground plans can be explained on the basis of the developmental cycle of Marakwet households, in particular on the grounds of changes in the lifecycle of Marakwet household heads (Moore, 1996). An example to identify persisting difference in a shared process of land rights negotiations is given by Robert Layton's analysis of Alawa landscape representations in northern Australia (Layton, 1997). He argues that Alawa and Westerners may fruitfully discuss some aspects of the environment (e.g., the occurrence of wild game) but that sacred sites, as features in Alawa representations of their environment, cannot completely be translated into Western legal representations because of differences in the underlying causalities (Layton, 1997). The premise that ancestral beings have created the land made Alawa representations of the landscape 'sensitive to features to which Western representations are blind' (Layton, 1997).

In other words, both the anthropology of settlement layouts and the anthropology of landscape seem to be well suited for the basic anthropological projects of raising awareness about difference and patterns of similarity. In fact, given the fruitfulness of analyses in these two subfields of anthropology it is somewhat surprising that the separation of the two fields continues to be taken for granted. In this chapter I argue that research has reached a point that allows us to pursue methods and theories that are no longer limited to either the analysis of settlements or that of the wider landscape. There are two main reasons for this endeavour. First, an analysis that can account for a wide range of phenomena is to be preferred against two separate theories that meet, as it were, at the doorstep to a homestead or a house but do not allow us to cross the threshold. Second, the separation into the 'domestic' space of settlement and the 'greater' space of landscape is not a universal one so that a separate analysis of the two spaces privileges one particular folk model that may not be adequate for comparative research.

15.2.1. Hunter-Gatherers and Their Environment

Hunter-gatherer studies have provided an arena for fundamental discussions about the relationship between humans and their environment. Two extreme positions are to be found: an adaptationist and a culturalist paradigm. Especially in the early literature the absence of elaborate settlements (in the archaeological record and in the ethnographic present) led to the image of hunter-gatherers as adapting to their environment but not as actively changing it. This gave rise to a view that considered foragers to be either completely encompassed by nature or as indeed being a rather insignificant part of it. Their flexible settlements and their movements in space seemed to be without any cultural logic and simply responses to environmental needs. Nevertheless, the ingeniousness and inventiveness of the diversity of forager adaptations has been emphasised and ever more complex simulations of these practices in models are being developed in this strand of research.

At the other end of the spectrum a completely different approach has emerged based on the growing recognition that foragers, too, change the world they live in (e.g., through the use of fire) and that their movement in space follows social practices (e.g., conflict resolution) and cultural ideas (including religious ideas) and not only environmental pressures (see Figure 15.1). The fact that many foragers recognise personal agency not only in humans, but also in other living beings had for a long time been discussed under the label 'animism' but has more recently received new attention. In this strand of research the cultural meanings that are attached to features of the environment are being emphasised instead of its 'physical' properties (resource availability, climate, etc.). Most recently anthropologists with a background in phenomenology and ecological psychology have sought to once again include the physical characteristics, such

Figure 15.1. ≠Akhoe Hai//om hunter-gatherers in northern Namibia moving camp (*See also Color Plates*)

as the inclination of the land, but in very different terms than those discussed in the adaptationist paradigm.

The separation into 'settlement' (read: civilisation) and 'landscape' (read: wilderness) is deeply entrenched in European thought but is also in the worldview of other agrarian societies. West African farmers, for instance, clearly cultivate this distinction and use it as an identity marker that distinguishes them from nomadic-pastoralist people in the area (see Dafinger, 2004). In southern Africa the use of the allonym 'Bushmen', and the discrimination against the people labelled this way, is also based on a strong underlying separation between the cultivated land of agriculture and livestock-raising and the wild land of the bush (see Widlok, 1999a). Although the Europeans in Africa, at least today, may see the bush much more positively than the African farmers, they both take this separation for granted.

However, as Dafinger (2001) has observed in Burkina Faso, in many parts of Africa the separation is anachronistic in the sense that it is increasingly possible and easy to move through the whole country by passing from one domestic homestead to another because the stretches of 'wild land' in between are disappearing. But even where there are still 'wastelands', they are not unstructured or untouched by the practices of the people living in the area, and this is probably true for a much longer period than we are inclined to think (see Hayes and Dieckmann, this volume).

As it has been demonstrated for Australia the use of fire by a handful of mobile residents clearly shaped the whole continent and the distribution of species that inhabit it (Pyne, 1991; Latz, 1996). There is no reason to assume that this was different in Africa which in all likelihood has a longer history of human occupation. Moreover, one of the decisive features of space is that it can be described in terms of its material characteristics for the people moving about within it, no matter whether it is the small space of a house or homestead or the somewhat wider space of the surrounding landscapes.

The method that I am discussing below focusses on one particular dimension of the characteristics of space, namely permeability. The case is particularly clear with regard to the built domestic or urban environment and its physical characteristics. Walls with predefined passageways define the permeability of space, including the number of rooms or places one has to cross in order to get to another place (or room) in the system and the number of possible routes that one may take when entering these systems and when moving about in them. Not surprisingly, therefore, architects have been exploring these features of space which allow them to provide simplified models of complex spaces. On this basis they are able to quantify the relative permeability of a system (a homestead, a building, a housing estate) either with reference to another system or with reference to another point within the system (see Hillier & Hanson, 1984).

In other words it is possible to compare not only the permeability of particular buildings (see Hillier, 1991) or homesteads, but also to compare the implications for being located at different places within these systems, for example, in the visitors' hut rather than the chief's hut, the secretary's office rather than the office occupied by the chief executive. Moreover, the underlying strategy, namely to look for the permeability features of space and to correlate them with features of the social permeability or political relations between the various occupants that use this space, can be applied to larger space as well as to domestic space (Hillier, 1991; see Widlok, 1999b,c). It can thereby help us to overcome a description predicated on the divide between 'civilised' and 'wild' space. But before elaborating this point further, it is useful to give a short summary of how a permeability analysis of this sort can be applied to a concrete case study from Africa.

15.3. RECORDING THE PERMEABILITY OF SPACE

The basic spatial relationships depicted in permeability maps have been theoretically conceptualised by Hillier and Hanson (1984) with the help of a specialised terminology that distinguishes places according to the number of access routes and their connectedness. Unipermeable (single-access) places are distinguished from multipermeable (multiple-access) places (see Figure 15.2). For simplicity the latter are here depicted as bipermeable places (i.e., with two connecting lines). Where space is distributed, all places tend to have the same number of access routes.

Correspondingly, space is nondistributed when access routes are unequally distributed with some places being better connected than others. Where space is completely symmetrical no place controls the access to another place; that is, there is no need to pass through one place in order to get to a second place without the second place being at the same time a necessary passage to the first place. With nonsymmetrical space, by contrast, one place controls access to a second place. These properties can be related either to a spatial structure on the whole or to different parts of it. Although these are materially given distinctions which can be used as modules that make up the map of a camp, homestead, or entire village, they do depend on a position taken within or outside the settlement for the calculation of distributedness and symmetry.

15.3.1. 'Permeability Maps'

Ways of mapping the 'permeability' of space were initially explored by a group of architects, led by Bill Hillier and Julienne Hanson (1984), and are further

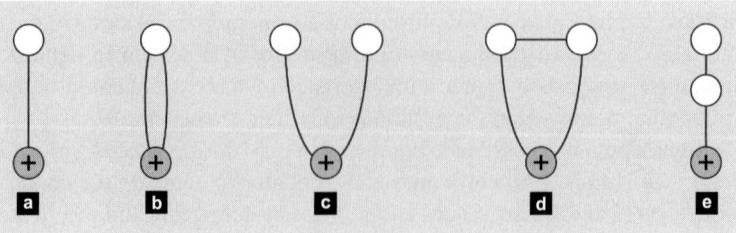

Figure 15.2. Basic elements of permeability maps

developed in regular conferences on 'space syntax' (see www.spacesyntax.net). One of the original core ideas was to develop a tool that would explain why building estates, even though they may have the outward appearance of 'villages' made occupants feel lost and insecure because they prevented occupants from moving or meeting appropriately in their built environment. With the help of a new graphical way of mapping, a set of descriptive technical terms, and a calculus to quantify spatial relations of permeability, an analytic tool was established that could explain why residents felt insecure in large estates and conversely why visitors felt attracted to 'villages'.

Although based on a highly technical Western understanding of space, these basic modules for drawing permeability maps have also been used to describe spatial structures comparatively across cultural contexts. These maps distinguish unipermeable (a) from bipermeable (b) space, depending on the number of access routes to the place (room, house) in question from the position of a 'carrier' entering the structure from the outside (marked ' + ' in the maps). The relationship amongst a number of places is distinguished firstly in terms of the number of connections that link these places. Spatial structures in which all places have the same number of connections are called 'distributed' (c), and distinguished from 'nondistri-buted' (d) constellations in which some places are better connected than others. Secondly, we may distinguish symmetric relations (as in (c)) from asymmetric relations (as in (e)) insofar as one space can regulate the access to another (such as a room or a passage leading to another room). A combination of these modules, to which quantitative factors may be attached, can then help to characterise places (rooms within a house, houses within a settlement, settlements in a region) comparatively and with reference to its permeability rather than to its outward form or size.

Permeability maps are designed as graphical maps which do not depict cultural ideal types. In their original conceptualisation, they incorporate the material reality of spatial layouts in order to show to what extent cultural ideas and practices can be traced to spatial conditions or at least be explained by the way they are supported by built forms. The continuity of spatial layouts is not necessarily established through sharing and elaborating cultural norms but also through regular practices that continually re-create similar patterns of permeability across time and settings. Permeability maps complement ground plans because they help to show that spatial layouts and built forms not only reflect cultural ideas but also produce certain ideas and practices. Permeability maps replace the (fictitious) bird's-eye view with the perspective of a 'carrier' (a visitor or resident) striding through the camp (see Figure 15.3). They distinguish the relation between points in a settlement by the number of possible access routes and the number of the steps required in order to reach one place (or all places) from another (or all others). This creates a descriptive tool for comparing variations and changes in settlement layouts (for a more detailed discussion of the use of permeability maps in anthropology see Widlok, 1999b,c).

To illustrate the usefulness of permeability maps in a concrete case, we may turn to the map of a homestead as it is built by Owambo agropastoralists in the

Figure 15.3. ≠Akhoe Hai//om dry-season camp (mapped in Figure 15.6)

north of Namibia (see Figures 15.4 and 15.5). In an Owambo homestead the various occupants of a homestead (household head, first wife, other wives, unmarried
boys, and visitors) occupy distinct places in the spatial structure. The *olupale*
(number 9 on the map) is a central place where visitors are welcomed and served
a meal (see Widlok, 1999c). As the permeability map of the homestead shows,
the *olupale* is highly connected (or 'distributed') with regard to other parts of the
palisade structure that make up the settlement as a whole.

The particular homestead mapped here is the homestead of an Owambo chief
and the permeability map provides a systematic and rigorous tool for establishing
differences between the homesteads of different homestead owners. Ultimately,
they also help us to describe the features that bring about this diversity. In this
particular case the distributedness of this meeting place, which emerges from the
graph in terms of many connecting lines, is particularly marked. It is far greater
than that of the average place in this spatial structure and particularly great in
comparison with the sleeping quarters of the homestead head (Figure 15.5, 33).
This makes it a privileged place for social control and for observing the movements of the inhabitants.

From the visitor's perspective the meeting place is already far inside the
homestead, which presents itself to the visitor as a rather solid fortress of palisades (see Figure 15.10). Its position is more asymmetric and more significant
than other distributed places nearer the homestead's entrance, for instance, the
boys' sleeping huts (57–59). This relative asymmetry of the *olupale* increases
the noninterchangeability and significance of this place. To be admitted to the
meeting place is not trivial for a visitor to an Owambo homestead, because it
implies that the visitor is welcomed as a guest who can claim the attention of

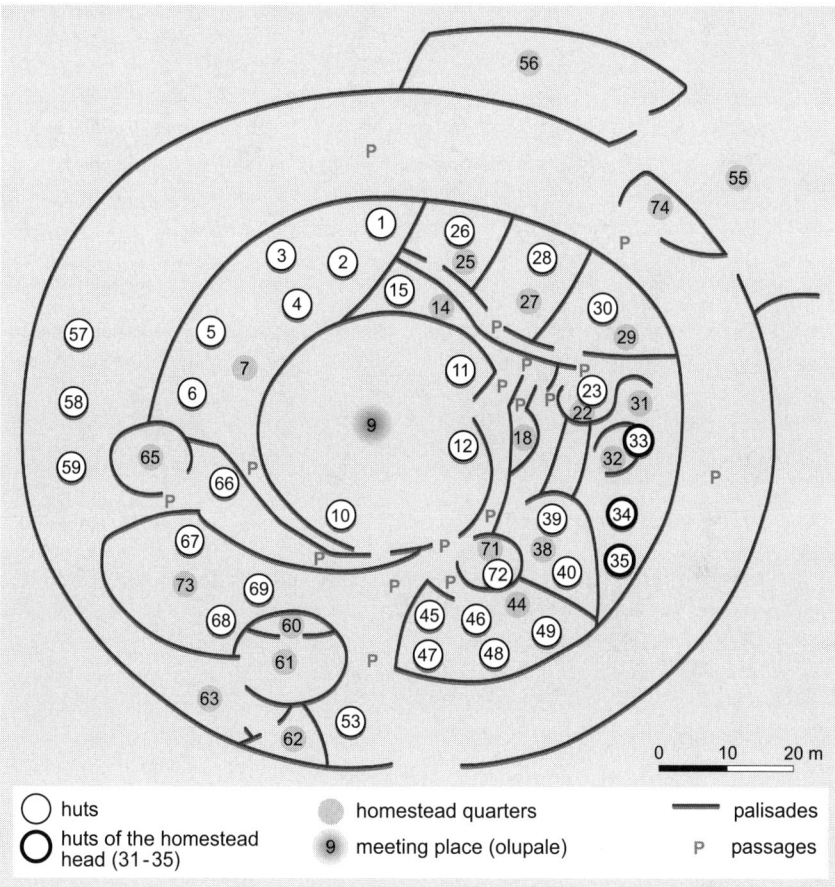

Figure 15.4. Ground plan of an Owambo homestead. P = Passage, 9 = meeting place (olupale), 31–35 = huts of the homestead head; all other numbers refer to quarters occupied by wives and children, visitors, and livestock (for more details see Widlok, 1999c)

the occupants. Permeability maps show how the connectedness of the *olupale* can imply asymmetry between visitor and inhabitant and at the same time imply distributedness between the homestead head and his fellow inhabitants.

Comparisons are not only possible for different positions within a single spatial structure such as that of an Owambo homestead but also for comparing the permeability of whole spatial structures that may otherwise seem to be noncomparable. For instance, a permeability map of a dry season camp (see Figures 15.3 and 15.6) constructed and occupied by hunting and gathering ≠Akhoe Hai//om – neighbours to the Owambo – has quite different characteristics than the Owambo homestead analyzed above but it can still be usefully compared in terms of its permeability.

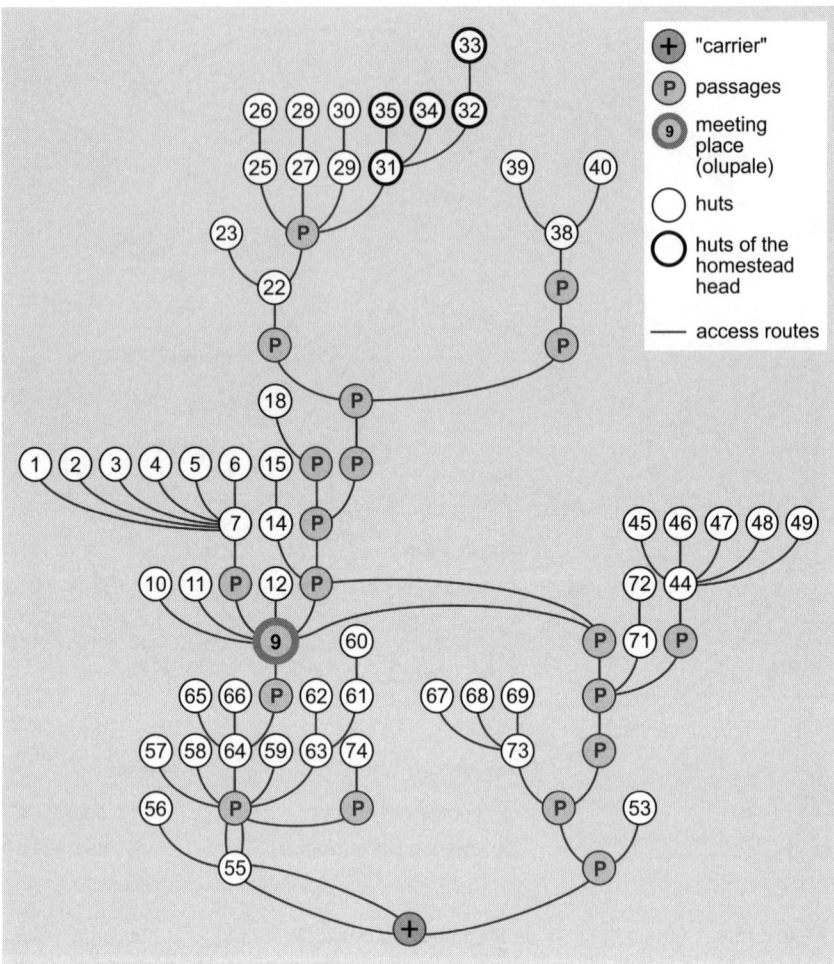

Figure 15.5. Permeability map of the homestead shown in Figure 15.4. + indicates a 'carrier' (e.g., a visitor) who enters the settlement. In order to reach the meeting place (9) the visitor has to go through a number of separate passages and entrance spaces in order to move on to the sleeping hut of the household head (33); there are more passages to pass through as well as an anteroom (22) and the daytime living rooms of the household head (31, 32). By contrast, there are a number of ways that can take an outside visitor to the meeting place (either via the boys' huts 57–59 and the cattle enclosures 60–63 or alternatively via waiting huts (67–69) and or past the wives' quarters (44–49 and 70–71); see also Hillier and Hanson (1984) and Widlok (1999c)

In this camp centrality, open access, and proximity to the central shady tree are the properties of the fireplace for the medicine dance (M). As with (almost) all fireplaces of the camp, this place is accessible from more than one direction and is therefore connected with double lines. Access to any of the huts in the camp,

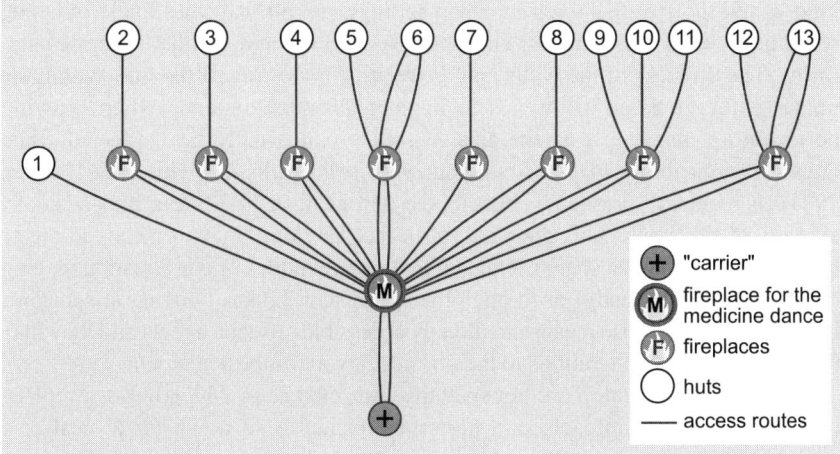

Figure 15.6. Permeability map of a ≠Akhoe Hai//om dry-season camp

except for the open shelters, is controlled by a fireplace (F) in front of the huts which any visitor has to pass in order to reach the interior of a hut. In some cases occupants of several huts share a fireplace, such as a couple (10), their unmarried daughters (11) and the husband's mother (9); in other cases a family has more than one fireplace (as with huts 7 and 8) All huts, except for a deserted 'fire-less' hut (1), at that stage used for utensils only, and a hut of the pregnant woman (7), have the same permeability. As the graph shows only one step is required to reach hut 1, instead of two steps for the other huts. Access to the fireplace of hut 7 is unipermeable only (only one route of access is possible). Seniority, gender, and social position (such as late arrival at the place) do not produce fixed spatial features in the Hai//om camp that would leave traces in the spatial layout. Or, to put it differently, the spatial order does not support a social differentiation according to age, sex, or social position.

Shelters in hunter-gatherer camps such as the one analyzed here do not have solid walls, but they have spatial features which may be regarded as having a very similar function. Of central importance are the fireplaces, especially the fireplaces in front of a hut or windbreak. Even if no fire is burning, these fireplaces have to be considered separate 'rooms' in terms of their permeability. Fireplaces are usually multipermeable; that is, they are accessible through a number of routes. But a visitor will never – under normal circumstances – disregard the presence of a fireplace and go into a hut or windshield without first stopping at the fireplace in front of it. Climbing through a window when there is a doorway, in the absence of special circumstances, would be a parallel behaviour in solid buildings.

The walls of Hai//om windbreaks and of huts which are made of wooden poles allow people to look – and certainly to talk and listen – through them, especially in the dry season. They are therefore different from walls between rooms in most urban environments, although the difference is one of degree rather than kind, given the fact that even in urban settings, not all walls are made of stone; some have windows, hatches, curtains, glass doors, and similar 'intermediate' forms. The thinness of the walls and the relative proximity of the huts within the hunter-gatherer camp allows us to neglect absolute distance when drawing permeability maps, just as the size of rooms can usually be neglected when drawing permeability maps of buildings with solid walls (see Hillier & Hanson, 1984). It requires no special effort to see and hear, or to be seen and heard, in all parts of the camp. This does not mean that Hai//om have no means to create boundaries in this open spatial setup, either through modes of conversation (calling out or conversing quietly) or through the use of their bodies (looking at someone, looking away) but these means are directly accessible to social agents and they may be changed without alterations to the physical layout of the settlement.

Visitors to the camp are not spatially restricted as to which hut or fireplace they may initially approach, or where they decide to sit down for a meal or a visit. However, the layout of the camp guides visitors to a central shady fireplace, where they are prompted (and expected) to sit down and wait before they move on to one of the other fireplaces. After the visitors have entered this central place they are expected not to venture farther without being encouraged to do so. The initiative lies with the residents who may turn to the visitors and who may come to greet them, inviting them to sit at one of the other fireplaces. This distinction between the central fire and other fireplaces corresponds to the activities carried out at the various places within the camp: open access to the central tree corresponds to the daylight activities (usually handicrafts or conversation) as well as to the night-time activities (medicine dancing and healing) that take place there. The more restricted access to the individual fireplaces corresponds to the main activities carried out at these places, namely eating and chatting.

Finally, access into a hut, which contains individual belongings and resting places, is possible only via the fireplace in front of it. The permeability maps presented here refrain from including normative aspects of permeability (the dos and don'ts of behaviour in space) unless they correspond to a physical feature in the environment (e.g., a fireplace or a path). It remains to be seen in the course of further anthropological adaptation of mapping permeability whether it proves to be useful to also 'map' the norms and values that define permeability for occupants and visitors in any given cultural context.

15.4. MAPPING PERMEABILITY AS AN ANALYTIC TOOL FOR COMPARISON

Permeability maps may help us to compare the spatial characteristics of settlements and landscapes where there are very different outward appearances, for instance, concerning the scale or shape of huts or landscape features (compare Figures 15.3 and 15.10). The other important feature of permeability maps, which is particularly important for our understanding of settlements as well as landscapes is that, conversely, we have a method of measuring and of testing whether we are misled by similarities in scale or outward shape. In built spaces and in landscapes we may find spatial features with similar form (types of huts in settlements, types of rock formation in the landscape, for instance).

The possibility of directly asking the occupants of the spaces in question about spatial forms is often restricted, typically so in the case of archaeology or history, but also where spatial structures are generated, not as an execution of a formal blueprint, but as the cumulative result of numerous individual decisions over time. Under this restriction, the observer may be inclined to draw parallels which may not be warranted. We know that the complexity of the built environment and the natural environment allows people who move around in it to have very different, selective perceptions of isolated features in this environment. Permeability maps direct our attention away from the isolated features and towards their connectedness, their position in a larger spatial structure. This can be illustrated by going back to the example of the *olupale*, the place where Owambo meet their guests in a homestead.

As is the case with other cultural features, the *olupale* is not only found in Owambo homesteads but also elsewhere. Today there are also Hai//om settlements that have a row of palisades that look like an Owambo *olupale* (see Figure 15.7) and that are even given a name (*olupare* or *orupare*) that is clearly derived from the Owambo term. The Hai//om settlement depicted in Figures 15.8 and 15.9 in fact has two '*olupare*' (O), a resting place where palisades protect visitors and residents from the hot afternoon sun from the west. However, a comparison of permeability maps shows that the position of these *olupare* in the overall structure of the settlement is quite different from that which is found in an Owambo homestead. The two *olupare* in the Hai//om settlement shown here have no special features with regard to their position in the overall layout. As the permeability map shows (see Figure 15.9), they differ from the other fireplaces only in that they do not control access to a hut.

Figure 15.7. The meeting place olupale/olupare

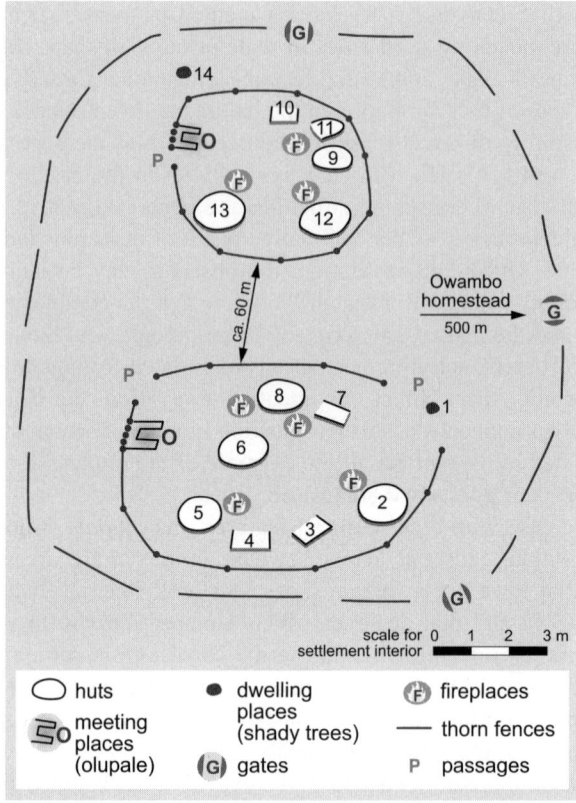

Figure 15.8. Ground plan of a ≠Akhoe Hai//om wet-season camp

The introduction of the *olupale* into Hai//om settlements is a cultural import that has not yet registered with the way in which domestic space influences daily practice. Hai//om visitors continue to eat at the hearths of their hosts, and they may occupy a hut in the circle of huts and move around the circular structure of the camp without being irritated by the cultural innovation of the *olupale*. However, its mere presence provides a spatial prompt which may develop into changing routines of welcoming visitors or of distributing occupational fields within the settlement. To a limited degree this is already the case in this particular settlement. The fireplace at the *olupare* is a preferred place for men and for visitors who take part in local medicine dances. As regards to Owambo visitors to the Hai//om camp, it seems to prompt the visiting agropastoralists to keep away from individual hearths and to take a seat at the *olupare*.

In terms of permeability another factor seems to be more important than the presence of an *olupare*. This settlement is surrounded by thorn fences that protect millet fields from cattle. It shares this feature with the poor Owambo homesteads that cannot afford a palisade fence and both are in contrast to a dry-season camp

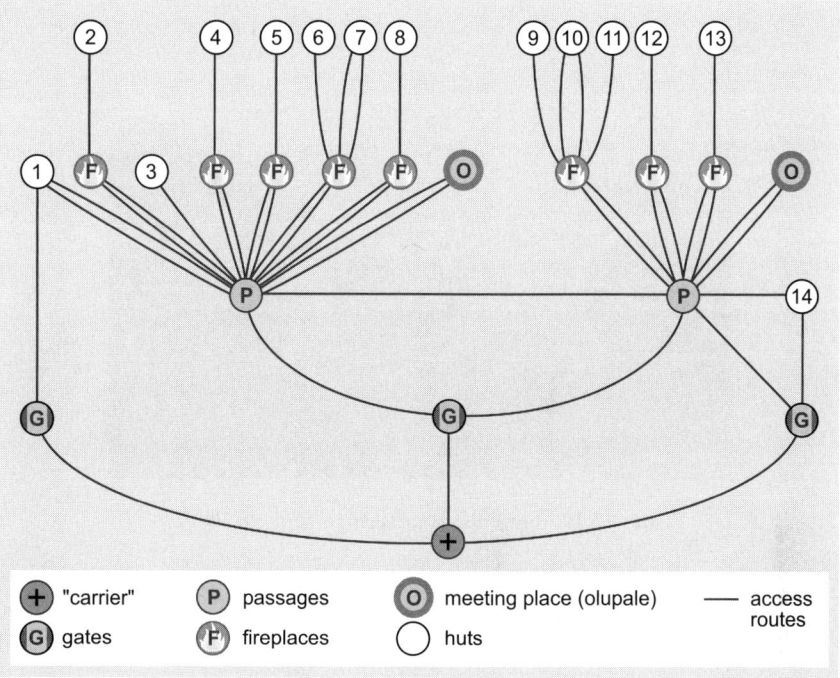

Figure 15.9. Permeability map of a ≠Akhoe Hai//om wet-season camp

as in Figure 15.3, which has no fences and no passages. The thorn fences of the
wet-season camp and palisade fences of permanent Owambo homesteads require
the establishment of clearly defined gates (G) for entering the fields and of less
clearly defined passages (P) through the garden. Unlike the case of the *olupale/
olupare,* these gates and passages are not culturally elaborated upon (neither
among Owambo nor among Hai//om), nor do the Hai//om use Owambo terms to
name them. We may even have difficulties in deciding whether these are features
of the built environment or of the landscape around it. Often these fences are
created simply by cutting down the encroaching bushes in and around fields. But
nevertheless gates and passages directly influence the permeability features of the
camp and its resting places.

The boundary of the camp thickens as visitors have to go through a number
of predefined steps before they approach the huts. Furthermore, activity areas
under shady trees in the field form another stopping place before reaching the
huts. Within the settlement the huts are no longer accessible through a central
space as families create two clusters with fields in between. Unlike the inhabitants of a large Owambo homestead (see Figure 15.10), Hai//om do not cultivate a
communal field; each couple or individual has a distinct stretch of land on which
to grow millet. Accordingly, Hai//om wet-season camps are frequently made up
of two or three clusters of huts. The dry- and the wet-season camps compared

Figure 15.10. Owambo homestead in northern Namibia

here are occupied by the same local group, which forms a single circle of huts under the conditions of the dry-season camp (see above) and it breaks up into clusters under the conditions of the wet-season camp.

Therefore, whereas the addition of an *olupale* to a settlement need not have any immediate effect on the permeability of space, it emerges from the permeability maps that the addition of surrounding fields (fences, gates, and paths) has indeed structurating implications. In places where there are many homesteads and many gardens these fences also directly influence the permeability of the landscape at large as I experienced several times when, what appeared to be a road fit for vehicles, abruptly ended at a newly erected fence surrounding an Owambo homestead. The last decades have seen a continual shrinking of communal lands for the use of transhumant herders and of foragers. Fencing, especially illegal fencing, has been a major factor in this process and it has affected mobility patterns and the scope for movements both at a microscale and in the region at large.

15.5. THE MERGING OF LANDSCAPE AND SETTLEMENT

Having outlined the method of permeability analysis above we may now consider how the same framework can be used to include, not only the domestic built space, but also the landscape around it. To begin with it is important to recall that some features of the landscape are already included in the analysis of domestic space. This includes, for instance, the presence (or absence) of big shady trees in the settlement or the fact that in the hunter-gatherer camps the difference between a ('natural') bush and a ('cultural') windbreak or a hut is not a difference of kind but one of degree. Anything within the spectrum between naturally growing bushes and carefully made constructions may be used as a windbreak and for storing things and food at a high-up place out of the reach of most animals (and children). In a similar vein some built spaces may be considered to be like natural landscapes (urban landscapes, inner-city jungle, picturesque villages). Certainly the Hai//om whom I accompanied to Windhoek considered the capital city to be a wild place. Conversely, for occupants of the bush, spaces considered 'natural' or even 'wild' by others are often straightforward 'cultural' landscapes (as exhibited by the practices of naming places).

Thus, the ethnography provides numerous examples in which the boundary between 'natural' and 'cultural' landscapes is blurred or fuzzy at best, and outright misleading at its worst. We may now consider how to draw permeability maps for the larger landscapes surrounding the settlements. In southern Africa, in particular, the first features to be mapped in such large-scale permeability maps are fences and cutlines (linear clearances, introduced primarily by the army to control the land). Both features have interesting implications in the colonial and postcolonial history of countries such as Namibia (see Widlok, 2003 and Dieckmann, this volume).

Boundaries not only exist on paper or in peoples' minds. In this part of the world at least, they also structure the landscape on the ground. This is true for international borders but also very much so for boundaries such as the 'red line' which used to separate the colonial police zone from the communal areas of the north (see Widlok, 1999a). Today the red line still exists as a gameproof fence and a series of guarded gates cutting across Namibia and the Hai//om settlement area (see Widlok, 2003). It effectively prevents the movement of livestock and of game from the north to the south. It therefore has an indirect effect on the landscape through the impact of animals (usually overgrazing on the so-called 'communal' side) as well as through the suppressed impact of animals (elephants and other animals can no longer move seasonally between Etosha and the Kavango).

Fences have become an entrenched feature of the landscape itself. Humans may, theoretically, cross any fence (just as they may – theoretically – open any door or window and pass from one room to another). But sanctions for doing so are very explicit, especially in the so-called commercial farming area but also increasingly within both the so-called 'communal' and the 'commercial' land. Taking shortcuts by moving from one place more or less directly to another is inhibited as farmers threaten to sanction this kind of 'trespassing'. Fences very directly influence what one is allowed or prohibited to do on either side of the fence. For the Hai//om fences restrict not only the right to settle or to hunt and gather but they also restrict the simple crossing of a fence that separates the road from the farms because farmers do not want any travellers to use the bush as a toilet because they fear the spread of diseases to their livestock. Thus, fences not only affect livestock or symbolise ownership, they regulate people's movements and directly affect the permeability of space. The same holds true for cutlines and the roads that are usually found on these cutlines. Given that free movement across the land is often no longer possible and long detours are required, all residents in the area increasingly depend on transport opportunities along the roads and cutlines which have now replaced the long linear open spaces of the dry riverbeds. The *omurambas* served for orientation and transportation purposes in the early colonial era but are now cross-cut by fences.

Therefore, fences may be considered to be walls in the landscape; at least they have equivalent entailments with regard to permeability. Just as not all walls are the same (some are more perforated or permeable than others) so fences, too, differ. But what about features of the environment that are much less clearly a product of human intervention? Water sources are a good case in point. Here, again, the boundary between a 'natural' and a 'manmade feature' is a fuzzy one. The spectrum reaches from the concrete water reservoirs with diesel pumps, across water sources such as windpumps, manually operated wells, dug-out wells, to open water sources and hollow trees. The amount of human work invested into these sources may differ, but it is a difference of degree and all these sources are equally features of the landscape as well as integral features of human settlements (at least potentially). And, as with settlement features, it is the pattern of water points, more than the design of the individual water point, which regulates permeability. The pattern is influenced by the availability of water (seasonal or permanent) but in other regions also the ability to cross water where there are forts (just as passes across mountains for that matter).

There are other features of the land that also come to mind. In some cases the spectrum also includes animal-induced changes of the environment, thereby further softening the boundary between 'natural landscape' and 'settlement'. When looking at aerial photographs of my research area in northern Namibia it was notable that certain trees were growing in groves, in particular the prized Mangetti nut (see Widlok, 2006) but there were also thin lines of Mangetti trees which – like roads – connected these groves, and the groves with other places such as water points in the area. These trees were not planted, they had grown either as humans moved around in regular patterns and dropped nuts or – more likely – as wild animals (ruminants in particular) had moved along their paths, chewing the outside of the Mangetti nuts, leaving the intact kernel behind. The 'avenues' that were created in this way now continue to be used by animals and humans because the trees provide food and shade (and probably orientation, as well). This is a clear case in which landscape and settlement patterns merge in terms of their permeability.

Similarly, changes in soil type, another landscape feature, also influence permeability. Soft sand and densely overgrown soils make it difficult to move whereas places linked through stretches of hard soil are much better connected to one another. Consequently, the semantics of landscape descriptors in the language of the ≠Akhoe Hai//om who live in this area highlights features of the ground. As I have pointed out elsewhere (Widlok, 1997) it is important to note in this context that landscape terminology is used in spatial language situations ranging from large-scale (orientation in the bush) to small-scale (orientation of objects right in front of the speaker). Furthermore, landscape terms are not only used for the larger landscapes around but also within larger settlements. The settlement at /Gomais, for instance, was divided into three 'quarters', associated with the residents' different places of origin and which highlight features of the soil, namely 'red mound' (*/aba!anis*), 'black valley' (*≠nu!hoas*), and 'white hill' (*!uri!hums*), although these quarters were only a few metres apart from one another.

The merging of landscape and settlement may be particularly marked in the case of African hunter-gatherers who seem to be less inclined than their neighbours to subscribe to a rigid divide between nature and culture, between human and nonhuman, between the agency of humans and other forms of agency (see Ingold, 2001). However, I suggest that upon closer inspection there may be more cases of this than we may initially think. Other groups in the area may conceptually distinguish large significant features of the landscape (e.g., the Waterberg for Namibian Herero) from the small but significant features of their settlements (e.g., the place of ancestral fires and graves). But recent research on the importance of small-scale feature for large-scale landscape management suggests – for instance, the importance of graves for dam construction (see Bollig, 1997, this volume) – that the two are often mutually constitutive. Anecdotal evidence from sites such as the Tsolido hills (see Gall, 2001) also suggest that these connections are not recent phenomena but are likely to go back a long way and therefore are relevant to archaeological reconstructions as well (see Widlok, 2005).

The continuities between landscape and settlement in these cases are largely implicit in the practices of moving around on the land, by using spatial language, by visiting graves, or by attending to special fireplaces. In other cases, such as that of the Australian Aborigines, the continuities are made much more explicit; they even become the orthodox view. However, I do not think that we are dealing with a feature that is peculiar to hunter-gatherer societies or to non-European societies for that matter. Some of the most sophisticated forms of landscape gardening similarly connect domestic space with the landscape 'out there'. In classical Japanese gardens the views from buildings into the garden as well as from various points in the garden to the 'borrowed landscape' beyond the confines of the garden are well known and have been cultivated practices for a long time (Ito, 1999). European landscape gardens have been preoccupied with axes of views for centuries (v. Pückler-Muskau, 1996) so that any present-day reconstruction of these gardens usually relies on a representation that is very similar to the permeability maps presented here. There are, therefore, a number of possible starting points for a more systematic connection between the analysis of landscape and the analysis of settlements. Permeability maps will only be one feature of this kind of analysis.

15.6. CONCLUSION: THEORETICAL AND METHODOLOGICAL IMPLICATIONS

What are the theoretical and methodological implications of the type of analysis presented here? Firstly, the question as to whether social agents separate a settled world from 'the landscape' out there is an empirical one which has to be investigated in each case. We cannot assume that such a separation necessarily exists, that it is defined in the same way across cases, and that it remains unchanged over time. Secondly, no matter whether there is such a separation in emic terms, it is useful to consider which features influence spatial permeability in settlements and beyond the domestic space. Some may be found across the spectrum; others may be limited to a micro- or a macro-scale.

Thirdly, cross-fertilisation between two hitherto separate subfields of research is enhanced. Some of the ideas (and methods) that have been developed in the anthropology of landscape may prove fruitful when they are carried over into the analysis of settlements and vice versa. Fourthly, there are already elements that can lead towards an integrated theory of space as outlined with regard to the definition of spatial permeability and the ways in which it relates to social permeability.

Fifthly, the methodological emphasis on bird's-eye view maps (whether of homesteads or of the world at large) can be usefully complemented by the application of permeability maps as research tools. One of the main advantages of these maps is that they are practice-oriented; that is to say they allow us to include the relative positions of social agents in spatial structures, whether they are manmade, natural, or most appropriately considered an integral product of both.

Lastly, developing tools and theories that bridge the divide between the anthropology of built forms and the anthropology of landscape not only helps to redress a bias based on a specific cultural background (that of sedentary agriculture). It also helps us to address the current and future changes of humans living their lives in an environment in which landscapes increasingly resemble the built environment, either because of increased population pressure or because they become particularly marked and intertwined as enclaves in a global built environment. The notion of a 'global village', misleading as it may be, is now supplemented by a growing notion of a 'global garden' (Harrison & Harrison, 1999) and it will be anthropology's role to critically assess this notion and to identify the social relations that lead to a situation in which some individuals and groups occupy the centre of this 'garden' and others are being placed outside, or in the backyard as it were. Further research will need to show, not only the permeability of places in comparative perspective, but also how permeability is being altered over time.

REFERENCES

Brody, H. (2001). *The Other Side of Eden. Hunter-Gatherers, Farmers and the Shaping of the World.* London: Faber & Faber.

Dafinger, A. (2001). An anthropological case study on the relation of space, language, and social order: The Bisa of Burkina Faso. *Environment and Planning*, 33, 2189–2203.

Dafinger, A. (2004). *Anthropologie des Raums: soziale und räumliche Ordnung im Süden Burkina Fasos.* Köln: Köppe.

Gall, S. (2001). The Bushmen of Southern Africa. Slaughter of the innocent. London: Ghatto & Windus.

Gell, A. (1998). *Art and Agency. An Anthropological Theory.* Oxford: Oxford University Press.

Harrison, H.M. & Harrison, N. (1999). *Grüne Landschaften. Vision*: Die Welt als Garten. Frankfurt/ Main: Campus.

Hillier, B. (1991). Visible colleges: Structure and randomness in the place of discovery. *Science in Context*, 4, 23–49.

Hillier, B. & Hanson, J. (1984). *The Social Logic of Space*. Cambridge: Cambridge University Press.

Hirsch, E. & O'Hanlon, M. (Eds.) (1995). *The Anthropology of Landscape: Perspectives on Place and Space*. Oxford: Oxford University Press.

Ingold, T. (2001). *The Perception of the Environment. Essays in livelihood, dwelling and skill*. London: Routledge.

Ito, T. (1999). *Die Gärten Japans*. Köln: DuMont.

Kent, S. (Ed.) (1993). *Domestic Architecture and the Use of Space: An Interdisciplinary Cross-Cultural Study*. Cambridge: Cambridge University Press.

Latz, P. (1996). *Bushfire and Bushtucker. Aboriginal Plant Use in Central Australia*. Alice Springs: IAD Press.

Layton, R. (1997). Representing and translating people's place in the landscape of northern Australia. In A. James, J. Hockey & A. Dawson (Eds.), *After Writing Culture. Epistemology and Praxis in Contemporary Anthropology* (pp. 122–143). London: Routledge.

Moore, H.L. (1996). *Space, Text, and Gender: An Anthropological Study of the Marakwet of Kenya*. New York: Guilford Press.

Pückler-Muskau, H.F.v. (1996). *Andeutungen über Landschaftsgärten verbunden mit der Beschreibung ihrer praktischen Anwendung in Muskau*. Frankfurt/Main: Insel.

Pyne, S. (1991). *Burning Bush. A Fire History of Australia*. New York: Henry Holt.

Widlok, T. (1997). Orientation in the wild: The shared cognition of Hai//om Bushpeople. *Journal of the Royal Anthropological Institute*, 3, 317–332.

Widlok, T. (1999a). *Living on Mangetti. 'Bushman' Autonomy and Namibian Independence*. Oxford: Oxford University Press.

Widlok, T. (1999b). Die Kartierung der räumlichen und sozialen Durchlässigkeit von flexiblen Lokalgruppen. *Zeitschrift für Ethnologie*, 124, 281–298.

Widlok, T. (1999c). Mapping spatial and social permeability. *Current Anthropology*, 40, 392–400.

Widlok, T. (2003). The needy, the greedy and the state. The distribution of Hai//om land in the Oshikoto region. In T. Hohmann (Ed.), *History, Cultural Traditions and Innovations in Southern Africa*: vol. 18. *The San and the State. Contesting Land, Development, Identity and Representation* (pp. 87–119). Köln: Köppe.

Widlok, T. (2005). Theoretical shifts in the anthropology of desert hunter-gatherers. In P. Veth (Ed.), *Desert Peoples. Archaeological Perspectives* (pp. 17–33). Malden, MA: Blackwell.

Widlok, T. (2006). Two ways of looking at a Mangetti grove. In J. Maruyama, L. Wang, T. Fujikura & Ito M. (Eds.), *Proceedings of Kyoto Symposium: Crossing Disciplinary Boundaries and Re-Visioning Area Studies: Perspectives from Asia and Africa* Kyoto: University of Kyoto (pp. 12–24).

Part IV

Language and the Conceptualisation and Epistemics of African Arid Landscapes: Perspectives from Linguistics and Oral History

Chapter 16

Two Ways of Conceptualising Natural Landscapes: A Comparison of the Otjiherero and Rumanyo Word Cultures in Namibia

WILHELM J.G. MÖHLIG

16.1. INTRODUCTION

Landscapes, as natural phenomena of the environment, are not stable/unstable during the course of time. As seen from the perspective of those living in a specific environment, the dynamics of landscape are caused by inner and outer factors. Inner factors are permanent or temporary changes due to the global or local climate and degradation. By outer factors, we understand political or government interference and processes of migration. Whatever the reasons may be, according to our experience, the speakers of specific word cultures always tend to adapt their former systems of conceptualising landscape to new situations. For a historical linguist, the processes of adaptation remain visible in the records of landscape terminology and the underlying historical processes can be reconstructed on this basis. Therefore in this chapter we try to show, not only the contemporary dimensions

M. Bollig, O. Bubenzer (eds.), *African Landscapes*,
doi: 10.1007/978-0-387-78682-7_16, © Springer Science+Business Media, LLC 2009

of landscape conceptualisation of two Namibian word cultures, but also, in a historical perspective, the dynamism of adaptation to new environments, active in these languages.

16.2. METHODOLOGICAL REMARK

16.2.1. The Problem of Incompatibility

In comparing several word cultures with respect to their systems of landscape conceptualisation, we are confronted with the problem that the analytic results are often so individualistic that they finally become incompatible. If the word cultures to be compared belong to the same linguistic subfamily, the probability of discovering some general characteristics showing their dependencies on shared experiences with regard to history, economy, or environment increases. For such a case study of 'genetically narrow' comparison, we have chosen the word cultures of the Herero living in Kaokoland, central and east Namibia, and the Manyo (Gciriku and Shambyu) living along the central Kavango river. Both languages, Otjiherero and Rumanyo, belong to the Bantu family, which covers the whole subcontinent of Africa starting from Cameroon, Congo, Uganda, and Kenya in the north and going down south to Namibia, Botswana, and the South African Republic. According to Guthrie's still valid classification of Bantu languages (Guthrie, 1971, vol. 2, pp. 28 ff.), they belong to two different zones. Otjiherero, the language of the Herero, is a member of zone R together with the Owambo languages and Umbundu in Angola. Rumanyo, the language of the Gciriku and Shambyu, is allotted to zone K together with some other Kavango languages and a majority of languages spoken in central and east Angola as well as in Zambia.

The linguistic inventories of the Bantu languages in general show many similarities that regularly correspond in form and meaning. The British Bantuist Malcolm Guthrie systematically recorded and compared these correlations. On this basis, he constructed a thesaurus of shared features, which he called *Common Bantu* (Guthrie, 1967–1971). We use this thesaurus as a reference to compare the sound inventories and some grammatical items of Otjiherero and Rumanyo. In this way, our comparative analysis is in fact threefold, namely amongst Common Bantu, Otjiherero, and Rumanyo. The advantage of this procedure is a greater certainty in drawing conclusions on a comparative level.[1]

---

[1] In this chapter, we leave aside the problems of cultural outsiders when dealing with the conceptual systems of insiders. The interested reader will find further details concerning this question in Kathage (2004).

16.2.2. Semantic Fields

The concepts of landscape together with their linguistic representations in a specific word culture form so-called semantic fields.[2] For nonlinguists, the analogy may be a bit misleading. Unlike fields as phenomena of the natural surface, semantic fields are not defined by their boundaries but only by their centers. Consequently, at their peripheries, they overlap with other semantic fields that are similar and with which they share some properties.

Concepts, in each word culture, form multidimensional networks in the sense that their semantic components show all sorts of relations such as overlapping, contrast, antagonism, complement, redundancy, and so on. These networks are unstable. In a historical dimension, it can be observed that they are in constant alteration. New concepts have to be named to meet the communicative necessities of daily life; old concepts fall into disuse. They are no longer communicated, and their names become obsolete. To cope with this endless and dynamic nature of conceptualisation and coining new notions, semanticists 'plant their stakes' wherever they have a thematic interest and take the perspective chosen at random as the center of a more or less circular 'horizon' of other adjacent and interrelated concepts. This arbitrary section of a semantic whole is called a semantic field. In this sense, in this chapter we speak of the semantic field of landscape.

16.2.3. The Structure of Semantic Fields

Despite the arbitrariness on which the first methodological step is based, semantic fields universally show hierarchical (vertical) and horizontal structures. Hierarchies[3] are governed at all levels by the principle of inclusiveness. An item at a higher level always includes items of a lower level. In a horizontal dimension, the semantic relationships between different items are more complex. They may be contrastive, complementary, adjacent, overlapping, and so forth. For instance, the English concept and lexical term for 'tree' includes the more specific terms such as 'oak', 'beech', and 'fir', but it excludes 'shrub', 'bush', and 'grass'. If we follow the taxonomic tree of the item 'tree' and compare the items at the level of individual species, we find that 'oak' and 'beech' belong to the subcategory 'deciduous trees' and 'fir' to the 'coniferous trees'. In other words, 'oak' and 'beech' are horizontally in complementary relationship, whereas 'fir', in the same dimension, has a contrastive relationship with the other two items.

---

[2] The term 'semantic field' was introduced by Trier, Portzig, and others (Lyons, 1977, p. 250) in the 1920s and 1930s. Since then the theory has further been developed. However, the scholarly discussions on certain aspects of the field theory are still going on. Anyway, for the description of thematically related concepts and their denotations, the instrument of semantic field has proved its heuristic value. This is the reason why we use it in this chapter.

[3] For the principle of hierarchy in taxonomies, see Berlin et al., (1973).

For our descriptive purposes, focusing on landscape, we distinguish the following levels of taxonomic hierarchy.

1.   Type of landscape
2.   Subtype of landscape
3.   Individual landscape
4.   Components of landscape

– General components
– Complex components
– Single components

5.   Elements of components
6.   Surface

For instance, in Otjiherero,[4] a landscape type is *okutí*, 'uncultivated land'. At the same hierarchical level, it is opposed by another type called *otjirongó*, 'inhabited land'. An Otjiherero example of a subtype is *okutí ongaángo*, 'unpopulated land'. As the header noun shows, it falls under the type *okutí*. The appositional noun *ongaángo* denotes 'waste country, wilderness'. Thus it underlines and specifies the inherent distinctive feature of the head noun *okutí*. At the same hierarchical level of concepts, it is complemented by another subtype called *okutí kozóndundú*, 'mountainous region'. An example of an individual landscape in Otjiherero one level below *okutí ongaángo* is *ehándjá* or *otjáná*, 'flat grassland, treeless plain'. At the same hierarchical level, we find *otjihwa*, 'thick forest area, bush country' and *onamiva*, 'treeless sand desert'. Neither in Otjiherero nor in Rumanyo, do we find specific notions with the abstract meaning 'landscape'. However, the concept as such appears to exist, because adult indigenous speakers can usually name distinct coherent areas that an outsider would also recognise as individual landscapes.

Some components of landscape belong to several hierarchical levels. They may either constitute individual landscapes or contribute to several landscapes. In Otjiherero, such an example is the term *ozondondú*, 'mountains'. At the level of subtypes, it functions as a qualifier *okutí kozóndundú*, verbally 'uninhabited region of mountains'. At a lower level, its plural form denotes a landscape, and one level further below, its singular form may refer to a general component of a landscape. The same hierarchical level comprises a variety of complex and single components such as the notions for 'table-mountain', 'conical hill', 'cliff', and so on. (see Table 16.4). Complex components are conceptualised as individual wholes of landscape although they are visibly composed of several elements such as, for instance, valleys. A valley consists of at least a V- or U-shaped profile with slopes on two sides, a floor, and a long stretched ground. Sometimes it may

---

[4] We follow the established rules of orthography, but deviate from it in marking the distinctive tones. High tones are marked by an acute over the bearing syllable: *á*, and falling tones by a circumflex: *â*. Low tones remain unmarked: *a*.

become difficult to decide whether a component is complex or single. In a conceptualisation analysis, the question does not matter, because differences become visible at the level of features (see Section 16.2.4).

Elements of landscape are the geological constituents of complex or single components such as 'slopes', 'riverbanks', 'paths', and the like. The same elements may constitute various landscapes. Some, for example, 'paths' and 'trails', even connect or run across different landscapes. Finally, the term 'surface' normally refers to the visible, not the hidden ground. As compared to the great variety of types of soil that Kanuri[5] differentiate (Platte & Thiemeyer, 1995), it is amazing that the Otjiherero and Rumanyo language cultures possess only small catalogues of names referring to the characteristics of the ground. The same surface notions are used in the context of different landscapes.

16.2.4. Prototypical Features and Concrete Values

Hierarchical concepts at all levels have prototypical features that are filled by concrete values.[6] The features referring to size, situation, shape, vegetation, ground conditions, and the like are more or less 'objective' properties that can also be observed and measured by outsiders, for instance, by geographers. The concrete values, however, refer to properties that only insiders attribute authentically to specific concepts. Whether a certain solitary hill in Otjiherero is referred to as *okarundú*, 'hillock', *otjivándá*, 'head shaped hill', or *oruvándá*, 'long hill with flat top' depends on how an insider evaluates the natural phenomenon according to the parameters she has at her disposal. To find out the insiders' evaluations, there are three methods at hand.

- Firstly, the whole set of landscape terms is internally compared one by one.
- Secondly, several insiders are independently interviewed with regard to the meaning and application of single terms, and these texts are later analysed by an outsider, eventually with the support of an insider.
- Thirdly, texts that have been created for other purposes but contain landscape terms are analysed with text linguistic methods. This applies to cases where amongst two or more landscape elements the surface features are the same, but in the minds of the people they are treated as being different. Place-names may provide further insights here.

For instance, in the Gciriku conceptual system, a particular waterhole is called *Shamahího*, 'place of women's headdresses or wigs'. Its water is not used. The conspicuous name goes back to the first decades of the twentieth century,

---

[5] Nilo-Saharan language spoken near Lake Chad.

[6] In this chapter, we do not use the concepts of feature and value in the same sense as the so-called frame theory (see Barsalou, 1992), although there may be overlaps. Again, our definitions and applications of these concepts are governed by heuristic considerations.

when two women were killed and swallowed by a lion there. Only the indigestible remnants of their headdresses were later found in the lion's droppings so that their fate became known. Because of this event, the waterhole is affected by a taboo that distinguishes it considerably from neighbouring waterholes. For an outsider, all the waterholes of the region appear to belong to the same geographical type. However, insiders know of its particular properties. In their eyes, it bears an emotional feature that is distinctive.[7]

16.2.5. Etymologies

In Otjiherero and Rumanyo, landscape concepts are usually expressed by nouns or nominal phrases, that is, nouns and complements (other nouns, adjectives, nominal possessives[8]). Most of these names show formal features that give the expert, not necessarily also the insiders, hints about their history. For instance, the Rumanyo term *rukâmbero*, 'outlook' is composed of four morphemes: the class prefix *ru-* 'cl. 11 for long objects or processes of long duration', the verbal root *–kâmb-* 'look out to the far distance', *-er-* 'verbal derivation of applicative', and *–o* 'nominal derivation expressing the means or instruments of performing the action of the verbal stem or root'.[9] When we put these semantic components together, we arrive at the following paraphrased meaning: 'an elevated place from where one can look out to the far distance'.

16.3. LINGUISTIC PROPERTIES SHARED BY THE OTJIHERERO AND RUMANYO LANGUAGES

In this section, we discuss some linguistic properties that both word cultures have in common and that are important for the etymological analysis of the language material offered in this chapter.

16.3.1. Comparative Sound Inventories

In Table 16.1, the first column shows the so-called starred sounds of Common Bantu (CB) as compiled by Guthrie (1967–1971) at the diaphonemic level. The symbols by which Guthrie represents the starred items may be confusing inasmuch as they do not represent, unlike orthographical symbols, real sounds, but more or less long series of corresponding sounds in a multitude of related languages. In

---

[7] I first learned about the story through a text by the indigenous writer Herbert Ndango Diaz (1994, pp. 12 ff.).

[8] A nominal possessive qualifies a header noun by its possessor. This syntactical figure corresponds to the genitive in Latin, French, English, or German.

[9] Further details see Möhlig, (2005, Section 4.2.1).

other words, these starred phonetic symbols are stereotyped entities at the abstract dialevel of comparison. Their regular and real correspondences in Otjiherero (Her) and Rumanyo (Ma) are shown in the second and third columns.

16.3.1.1. Comparison of the Vowels

In Rumanyo, the close vowels *î and *û merge with their nearest neighbors in the vowel system, that is, *i and *u. However, in Otjiherero *î merges with *e, and only *û merges with its nearest neighbor in terms of tongue height position. For instance, CB *-kíngò, 'neck' = Her *osengo*, Ma *ntîngo*; CB *–céké 'sand' = Her *ehéké*, Ma *mushêke*; CB *-gÙbÚ 'hippo' = Her *ondúú*, Ma *mvúghu*, *-júdù 'nose' = Her *eúrú*, Ma *liyûru*.[10]

Table 16.1. Comparative List of Vowels

Comparative Bantu (CB)	Othiherero (Her)	Rumanyo (Ma)
*a	a	a
*e	e	e
*Iᵃ	i	i
*o	o	o
*u	u	u
*U	u	u

ᵃThe symbols *I and *U stand for high close vowels that often have palatalising or labialising effect on the adjacent consonants. These effects differ from the phonetic effects of "ordinary" *i and *u and therefore have to be written differently.

16.3.1.2. Comparison of Consonants

The phonetic values of the characters are almost the same in Otjiherero and in Rumanyo. The following characters need some explication: *tj* = voiceless, palatal affricate; *v* = voiced, bilabial approximant; *vh* = voiced, dentilabial fricative; *gh* = voiced, velar fricative or approximant; underlined characters mark dentals; *y* = voiced, palatal fricative; nasals in combination with other consonantal characters mark prenasals.

The consonantal system of Otjiherero, in comparison with Rumanyo and other Bantu languages, possesses two particular features. Firstly, it has a class of addental consonants – namely *s*, *z*, *t*, and *n*, forming a distinctive contrast with homorganic alveolars, namely *s*, *t*, and *n*. Secondly, it has only voiced prenasals, namely *mb*, *nd*, *ng*, and *ndj*. The peculiarity of Rumanyo is its behavior in

---

[10] All the other divergent consonants between Otjiherero and Rumanyo are based on regular sound correspondences.

Table 16.2. Comparative List of Consonants

Comparative Bantu (CB)	Othiherero (Her)			Rumanyo (Ma)		
		In front of			In front of	
		Close vowels			Close vowels	
	Ordinary vowels	I_I	I_U	Ordinary vowels	I_I	I_U
*p	p	s	s	p	f	f
*t	t	s	t ~ s	t, sh~tj /_i	t	t
*k	k, tj /_i/e[a]	s	t ~ s	k, sh /_i	t	f
*c	h	h		sh ~ h		
*b	v ~ w	z		v	vh	vh
*d	r	z		l ~ r	d	
*g	y ~ Ø	z		gh, Ø	d	
*j	y			y		
*m	m			m		
*n	n	n		n	n	
*ny	ny			ny		
*mp	mb			mp		
*nt	nd			nt		
*nk	ng		nd	nk		
*nc	h			ntj ~ h		
*mb	mb			mb		
*nd	nd			nd		
*ng	ng	nd		ng		
*nj	ndj			ndj		

[a] *The symbol /_ means 'in front of'*

front of close vowels. Whereas most Bantu languages, such as Otjiherero, show affrication, Rumanyo, at least in the domain of *t* and *k*, retains the plosive character of these sounds. Yet, both languages show affrication in front of *i*, Otjiherero additionally in front of *e*. All sounds of the Otjiherero inventory find their counterparts in Common Bantu. Rumanyo, in addition to this catalogue, has five click sounds, which are probably borrowed from Khoisan languages. However, with all the progress made in the documentation of the diverse Khoisan languages, even after 40 years of research, we are still unable to trace the donor language of the Rumanyo click sounds. The landscape terminology of Rumanyo is also affected and, furthermore, many place-names, particularly in the south of the Kavango valley, bear click sounds.

16.3.2. Strategies of Word Formation

Under the aspect of word formation in the domain of landscape terminology, four grammatical processes are relevant for the two languages compared. These are: (1) nominal class shift, (2) compounding of different word stems, (3) adding qualifiers at phrase level, and (4) verbal derivations. We deal with these grammatical topics one after the other.

1. Shift of Noun Classes

Perhaps the most prominent feature of all Bantu languages is the nominal class system. It governs the syntactical structures of sentences and is important for word-building processes. For comparison, we list the Otjiherero and Rumanyo systems side by side (see Table 16.3).

Table 16.3. Comparative List of Noun Classes

CB Class number	Otjiherero (Her) Class prefix	Example/function	Rumanyo (Ma) Class prefix	Example/function
01	*omu-*	*omundu* "person"	*mu-*	*muntù*[a] "person"
02	*ova-*	*ovandu* "persons, people"	*va-*	*vantù* "persons, people"
01a	Ø	*maamá* "my mother"	Ø	*nná* "my mother"
02a	*oo-*	*oomaamá* "my mothers"	*va-*	*vanáne* "my mothers"[b]
03	*omu-*	*omutl* "tree"	*mu-*	*mutávi* "branch"
10+03			*dlmu-*	*dimutávl* "branches"
04	*omi-*	*omitl* "trees"		
05	*e-*	*ewe* "stone"	*li-*	*llwé* "stone"
06	*oma-*	*omawe* "stones"	*ma-*	*mawé* "stones"
07	*otji-*	*otjihávéro* "chair"	*shi-*	*shipúna* "chair"
08	*ovi-*	*ovihávéro* "chairs"	*vi-*	*vipúna* "chairs"
09	*o(N)*[c]	*ongombe* "cow"	*(N)*	*ngómbe* "cow"
			Ø	*hóve* "ox"
10	*ozo (N)*	*ozongombe* "cows"	*(N)*	*ngómbe* "cows"
02+09			*va (N)*	*vahóve* "oxen"
11	*oru-*	*oruvyo* "knife"	*ru-*	*rufúro* "dagger"
06+11			*maru*	*marufúro* "daggers"
12	*otu*	*otuvyó* "knlves"		
13	*oka-*	*okakambe* "horse"	*ka-*	*kakâmbe* "horse"
				kakúru "owl"
12			*tu-*	*tukâmbe* "horses"
02+13			*vaka-*	*vakakúru* "owls"
14a	*ou-*	*oukambe* "horses"		
14	*ou*	*outá* "bow"	*(gh)u-*	*utè* "gun"
06+14a	*omau-*	*omautá* "bows"	*ma(gh)u-*	*mautà* "guns"
15a	*oku-*	*okurama* "leg"	*ku-*	*kufù* "cold season"
		okulyá "thom"		
02+15a				*vakufù* "cold seasons"
06	*oma-*	*omarama* "legs"		
06+15a	*omaku-*	*omakulyá* "thoms"		
16	*pu-*	*pondjúwó* "at a house"	*pa-*	*pandjúgho* "at a house"
17	*ku-*	*kondjúwó* "towards a house"	*ku-*	*kundjûgho* "towards a house"
18	*mu-*	*mondjúwó* "inside a house"	*ku*	*mundjûgho* "inside a house"

[a] For Rumanyo, we observe the following conventions of marking the distinctive tones: á = high tone, â = falling tone, à = low tone that causes the preceding syllable to be high. Low tones and predictable tones remain unmarked.

[b] The term also includes the sisters of my mother.

[c] The symbol (N) stands for any homorganic prenasal such as *mb, nd, nz, ng,* etc.

Due to their semantic properties, nouns usually belong to specific pairs of nominal classes (Möhlig et al., 2002, pp. 34 ff.). Sometimes, however, their affiliations to the class system appear opaque to us. Both languages compared are familiar with the mechanism of shifting nouns from one class to another to underline certain prototypical properties such as size, shape, location, or even esteem. For instance, in Otjiherero, the general word for 'mountain' is *ondundú* belonging to class 9/10. To express the smallness of an elevation in the sense of 'hillock' the underlying stem, *–rundú* is shifted to class 13/14a, which usually contains small objects. In other words, the form becomes *okarundú*, plural *ourundú*.

In Rumanyo, we find a comparable case. The general word for 'mountain' is *ndúndu* belonging to class 9/10. To express the particular steepness of an elevation in the sense of 'cliff, steep riverbank', the underlying nominal stem, *–rúndu* is shifted to class 14/6; that is, the form becomes *urúndu*, plural *marúndu*. In a general way class 14 expresses, among other concepts, something that is long and thin or flat. This can also be demonstrated by the Otjiherero term *oruwe* 'cliff', which is derived from the basic word *ewe* 'stone'. The shift of the nominal stem *–we* 'stone' from class 5 to 11 adds the meaning of 'length and thinness'. In a way, this device replaces adjectives in other languages.

2. Compounding of Different Word Stems

In both languages, it is possible to combine two or more word stems (nominal, verbal, or adjectival) within one class prefix. For instance, in Otjiherero, the denotation for 'island' is *okakondwáhí*. This form is composed of the class prefix (cl. 13), the passive verb stem *–kondwá*, 'be cut off' and the nominal stem *–hí*, 'ground'. The etymological meaning is: 'piece of land that has been cut off'. In the perspective of history, this type of formation suggests a rather recent origin, for in the inventory of Common Bantu we even find two entries with this meaning,[11] which shows that the concept as such belongs to the Bantu cultural heritage. The Herero may have developed the concept of 'island' only when they moved into their present settlements. In other words islands did not exist in their former settlement area.

3. Rendering Concepts by Complex Forms at Phrase Level

In all word cultures, there are more concepts than there are denotations for concepts. Lyons (1977, Section 9.6) discusses this question comprehensively in the context of lexical gaps. As a practical example for this general phenomenon is the conception of colors that an individual is neurologically capable of discerning and the terms that are at his disposition to render these concepts in a one-word

---

[11] Compare C.S. 289 *-càngà* with a western distribution and C.S.676 *-dúá* with an eastern distribution. (Guthrie, 1967–1971, vol. 3, pp. 86, 182).

manner in any given language. If there is no one-to-one correspondence between concepts and their expression in words available but felt to be needed in communication, the denotation may be derived from a different, more basic concept, as we have seen in the preceding sections.

Another strategy would be to paraphrase these concepts by an ad hoc invented phrase or a stereotyped formula. Hence, some landscape concepts are rendered at the syntactical phrase level by complex forms, that is, by forms consisting of several words with their own class prefixes. A complex form that both languages use are nouns followed by an adjective. For instance, in Otjiherero, the concept of 'main road' is named, almost as in English, *ondjira onéné*, verbally 'way big'; in Rumanyo, a crossroad is named *mahánga ndjíra*, verbally 'mixture way'. For specifying concepts at a lower hierarchical level, both languages use the grammatical device of nominal possessive (genitive). In taking the higher-ranked word in the position of the item possessed (*possessum*) and another specifying concept in the position of possessor a new name for a concept at a lower hierarchical level may be coined. For instance, in Otjiherero, *okutí* means 'uncultivated land in general'. On that basis, *okutí kozóndundú* 'mountainous region' is formed. It means verbally 'uncultivated land of mountains'. In Rumanyo, we find a parallel example, *shiróngo shándúndu*, verbally 'country of mountains'.

Nominal possessive constructions are also used to express a componential relationship: 'part of a higher ranked concept'. For instance, in Otjiherero, the parts of a mountain are named in this way: *ohongá yondúndu* 'top of a mountain' and *kehí yondúndu* 'base of a mountain'. Specifying landscape units by using the syntactical instrument of nominal possessive is very common all over the world. Because of its casual and partially improvisational character, word formation of this kind is historically always young. If it is not a loan translation copied from another language (European or, as in the case of Otjiherero, also Nama), it may be even younger than word compounding, because historical linguists can often show that a compounded notion has developed from a former nominal possessive. We find that the strategy of nominal possessive formation often serves actual intellectual needs, when, for instance, gaps in the hierarchical system of landscape terminology suddenly become apparent in a discourse.

4. Derivations from Verb Stems

Bantu languages have developed elaborated systems of forming nouns by derivation from verb stems. Usually vowel suffixes with particular stereotyped functions are attached to the verbal roots (Kähler-Meyer, 1967, for Bantu in general; Möhlig et al., 2002, pp. 40 ff., for Otjiherero; Möhlig, 2005, pp. 135 ff., for Rumanyo), and the derived stems are subsumed under the noun classes that correspond best to the stereotyped properties of the newly coined nouns. Otjiherero as well as Rumanyo also use this device in their landscape terminologies.

For instance, in Rumanyo, a watering place for cattle is called *shinwêno*. This noun is composed of three elements: the class prefix *shi-* for class 7, the applied verb stem *–nwên–* 'drink from', and the derivational suffix *–o* denoting 'instruments or other devices to perform the action expressed by the verbal root'. Thus, *shinwêno* means verbally 'container or place to drink from'. In this case, it has taken the special meaning of 'watering place for cattle'. A similar example in Otjiherero is the term *otjikúnino*, 'garden, sowing ground'. It is composed of the class prefix (cl. 7) *otji-*, the applied verb stem *–kúnin–* 'to sow at', and the derivational suffix *–o*. Etymologically it means 'a place where to sow at'. We consider such words as recent formations. It is certainly not accidental to find forms such as *shinwêno*, 'watering place for cattle' and *otjikúnino*, 'garden, sowing ground', because, by tradition, the Manyo are not cattle breeders and the Herero are not agriculturalists. Therefore, the Manyo did not need watering places for cattle, and the Herero had no need of gardens. In this way, both terms can be interpreted as indicators of shifting subsistence strategies in both cultures.

16.4. THE OTJIHERERO SYSTEM OF CONCEPTS ON LANDSCAPE

16.4.1. Basic Structure

Going from top to bottom, *evávêrwá*, 'the universe' is divided into *eyûrú*, 'the sky' and *ouyé*, 'the earth'. The earth comprises of two units: *okuvare*, 'the ocean(s)' and *ehí*, 'the land'. Following the line of 'land', two types of landscape are distinguished by specific terms:

okutí	'uninhabited area, wilderness'
otjirongó	'inhabited area'

Following the line of *okutí* 'uninhabited area', Otjiherero distinguishes three landscapes:

okutí kozóndundú	'mountainous region'
onamiva	'sand desert'
otjihwa	'forest area'

Following the line of *otjirongó*, 'inhabited area' we find the following notions.

otjihúró	'village'
otjikúnino	'cultivation area, garden'

In Otjiherero, three more landscapes are notionally distinguished. These are:

orutjândjâ	'flat grassland'
omaryó	'grazing ground'
omuronga	'river'[12]

They can neither be subsumed under *okutí*, 'uninhabited area, wilderness' nor under *otjirongó*, 'inhabited area', because they cannot be associated with the feature of inaccessibility or hostile nature as in the case of *okutí*, and on the other hand they underlie certain restrictions with respect to human settlement. It is

---

[12] The term refers particularly to the Ókunéné River landscape.

likely that they form a third type of landscape characterised by accessibility and intensive exploitation of their natural resources, without being an area of permanent settlement. This case needs further investigation. So far, it could not yet be ascertained with the help of local experts.

16.4.2. Focus on Mountainous Areas

In order not to overburden this essay with long lists of landscape terms, we restrict ourselves to a componential description of the elaborate semantic domain of mountainous areas, where Otjiherero has a clear conceptual focus. Going from west to east, the vast area of Herero settlement in the northern and northwestern parts of Namibia is characterised by high and low mountains, adjacent plains with scattered hills and inselbergs, and by elevated grass plains that expand over a vast area. Thus, mountains are an important element in Herero perception, particularly in Kaokoland (see Table 16.4).

Table 16.4. Semantic Structure of Mountainous Areas in Otjiherero

	Terms	Etymologies	Features	Values
Subtype of landscape	*okuti kozóndundú* "mountainous area"	Nominal Possessive	1. Anthropo-centric factor 2. Ground relief	1. Not inhabited by human beings 2. Mountainous
Landscape	*ozondundú* "mountains"	Genuine	1. Elevation 2. Size 3. Profile 4. Number	1. Above level 2. Diverse 3. Diverse 4. Many
General component	*ondundú*	Genuine	1. Elevation 2. Size 3. Profile	1. Above level 2. Diverse 3. Diverse
Complex or single components (choice of 10 Items)	*etaka* "table mountain"	Genuine	1. Elevation 2. Profile	1. Tall 2. Table like
	ohúngú "conical hill"	Genuine	1. Elevation 2. Profile	1. Medium 2. Conical
	oruúwá "flat rock"	Genuine	1. Constellation 2. Elevation 3. Size 4. Profile 5. Vegetation	1. Solitary 2. Medium 3. Longish 4. Flat top 5. Bare
	otjivándá "koppie"	Genuine?	1. Constellation 2. Elevation 3. Profile	1. Solitary 2. Medium 3. Round, head like
	oruvándá "long small hill with flat top"	Class shift	1. Constellation 2. Elevation 3. Profile	1. Solitary 2. Medium 3. Longish, flat top
	epúngúwe "solitary tall rock"	Compound	1. Constellation 2. Elevation 3. Profile	1. Solitary 2. Tall 3. Tower like

(continued)

Table 16.4. (cont'd)

	Terms	Etymologies	Features	Values
Elements (choice of 17 items)	*ohongá yondúndu* "peak of mountain"	Nominal Possessive	1. Elevation 2. Size	1. Summit 2. Tall
	kehl yondúndu "base of mountain"	Nominal Possessive	1. Elevation 2. Size	1. Base 2. Big
	ozondotó "chasm"	Genuine?	1. Inclination 2. Size	1. Steep 2. Tall
	orutjené "precipine"	Genuine?	1. Speaker's perspective 2. Inclination	1. Seen from top of mountain 2. Steep
	ombéró "pass, gateway between two summits"	Verbal Derivation	1. Elevation 2. Constellation 3. Size	1. Summit 2. Region between two summits 3. Broad
	otjikoyo "rock shelter"	Genuine	1. Elevation 2. Inclination (profile) 3. Anthropo-centric factor	1. Tall 2. Protruding slope leaving small cave underneath 3. Refuge
Surface (choice of 8 items)	*ehí* "ground"	Genuine	Substance	General
	ehéké "sand"	Genuine	1. Substance 2. Size 3. Consistency	1. Mineral particies 2. Small 3. Solid
	omunókó "mud"	Genuine	1. Substance 2. Size 3. Consistency	1. Mineral particies 2. Small 3. Mixed with water
	ewe "stone"	Genuine	1. Substance 2. Size 3. Consistency	1. Mineral particies 2. Medium to large 3. Solid

16.4.3. Otjiherero Landscape Terms in General

With respect to the linguistic structure and the etymology of the landscape terms collected, over 50% are genuine terms that have to be tentatively classified as inherited. The figure may be too high because of undiscovered loans. Up until now, we have only discovered less than 10% of the modern loans amongst this collection. Otjiherero loans particularly from Nama or other Khoisan languages are difficult to ascertain, because Otjiherero has almost perfectly adapted such items to its own phonetic characteristics. About 10% of the items are derived terms that go back to verbal roots or to other nouns. It is highly probable that the majority of them were coined after the Herero settled in Kaokoland. Compounding with over 10% occurs comparatively frequently. Although, in terms of language history, compounded notions can often be regarded as fossilised syntactical clauses, such as nominal possessives, adjectival clauses, relative clauses, and so forth, in the case of Otjiherero we also have to think of loan translations from Nama, where nominal compounding is still an operative instrument of derivation (Haacke, 1999; Haacke and Eiseb, 2002). In such a case, they would be historically younger formations.

16.5. THE RUMANYO (GCIRIKU, SHAMBYU) SYSTEM OF CONCEPTS ON LANDSCAPE

16.5.1. Basic Structure

In Rumanyo, the universe is divided into *liwîru*, 'heaven' and *livhù*, 'earth'. The earth is divided into *limudíva*, 'ocean' and *liróngo*, 'land'. As in Otjiherero, the land comprises two types of landscape: *mâmbo*, 'uninhabited wilderness' and *utûro*, 'inhabited area'. Following the line of 'wilderness', no conceptual subtypes appear to exist. One step below, at the level of individual landscapes, we find three notions:

In the line of 'inhabited area', the following four notions are to be found.

mbúrundu	'sand desert (Kalahari)'
wíya	'savannah'
mutîtu	'forest area'
In the line of 'inhabited area', the following four notions are to be found.	
shiróngo shámukúro[13]	'river (landscape)'
mundí	'settlement'
mafûva	'cultivation area'
liyâna	'flat meadows along the rivers'

The Kalahari sand desert stretches to the south of the Manyo settlements. It is almost waterless and covered with little vegetation. In the perspective of the Manyo people, it is the home of small groups of !Khung[14] hunter-gatherers. In the olden days, the Manyo did not enter this hostile area. Nowadays, it has been opened up for human settlement by the drilling of boreholes, which make ranching possible.

The savanna area is far-reaching and stretches to the north of the Kavango river oasis. It has rich vegetation and, until recently, was abundant in game. The water resources of the savanna were perennial depending strongly on the productiveness of the rainy seasons. An elaborate terminology exists at the level

---

[13] The terms *shiróngo* and *liróngo* belong to the same nominal root, but to different classes.

[14] The combination of the letters *!kh* refers to a voiceless aspirated postalveolar click sound with an aspirated efflux.

of components and elements to denote the geomorphology of the savanna. Furthermore, we were able to collect over 200 names for trees and shrubs growing in the savanna. Most of them are traditionally exploited for human purposes (diet, medicine, timber, hunting weapons, household utensils, decoration, firewood, etc.). Up to the 1950s, the collection of honey, wild fruits, roots, and plants in the savanna was an important factor for the Manyo economy. Hunting was important too (Fisch, 1994). Linguistic traces of this former economic focus are special terms referring to elements of the savanna landscape such as *lirándo*, 'antelope track', *matênde*, 'elephant footprints', *muvâva*, 'trail of a snake', *lilíra*, 'path of footprints of big game', and so on.

The terminology of the forest area is comparatively simple. The thick forest, called *ucó*, was considered hostile and indigenous people tried to avoid these areas. Nowadays, when the trees of the forest are cut and the area, with the help of artificial boreholes, is turned into cultivation land, the savanna terminology is still applied to the newly cultivated areas.

The riparian landscape shows an elaborate structure of conceptual components and elements. Some denotations refer particularly to the size of the river. The flexible system of the nouns with its derivational prefixes and suffixes allows the expression of these semantic variations, for instance, *mukúro*, 'river in general', *limukúro*, 'big river, stream', and *mukúroghona*, 'brook'. In the case of 'big river', the basic noun *mukúro* consisting of the nominal prefix *mu-* for class 3 and the nominal stem *–kúro* is pre-prefixed by an additional prefix *li-* for class 5, which usually alludes to bigness. In the case of 'brook', the basic noun is affixed by the morpheme *–ghona*, which etymologically means 'child', but as a suffix to nouns, it usually expresses diminishment or smallness. Other elements that concern sections or parts of the river are found in Table 16.5.

The last item has a strong emotional connotation in that it is a favored burial place for people that have been murdered. A corpse dumped there will disappear without leaving any traces behind, because crocodiles will soon devour it.

The Kavango floods annually from December to March. Then large parts of the flat embankment are under water. Perhaps due to the dynamism of the river, the Manyo distinguish many kinds of embankment as shown in Table 16.6.

It is amazing how many water landscape terms refer to the activities of hippos *mvhúghu* as shown in Table 16.7.

Table 16.5.   Terms Denoting Parts of River

runóne	"spring, source"
mpûpa	"waterfall, raplds"
magwánekararo	"confluence"
litóndo	"deep water"
lirúol	"Island"
lintêta lyámukúro	"bend of the river"
linónge	"flcating grass field"

Table 16.6.   Terms Denoting Parts of River Embankment

rugwà	"any visible section of the river bank"
rukénka	"steep embankment"
urúndu	"cliff"
mushêli	"area along the river"
ntére	"water line"
mutambo	"river bank line"
liyâna	"vast plain near the river after the high waters have fallen; a preferred pasture ground"
uwáro	"fjord"

Table 16.7.   Terms Denoting Hippo Activities

mpáre	"place where hippos bathe in the sun"
maghómbe	"pasture of hippos"
lipínda	"resting place of hippos in the reed grass near the riverbank"
livúgho	"resting place on a sandbank"[a]

[a] FISCH (1994:22) lists some more items.

The lexicon contains hundreds of terms denoting the abundant flora of the river landscape. Many of the plants are used for human purposes, be it as food, the fabrication of utensils and weapons, the thatching of houses, and so on.

The catalogue of surface or substance terminology referring to the lowest level of our conceptual hierarchy is very short. As in Otjiherero, we collected less than ten genuine terms, for example, *muvhù*, 'soil, earth', *mushéke*, 'sand', *liróva*, 'clay, mud', *utâre*, 'iron ore', *liwé*, 'stone', *mpéyo*, 'calcrite', and *katirá*, 'red soil powder'.

16.5.2. Focus on River Landscape

Because of limited space, we restrict ourselves to the componential analysis of some selected terms relating to river landscape (see Table 16.8). The conventions are the same as applied in the analysis of Otjiherero terms.

16.5.3. Rumanyo Landscape Terms in General

With respect to the linguistic structure and the etymology of the landscape terms, over 55% are genuine terms that have to be classified as inherited. About 25% are derived terms going back to verbal roots or to other nouns. Because of their periphrastic casual character, it is highly probable that the majority of them were coined after the Manyo settled in the Kavango region. More than 20% are clearly marked as loans. When they contain click sounds, their Khoisan origin is rather obvious. Other notions are recognisable as loans by their prosodological shape deviating from the usual rules of word architecture. Sometimes, even Otjiherero

Table 16.8. Semantic Structure of River Landscape in Rumanyo

	Terms	Etymology	Features	Values
Landscape	*shiróngo shámuróngo* "river area"	Nominal Possessive		1. Inhabited 2. Perennial
General component	*mukúro* "river"	Genuine		1. Broad, long 2. Flowing 3. Perennial 4. Seasonal floods
Complex or single components	*mpûpa* "rapids"	Genuine		1. Falling
	mpáre "resting place of hippos ashore"	Genuine		1. Outside water 2. Resting place of hippos
	urúndu "cliff on river bank"	Class shift		1. Transitional 2. Steep 3. Tall
	magwánekerero "confluence"	Verbal derivation		1. Meeting of two rivers 2. Uniting
	lirúdl "island"	Genuine		1. Inside 2. Above water level
	linónge "floating grass field"	Genuine		1. Inside 2. Floating 3. Grass spec.
	mbámbo "embankment"	Genuine	1. Constellation 2. Constellation	1. Transitional zone 2. Outside riverbed
Elements	*litóndo* "deep water"	Genuine	1. Constellation 2. Size	1. Inside 2. Deep
	licwà "shallow water"	Loan from Khoisan	1. Constellation 2. Size	1. Inside 2. Shallow
	livûndu "thicket near embankment"	Verbal derivation	1. Constellation 2. Vegetation	1. Transitional 2. Thicket
	rukénka "steep embankment"	Genuine	1. Constellation 2. Inclination 3. Size	1. Transitional 2. Steep 3. Medium
	litatá "swampy area"	Loan, unknown origin	1. Substance 2. Size 3. Consistency 4. Vegetation	1. Mineral particles 2. Small 3. Mixed with water 4. Paludal
	lintêta lyámukúro "bend of the river"	Nominal Possessive	1. Ground plan 2. Size	1. Curved 2. Degree of bend
	ntére "water line"	Genuine	1. Constellation 2. Dynamics	1. Inside 2. Separating
Surface (choice of 10 items)	*muvhù* "ground"	Genuine	Substance	General
	mushêke "sand"	Genuine	1. Substance 2. Size 3. Consistency	1. Mineral particles 2. Small 3. Solid
	liróva "mud"	Genuine	1. Substance 2. Size 3. Consistency	1. Mineral particles 2. Small 3. Mixed with water
	liwé "stone"	Genuine	1. Substance 2. Size 3. Consistency	1. Mineral particles 2. Medium to large 3. Solid

appears to be the donor language, as for instance for *murambá*, 'drainage ditch'. In a few instances, European donors (Afrikaans, English, German) are visible as, for instance, in *dóropa*, 'village, small township' from Afr. *dorp*. However, in many cases, a donor language could not yet be identified. There is no doubt that many loans were acquired when the Manyo ancestors began to settle along the Kavango and Kwito rivers. Other loans, particularly those of European origin, entered the language even later on.

16.6. COMPARISON OF THE OTJIHERERO AND RUMANYO CONCEPTUAL SYSTEMS ON LANDSCAPE

16.6.1. General Differences

As we have shown above, Otjiherero and Rumanyo have developed different semantic domains in the field of landscape terminology that they put into prominence. In Otjiherero, this is the mountainous domain; in Rumanyo, there are even two semantic domains selected in this way, the savanna and the habitat of the river oasis.

16.6.2. Comparative Results at the Structural Level

The hierarchical structure of landscape concepts and terms at seven levels is sufficient to deal with all conceptual relationships of inclusiveness and exclusiveness. The two systems compared do not show greater differences in this respect.

What we did not expect right from the beginning of our semantic analysis is the fact that a rather restricted catalogue of abstract features suffices to describe the values (see Section 16.2.4) of both systems. These features can roughly be subsumed under three categories: general geographical, special geographical, and anthropocentric properties. The geographical factors correspond partly to the general features that geographers would also use,[15] such as constellation, profile, ground relief, size, elevation, inclination, substance, and consistency. The special geographical factors include the dynamics in nature, animal orientation, and vegetation. The third category of anthropocentrism refers to the human perspective, emotional values, and history. When contrasted with European systems of landscape conceptualisation, our findings suggest that not only the hierarchical principle, but also the catalogue of geographical features has a high degree of universality.

Already our experience with two word cultures that are genetically related reveals that the number and composition of elements and properties defining individual landscapes differ considerably. For instance, in Otjiherero, the mountainous region has a choice of at least ten single or complex components, whereas

---

[15] Leser, (1998, p. 700), Hendl and Liedke (1997, Chap. 2).

Rumanyo has only two units for the same type of landscape. One can easily imagine that the number of such differences increases when the degree of linguistic relationship decreases, leading straight into incompatibility. In our opinion, an interdisciplinary effort of at least semanticists and geographers is required to overcome this problem.

16.6.3. Comparative Results at the Etymological Level

Both word cultures have the basic distinction between 'uninhabited area' and 'inhabited area'. This is also found outside the Bantu languages, at least within the wider genetic context of Niger, Congo languages. For these languages, it may therefore tentatively be considered a universal feature of landscape conception.

Conceptual differentiation starts at the hierarchical level of individual landscapes. At this lower level, the concepts and their denotations are completely dependent on the environmental circumstances. Nevertheless, etymological analysis of the terms can reveal whether the terms used for denoting the various components of landscape belong to the old cultural heritage or whether they have been coined in comparatively recent times. The results of such a historical analysis allow us to draw conclusions on the environment of former areas of settlement or even of economic changes that have taken place in the past.

In Rumanyo, the etymology of the three landscape names in the hierarchical line of *mâmbo*, 'wilderness' is an example of this. We tentatively interpret the first term *mbúrundu*,' sand desert (Kalahari)', which also occurs in Rukwangali (spoken to the west of Rumanyo), as a compound of the two nouns *Mburú*, 'Boer, Afrikaander' and *ndúndu*, 'mountain'. In this interpretation, the term refers to the Boer country to the south of the Kavango. Hence, it is a relatively recent coinage. The second term *wíya*, 'savanna' reminds one of the Common Bantu entry *-yìkà*, 'grassland' (Guthrie, 1967–1971, vol. 4, p. 166, C.S. 2002), which has a wide distribution through Eastern Savanna Bantu. However, this correlation is problematic because in Rumanyo the second consonant *k regularly corresponds to k and not to y as in the present term *wíya*. The case can only be explained by borrowing either from another Bantu language where the correspondence *$k > y$ is regular[16] or, what is more likely, from a non-Bantu language as a mediator. Because we do not know of any donor, we assume that the term came into the language during a phase of language contact, before the Manyo migrated to the Kavango region. Only the third landscape term in this line *mutîtu*, 'forest area', has a direct link via Common Bantu (Guthrie, 1967–1971, vol. 4, p. 114, C.S. 1765) to other Bantu languages spoken all over the Bantu area. Evidently, this kind of landscape was present wherever the Manyo settled in the course of their ethnic history.

---

[16] In fact, we do not know of such a language.

16.7. CONCLUSIONS

The settlement area of the Herero has only one perennial river at its northern frontier. Therefore, in Otjiherero, denotations for components or elements of a river landscape do not play an important role. Nevertheless, water supply is vital for any society. As elements or components of other landscapes, Otjiherero distinguishes diverse sources of water. The Manyo, on the other hand, have been living in the river oases of the Kavango and the Kwito for several generations. This is reflected by an elaborate system of terms referring to this kind of landscape, whereas mountains are rare in their area. Therefore, in Rumanyo, the terminology referring to that kind of landscape is rather rudimentary.

In Otjiherero, the focus on mountain-related terms indirectly reflects the cattle-oriented economy of its speakers. Interviews have shown that they use the topographical features of the mountainous landscape as landmarks to orient themselves with their herds outside the settlement areas. The mountainous system of orientation is evidently more effective than the system of cardinal points, by which the Herero distinguish two main directions: *Okumuhûká*, 'east', literally 'sunrise' and *Ongurôvá*, 'west', literally 'evening'. Following the course of the sun, two secondary directions can be defined: the south is called *Eyúvá kókumohó*, 'the sun to the left side' and the north is called *Eyúvá kokúnéné*, 'the sun to the right side'. The anthropocentric system of defining the cardinal directions with reference to the daily course of the sun works only under two conditions: First, the person who indicates a direction has to take a more or less stable position within the region. Second, the cardinal system should allow directions between the cardinal points of the sun. For people keeping cattle in big herds with the necessity to migrate with their herds, both conditions are not fulfilled. Also, for an outsider's perspective, a decentralised system of stable landmarks would have many advantages as opposed to this vague system of sun orientation.

For an etymologist, the landscape terms of Otjiherero clearly reflect the history of the Herero. The names of the four main landscapes are all comparatively young. The denotation for 'mountainous region', *okutí kozóndundú*, represents, in linguistic terms, a nominal possessive, which per se indicates a recent origin. Such forms could easily be invented at any time, if the communicative necessities demand it. The name for 'sand desert', *onamiva*, proves to be a young acquisition from Nama *nami-b*, 'desert'. In other words, both terms were most probably coined when the Herero began to settle in Kaokoland. The terms *otjáná*, 'flat grassland' and *otjihwa*, 'thick forest area' can etymologically be traced back to old Bantu roots, yet with slightly different meanings. The term *otjáná* is probably derived from an obsolete intransitive verbal stem *-yánà*, 'spread out'. In Rumanyo, the corresponding term *liyâna* has the special meaning of 'flat plain emerging along the river, when the annual floods go down'.

As we have already pointed out, landscape terms coined by a process of verbal derivation are usually of younger origin, although not necessarily recent. The term could have been brought along, when the forefathers of the Herero came to Kaokoland. Yet, it does not belong to the ancient cultural heritage. They may have coined it for a landscape where they previously settled. Finally, the term *otjihwa* is a landscape term also in other remote Bantu languages [17] with a broader spectrum of meanings: 'grassy place, shrub, garden'. The special meaning of 'thick forest' is certainly one adapted to the circumstances of the present habitat in comparatively recent times. It could be a loan from Oshindonga *oshihwa* (Viljoen et al., 1984, p. 13), where it is a general term for 'shrub, bush' denoting an area that is densely covered with bush vegetation. The interpretations of the last two terms taken together, lead to the conclusion that the Herero came to Kaokoland from an area that was characterised by tree savannas, as found in the eastern section of central Angola.[18]

The Rumanyo system of landscape terminology is highlighted by the elaboration of two semantic fields. In our experience, this case is rather unusual. However, it can be explained by the reported history of the Rumanyo speakers (Fleisch & Möhlig, 2002, pp. 57 ff.). Before the Manyo settled in the river oasis of the Kavango about 250 years ago, they lived in the flat and sandy savanna area between the Mashi River in Zambia and the Kwito River in southeast Angola. Their economy was based on hunting and gathering. The close connection with the savanna landscape and the exploitation of its resources is well reflected by the terminology concerning the geographic properties of that geomorphologically monotonous landscape and by its rich vegetation and wild life.

When the forefathers of the Manyo settled in the lower Kwito and the central Kavango valleys, they evidently expanded their subsistence to a water-oriented economy with fishing and garden cultivation on the moist river banks. The existence of a second focus in Rumanyo landscape terms clearly witnesses this event. Oral tradition is silent about how this economic expansion took place, although the historical reports do cover the period before the migrations. However, many non-Bantu click words culminate in the semantic domain of riparian landscape and river activities. This also refers to riverbound animals and aquatic plants: *macâva*, 'crocodile', *gcérembe*, 'hippo bull', *gcó*, 'smell of fish', *kacúru*, 'tortoise', *licwà*, 'shallow water', *licì*, 'otter', *ligcù*, 'belly of a fish', *muncúngu*, 'water lily', *linchè*, 'sardine', and *mancé*, 'shallow section of the river where reed grows'. These words allow us to assume that larger portions of an earlier Khoisan-speaking population living already in the river oases were absorbed into the ethnic community of the Manyo. For the Shambyu territory, we possess the Tjaube chronicle (Fleisch & Möhlig, op. cit., pp. 29 ff.), referring to a pre-Manyo population at the Kavango. It tells us how a group of Bantu immigrants deprived them of their power. This event can be extrapolated as an example of Bantu gaining supremacy over riverine Khoisan for the whole central Kavango, respectively the lower Kwito.

---

[17] Compare Guthrie, op. cit. vol. 3, p. 116, C.S. 393 *-cúá* grass with a distribution mainly in Eastern Savannah Bantu.

[18] On the basis of other sources, this area was also defined as a former settlement of the Herero. Compare: Möhlig, (2000, pp. 135 ff.).

A variety of modern landscape terms, either loans or coinages based on operative word-building strategies, can be found in both word cultures in the semantic domains of modern settlement and the development of former infertile areas. The terms for villages, townships, farms, boreholes, water pumps, dams, and so on fall under this category. They do not only reflect the processes of how modern landscapes develop, but also demonstrate the creative capacities of Otjiherero and Rumanyo to respond to the modern communicative demands.

Acknowledgements Jekura U. Kavari collected the major part of the Otjiherero data in this article. Likewise, Karl Peter Shiyaka Mberema contributed to the Rumanyo material. I have to thank both counterparts for their valuable support. Nina Gruntkowski and Michael Bollig read an earlier version of this article. I am grateful to them for their valuable criticism and suggestions. All shortcomings in the handling of the data are, of course, my exclusive responsibility.

REFERENCES

Barsalou, L. (1992). Frames, concepts, and conceptual fields. In A. Lehrer & E.F. Kittay (Eds.), *Frames, Fields, and Contrasts. New Essays in Semantic and Lexical Organization* (pp. 21–74). Hilldale, NJ: Lawrence Erlbaum.

Berlin, B., Breedlove, D.E. & Raven, P.H. (1973). General principles of classification and nomenclature in folk biology. *American Anthropologist*, 75, 214–242.

Diaz, H.N. (1994). *Kashuta Mukokera Mbanga* [Beggars can't be choosers]. Windhoek: Gamsberg Macmillan.

Fisch, M. (1994). *Die Kavangojäger im Nordosten Namibias. Jagdmethoden, Religiösmagische Praktiken, Lieder und Preisgedichte*. Windhoek: Namibia Wissenschaftliche Gesellschaft.

Fleisch, A. & Möhlig, W.J.G. (2002). *The Kavango Peoples in the Past. Local Historiographies from Northern Namibia*. Cologne: Köppe.

Guthrie, M. (1967–1971). *Comparative Bantu. An Introduction to the Comparative Linguistics and Prehistory of the Bantu Languages*, 4 vols. Farnborough: Gregg.

Haacke, W.H.G. (1999). *The Tonology of Khoekhoe (Nama/Damara)* research in Khoisan Studies, vol. 16. Cologne: Köppe.

Haacke, W.H.G. & Eiseb, E. (2002). *A Khoekhoegowab Dictionary with an English-Khoekhoegowab Index*. Windhoek: Gamsberg Macmillan.

Hendl, M. & Liedtke, H. (Eds.) (1997). *Lehrbuch der allgemeinen physischen Geographie*. Gotha: Justus Perthes.

Kähler-Meyer, E. (1967). Die Beziehungen zwischen Klassenpräfix und vokalischem Suffix am Nomen in den Bantusprachen. In *La classification nominale dans les langues négro-africaines* (pp. 313–336), Actes du Colloque International du C.N.R.S., Aix-en-Provence 3–7 juillet 1967. Paris.

Kathage, B. (2004). *Konzeptualisierung von Landschaft im Mbukushu (Bantusprache in Nord-Namibia)*. Cologne: Köppe.

Leser, H. (Ed.) (1998). *Diercke-Wörterbuch Allgemeine Geographie*. München: Deutscher Taschenbuch Verlag.

Lyons, J. (1977). *Semantics*, vol. 1. Cambridge: Cambridge University Press.

Möhlig, W.J.G. (2000). The language history of Herero as a source of ethnohistorical interpretations. In M. Bollig & J.-B. Gewald (Eds.), *People, Cattle and Land. Transformations of a Pastoral Society in Southwestern Africa* (pp. 119–146). Cologne: Köppe.

Möhlig, W.J.G. (2005). *A grammatical sketch of Rugciriku (Rumanyo)*. Cologne: Köppe.

Möhlig, W.J.G., Marten, L. & Kavari, J.U. (2002). *A Grammatical Sketch of Herero (Otjiherero)*. Cologne: Köppe.

Platte, E. & Thiemeyer, H. (1995). Ethnologische und geomorphologische Aspekte zum Bau von Brunnen und Getreidespeichern in Musene (Norost-Nigeria). In K. Brunk & U. Greinert-Byer

(Eds.), *Mensch und Natur in Westafrika. Eine interdisziplinäre Festschrift für Günter Nagel* (pp. 113–129). Berichte des Sonderforschungsbereichs 268, vol. 5. Frankfurt.

Viljoen, J.J., Amakali, P. & Namuandi, M. (1984). *Oshindonga/English English/Oshindonga Embwiitya Dictionary*. Windhoek: Gamsberg Macmillan.

Chapter 17

Landscape Conceptualisation in Mbukushu: A Cognitive-Linguistic Approach

Birte Kathage

This chapter outlines a cognitive-linguistic approach to describing conceptualisation of landscape. It is based on the findings of a cognitive-linguistic investigation into Mbukushu, a Bantu language that is spoken mainly in the semiarid northeast of Namibia. Taking language as the main source this chapter highlights some of the main universal principles that underlie conceptualisation. Culture-specific principles are illustrated by examples from Mbukushu. In order to provide means for further cross-cultural research into landscape conceptualisation this chapter proposes descriptive parameters which are not culturally biased.

The findings substantiate common cognitive theories that universal principles of conceptualisation are based on the human physical endowment and on basic human needs. Culture-specific principles reflect how the Mbukushu people adapt culturally to specific features of their natural environment.

17.1. INTRODUCTION

Conceptualisation is a fundamental cognitive process which transforms human experience and the world external to human beings into mental representations–that is, into concepts.[1] But as conceptualisation is not 'observable' it is one of the

---

[1] Following the basic assumptions of cognitive theories I distinguish between perception and conceptualisation as two components of cognition. Perception is purely a neuro-physiological process involving sensory systems such as the visual, auditory, tactual and so on.

M. Bollig, O. Bubenzer (eds.), *African Landscapes*,
doi: 10.1007/978-0-387-78682-7_17, © Springer Science+Business Media, LLC 2009

'black-box problems' of the human mind. Language, however, being the output of cognitive processes that comprise conceptualisation, is 'observable'. Thus a cognitive-linguistic investigation is suited to bringing about information on conceptualisation.

'Landscape' – the natural environment which consists of physical constituents – is one of the basic domains of human experience as it is a fundamental constituent of daily life. Therefore landscape terminology is particularly suitable for a cognitive linguistic investigation into conceptualisation. The analysis of landscape terminology reveals information on lexical structures and linguistic strategies; these are linked to conceptual structures and conceptual strategies which are applied to conceptualise landscape.

Conceptual structures and conceptual strategies are based on cognitive principles which are universal and others which are culture-specific: universal are those principles which do not differ cross-culturally because they are inherent parts of human cognition. Culture-specific principles, on the contrary, differ because they are due to culturally biased knowledge.

In the following I outline some of the main universal principles of conceptualisation drawing on cognitive theories. Culture-specific principles of landscape conceptualisation are illustrated by examples from Mbukushu.

In addition I present some methods which were applied to the collection of data on landscape conceptualisation. These methods serve to build a comparative database by means of parallel research in different languages and cultures. They make use of linguistic, visual, and three-dimensional stimuli in order to elicit linguistic and nonlinguistic data on landscape conceptualisation.

Furthermore, I present systems of spatial orientation in Mbukushu in order to make apparent that universal and culture-specific strategies of landscape conceptualisation are not only reflected in linguistic but also in nonlinguistic abilities.

Finally, a proposal is elaborated emphasising that descriptive parameters, which are not culturally biased, are required in order to provide means for a cross-cultural comparison of landscape conceptualisation.

17.2. THE SEMIARID ENVIRONMENT OF THE HAMBUKUSHU

The Hambukushu, the speakers of Mbukushu, live in the northeast of Namibia in a semiarid area bordering Angola and Botswana. From a physical geographical point of view the three major items of the landscape are the Kavango River (*rware*), the river terraces (*ndundu*), and the sandveld covered in thick deposits of Kalahari sands (*ndundu* as well) (Figure 17.1). Floodplains (*diyana*) associated with the river are dominated by grasslands (*divayi*), arms of the river (*rwareghana*), and water pools (*diviya*). The annual floods of the Kavango begin in December and peak in March or April. The river terraces are extremely fertile and preferred settlement sites. The sandveld is partly covered by woodlands (*muthitu*).

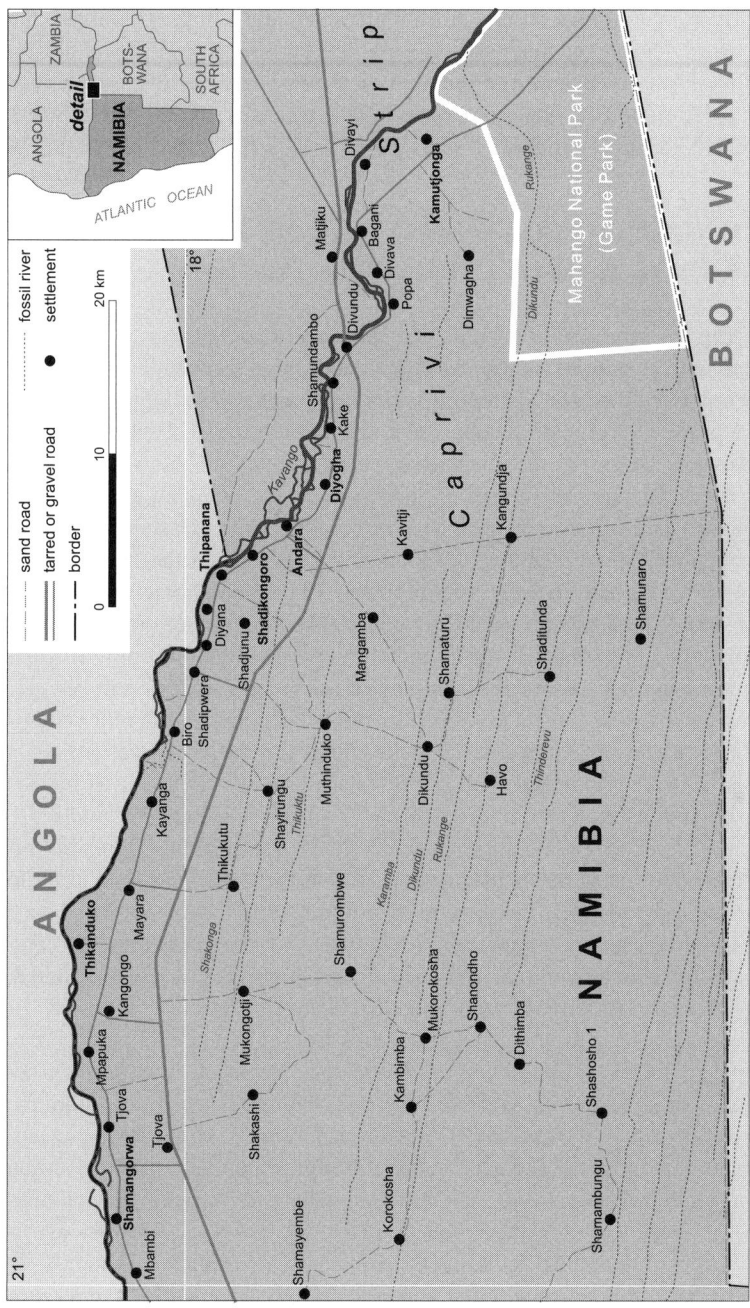

Figure 17.1. Recent Hambukushu settlement area (Graphic: Monika Feinen)

Parallel drainage channels of fossil rivers (*muronga*) run through the veld. In general they run from northwest to southeast. The main source of water is the Kavango River but there are several other sources available. In the woodlands the water is – mainly during the rainy season – found in pans (*dikorokoro*) and natural wells (*dithimanyambi*). Because of the sandy loam soil in the *mironga* (plural of *muronga,* 'fossil river') the water does not drain away easily and the groundwater resources there are fairly reliable.

17.3. CONCEPTUALISING LANDSCAPE IN MBUKUSHU: UNIVERSAL AND CULTURE-SPECIFIC PRINCIPLES

By the cognitive process of conceptualisation human experience is transformed into mental representations of knowledge, that is, into concepts. In order to achieve communication, concepts are codified as lexemes. As lexicalisation is a linguistic process of encoding conceptual information the analysis of linguistic data provides access to the 'cognitive world'. Therefore this investigation is mainly based on linguistic material and involves the compilation of lexical items related to landscape as well as the compilation of expressions related to spatial orientation. The collection of data was carried out on the basis of ethnosemantic methods making use of linguistic, visual, and three-dimensional stimuli for detailed information see Kathage, 2004). Combining both linguistic and anthropological approaches these methods serve to obtain not only linguistic but also nonlinguistic data. The nonlinguistic data on spatial orientation, for instance, offers different perspectives on conceptualisation because other cognitive abilities, apart from language, are taken into account (see Section 17.4). The interviews were conducted with informants from various places within the Hambukushu area using a standardised structured interview format to minimise the problem of obtaining noncomparable data across informants: Every informant was exposed to the same set of questions and to the same stimuli in order to control the input that triggers the informant's responses. Thus the output of each interview could be reliably compared. The interview data were analyzed – inter alia – with methods of a cognitive approach on text linguistics (de Beaugrande & Dressler, 1981; for detailed information see Kathage, 2004).

When investigating languages and cultures from an outsider's perspective, conventional linguistic methods that are based mainly on introspection are not adequate enough to gain access to the insider's culturally specific 'cognitive world', or in this case to the insider's 'cognitive landscape': the access to conceptual information is indirect. The access to cultural information is likewise: when investigating cultural issues one is confronted with the problem that cultural knowledge is intuitive and/or taken for granted. As a result such knowledge is not readily accessible and difficult, even for insiders, to articulate explicitly. In order to 'sift out' cultural knowledge that is shared by the members of a community the methods of data collection should yield data that can be compared across informants and – as a long-term goal – across cultures. In addition the methods should yield information on linguistic strategies and lexical structures which must be due to conceptualising processes and conceptual structures. To meet these requirements the collection of data was designed as follows.

In the first phase, terms relevant to landscape were elicited either from available language documents and/or from a native-speaking informant. These lists served as preliminary databases for further interviews with native speakers. In order to elicit categorisations of landscape terms I applied 'pile sorts', a method adopted from cognitive anthropology (see Weller, 1998). The informants were asked to sort (1) a pack of cards each of which contains a term of the basic landscape vocabulary and (2) a pack of photos each of which depicts a certain landscape entity according to whatever criterion makes sense to them. As preparation the photos were selected by a native-speaker and thus from an insider's perspective. After the first sorting the informants were asked to subdivide the piles. Finally they were asked if there was a title or a phrase that described each pile. As lexical structures reflect the mental categories and hierarchies which people have formed to conceptualise a domain, it can be assumed that the title of a pile is the superordinate general term, whereas the subordinate cards represent the members of the category. The elicitation of data on systems of spatial orientation was carried out on the basis of standardised 'Space Kits' developed by the Cognitive Anthropology Research Group at the MPI of Psycholinguistics in Nijmegen, The Netherlands. The first Space Kit 'Animals in a Row' is a nonlinguistic task. It involves memorising a transverse sequence of three different three-dimensional toy animals all facing the same direction. After a 180° rotation the informant has to reconstruct the memorised array. Two systems of spatial orientation can be demonstrated by the directional placement of the animals (for information on systems of spatial orientation see Section 17.4). Linguistic data was elicited by the second Space Kit 'Man-and-Tree'. It involves describing a standardised set of 12 photos each of which depicts static arrays of objects. The descriptions provide data on spatial terms.[2]

In the second phase, open-ended questions were used to obtain descriptive information on landscape entities. Using the preliminary list as linguistic stimuli, and photos – which depict landscape entities of the respective natural environment – as visual stimuli, the informants were asked to describe the landscape terms (and the photos, respectively) in their own words (and of course in their native language). As preparation the photos were selected by a native-speaker and thus from an insider's perspective. The descriptions served firstly to scrutinise the preliminary list of landscape terms, secondly to collect additional terms related to landscape, actually due to the fact that these terms are linked in a semantic network with the landscape term in question and therefore mentioned in the descriptions, and thirdly the descriptions served as a database to 'sift out' the terms which belong to the basic vocabulary. Assuming that such a landscape term is common, generally known, and frequently used, one of its indices is that it must be mentioned by the majority of the informants.

---

[2] See Senft, (1994).

In the third phase, the results of the second phase were used to develop new interview stimuli. Again linguistic as well as visual stimuli were used to take different cognitive abilities into account.

17.3.1. Conceptual Transfer

Cognitive strategies applied to conceptualisation consist of universal and culture-specific principles. They are reflected in linguistic strategies which are applied to the lexicalisation of conceptual information.

Employing concrete entities in order to understand and explain less concrete phenomena seems to be a universal principle of conceptualisation (Heine et al., 1991). In the following I refer to this strategy as 'conceptual transfer'. It can be described in terms of different domains of entities. With regard to their degree of abstraction they can be arranged along a chain (Figure 17.2).

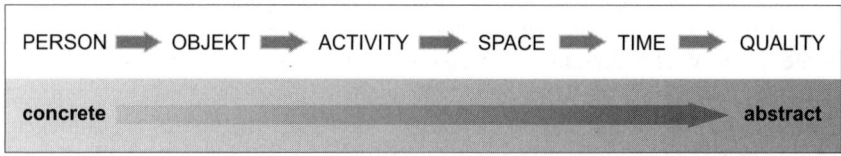

Figure 17.2. Conceptual transfer from concrete to abstract entities (Heine et al., 1991:48)

Because humans think of their own body as the most concrete entity it serves as a 'source' to conceptualise more abstract entities. To give an example: in Mbukushu the body part 'navel' is used as a metaphorical vehicle to conceptualise an element of landscape, a pan, which appears to the human mind as being more abstract than a part of the body. Linguistically this conclusion can be drawn from the lexical representation as the lexeme for 'pan, pool', *dikorokoro*, is derived from the lexeme for 'navel', *dikoro*.

17.3.1.1. Metaphor and Metonymy

The relation of the domains mentioned above is metaphorical in nature. It involves a 'target entity' (here: 'pan, pool'), and a 'source entity' (here: 'navel') that is projected onto the target entity (Lakoff, 1987). Along with metaphor, conceptual strategies also involve metonymy. A metonymy occurs within a single domain of human experience. The source entity stands for the target entity that is in some way contiguous to the source entity. In Mbukushu, for instance, an anthill is used as a source entity to conceptualise particular areas in the woodlands covered with a special kind of soil (clay-loam). An increased number of termite hills is one of the main characteristics of these areas. The kind of metonymy applied is called 'part-whole-metonymy' as the target entity is conceptualised by one of its parts. Linguistically this is reflected in the naming strategy: the lexical item of the part, *thighuru* ('termite hill'), denotes the whole, *rughuru* (see also Section 17.3.2: shift of noun classes).

Linguistically metaphor and metonymy are likely to give rise to polysemy: where both the lexeme for the source entity and the lexeme for the target entity coexist in a given language different meanings are associated with a single lexeme only. Accordingly the presence of a linguistic form with several different meanings suggests conceptual transfer. A conceptual transfer in which a lexical form was used to denote one entity before it was extended to designate another entity is a historical process which may serve as an indicator for ethnohistorical processes. Moving from one habitat to another, for instance, a speech community might extend the meaning of a lexical item in order to denote landscape entities in the new environment.

In Mbukushu the lexeme *ndundu* has several meanings denoting 'hill', 'riverbank', and 'raised area, up hill'. The polysemy of *ndundu* can be explained as being due to a conceptual transfer of the following kind. The root *-dundu* corresponds to Guthrie's (1971) Common Bantu C.S. 707 **-dundu* which has the meaning 'hill'. Accordingly it can be concluded that 'hill' is the original meaning of *ndundu* which was later extended to designate the stretch of land bordering on the river ('riverbank'). This conceptual transfer is prompted by the conceptual focus on river landscape.[3] As illustrated in Figure 17.3 the Hambukushu seem to conceptually position themselves close to the river. From this line of vision the river bank has features similar to those of a hill-for example, MUVE WAYEYUKA ('the ground rises').

In addition to 'hill' and 'riverbank' the meaning of *ndundu* was extended to denote a greater landscape unit: from the line of vision just mentioned it is the elevated area which I refer to as 'raised area, up hill'. The relation of the source entity 'hill' and the target entity 'raised area, up hill' is metonymical in nature (part-whole-metonymy).

As it is a matter of intuitive cultural knowledge speakers of Mbukushu are unaware of the fact that the entities named by *ndundu* belong to different degrees of conceptual abstraction (see also Figure 17.5).

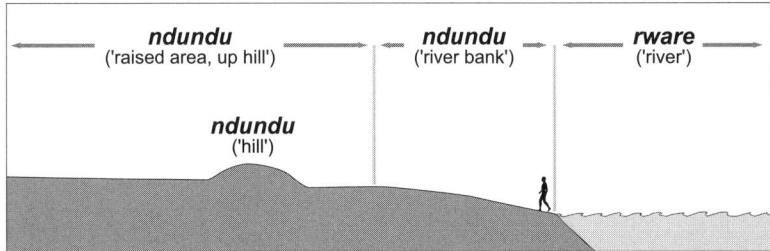

Figure 17.3. Conceptual transfer: the case of *ndundu* (Graphic: Monika Feinen)

---

[3] Culturally significant domains are coded with extensive sets of lexemes. Thus the significance of a particular domain can be measured by the size of the semantic field in a lexicon (Dimmendahl, 1995). In Mbukushu this applies to the river landscape. The lexical focus accounts for a conceptual focus: it is the river the Hambukushu put into prominence conceptually.

17.3.2. Assigning Features to Entities

Assigning sets of (abstract) 'features' to entities seems to be another universal principle of conceptualisation. By 'features' I mean attributes that are cognitively assigned to an entity. These 'conceptual' features are considered as characteristic and as naturally or necessarily belonging to it, but they are not necessarily features in the real world.

The assignment of features as such is universal, but the selection of features considered characteristic for a given entity is culture-specific: The assignment of features to one and the same entity may vary across cultures reflecting culture-specific differences in conceptualisation. For instance UNINHABITED[4] and UNCULTIVATED are (in English) features of 'the wilderness'. In Mbukushu the culture-specific feature YINYUNYI ('birds') is assigned. In the perspective of the Hambukushu the birds – the audibility of the birds singing in particular – indicate that a person is left to cope on his own in *mamboreya* ('the wilderness').

Features play an integral role in conceptual transfers: The choice of a source entity is due to the feature(s) it has in common with the target entity. Two examples help to illustrate this point:

1. In Mbukushu 'threshing floor' (*thindanda)* serves to conceptualise 'riverbed' (*rundanda*). The feature that is responsible for this conceptual transfer is the FIRM GROUND of the source entity.[5]
2. In Mbukushu a particular feature of a tree (*thitondo*) is considered to bear resemblance to the deep part of a river (*ditondo*): Through the Hambukushu's perspective a big tree could be put there without being seen because, due to the depth of the river, it would be totally covered with water. Precisely, the features in question are the HEIGHT of a big tree (*ditonto*: 'big tree') and the DEPTH of a river (Figure 17.4).

Figure 17.4. Conceptual transfer: the case of *ditondo* ('deep part of the river')
(Graphic: Monika Feinen)

---

[4] In the following features will bare highlighted by CAPITAL LETTERS.

[5] *Thindànda* ('threshing floor') bears a descending and *rundánda* ('riverbed') an ascending tone. Because the interaction of semantic shift and tonal shift has not yet been systematically investigated, the matter cannot be further discussed in this chapter.

Linguistically the above-mentioned conceptual transfers are reflected in the shift of noun classes. Like all Bantu languages Mbukushu has a nominal class system which supplies the means for word formation in terms of class shifting (see Möhlig, this volume). The noun classes, 11 in number, represent to some extent semantic properties. Thus, shifted from one class to another, nouns adopt different meanings according to the semantic properties of the class. In order to express the depth of the river *thitondo* ('tree') is shifted to class 3 (*di-/ma-*)[6] which indicates among other things the big/large size of entities. The vast majority of landscape terminology belongs to class 3 (*di-/ma-*) followed by class 6 (*ru-/maru-*) and class 4 (*thi-/yi-*). Nouns of this class denote inter alia long entities. This applies to *rughuru* ('area in the woodlands covered with clay-loam') which is conceptualised by one of its parts (*thighuru* ('termite hill')) (see Section 17.3.1). The nominal class shift from class 4 (*thi-/yi-*) to class 6 (*ru-/maru-*) indicates that the Hambukushu think of LENGTH as a characteristic attribute (i.e., feature) of *rughuru*. In fact the *rughuru*-areas run through the sandveld just as the drainage channels of fossil rivers (*muronga*).

Some nouns grouped into class 4 (*thi-/yi-*), like *thirudhi* ('island') and *thighuru* ('termite hill'), are conceptualised as containers having features such as INTERIOR, BOUNDARY, and EXTERIOR. In this example it can be seen that not only one feature but bundles of features are assigned to entities. In the following I refer to a feature bundle as a 'prototype' because it is taken as a representation of the prototypical attributes of an entity's mental representation (i.e., its concept). Note that a prototype is a mental abstraction and not an entity of the real world. The approximation of entities in the real world to the prototypical representation in the mind is defined in terms of shared (prototypical) features. A prototypical 'bird' (in English), for instance, has the features FEATHERS, WINGS, LAYS EGGS, and CAN FLY. But birds which don't fly, such as ostriches, are still birds, albeit not typical ones.

Features[7] supply (culture-specific) conceptual information on entities – descriptive as well as definitional (Table 17.1). Thus one approach to describing conceptualisation is to describe features. For descriptive purposes I propose to establish different classes of features which I refer to as 'feature-types'. Following a classification of de Beaugrande and Dressler (1981) I suggest feature-types such as 'PART', 'MINUS-PART', 'EQUIVALENT', 'LOCATION', 'QUALITY', 'STATE', 'TIME', and 'FORM'.[8]

---

[6] In the parenthesis, the first is the singular prefix, the second the plural prefix.

[7] Note that we are dealing with conceptual features which are not necessarily features in the real world.

[8] This list is but a selection of feature-types. For detailed information see Kathage (2004).

Table 17.1. Feature Type

Feature type	Represents	Example
'PART'	An entity which is considered to be an integral component of another entity (the whole)	MEYU (water) is a PART of *rware* ('river')
'MINUS-PART'	An entity which is explicitly not part of a given entity	HAGUVA ('people') are a **MINUS-PART** of *mamboreya* ('the wilderness')
'EQUIVALENT'	An entity which is equal or similar to another entity	DIYANA ('river meadow') is an **EQUIVALENT** of *muronga* ('fossil river')
'LOCATION'	The canonical position of an entity	KWISHI DHOMEYU ('under the water') is a **LOCATION** of *rundanda* ('riverbed')
'QUALITY'	A characteristic property of an entity	MUVE WAYEYUKA ('the ground rises') is a **QUALITY**, of *ndundu* ('hill', 'river bank', 'raised area, up hill')
'STATE'	A temporary property of an entity	KUYARA MEYU ('be full of water') is a **STATE** of *dikorokoro* ('pan, pool')
'TIME'	A temporal frame. It often correlates with feature-type STATE	PARUVEDHE ROMVURA (lit. 'at the time of rain') is a **TIME** of KUYARA MEYU ('be full of water') which is a STATE of *dikorokoro* ('pan, pool')
'FORM'	The shape of an entity	Note, that the Hambukushu do not explicitly conceptualize landscape entities by assigning features of the type **FORM**

In Mbukushu it is the feature-type PART that is assigned to landscape entities with the highest frequency. In the following examples I focus on this type.

Prototypical features – often belonging to different feature-types – constitute the mental model, which is the prototype of an entity. Due to culture-specific notions a certain feature of a prototype can be conceptually emphasised in order to indicate its relative importance. For example, YITONDO ('trees'), MBUYO ('fruits'), and YIYAMA ('animals') are culture-specific prototypical features (type PART) of *muthitu* ('Bush, forest'), but the Hambukushu put emphasis on the former: YITONDO ('trees') are conceptualised as being integral to *muthitu* ('Bush, forest') having a more or less determinate function with respect to this entity.[9]

The list below comprises conceptually emphasised prototypical features of the type PART. The features were ascertained by the majority of informants. In Mbukushu the inventory of PARTS assigned to landscape entities is limited mostly to MEYU ('water'), YITONDO ('trees'), and MUVE ('ground, soil'). Apparently – serving as natural resources – they are selected due to their usability (Table 17.2).

---

[9] Notwithstanding *yitondo* ('trees') are also found in other places –, however, they are not always integral but optional parts.

Table 17.2. Emphasised prototypical features (type PART) assigned to landscape entities in Mbukushu

Landscape entities		Emphasized prototypical features of the type PART
didovongiri	('(water)guily')	MEYU ('water')
difakateya	('high woodland')	YITONDO ('trees')
digcu	('thicket')	YITONDO ('trees')
dike	('(salt)pan')	MUVE ('ground, soil')
dikorokoro	('pan, pool')	MEYU ('water')
dikorongwa	('pool')	MEYU ('water')
dikutheya	('grove of ghukuthi')	YITONDO ('trees')
dimbombwera	('cove')	MEYU ('water')
dimbwerenge	('depression: ditch')	DIDHAMENA ('valley')
dirundu	('big hill')	MUVE ('ground, soil')
dingengera	('edge of steep riverbank')	MEYU ('water')
rungengera	('edge of steep riverbank')	MEYU ('water')
dinonge	('swamp')	MVU ('hippo'), MUNGCIDI ('Hippo grass: Vossia cuspidata')
dipembe	('field in *muthitu*')	HANU ('people')
dipenga	('rapids')	MEYU ('water'), MAWE ('stones')
dipya	('field')	HANU ('people')
dithimanyambi	('natural well')	MEYU ('water')
ditondo	('deep part of river')	MEYU ('water')
diviya	('pool, pond, lake')	MEYU ('water'), THI ('fish')
divundu	('marsh')	MUVE ('ground, soil')
divava	('bushes, brush')	YIPUMBU ('shrubs')
divayi	('river meadow')	MEYU ('water'), MVU ('hippo')
diya	('moorings')	WATO ('boat'), MEYU ('water')
diyana	('(flood)plain')	MUHONYI ('grass')
ghudhungi	('depth')	MEYU ('water')
ghukandami	('shallow')	MEYU ('water')
kadhamena	('shoal')	MEYU ('water')
kapupo	('torrential water')	MEYU ('water'), MAWE ('stones')
kaye	('world')	YOYIHEYA ('everything')
mamboreya	('the wilderness')	YINYUNYI ('birds')
muronga	('fossil river')	MUHONYI ('grass')
mughambo	('wood- and shrubland')	YITONDO ('trees')
muthitu	('Bush, forest')	YITONDO ('trees')
ndundu	('hill', 'river bank', 'raised area, up hill')	MUVE ('ground, soil')
rware	('river')	MEYU ('water')
rundanda	('riverbed')	MEYU ('water'), MUVE ('ground, soil')
rughuru	('area covered with clay-loam in *muthitu*')	THIGHURU ('termite hill'), DINGANDA ('firm ground')
thivagho	('ford')	MEYU ('water')

All in all the different feature types are represented by entities of the following domains.

1.   Water
2.   Vegetation
3.   Condition of ground (soil, respectively)
4.   Morphology of the earth's surface[10]

By assigning particular features to entities each society designs a culture-specific conceptual model of its natural environment.[11]

One approach to a problem-solving attempt of describing culture-specific aspects of conceptualisation is to employ feature-types. As they are not culturally biased they provide descriptive parameters for a cross-cultural comparison of several systems of landscape conceptualisation.

17.3.3. Categorising Entities

Categorisation is another universal principle of conceptualisation. Mental representations of entities are organised in terms of categories (Lakoff, 1987; Taylor, 1989). By 'category' I mean a number of entities that are considered to belong together due to the features they have in common with the best example of the category-which is the prototype-a mental model representing prototypical features that (can) belong to different feature-types (Rosch & Mervis, 1975).

Categorisation differs (just as the assignment of features) from culture to culture. In some cultures, for example, the category for 'bird' includes bats, in spite of the fact that bats don't lay eggs. Accordingly LAYING EGGS is not a prototypical feature for 'bird' but ABILITY TO FLY is.

Conceptual Hierarchies

Categories are organised hierarchically in conceptual structures that consist of taxonomies and partonomies (Berlin et al., 1973). Linguistically these hierarchies are reflected in semantic fields that are structured in a similar way. A 'semantic field' consists of lexemes which are applied to a specific domain, for example, to 'landscape terminology'.

Taxonomic structures of a semantic field are based on 'kind of relations' – for example, *dipembe* ('field in the woodlands') is 'kind of' *dipya* ('field') – whereas partonomic structures are based on 'part of relations' (Schladt, 1997). In Mbukushu landscape entities are considered to be component parts of greater landscape units. Thus their lexical representations are organised in terms of partonomies. On a rather abstract level the above-mentioned *ndundu*, for instance,

---

[10] Note that 2–4 correspond to the nomenclature of modern physical geography.

[11] From a historical point of view it is interesting to investigate which features a speech community selects and what this reveals about their history (Dimmendahl, 1995).

which I referred to as 'raised area, up hill', 'hill', and 'riverbank', denotes a greater landscape unit. The concrete entity 'hill' (*ndundu*) is, along with other things, a prototypical PART of it.

Figure 17.5 visualises the conceptual hierarchy of landscape entities in Mbukushu. It comprises five levels standing for different degrees of abstraction. The degree of abstraction is illustrated in an ascending order[12]: levels (3) and (4) represent the most concrete entities, whereas level (0) represents the most abstract entity.

Level (0) represents the entirety of landscape entities
Level (1) represents the conceptual contrast of 'water area' and 'land area'
Level (2) represents four types of landscape
Level (3) represents elements of landscape types (i.e., the items listed in Figure 17.5)
Level (4) represents segments of the entities on level (3)

The vertical relations in the hierarchy consist of part-of relations: the entities on a lower level are features (type PART) of entities from the level above. As already mentioned, they partly serve as source entities to conceptualise more abstract concepts on higher levels.

Figure 17.5 comprises the most common Mbukushu lexemes related to landscape. The hierarchical levels are listed in reverse order.

17.4. SYSTEMS OF SPATIAL ORIENTATION IN MBUKUSHU

Systems of spatial orientation are closely related to the conceptualisation of landscape. The following section deals with spatial orientation in Mbukushu in order to make apparent that the conceptual strategies described above are not only reflected in linguistic but also in nonlinguistic cognitive abilities.

I follow Levinson (1996) proposing three general types of spatial orientation: intrinsic, relative, and absolute systems. Intrinsic systems are associated with features such as FRONT, BACK, TOP, BOTTOM, LEFT, and RIGHT which are assigned to objects or persons. The assignment of these features is culture-specific. In some languages, for instance, objects such as stones are construed as having an intrinsic FRONT and BACK (Levinson, 1996, p. 367). Relative (or deictic) systems are 'relative' with regard to the deictic centre. In English, for example, the primary deictic centre is based on the location of the viewer. Under a 180° rotation, however, the deictic centre can be assigned to another entity. Relative systems are culturally biased as they depend on intrinsic notions.

Linguistic as well as nonlinguistic data provide strong evidence that the Hambukushu mainly employ absolute systems of spatial orientation. These systems

---

[12] I follow Berlin et al. (1973) in numbering the levels from (0) to (4).

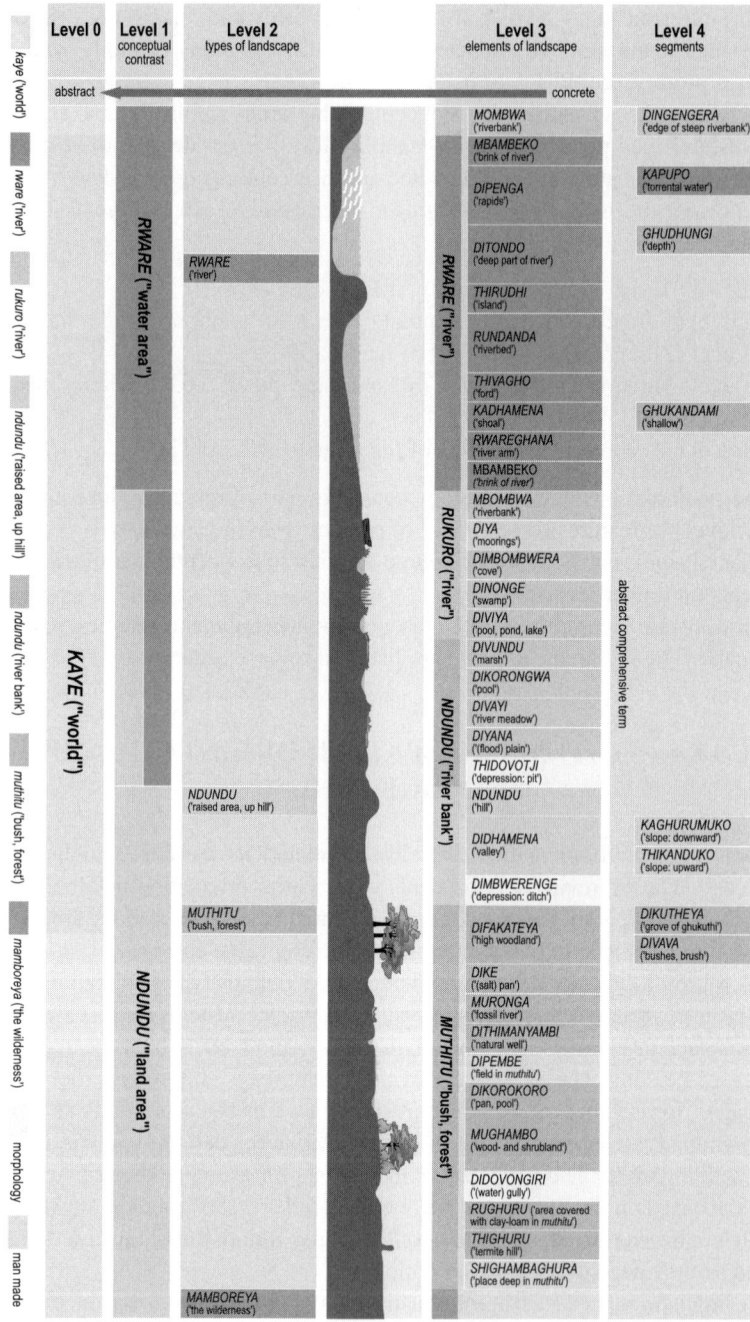

Figure 17.5. Conceptual hierarchy of landscape entities (*See also Color Plates*)

are independent of the deictic centre, that is, the position assumed by the viewer. In Mbukushu they are based on two different types of culturally fixed reference points:

1. On two landmarks: *Rware* ('river'), on which the Hambukushu focus conceptually, and *ndundu* in the sense of 'raised area, up hill'
2. On cardinal directions: *Diva* ('east'), *ditokera* ('west'), *mbunda* ('north'), and *ghucuma* ('south')

As landmarks are rooted in salient entities of landscape they are restricted to the particular local environment and highly culture-specific. This applies to the system based on *rware* ('river') and *ndundu* ('raised area, up hill') as it is limited to the rather small Mbukushu settlement area.

Landmarks are defined in terms of fixed reference points that can be reached, whereas cardinal directions cannot be reached: *Diva* ('east') and *ditokera* ('west') are derived from verbs that denote the rising and setting of the sun. As the course of the sun varies throughout the year, 'sunrise' and 'sunset' do not determine fixed bearings (Levinson, 1996). Accordingly spatial terms derived from the course of the sun refer rather to the eastern and western half of the horizon. This can be seen to be confirmed by the lexical representation because *diva* ('east') and *ditokera* ('west') belong to class 3 (*di-/ma-*) that contains, along with other things, big/large-sized entities and entities covering large/extensive areas. *Mbunda* ('north') is derived from the name of a region in southeastern Angola which is north of the recent Hambukushu settlements. *Ghucuma* ('south') is probably derived from an ethnic name for a Khoisan group (*mà/úmá kxòè*) (Heine, 1997b) both having a voiceless dental click. I assume that the 'east–west orientation' was established first inasmuch as the notion for 'north' and 'south' must have been lexicalised after the Hambukushu's arrival to the present-day area of settlement.[13]

The (universal) conceptual strategy looked at here is the same as described in Section 17.3.1: concrete entities serve as reference points (e.g., a river, the sun, a certain region). They are employed in order to conceptualise the less concrete phenomenon 'space'. Cultural differences in conceptualisation occur due to the variation in concrete entities that serve as reference points.

17.5. CONCLUSION: UNIVERSAL CONSTRAINTS AND CULTURE-SPECIFIC VARIATION

As the conceptual system is reflected in language an investigation into landscape terminology reveals universal as well as culture-specific principles of landscape conceptualisation. The findings presented in this chapter substantiate

---

[13] This chronology corresponds with Brown's findings on establishing systems of spatial orientation (Brown, 1983, p. 144).

cognitive theories purporting that conceptualisation is simultaneously constrained by universal cognitive predispositions and specified by culture-specific notions. Considering that human needs and interests are at the center of attention, anthropocentricity appears to cause the rise of universal principles and thus constrains the range of possible solutions to conceptualising landscape. Yet there are considerable differences across cultures in the way landscapes are conceptualised (Levinson, 1999[3]): Culture-specific strategies of landscape conceptualisation reflect how people culturally adapt to specific features of their natural environment. Universal strategies of landscape conceptualisation are based on the human physical endowment and on basic human needs: they are partly determined by the bodily nature of humans and affected by the quest for a means of living. Entities, for instance, which serve as conceptual features are often chosen due to their usability. In Mbukushu this especially applies to features of the type 'PART'.

Empirical work on landscape conceptualisation that is grounded on cognitive-linguistics is still at an early stage as there are only few empirical studies addressing the issue and research has been done mainly within a single language. Therefore, further cross-cultural research is required, not only in order to develop comprehensive models of landscape conceptualisation, but also in order to solve the important question on parameters of culture-specific variation in conceptualisation: to what extent is conceptualisation universal and to what extent is it variable across cultures? In order to come closer to the answers to these questions, there is a need to build comparative databases by means of parallel research in different languages and cultures. To obtain comparable data, it is essential to apply standardised procedures and methods of data collection as those described above.

In addition aiming at cross-cultural comparison of several systems of landscape conceptualisation requires a set of descriptive parameters (such as the above-mentioned feature-types) that are not culturally biased. Universal predispositions-for instance, the human physical endowment on which universal strategies of landscape conceptualisation are based-can serve as a frame of reference by which variables-that are, culture-specific strategies of landscape conceptualisation-can be identified. Following this procedure we can expect to come across strategies of landscape conceptualisation that are characteristic, not only of a certain cultural group, but also of a certain cultural and geographic area.

REFERENCES

Beaugrande, R.-A. de & Dressler, W.U. (1981). *Einführung in die Textlinguistik.* Tübingen: Niemeyer.

Berlin, B., Breedlove, D.E., & Raven, P.H. (1973). *Principles of Tzeltal Plant Classification.* New York: Academic Press.

Brown, C.E. (1983). Where do cardinal directions come from? *Anthropological Linguistics,* 25(2), 121–161.

Dimmendahl, G.J. (1995). Studying lexical-semantic fields in languages: Nature versus nurture, or where does culture come in these days? *Frankfurter Afrikanistische Blätter,* 7, 1–29.

Guthrie, M. (1971). *Comparative Bantu. An Introduction to the Comparative Linguistics and Prehistory of Bantu-Languages*. (Vol. 2). Farnborough: Gregg Press.

Heine, B. (1997a). *Cognitive Foundations of Grammar*. Oxford: Oxford University Press.

Heine, B. (1997b). On spatial orientation in Kxoe: Some preliminary observations. *Khoisan Forum*. Working Paper 6.

Heine, B., Claudi, U., & Hünnemeyer, F. (1991). *Grammaticalization: A Conceptual Framework*. Chicago: University of Chicago Press.

Kathage, B. (2004). *Konzeptualisierung von Landschaft im Mbukushu (Bantusprache in Nord-Namibia)*. Cologne: Rüdiger Köppe Verlag.

Lakoff, G. (1987). *Women, Fire and Dangerous Things. What Categories Reveal About the Mind*. Chicago/London: University of Chicago Press.

Levinson, S.C. (1996). Language and space. *Annual Reviews Anthropology*, 25, 353–382.

Levinson, S.C. (1999[3]). Relativity in spatial conception and description. In J.J. Gumperz & S.C. Levinson (Eds.), *Rethinking Linguistic Relativity*. Studies in the social and cultural foundations of language 17 (pp. 177–202). Cambridge: University Press.

Rosch, E. & Mervis, C.B. (1975). Family resemblances: Studies in the internal structure of categories. *Cognitive Psychology*, 7, 573–605.

Schladt, M. (1997). On describing conceptual structure. Examples from Kxoe. *Khoisan Forum*. Working Paper 5.

Senft, G. (1994). Ein Vorschlag, wie man standardisiert Daten zum "Thema Sprache, Kognition und Konzepte des Raumes" in verschiedenen Kulturen erheben kann. In G. Grewendorf & A. von Stechow (Eds.), *Linguistische Berichte. Forschung, Information, Diskussion* (pp. 413–429). Opladen: Westdeutscher Verlag.

Taylor, J.R. (1989). *Linguistic Categorization. Prototypes in Linguistic Theory*. Oxford: Clarendon Press.

Weller, S.C. (1998). Structured interviewing and questionnaire construction. In R. Bernard (Ed.), *Handbook of Methods in Cultural Anthropology* (pp. 365–409). Walnut Creek/London/New Delhi: Alta Mira Press.

Chapter 18

Otjiherero Praises of Places: Collective Memory Embedded in Landscape and the Aesthetic Sense of a Pastoral People

J.U. Kavari and Laura E. Bleckmann

This chapter allows insight into the intricate relation between landscape and memory. The *omitandu* – this is the Otjiherero term for these praises – are part of Ovaherero, Ovahimba, and Ovatjimba oral traditions. They are highly valued among the pastoral people of Kaokoland[1] in northwestern Namibia for their aesthetic as well as for their historical and political content[2] (see Figure 18.1). Otjiherero praises allude to the collective memory of the community and enshrine

---

[1] The term 'Kaokoland' became commonly used during South African rule when it referred to the homeland 'Kaokoland' stretching over an area that nowadays comprises the northern administrative unit of the Kunene Region, divided up into different constituencies (Epupa, Opuwo, and Sesfontein). As the inhabitants of the Northern Kunene Region refer to themselves as 'Kaokolanders', as inhabitants of 'Kaokoland', or simply 'Kaoko' this chapter likewise refers to this self-chosen label (see Friedman, 2004).

[2] The prefix 'omu-' determines the singular form of the first class in Otjiherero whereas 'ova-' is prefixed to express the plural form of this class that mostly encompasses human beings. 'Otji-' belongs to the seventh class and determinates mainly things.

M. Bollig, O. Bubenzer (eds.), *African Landscapes*,
doi: 10.1007/978-0-387-78682-7_18, © Springer Science+Business Media, LLC 2009

Figure 18.1. The Omberangura mountains in Opuwo – 'where Kazu went through'
(Bleckmann, 2005)

what the community has estimated is of importance to remember. As such they constitute an essential part of oral history. They refer to historic events and persons and link these to specific places and landscape features. They further condense information on genealogies and resources and although on the surface they address relations between people and land, the praises have strong political connotations through their reference to the past tenure system. Last but not least they provide insight into the perceptions and the aesthetic sense of a pastoral people. With a special focus on praises of places this chapter aims, in the first instance to illuminate the special structure of the *omitandu* and to highlight their content. A second aim is to draw upon the representation of landscape features as well as on the poetry in such praises. How are landscape and other features remembered and represented by Kaokoland's pastoralists? Furthermore, the ways in which memory is linked with places and landscape features are discussed. How far is the recited collective memory spatialised and in which way is the reference to spatialised collective memory of importance for recent debates in the Otjiherero speaking community of Kaokoland?

18.1. INTRODUCTION

When the oral poet starts reciting a praise that is embedded in a speech or a song in a sometimes lower or higher voice, the audience remains silent and tries to follow the quick flow of words. These praise poems of the Otjiherero-speaking community are, beside their emotive meaning and aesthetic value, an important source of collective memory and are essential for the coding and legitimisation of kinship, ownership, and belonging and are thus cited in religious, political, and everyday contexts alike. With the examination of the structure and content of the *omitandu* (Sg.: *omutandu*) this chapter contributes to oral literature and oral history studies, respectively, by focusing on the *omitandu* as an important store of collective memory in oral societies (see Ricard, 2004). The aesthetic sense of a pastoral people and their reflections on the past via the landscape is at the centre of the analysis. A further matter of investigation is in how far the *omitandu* comprise a spatialisation of collective memory. Finally, the role the *omitandu* play in land tenure conflicts and identity politics and how far they are of importance concerning spatial orientation and belonging are investigated.

Halbwachs (1925, 1941, 1950) introduced the term 'collective memory' and discussed it in a sociological perspective emphasising its meaning for the identity of a group. Whereas Halbwachs seeks to stress the formation of a collective memory that goes beyond individual memory and is transmitted over generations, Assmann (1995, 1999b, 2000) goes one step further and sets up the term 'cultural memory'. With this he aims 'to relate all three poles – memory (the contemporised past), culture and the group (society) – to each other' (Assmann & Czaplicka, 1995, p. 129). He discusses the store of cultural memory in oral societies and states the following elements as essential for the survival of collective memory: recalling and transmission of the memory, social participation, its meaning for the present, and its saving in a 'stable' formation (see Assmann, 1999b; Halbwachs, 1950).

The recalling emphasises the continuous reproduction, the transmission refers to the transfer from generation to generation, social participation is directed towards the audience by the members of the community, and the meaning for the present concerns the implications that the collective memory has for the present debates. The importance of this last feature for collective memory and oral history, respectively, is strengthened by Vansina (1985 , p. 92) who argues: 'All messages have some intent which has to do with the present, otherwise they would not be told in the present and the tradition would die out'. The poetic and thus stable form of the *omitandu* enables a better preservation of the collective memory and their passing it on orally.

The binding of memory to specific places or a landscape in general has been examined in a variety of studies. Assmann (1999a,b) differentiates here between the binding of memory to places or things that serve as mnemonic devices such as was common in the praise poetry of the ancient Romans and memory that is embedded in landscape due to its remembrance of events, people, or any kind of other 'lived' memory. The so-called 'art of memory' of the first approach which has been exhaustively examined by Yates (1966) concentrates on the capacity of the human mind to remember and discusses places and things as mnemonic aids in rhetoric. The latter approach underlines the symbolising of landscape connecting memory to where 'it has taken place'. Nora (1990) introduced the term memory-place[3] for this and Assmann (1999a) as well as Gordillo (2004) examined the 'spatialisation of memory'.[4]

---

[3] Translated from French (*'lieux de mémoire'*) into English ('memory-place') (Flores, 1998).

[4] When differentiating the two terms 'place' and 'space', the latter is mostly defined according to the Cartesian concept as 'infinite' and 'empty' bearing meaningful 'places' which are thus 'concrete' and 'filled'. There are, however, other approaches which define the terms in another way and thus 'place' as 'general' from which 'space' is derived. The article chapter makes use of 'place' as 'meaningful' and 'concrete' as treated in the here- mentioned theoretical approaches and to which the chapter refers. Although it has to be mentioned that the term 'spatialization' is not best chosen and 'placializsation' keeps to the point better. For discussions on the differentiation of 'place' and 'space' see, for instance, Feld & Basso (1996) or Low & Lawrence-Zúñiga (2003).

One way of examining the relationship of landscape and memory is to make use of the metaphoric concept of landscape as a text underlining the 'inscription' of memory into landscape which thus can be 'read'. P. Carter (1987) makes reference to landscape as text and discusses the 'inscription of memory' into landscape, focusing mainly on the inscription of colonial power onto the landscape through European naming. Several authors examined the relationship of landscape and memory offering different views onto its interplay (see, e.g., Cosgrove, 1998; Hirsch, 1995). Schama (1995) gives insight into memory embedded and experienced in different 'landscapes' that he divides up into the three elements 'wood', 'water', and 'rock'.

The Ovaherero, Ovahimba, and Ovatjimba who form the main groups of the Otjiherero-speaking pastoralists inhabit Kaokoland's semiarid savanna in the northwestern corner of Namibia where due to the harsh environmental factors, they practice a (semi-) nomadic lifestyle. The different labels 'Ovaherero', 'Ovahimba', and 'Ovatjimba' are mostly used nowadays to underline the identity of one or the other group although most of the Otjiherero-speaking Kaokolanders emphasise common descent, common language, and to a great extent common tradition and shared history by the uniting label 'Ovaherero' whereas 'Ovahimba' or 'Ovatjimba' can be regarded as differentiating ones. Migrations during war times have been of great importance for the evolvement of the different labels and groups. Whereas the Ovahimba belonged to those people who migrated to Angola between roughly 1860 and 1890 from where they returned with large herds of cattle, the label 'Ovatjimba' used to be in earlier times a derogatory label for those who stayed in Kaoko, lost their herding animals, and became impoverished.[5] Nowadays tourism and political factors favor the trend of ethnic labeling that often serves economic or political aims (see Rothfuss, 2004; Bollig & Heinemann, 2002).

The *omitandu* present the most important source of oral literature in these Otjiherero-speaking communities of Kaokoland as there is little other prose or poetry that is passed on orally. They are part of a genre that is very popular in southern Africa: praise poetry in southern Africa has been of interest for a number of studies: The early work of Shapera (1965) sheds light on praises among the Tswana and concentrates on the identification of the royal lines of the different Tswana kingdoms. Cope (1968) examines the so-called *izibongo*, the heroic praise poems of the Zulu, and Kunene (1971) offers a linguistic analysis of the comparable Basotho praise poems. Opland (1983) analyses Xhosa praise poetry from an interdisciplinary perspective whereas Gunner (1996) researches the relation between belonging and land in a comparative study of Zulu, Xhosa, and Sotho praise poems.

---

[5] The term 'Ovatjimba' is allegedly derived from the Otjiherero noun '*ondjimbandjimba*' ('antbear'), because the Ovatjimba were said to dig their food from the soil like the antbear does. For early sources about this labelling see, for example, van Warmelo (1962).

From a linguistic approach Otjiherero praises have been analyzed by Ohly (1990), and Kavari (2002) contributed a detailed linguistic but historically sensitive analysis of Otjiherero praises. Alnaes (1989), Dammann (1987), Henrichsen (1999), and Förster (2005) concentrate on an ethnohistorical analysis whereas Henrichsen and Förster underline the spatiopolitical dimensions of Otjiherero praises in the context of the Namibian land question. Bleckmann (2007a) gives insight into ethnohistoric aspects of northwestern Namibia's praise poetry concentrating on aesthetics and spatial aspects of the recited memory. Hofmann (forthcoming) examines the relation of landscape and identity in Namibian Ovaherero communities with praise poetry being an element of the figuration of this interplay.[6]

Southern African praises cannot always be said to 'praise' in a proper sense. The creation of a praise is certainly in order to extol the topic but the praise as such can also include pejorative or amusing attributes and give critical comments. A further characteristic of the genre is the spontaneous creation of praises in oral performances. Rothenberg and Rothenberg (1983) give a vivid description of performing Zulu praise poetry: 'The praiser recites the praises at the top of his voice and as fast as possible. These conventions of praise poem recitation, which is high in pitch, loud in volume, fast in speed, create an emotional excitement in the audience as well as in the praiser himself. . . .' (Rothenberg & Rothenberg, 1983, p. 127).

Furthermore, the typical structure of praises is a linking one consisting of diverse linkages that can comprise different levels. Concerning the *omitandu* the following types of linkages can be accounted for: the creation of a praise by linking and combining old and new or simply different praise verses, the phonetic structure that strings words together and gives the praise its melodic flow and links of content that intertwine different topics, set up genealogies or social relations, and that links people, cattle, events, and landscape features to places often bound in metaphorical expressions. Kavari (2002) describes the linking structure with the following words: 'The entities in a praise do not function in isolation, but in relation to entities that precede and follow. Thus a link exists between entities which, in many instances, are realised by a linking element.' Kavari (2002, p. 65). He therefore differentiates between the linguistic terms 'referent' and 'epithet'. Whereas the referent points to an element that is praised, the epithet is the element that serves to praise the referent. A further description of the epithet is realised by the subordination of 'key words' and 'complements'.

When different praises or praise verses of origin or sujet are linked up the *omitandu* create a structure that resembles those of hyperlinks. A praise about a certain hero can therefore include the praise of the lineage of this hero, or the

---

[6] Although most of the studies are based on Otjiherero praises in Central Hereroland, Alnaes (1987) devotes her paper to the Ovaherero community in Botswana and Kavari (2000) and Bleckmann (2007a) concentrate on praises of Kaokoland.

praise of his place of birth. This linking structure lays ground for the special structure of Otjiherero praises and is one of their main characteristics, making them additive and thus open for new creations that are added and produce manifold variations that vary due to the selection of verses the oral poet makes. The individual selection is of great importance as it influences what is said in total and how it is said. The oral poet makes his selection according to context, audience, aim, and individual knowledge of praises.

Thus *omitandu* can be considered to be infinite in their combinations and in duration without beginning and end in form and content. But in fact the performance has a beginning and an end and the verses of, for instance, a certain village don't vary a lot although the number of verses and the way it is linked with other praises differs when recited in different contexts and by different oral poets. To a lesser extent regional variations have been considered (see Bleckmann, 2007a).

Together with their rich imagery which makes them beautiful but also difficult to access they are not easy to understand even for native speakers. The main subjects that are interconnected in the *omitandu* are people, clans, lineages, events, cattle, and landscape or places. These topics may be further classified into historical, sociocultural, and landscape features as the three main categories which are closely intertwined in metaphoric expressions.

There are *omitandu* that refer to Okarundu kaMbeti, a place near to the Angolan border which oral history refers to as the first settlement of the Otjiherero-speaking pastoralists on their arrival in Kaokoland after migrating from the Interlacustrine Area;[7] others reflect the catastrophic events of the last two centuries when turmoil, cattle raids, wars, and escape hallmarked Kaokoland. Nonrecurring migrations that were fuelled by certain economical or political circumstances are often remembered in the *omitandu* by highlighting special landscape features such as fjords or passes that the groups found on their way (see also Figure 18.1). There are praises about the crossing of the Kunene towards the Angolan side where the pastoral Ovaherero and Ovahimba were looking for a refuge in order to escape the cattle raids of the Ovakwena, or about their remigration a few decades later (see also Bollig, 1997a).

The label 'Ovakwena' refers to Nama-speaking groups that originated from the central areas. From the beginning of the 1840s they penetrated into Kaokoland where they – well armed with guns from the capebound trade – undertook several raids whose brutality is still vividly remembered. There are further praises about heroic personalities, such as Vita Thom, who was an important leader during the first decades of the twentieth century. Others consist of a long stringing together of names that present genealogies.

Time in the *omitandu* is evanescent. Their structure is more a descriptive than a narrative one although the praises often refer to shorter or longer narrations that the informant gives in added explanations. Different time periods flow into each other and events or people can seldom be temporally fixed but easily

---

[7] Möhlig (2000) gives linguistic evidence for this migration of Otjiherero pastoralists.

localised and it seems that places take over for time. Time is not used as an important point of reference whereas places are and the identification of somebody or something is mainly operated along places and kinship than along time.[8] If the name of a person is known in Otjiherero society and his genealogical embedment is possible, one can accord the name to a certain timeframe but the rest of the content probably remains lost in time apart from a very rough dating covering mostly the period from around 1840–1940.[9] With this negation of time it can be stated that the concept of history of Kaokoland's pastoralists differs from the European concept by absorbing time into space. [10]

An *omutandu* is created by an individual, the oral poet. During the performance he combines a praise by taking bits and pieces of different praise verses. A new praise or praise verse is added to the existing store of knowledge and becomes a communal product when it is accepted by the community. In other words, it is transferred from individual to collective memory and nobody will claim the ownership of its creation. This means that a praise might be a product of more than one individual, as many have contributed pieces to the whole. As such an *omutandu* is the common good of the community which is attached to the memory cited in the praise, mostly by kinship or spatial ties, and which thus can be regarded as the 'owner' of the *omutandu*. Therefore, a kind of ownership comes up where different *omitandu* belong to different communities. This aspect is of special importance concerning land rights: here an *omutandu* of a certain place can be used to underline the belonging of a certain group to this place and thus claim exclusive rights.

18.2. METHODS

Omitandu were recorded both with specialists and those who only knew a few verses on their place of birth or on their family in different places in northern and southern Kaokoland. Because of the special word code and rich imagery the content of the *omitandu* can only be decoded and translated after further

---

[8] In Yolngu ontogeny Morhpy (1995) states a similar precedence of place over time which was created 'through the transformation of ancestral beings into place, the place being for ever the mnemonic device of the event'.

[9] Traditionally, the Otjiherero- speaking pastoralists named different time periods according to important events such as droughts, raids, or the death of important personalities. This 'calendar' thus can be considered as a conceptual time frame which beside in addition to genealogical sources serves as a point of reference. Thanks to the work of Gibson (1977) and van Warmelo (1962) who set up a chronology of the 'named years' and accorded them to the Western time frame, it can sometimes be figured out when somebody lived or something happened.

[10] This absorption of time into space is a special characteristic of the *omitandu* although it is also noticeable in the way that the people talk about the past in other contexts when past events are summed up under '*rukuru rukuru*' ('long long ago').

explanations by the oral poet.[11] A 'simple' taping, transcription, and translation has therefore turned out to be insufficient.[12]

After recording, a first transcription, and translation, further interviews were conducted to correct the transcription and translation, to understand the content and last but not least to deepen the narrative and interpretative context. For this the informants were asked for explanations of words or word combinations that don't exist in the Otjiherero everyday vocabulary, expressions, and contexts as well as the interpretation of metaphors and metonyms. They further embed events and people in their wider contexts; they can give temporal references and help in drawing up genealogies linking these to names mentioned in the praise poems. These explanations are another part of the collective memory and are mostly transmitted with the passing on of an *omutandu* itself. However, it often happened that the meaning of a word could not be deciphered and that the deeper or transferred meaning of the memory was thus lost. Whereas the following analysis is mainly based on the explanations that were given by the informants, it is enriched by a linguistic examination and further field notes.

The following questions stand in the centre of the analysis. What does the oral poet say and mean? How does the oral poet say what he wants to express? Why does the oral poet say what he says, and in the way he does? How are landscape features and places linked to memorable events and persons? In order to examine in how far the praises allude to a spatialisation of collective memory most of the different places that are referred to have been visited and mapped, and most of them have been photographed. By visiting the different places and landscape features the way a certain place or landscape feature is remembered in an *omutandu* could be related to its physical being and appearance. The mapping helped to localise the different places and landscape features and served to make statements on the distribution of praised places and landscape features.

18.3. PRAISES OF PLACES

The Ovaherero and Ovahimba create praises for nearly every place they have lived in and herded their livestock. In praises of places a place can acquire its name or praise because of a memorable incident or event or because of its

---

[11] As shown in the next section there are some expressions that are very common in *omitandu* and whose meaning can be 'decoded' in a proper sense. Apart from these there is no proper 'code' that can be decoded although a deep understanding of Kaokoland's pastoralists's tradition and the informants' explanations are necessary to get a grip on the content.

[12] One author is a native speaker of Otjiherero and long accustomed with the local tradition as he was born and bred in Kaokofeld. He started the systematic collection and analysis of *omitandu* in 1998 during a three month fieldwork, a process that is still continuing today. The other author conducted anthropological fieldwork from September to December 2005 and March to June 2007 and also still continues to collect and analyze *omitandu*.

physical characteristics such as well-shaped mountains in an aesthetic sense or a particularly characteristic topography (Figure 18.2). Some places are liked and others disliked for a variety of reasons, as some are bad because of livestock diseases and others are free of them. In some places livestock multiplied rapidly, and

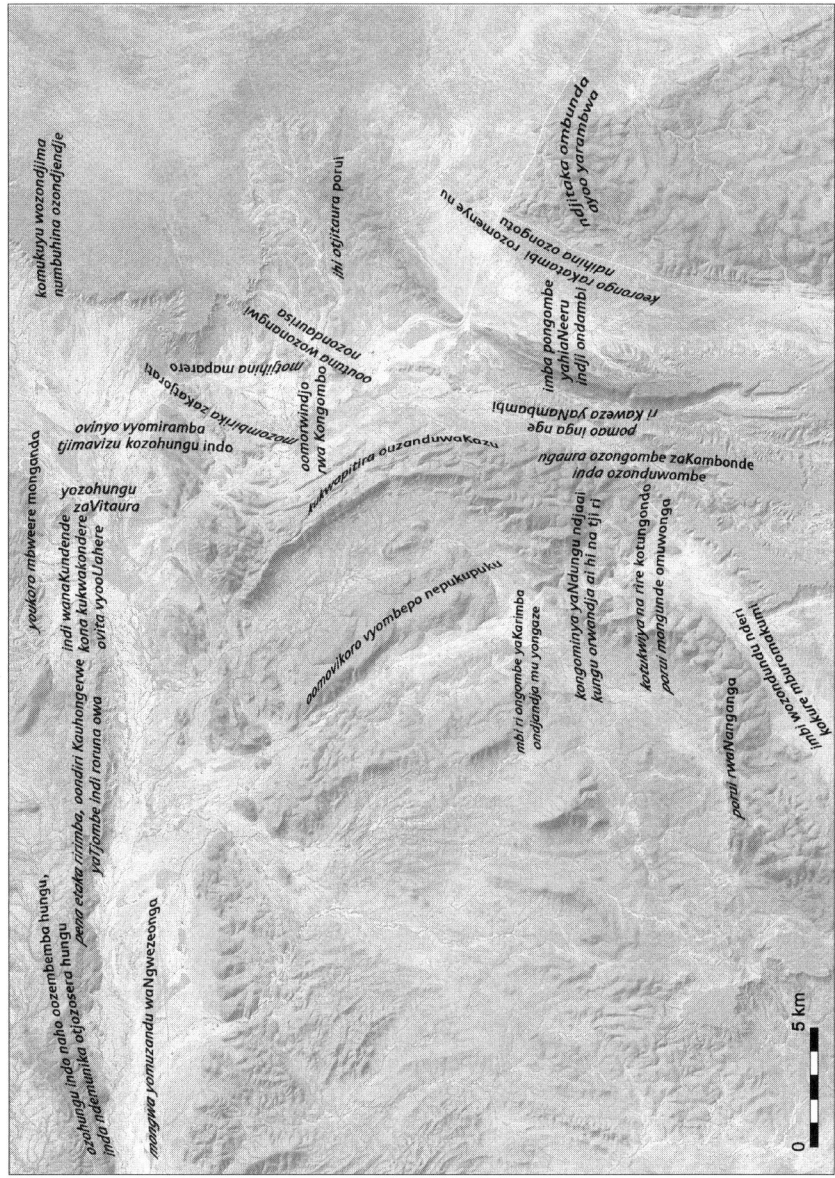

Figure 18.2. Praises of places (*See also Color Plates*)

some places became memorable because of historical incidents. These incidents become closely associated with the specific locations that can become historic monuments. The most important features in praises of places are topographical features of a place, the localisation of resources, and remarkable events which took place there. Births and burials as well as battles are often linked to a certain place or landscape feature. Before the analysis turns to a detailed examination of two *omitandu*, the way in which these features are set up in the praises are highlighted along with some examples.

18.3.1. Structure of Praises of Places

Praises of places consist not only of their formal aspects of a linking structure which interweaves praises with praise verses. Also concerning their content and meaning the linking aspects stand in the centre as persons are linked to other persons and thus set up genealogies and connect memorable events and persons with the place to which they allude. The topic of the *omutandu* in praises of places is always a village or a landscape feature. But to give honor to the place and a sufficient characterisation of it, its physical appearance is given and it is linked to the memory of mainly events, resources, people, or cattle. A praise about a village can therefore contain a longer passage where a certain hero or lineage is praised. The creation of the *omutandu* as such contributes to the subject's credit, whereas the exact way the topic is praised does not always do so. An *omutandu* aims to give a characterisation of the topic and seeks to capture its essence and therefore counts its most remarkable characteristics as far as they are estimated to be of importance for the memory of the place.

To give an example people are remembered according to their physical appearance, by recounting remarkable characteristics, by linking them to acts or events, or by embedding them in their genealogical or wider sociocultural context. The *omutandu* of Omanguete, a village approximately 70 km north of Opuwo, is praised with the following words: '*indji ndjiri komuvi wetu ingwi ngutuyereka euru*' ('this is at our ugly person whose big head we used to joke about'). The village is praised by making reference to somebody whose name is not known but who is described by his most remarkable characteristic, his ugly physical appearance and the fact that the people used to joke about his big head. This verse further points to the descriptive character of the *omitandu* and the stylistic device of the omission of names. In this case the name was lost as the informant didn't know it only referring to the person's roughly familiar context.

Descriptions of the topography are further very common in praises of places. Ohandungu[13] is praised with reference to its topographic position in a valley surrounded by mountains: '*Ihi otjiwa kovikuro*' ('The nice one on its sides'), whereas 'the nice one' refers to the valley itself, the position of the mountains is given by 'its sides' as the mountains surround the valley. In addition to

---

[13] Ohandungu is a village that is to be found 40 km northeast of Opuwo.

characteristic descriptions of the topography, certain plants, water sources, or fruits are mentioned that contribute to enrich characteristic vegetational and hydro-logical features. Another praise verse of Ohandungu contains this reference to resources by indicating a former pasture of the place which is further linked to somebody whose cattle grazed there: '*motjohozu tjitjira ongombe yaNdungu yaKatwe kondura*' ('at the thick grass that has been eaten by the cow of Ndungu of the fat Katwe'). Concerning the linking of events with a place an example is given in the praising of Otjikotoona, a popular mountain pass in oral history where warrior groups used to pass through (Figure 18.3). It is therefore praised with the following verse: '*ombero yovita nozondjou*' ('mountain pass of elephants and war'), giving further reference to the former abundant wildlife of the area.

There are certain linking metaphors that connect people with places and at least one but mostly several turn up in every praise. People who inhabit a place, who were born or who were buried there are alluded to and in this way localised. Not only heroes or popular personalities are honored by integrating them into a praise but also people about whom often nothing more is known. The simple possessive concord 'of' and the name of the person point to a person's inhabit-ance of a place. The reference to a person's place of birth or grave is given in a more metaphoric manner. The metaphor 'at the umbilical cord of' with an assigned name points to the place of a person's birth. An example is given in the *omutandu* of Otjikango that is praised by referring to the umbilical cord of someone who is characterised as wearing the typical necklace for twins: '*pongwa*

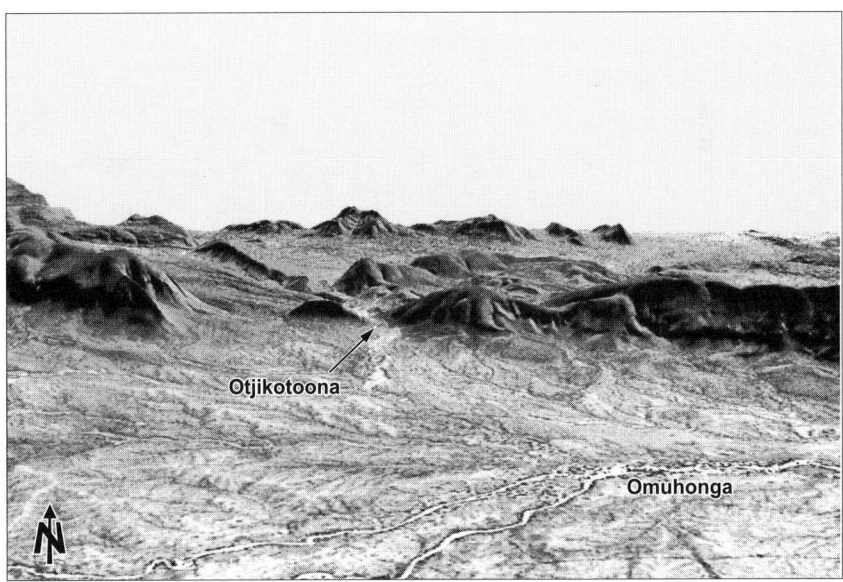

Figure 18.3. Mountain pass of Otjikotoona (3D-satellite image [Landsat, SRTM] subproject E1, CRC 389) (*See also Color Plates*)

yamuyambwa ovitjuma otjowepaha' ('at the umbilical cord of the one wearing
the twin-necklace). Even more common in praises is the reference to somebody's
gravesite.[14] The preposition 'at' and a following name or reference to an oxen
or the horns of an oxen express that a grave is to be found in the praised place.
Horns or oxen mostly refer to the grave of prestigious people, so-called *'ovahona'*, because the horns are placed on their graves. In some areas the horns on
the grave are renewed once within a period of about two to three years during
commemorations. Oxen are slaughtered and the old horns are replaced with new
ones. A possessive concord between two names is mostly a reduction referring
to the meaning of 'who is the child of'. It thus determines the father of another
person. In the *omitandu* it is the most common form to express genealogical relationships and in the Otjiherero-speaking pastoralists' traditions this genealogical
reference not only serves to identify but also to honor as it gives credit to a father
to be praised by his son and vice versa.

18.3.2. Analysis of Praises of Places

18.3.2.1. Ehomba Mountain

The Ehomba mountain is approximately 20 km from Kunene River, south of
Swartbooisdrift (Figure 18.4). It is one of the highest elevations in Kaokoland,
with an approximate height of 1900 m. In former times the Ehomba mountain

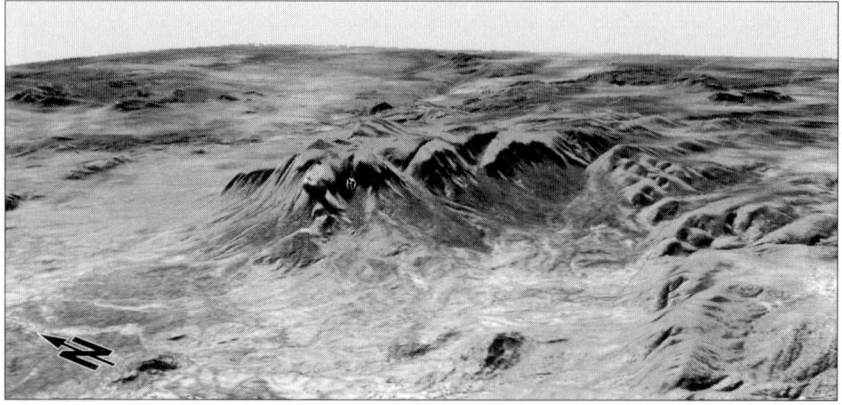

Figure 18.4.   Ehomba mountain (3D-satellite image [Landsat, SRTM] subproject E1,
CRC 389) (*See also Color Plates*)

---

[14] In the ancestral belief of the Ovaherero, Ovahimba, and Ovatjimba, respectively, graves
play an important role in various rites and ceremonies (Bollig, 1997b,; 1998,; 2001, this
volume; Kavari, 2004; Van Wolputte, 2002). The distribution of graves on the ancestral land
can further have implications for belonging and land ownership (see Bollig, this volume).

used to be a habitat for abundant wildlife. During times of war or scarcity when some groups in Kaokoland practiced a hunting and gathering lifestyle due to the loss of cattle, the abundance of plants, vegetables, and game that used to be found on the Ehomba mountain offered an important source of food and due to its seclusion a refuge where the people could hide from war. These groups of people that made their living mostly from hunting and gathering on the Ehomba mountain were labeled 'Ovatjimba'. The inhabitants of the surrounding villages of the Ehomba mountain, especially the Ehomba village itself, nowadays strengthen their ethnic identity as being 'Ovatjimba'. The label 'Ovatjimba' has its origin in the impoverishment of some groups who lost their cattle due to raids or to their escape into the mountains where they mainly practiced a hunting and gathering lifestyle. Although the label served in former times to underline an economic status it is nowadays mostly used as an ethnic label (see Bollig, 1998).

It is praised as follows.

Oyo ndjo yaTjipango omuhiringitita
There is the one of Tjipango the long one
Yoomama Tjitumaro kongwa yaKaemunika
Of the mothers of Tjitumaro at the umbilical cord of Kaemunika
YokoyooTjihende ndja ri nooVeongauri
Of the house of Tjihende that was with Veongauri and company
Ombonge ngu rir' ama kaeva
The ignorant one who cries while going to hunt

These first verses praise the Ehomba mountain by linking different people to it, Tjipango and the mothers[15] who probably lived there, Kaemunika who was born at Ehomba, and Tjihende with Veongauri and his company who settled there. The praise further links the Ehomba mountain to 'the ignorant one who was crying while going to hunt'. Characterizations of people in this manner expressing a special kind of humor are very typical for the *omitandu*. Despite giving honor in a proper sense the praising seems to capture the essence of somebody's character in depicting a caricature. The concept '*omuhiringitita*' in the first line refers to a 'very long thing', and it is used here to describe the elongated shape of the Ehomba mountain.

---

[15] The Ovaherero, Ovahimba, and Ovatjimba practice a double descent that divides into patriclan (*oruzo*) and matriclan (*eanda*).Whereas the inheritance of possessions is mostly operated along the matrilineage, religious duties and posts are inherited by the patrilineage. In this most of the cattle is inherited from the mother's brother. The brothers of a father are though addressed as 'father' and the sisters of a mother as 'mother'. For further information concerning the double descent see Van Wolputte (2004). Kavari (2004) and Bollig (2004, this volume) give extensive insights into inheritance in Ovaherero and Ovahimba tradition.

Ndja ri nooKarungu yongoro yaNeori

That was with Karungu of the zebra of Neori

Ondondo yaNauzemba

The black and white one of Nauzemba

Ngwa isan' omund' omukwao

Who called his colleague

Mbakuru yohoni yaNangoro ombinde yaTjituka

Mbakuru of the brown of Nangoro the spotted one of Tjituka

In these verses the Ehomba mountain is praised by linking it to several people's names and animals. The first and second lines mention a zebra that seems to have been hunted by Neori and Nauzemba, respectively. The second line contains a characterisation of the zebra without mentioning it directly. These omissions of names or references are very popular in the *omitandu* and are thus used in relation to human beings and animals alike. The last line gives another example of this stylistic device; here cattle are linked to their owners and are alluded to by drawing attention to their characteristic color pattern (see Kavari, 2004). These color and pattern terms that are of great importance among Kaokoland's pastoralists serve to differentiate between the manifold animal coats, especially cattle coats (see Eckl, 2002).

Ndjaai nyandis' ovitenda komarama aayo ozongweyuva

That played with irons on the legs as if it belonged to the Ongweyuva lineage

Ndji munik' otjomaze mekende nondana ai ri meondo

That looks like fat in a bottle and a calf amidst its age mates

The first line alludes to 'the spotted one' that is here further characterised and embedded, but interestingly like a human being. The Ongweyuva lineage is one of the lineages that consists of many patriclans belonging to the dominating ones in the area and is associated with hornless black and white cattle (*ozohungu ozombonde nozongonga*). The following clause draws attention to the aesthetics of a pastoral culture, as it metaphorically describes the beauty of the mountain through bovine imagery: butterfat (*omaze*) is of great value among Kaokoland's pastoralists. It is used as food as well as being mixed with different herbs and applied to the skin for perfume and protection. Symbolically *omaze* is strongly associated with fertility.[16] Put in a bottle and thus stored, this comparison here serves to describe the beauty of the Ehomba mountain. Anointing the body with butterfat is further said to recall the presence of the ancestors when they first brought cattle to the people and therefore alludes to the beginning of a pastoral life (Van Wolputte, 2002, p. 96). The second metaphor that is likewise drawn

---

[16] For an exhaustive examination of the symbolic meaning of butterfat and bodily praxis among the Ovahimba see Van Wolputte (2002).

from pastoral everyday life compares the mountain's beauty with a calf among its mates. Here the close relationship of the pastoralist with his cattle and the imagination of a well-fed calf among its mates that gives off warmth and is associated with fertility are of special importance. The two metaphors stand for utmost beauty and enable a vivid description of the mountain. They underline physical well-being and suggest the cultural continuity of a pastoral life with a calf pointing to the reproduction within the cattle herd. By ascribing these metaphors to the Ehomba mountain its importance as a refuge and reservoir of food is once again emphasised.

Kongwa yaKakoo Omutjimba ngwa yamenw' omuriro

At the umbilical cord of Kakoo the Omutjimba who was rescued by fire

Omuriro au ha yam' ovandu

While fire does not rescue people

Ona e ri ko kokarwi koumbao ku ku taukir' ovikora

It is at the water place with small calabashes where calabashes are broken

The first line praises the Ehomba as the place of birth of Kakoo, who was an Omutjimba. It thus links the mountain to Kakoo and he to the Ovatjimba, who are emphasised as the proper inhabitants of the Ehomba mountain. The person Kakoo can't be fixed in time but in space and perhaps in his familiar and certainly in his sociocultural context as being an 'Omutjimba'. The familiar context here refers to Kakoo as being the son or the father of somebody whose name is likewise known and that he is a descendent of a certain ancestor. In this way he can somehow be sorted out, but only in the Otjiherero-speaking pastoralists' way. He is associated with the Ehomba mountain which is praised with reference to him. The fire that is mentioned alludes to an oracle that foretold if somebody would die or not. It seems that Kakoo was rescued due to a prophesy that had been read in a fire that broke out. The next line 'while fire does not rescue people' underlines the specific feature of the fire as an oracle. Paradoxa such as these are very typical in *omitandu*. The last line refers to a spring from which a stream runs downhill and where people used to collect water in calabashes. In the olden days calabashes were used as water containers. During various raids, the enemies used to show up at the spring. Fights broke out and the calabashes were broken into pieces. 'Broken calabashes' became a general metaphor for war.

KoyaKaorori komuramba

At the place of Kaorori in the valley

Indji onene yOvatjimba vahi yaNambura womongo waMbingana

The big one of the Ovatjimba of the father of Nambura with the back of Mbingana

I nozoseu ozombwa zoviputuzumo ozombwa

It has delicious veldt onions with big round bellies

I nozombarurura ozombwa zoviputuzumo ozombwa

It has beautiful weaners with big round bellies

There is a valley at the foot of this mountain. During the rainy season water flows in the valley and at some places there are water pools. 'The big one' refers to the Ehomba mountain that is here again emphasised as being 'the one of the Ovatjimba'. Furthermore, there is reference to one of the most important veldt foods that the people used to gather on Ehomba mountain. Metaphorically it represents numerous veldt foods such as berries and the like and honey that are found on the mountain. The wild onions are alluded to in an anthropomorphic comparison that draws attention to the round form which is taken up again in the next verse. Here babies are described in their beauty having big round bellies and thus as being well-nurtured. Physical well-being is here again emphasised and associated with beauty. Recently weaned babies of the hunter-gatherers of the Ehomba mountain were well-fed as they could be nurtured on a variety of veldt foods.

The praise of the Ehomba links the mountain with the group of the Ovatjimba who found refuge in this remote and isolated place. The lushness and richness of the place is praised. The Ehomba is further linked to the different names of people who lived there or were born there. The *omutandu* of the Ehomba mountain emphasises the belonging of the Ovatjimba to this mountain that plays an important role in their constructions of identity.

18.3.2.2. *Omuhiva*

Omuhiva is a village situated approximately 25 km southwest of Opuwo (Figure 18.5). A village in Kaokoland always consists of several households that are

Figure 18.5. Omuhiva (3D-satellite image [Landsat, SRTM] subproject E1, CRC 389)
(*See also Color Plates*)

spread over the area. Although Omuhiva is close to Opuwo it is a remote place as it is not situated along the main road. Surrounded by a mountain range it was a popular place to settle as it offers several springs and thus plenty of water for people and cattle as well as for gardening. The main spring emanates from the mountain and its streams run downhill where they supply the different house-holds with water. The village is inhabited by people that identify themselves as Ovaherero, Ovahimba, or Ovatjimba alike.

It is praised with the following verses (Figure 18.6):

Porui rwaNanganga, koyaKasuko yaKatjandova,

At the spring of Nanganga, at the place of Kasuko of Katjondova,

Korui indwi rovizire mutenya

at the spring that has shadow during the day,

ku kwa tjirw' ozongombe ind' ozombambitaura,

where the *ombambi-taura* cattle spent the morning,

indji ombambitaura indj' orupoko rozongwe,

this cow of *ombambi-taura* colour this cow called the'gorge of the leopards',

At the entrance of the village when one crosses the pass in between the mountains the water from the spring offers a green and shady grazing area. It is skirted by several sweeping fig trees (*Ficus sycomorus*) and praised with reference to some*ombambi-taura* cattle that used to spend the morning there, drinking from the spring water. The cattle are praised by referring to their coat patterns.

Figure 18.6. Spring in Omuhiva: 'at the spring of Nanganga' (Bleckmann, 2005)

Figure 18.7. Cattle ombambi (*See also Color Plates*)

'*Ombambi*' is associated with a brown-reddish color and '*taura*' refers to a white coloring from the belly up to the tail (Figure 18.7).

The introductory verse 'at the water source of Nanganga' connects the place with the remembrance of Nanganga whose story is as follows. Nanganga lived in a house near the gardens. Her son, Tjikereta, was en route when she was raided by a group of Ovakwena (see Bollig, 1997a). When one of these raiding groups reached Omuhiva at an unknown time, they set fire to Nanganga's house, robbed the cattle, and murdered her. When Tjikereta found out what had happened he decided to avenge his mother and gathered together some young men. The fight took place at the same water source where afterwards so many corpses were scattered around that Tjikereta decided not to bury them but to leave them for the hyenas. Kasuko and Katjondova that are further linked to this place spent their lives in Omuhiva and are remembered in this aspect.

korui indu ku kwa tjirwa ingwi wa tjaka,
at the spring, where the child of Tjaka fought in the morning,
ingwi waKauez' ombande ngwa tukana Havinjanja,
the brave one of Kaueza who insulted Havinjanja,
wamama inaa Vahauri wozondondu zOvatwa
of my mother, the mother of Vahauri, of the rivers of the Ovatwa people

The first line mentions 'the brave one', Tjikereta, who is praised as the son of Kaueza who himself is praised and characterised by 'who insulted Havinjanja'. The next verse gives further genealogies. The last link connects Tjikereta – to whom the praise is still directed – to the rivers of the area around Sesfontein, a bigger village in the southwest of Kaokoland as well as to the Ovatwa. It is said that Tjikereta once fought on the banks of these rivers with the Ovatwa and proved his bravery. The Ovatwa are an ethnic minority that inhabits Kaokoland. Today the Ovatwa mainly live in the area around Okangwati in the northwest of

Kaokoland where they are popular for producing tools and jewelry whereas the biggest part of their group lives in Angola.

zomisepa nomingambu, ozondondu nda tukana ongombe omoro,

that has *omisepa* trees and mustard trees, the rivers that insulted our cattle,

omorui mOtjiundjambungu.

at the spring in Otjiundjambungu.

The pliable branches of the *omisepa* bush/tree (botanic classification not available) and of the *omingambu* bush (*Salvadora persica*) are favored resources for the production of bows. Nowadays, according to the informants, the *omisepa* bush/tree can't be found in the area anymore, due to the incisive drought of 1980–1981. 'Omoro' is one of the words whose meaning has become lost and which could therefore not be included in the translation. The last verse contains another name for Omuhiva, 'Otjiundjambungu', which can be translated with 'the place, where the corpses were left to be eaten by the hyena' (literally: 'the waiting place for hyena'). It thus makes reference to the story of Nanganga and how her son Tjikereta avenged her. The spring in Omuhiva plays an important role concerning the remembrance of the Ovakwena raids.

The *omutandu* of Omuhiva refers mainly to the water source and links it with the story of Nanganga and her son Tjikereta. The praise further devotes several lines to Tjikereta by embedding him genealogically, recounting attributes of his character, and links him to fights in which he took part. With the praise of Tjikereta which is included into that of Omuhiva the praise of Omuhiva gives a good example of how far praises resemble a hyperlink structure.

18.3.3. Metaphors and Landscape Aesthetics

The analysis of the two *omitandu* has given insight into the linking structure and into the poetry. This latter is here – with the main focus on landscape aesthetics – discussed by enriching the analysis with further examples. The praise of the Ehomba mountain underlines the historical importance of the Ehomba landscape as refuge for the Ovatjimba, its topography and richness of resources is given, and it is embedded in its sociocultural context by linking it with different names of people who lived or were born there.

The *omutandu* of Omuhiva alludes to one of its water sources as an important memory-place where an Ovakwena raid has taken place and further points to its favorable location by painting an image of the pasture around the water source where people and cattle are protected from the sun under sweeping fig trees while collecting water. As the aim of an *omutandu* is to praise and remember the most characteristic features and traits of a subject, they capture the essence of a person's character and delineate the character of a place by mentioning its inhabitants, giving the topography, localising resources, and recounting events that took place there. Therefore an *omutandu* does not always count positive attributes.

There are, for example, other verses that praise the Ehomba mountain by alluding to its antithetic character emphasising its utmost beauty but also its frightening attributes. In this it is praised with '*erero ya ri ondjandja, ndino ndi ya penduka ombiriona*' ('yesterday it was *ondjandja*, today it wakes up *ombiriona*'). While '*ondjandja*' decodes a light or brown color that has fine white dots, '*ombiriona*' does not simply refer to a color or pattern – although it is used in cattle coat conceptualisations – but to something that has been the reason for a conflict. According to Eckl (2002) the concept '*ombiriona*' is not based on a color or pattern concept but 'designates a piece of cattle which is object of a lawsuit', drawing back to '*ombiri*' that can be translated as 'matter of a dispute' (Eckl, 2000, p. 423). Although the verse can be interpreted as referring to the changing visual appearance, it can also allude to the unpredictability of the mountain that is not only beautiful and lush but also wild and dangerous. This aspect of its wildness that involves danger is further implicated in another verse '*onene ndji ri worukupo mozonda*' ('this big one that is like a marriage at a funeral'). This expression can be deciphered as follows. When somebody wants to make a marriage arrangement, he will hesitate to do this at a funeral. According to the explanations of the informant this hesitation is compared to the wavering of whether to climb up the mountain as the people fear dangerous animals.

The poetry in the *omitandu* expresses a very sensitive perception of Kaokoland's pastoralists. The imagery is mainly withdrawn from the pastoral everyday life closely intertwining nature, cattle, and people. It further reveals cultural immanence as the *omitandu* can only be deciphered by deepening the specific cultural setting of beliefs, customs, values, and perceptions. In fact it can only be fully deciphered by an expert, mostly the oral poet, who has kept not only the *omutandu* in mind but also the different explanations of metaphors, genealogies, and contexts. Cattle metaphors are frequent in praises and the linking of people with cattle is of great importance. People can be praised by descriptions of a cow or ox they owned and finally somebody's grave is identified and localised by (the horns of) the ox that has been slaughtered at this grave. References to resources, often those that are of importance for herding, are frequent. Otjiherero-speaking pastoralists are emotionally very attached to the landscape they inhabit as a valuable possession that maintains them and their livestock. This aspect is further underlined in the following example which draws attention to an anthropomorphism that is part of the pastoral perspective and which reinforces the close attachment of man to nature.

Anthropomorphic comparisons are very common in the *omitandu*; human features serve to describe landscape features and vice versa as, for example, given in the praise of an Omuhimba who is praised as having been beautiful 'like a forest and not a plain, like reed and not like *ondombora*'. The forest and the reed stand for abundance and diversity, however, the plain and the *ondombora* (a short growing grass) are associated with scarcity. A shepherd who is always in search of good pasture for his cattle will value the fertile grounds and abundance much more.

Furthermore, it is special for the *omitandu* that not only visual perceptions or the so-called 'seescape' are part of the imagery but the 'soundscape' plays an important role as well. The term 'soundscape' refers to the sensual perception of the environment by the sense 'hearing'. The verse of the Ehomba mountains gives an example where the soundscape is central: '*Onene ndji posa okatwezu kongombo aka ri oruteni*' ('a big one that sounds like a goat ram that ate spring'). This metaphoric expression contains an animalisation of the Ehomba mountain that is said to 'sound like a goat ram that ate spring'. In spring when fresh green leaves sprout on trees and bushes a goat will eat lots of them and its stomach, that is not used to such a lot of fresh green after scarce winter times, will make a noise. It is this noise that the Ehomba is said to produce alluding metaphorically to the plenty of animals that start grazing on the mountain. In a further explanation it is linked with the noises of joy that are made by the goats during mating.

Apart from their aesthetical value the *omitandu* are of great importance in their role of connecting collective memory with the places a community inhabits. The next section is devoted to this spatialising of memory and its links to identity construction and contemporary political debates.

18.3.4. Spatialisation of Collective Memory

To what extent does the linking of praise poems result in specific patterns of spatialised memory? Can such patterns be visualised using modern mapping devices? To answer these questions firstly the linking structure is looked at once more.

When Ohungumure, a small village about 15 km west of Opuwo, is praised with the following words, '*mongwa yaNakauwa*' 'at the umbilical cord of Nakauwa', the praise poem comprises the information that Nakauwa was born in Ohungumure. In this vein the memory of Nakauwa is spatialised and the village Ohungumure is praised by remembering Nakauwa. The metaphor can be regarded as twofold: the umbilical cord being anchored in the ground localises the birthplace. In a more interpretative explanation it expresses the person's attachment to this birthplace as the person is bound to it like a child to its mother by birth. Furthermore the descendants of Nakauwa are linked to this place. After the birth of a child the umbilical cord is buried near the homestead and in this way the metaphor of an umbilical cord indicates the place of a person's birth and the old homestead, respectively (see Van Wolputte, 2002).

Similarly graves of certain persons are linked to a certain place and in this vein localised. The *omutandu* of Omuhiva has given an example of a memory-place that in this case carries the remembrance of Nanganga and the Ovakwena raids. Resources are also an important aspect mentioned that are linked to the place and thus localised. Last but not least the named or described landscape features are an important part of this spatialisation of memory that works by linking different features to a place.

The importance of this kind of localised memory comes into being by the people's identification with and attachment to a certain place that has been inhabited for decades by their ancestors, where they were born and where they are buried, that nurtured the people and their cattle, and carries the memory of what life was like in former times, pointing to incisive and everyday events of the past by finally intertwining all these features with names of the deceased. The assigning of graves, mountains, or springs to a certain village is mostly fixed and this pattern is reflected in the praises. The 'borders' of a village or landscape feature are thus revealed in the collection of features that are assigned to the certain *omutandu* of this place: Memory-places, graves, springs, trees, passes, or mountains that spread over the landscape are linked in their assignment to special places.

The ensuing map has been designed to emphasise this aspect, further to give an 'overview' of the pattern of praised features in space and finally to connect the praise verses directly with the features they allude to and in this to indicate an inscription of memory. For the designing of the map,[17] the different places have been visited and fixed cartographically or by GPS. The praise verses are written onto the places they praise. The symbols indicate the places as they are praised in the *omitandu* and show the allocation of the different features. The assigning of places and landscape features to a certain praise poem – in most cases a village – is given in the same color as the different features that are cited in the *omutandu* and which in this are indicators of 'landmarks' for the 'border' of an area. Further explanations on the opposite page comprise – if given – the name or a description of the place. Finally, a closer view of the map reveals that in several examples collective memory can be rediscovered in the topography itself; see, for example, point 1, the plateau of Orotjitombo or point 14, the mountain pass of Omberangura (Figure 18.8).

Especially concerning the characterisation of landscape features it has been of interest to figure out how far the *omitandu* remember a landscape that can be recognised by a visitor. The photos that were taken further revealed this aspect of rediscovery in landscape itself so that it can be strengthened that the collective memory of landscape features is nonfictional. The settlement and waterfall of Epupa, an often-frequented tourist sight on the Angolan border, is praised by alluding to the baobab trees that line the river and the whitish and reddish hills whose soils and thus colors change as well as by referring to a special kind of aloe whose appearance is particular for the area around Epupa (Figure 18.9).

18.4. PERFORMANCE AND CONTEXT OF PRAISING

Are *omitandu* mental maps? Does Otjherero praise poetry result in song-lines as with Australian aborigines? When does an oral poet praise and what kind of importance do the *omitandu* have for the community?

---

[17] The maps that are represented in this article chapter have been designed and illustrated by Bleckmann and Bolten (2007). For further maps that illustrate the spatialization spatialisation of collective memory in Otjiherero praise poems, see Bleckmann (2007a).

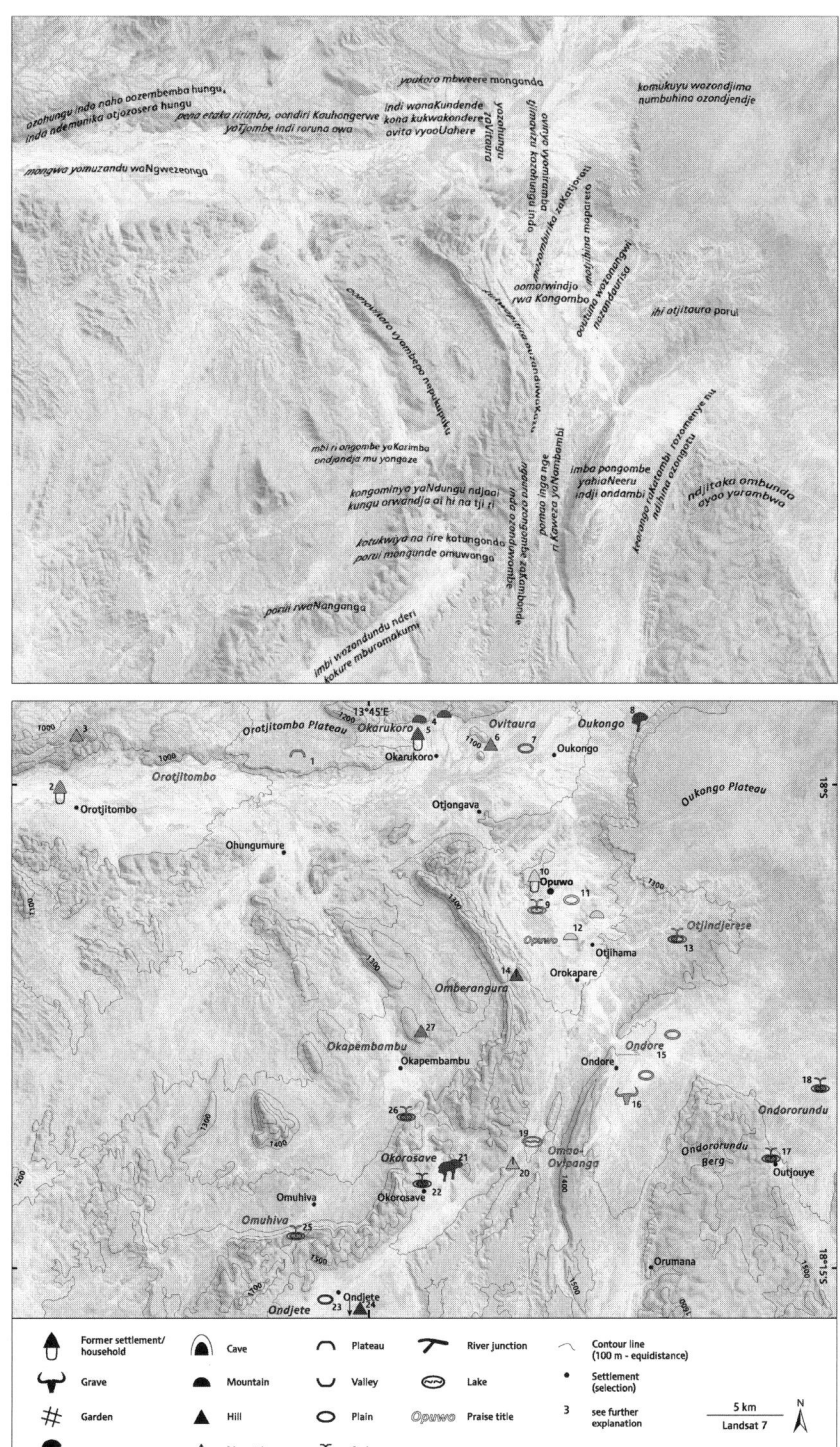

Figure 18.8. Map illustrating the spatialisation of collective memory (*See also Color Plates*)

Figure 18.9. Epupa: 'at the brownish and reddish hills' (Bleckmann, 2005)

In the performance the oral poet makes reference to the different places he praises by pointing with his arm in the direction of where the praised feature is supposed to be located or by facing it directly. A favorite manner of praising is 'to walk' from place to place. The order of *omitandu* in the recitation here is the same as in space itself. But the *omitandu* don't help the oral poet to orient in space and therefore the *omitandu* can't be called mental maps. However, the recitation of praise poetry is always spatially structured as the oral poet 'travels' through the landscape in a structured manner following the villages along the main road or going down a river. However, their pattern is not one that bears strong resemblance with the aborigines' song-lines in which the ancestors are embodied in the landscape features themselves and the singing ('dreaming') is much more a spiritual one alluding to the ancestors' spirits and decoding a whole mythological and cosmic framework (Morphy, 1995).

The gestures that the oral poet makes while praising relate the praise to the feature itself and thus express visualisation. In this manner it is then especially a pleasure for the oral poet to recite an *omutandu* when he is on the way to a place, when he has arrived at the place, or is actually there. Assmann (1999b) strengthens 'visualisation' as central in oral traditions as it relates the oral poet and audience to the matter of concern merging past and present.

omitandu are recited in personal and herding contexts when someone simply feels like singing or reciting *omitandu* mostly en-route and often while driving cattle from one pasture to another, when someone wants to enjoy, to overcome boredom, or feels like expressing his or her attachment to the place. Another occasion for the recitation of *omitandu* presents itself when people come together. The group then curiously listens to the floating words grasping some meaning and names of the shared history. Teaching *omitandu* is also an important aspect.

Children learn the most popular ones off by heart. Other favorite contexts for praising are ceremonies such as funerals or commemoration rites (see Bollig, this volume).

Praises in ceremonial occasions emphasise the shared history and tradition and honor the common ancestors. Their recitation awes the community that feels deeply and emotionally attached by these ancient linkages of the group with their ancestors and the land. *omitandu* are thus of great importance in the different group contexts concerning the identity of a group and underline a common memory by making reference to common traditions and beliefs and kinship ties as well as alluding to a pastoral way of being. They express the belonging of a group to the land of their ancestors. By visualising all these bondages in the common rite the recitation of the *omitandu* reinforces and renews the common identity and suggests cultural continuity. *omitandu* further strengthen the group identity by canonising how the world is perceived, understood, and aesthetically expressed.

A very important setting of praise poem recitation are political contexts. Especially meetings that discuss heritage and land ownership are of concern. Similar to a code-switching praises are interwoven with speech to underline the reciter's argument. Nowadays *omitandu* are, for instance, often cited concerning discussions about borders as have recently come up in the founding process of conservancies.[18] Here knowledge of *omitandu* or of a name that turns up serves to underline that a certain family or group has lived in the area for a long time and thus has therefore exclusive rights to it. In the traditional authority system knowledge of *omitandu* can be regarded as instruments of power, helping rights to prevail, especially land rights. Of great concern here is in how far a mountain, spring, or tree is part of one and the same or another praise as has been illustrated in the maps. The bondage of a certain family to the place of concern will be emphasised by making reference to the ancestors and by deciphering descent from this ancestor who is praised as having been born, lived, or buried at the place.

18.5. CONCLUSION

To conclude one can state that praises of places may be conceived as a source of a rich fund of landscape, genealogical, and sociohistorical knowledge and experiences as stored and communicated in the metaphorical and allusive language of praises. This knowledge as it is shared by the Otjiherero-speaking community and transferred from generation to generation thus represents an important part of the community's collective memory. In praises of places the collective memory is bound to the place that the praise alludes to and the collective memory is embedded in the landscape. A praise about a place connects people that stayed at this place, were born or are buried there, cattle, resources, or events with the

---

[18] A conservancy is a non-profit organization that is community based and aims to preserve the natural resources that are collectively managed in the registered area (RON, 1996).

place in a highly descriptive manner and in this way paints the character of the place and at the same time spatialises its memorable features.

The oral poet 'travels' during the performance from place to place linking one praise to another following the places along the main road or the way his ancestors took to arrive at his place of birth. During this 'traveling' the *omitandu* don't serve as an orientation in space although they can be said to be directional. Rather visualisation takes place where the oral poet relates to the praised feature. When praises of places allude to the memorable features of the place they refer to those that can be considered to be of historical momentousness and those that don't seem to be of great importance to an outstanding observer such as everyday experiences and long caricatures of people whose names are not even mentioned, but they are sometimes more in form than in content.

The structure of the *omitandu* reveals further aspects that allude to humor, paradoxical thinking, and appraisal of Kaokoland's pastoralists and as such give insights into the Kaokolander's worldview. Most important in this worldview are cattle and places, the latter sustaining cattle with pasture and water and cattle caring for the well-being of the Otjiherero-speaking society. The *omitandu* reflect this mobile pastoral everyday life and express a very sensitive perception of the pastoral environment in which humans, cattle, and landscape are closely intertwined. The *omitandu's* imagery and linking structure alludes to this interconnection.

The praises further emphasise the background of a society that has strong kinship and ancestral ties with the ancestors being 'the living dead'. The praises are a very important store of genealogical information. Furthermore, the fact that a praise of a place connects names with a place and assigns resources and landscape features to a certain village makes them important for political debates. Especially those that discuss land ownership and refer to the traditional land tenure system are of importance.

Finally it has to be emphasised that the praise poems serve as a storage of memory that is relevant in political debate but that they are, on the other hand, of importance concerning their aesthetic and emotional value. In so far the words of an informant capture this double meaning by stating: 'We carry the *omitandu* in our hearts and minds respectively'.

REFERENCES

Alnaes, K. (1989). Living with the past: The songs of the Herero in Botswana. *History in Africa*, 59(3), 267–299.

Assmann, A. (1999a). *Erinnerungsräume: Formen und Wandlungen des kulturellen Gedächtnisses*. München: Beck.

Assmann, J. (1999b). *Das Kulturelle Gedächtnis: Schrift, Erinnerung und politische Identität in frühen Hochkulturen*. München: Beck.

Assmann, J. (2000). *Kultur und religion*. München: Beck.

Assmann, J. & Czaplicka, J. (1995). Collective memory and cultural identity. *New German Critique*, 65, 125–133.

Bender, B. (1993). *Landscape: Politics and Perspectives*. Oxford: Berg.

Bender, B. & Winer, M. (2001). *Contested Landscapes. Movement, Exile and Place*. Oxford: Berg.

Bleckmann, L.E. (2007a). Zur Verräumlichung von kollektiver Erinnerung: landschaften in Preisgedichten der Herero/Himba im Nordwesten Namibias. In M. Casimir (Ed.), *Kölner Ethnologische Beiträge*. Cologne.

Bleckmann, L.E. (2007b). Poetry, landscape and the spatialisation of collective memory in Otjiherero praise poems. In O. Bubenzer, A. Bolten & F. Darius (Eds.), *Atlas of Cultural and Environmental Change in Arid Africa. Africa Praehistorica 21* (pp. 116–117). Cologne: Heinrich Barth Institut.

Bollig, M. (1997a). *"When War Came the Cattle Slept…": Himba Oral Traditions*. Köln: Köppe.

Bollig, M. (1997b). Contested places: Graves and graveyards in Himba Culture. *Anthropos*, 92, 35–50.

Bollig, M. (1998). Power and trade: In precolonial and early colonial Northern Kaokoland 1860s–1940s. In P. Hayes, J. Silvester, M. Wallace & W. Hartmann (Eds.), *Namibia Under South African Rule: Mobility and Containment 1915–46* (pp. 175–193). Oxford: Currey.

Bollig, M. (2000). Production and exchange among the Himba of northwestern Namibia. In M. Bollig & J.-B. Gewald (Eds.), *People, Cattle and Land: Transformations of a Pastoral Society in Southwestern Africa* (pp. 271–298). Köln: Köppe.

Bollig, M. (2001). Kaokoland: zur Konstruktion einer Kultur-Landschaft in einer globalen Debatte. In S. Eisenhofer (Ed.), *Spuren des regenbogens. Kunst und Leben im südlichen Afrika* (pp. 474–483). Stuttgart: Arnoldsche.

Bollig, M. (2002a). Koloniale Marginalisierung und ethnische Identität in Nordwest-Namibia: Ökonomie und Gesellschaft der Himba seit dem 19. Jahrhundert. In M. Bollig, E. Brunotte & T. Becker (Eds.), *Interdisziplinäre perspektiven zu Kultur- u landschaftswandel im ariden und semiariden Nordwest Namibia* (pp. 171–188). Kölner Geographische Arbeiten, 77.

Bollig, M. (2002b). Produktion und Austausch – Grundlagen der pastoralen Ökonomie Nordwest-Namibias. In M. Bollig, E. Brunotte & T. Becker (Eds.), *Interdisziplinäre perspektiven zu Kultur- u landschaftswandel im ariden und semiariden Nordwest Namibia* (pp. 189–204). Kölner Geographische Arbeiten, 77.

Bollig, M. (2004). Hunters, foragers, and singing Smiths: The metamorphoses of peripatetic peoples in Africa. In J.C. Berland & A. Rao (Eds.), *Customary Strangers. New Perspectives on Peripatetic Peoples in the Middle East, Africa, and Asia* (pp. 195–231). London: Praeger.

Bollig, M. (2005). Inheritance among the Himba of the Kunene region. In R. Gordon (Ed.), *The Meanings of Inheritance: Perspectives on Namibian Inheritance Practice*. (pp. 45–62). Windhoek: Legal Assistance Centre.

Bollig, M. & Gewald, J.-B. (2000). People, cattle and land: Transformations of a pastoral society. In M. Bollig & J.-B. Gewald (Eds.), *People, Cattle and Land: Transformations of a Pastoral Society in Southwestern Africa* (pp. 3–52). Köln: Köppe.

Casey, E.S. (1996). How to get from space to place in a fairly short stretch of time: Phenomenological prolegomena. In S. Feld & K.H. Basso (Eds.), *Senses of Place* (pp. 13–52). Washington, DC: University of Washington Press.

Cope, T. (1968). *Izibongo: Zulu Praise Poems*. Oxford: University Press.

Cosgrove, D. (1998). *Social Formation and Symbolic Landscape. With a New Introduction*. Wisconsin: University of Wisconsin Press.

Damann, E. (1987). *Was Herero erzählten und sangen, text, Übersetzung, Kommentar*. Berlin: Dietrich Reimer.

Eckl, A. (2000). Language, culture and environment. The conceputalization of Herero cattle terms. In M. Bollig & J.-B. Gewald (Eds.), *People, Cattle and Land: Transformations of a Pastoral Society in Southwestern Africa* (pp. 401–431). Köln: Köppe.

Feld, S. (1983). *Sound and Sentiment: Birds, Weeping, Poetics, and Song in Kaluli Expression*. Philadelphia: University of Pennsylvania Press.

Feld, S. & Basso, K.H. (1996). Introduction. In S. Feld & K.H. Basso (Eds.), *Senses of Place* (pp. 3–12). Washington, DC: University of Washington Press.

Förster, L. (2005). Land and landscapes in Herero oral culture: Cultural and social aspects of the land question in Namibia. *Analyses and Views*, 1. Windhoek: Konrad Adenauer Stiftung.

Friedman, J.T. (2004). Imagining the post-apartheid state. An ethnographic account of Namibia. Ph.D. dissertation, University of Cambridge, Cambridge.

Gibson, G.D. (1977). Himba epochs. *History in Africa*, 4, 67–121.

Gordillo, G.R. (2004). *Landscapes of Devils, Tensions of Place and Memory.* Durham, NC: Duke University Press.

Gunner, L. (1996). Names and the land: Poetry of belonging and unbelonging, a comparative approach. In K. Darian-Smith, L. Gunner & S. Nuttall (Eds.), *Text, Theory, Space: Land, Literature and History in South Africa and Australia* (pp. 115–130). London: Routledge.

Halbwachs, M. (1941). *La topographie legendaire des évangiles en Terre Sainte.* Paris: Presses Universitaires de France.

Halbwachs, M. (1950). *La mémoire collective, post. pubiliée par Mme Jean Alexandre, née Halbwachs.* Paris: Presses Univerisitaires de France.

Halbwachs, M. (1985). *Das Gedächtnis und seine soziale Bedingungen.* Frankfurt a. M.: Suhrkamp (french. orig. *Les Cadres sociaux de la mémoire,* 1925, Paris : Alcan).

Henrichsen, D. (1999). Claiming space and power in pre-colonial central Namibia: The relevance of Herero praise songs. *BAB Working paper,* 1/1999, Basel: Basler Afrika Bibliographien.

Henrichsen, D. (2000). Ozongombe*, Omavita and Ozondjembo* – The process of (Re-) pastoralization amongst Herero in pre-colonial 19th century Central Namibia. In M. Bollig & J.-B. Gewald (Eds.), *People, Cattle and Land: Transformations of a Pastoral Society in Southwestern Africa* (pp. 149–186). Köln: Köppe.

Hihuanguapo, M. (2000). How Opuwo got its name. In G. Miescher & D. Henrichsen (Eds.), *New Notes on Kaoko* (p. 9). Cape Town: BAB, Creda Communication.

Hirsch, E. (1995). Landscape: Between place and space. In E. Hirsch & M. O'Hanlon (Eds.), *The Anthropology of Landscape: Perspectives on Place and Space* (pp. 1–29). Oxford: Clarendon Press.

Kamupingene, T.K. (1985). *Ozondambo zaTjipangandjara.* Windhoek: Gamsberg Macmillan.

Kavari, J.U. (2000). Orature in Kaoko: Poetry and prose of Ovaherero and Ovahimba. In G. Miescher & D. Henrichsen (Eds.), *New Notes on Kaoko* (pp. 111–125). Cape Town: BAB, Creda Communication.

Kavari, J.U. (2002a). *The Form and Meaning of Otjiherero Praises.* Köln: Köppe.

Kavari, J.U. (2002b). *The Conceptualization of Landscape: Otjiherero.* (unpublished working document).

Kavari, J.U. (2005). Estates and systems of inheritance among Ovahimba and Ovaherero in Kaokoland. In R. Gordon (Ed.), *The Meanings of Inheritance: Perspectives on Namibian Inheritance Practice* (pp. 63–70). Windhoek: Legal Assistance Centre.

Kunene, D.P. (1971). *Heroic Poetry of the Basotho.* Oxford: Clarendon Press.

Low, S.M. & Lawrence-Zúñiga, D. (2003). *The Anthropology of Space and Place. Locating Culture.* Malden: Blackwell.

Malan, J.S. & Owen-Smith, G.L. (1974). The ethnobotany of Kaokoland. *Cimbebasia,* 2, 131–178.

Möhlig, W.J.G. (2002). The language history of Herero as a source of ethnohistorical interpretations. In M. Bollig & J.-B. Gewald (Eds.), *People, Cattle and Land: Transformations of a Pastoral Society in Southwestern Africa* (pp. 119–146). Köln: Köppe.

Möhlig, W.J.G., Marten, L., & Kavari, J.U. (2002). *A Grammatical Sketch of Herero (Otjiherero).* Köln: Köppe.

Nora, P. (1990). *Zwischen Geschichte und Gedächtnis.* Berlin: Wagenbach.

Ohly, R. (1990). *The Poetics of Herero Song: An Outline.* Windhoek: Gamsberg Macmillan.

Ohta, I. (2000). Drought and Mureti's grave: The we/us boundaries between Kaokolanders and the people of Okakarara area in the early 1980s. In M. Bollig & J.-B. Gewald (Eds.), *People, Cattle and Land: Transformations of a Pastoral Society in Southwestern Africa* (pp. 299–317). Köln: Köppe.

Opland, J. (1983). *Xhosa Oral Poetry: Aspects of a Black South African Tradition.* Cambridge: Cambridge University Press.

Ricard, A. (2004). *The Languages and Literatures of Africa.* Oxford: Currey.

RON (1996). Amendment of regulations of relating to nature conservation. *Government Gazette,* No. 1446, Windhoek.

Rothenberg, J. & Rothenberg, D. (1983). *Symposium of the Whole. A Range of Discourse Toward an Ethnopoetics.* Berkeley: University of California Press.

Rothfuss, E. (2004). *Ethnotourismus – Wahrnehmungen und Handlungsstrategien der pastoralno-madischen Himba (Namibia). Ein hermeneutischer, handlungstheoretischer und methodischer Beitrag aus sozialgeographischer Perspektive.* Passau: Selbstverlag Universität Passau.

Schama, S. (1995). *Landscape and Memory*. New York: Knopf.

Schapera, I. (1965). *Praise Poems of Tswana Chief*. Oxford: Clarendon Press.

Vansina, J. (1988). *Oral Tradition as History*. London: Currey.

van Warmelo, N.J. (1962). *Notes on the Kaokoveld (South West Africa) and its people*, Ethnological Publications, No. 26. Pretoria: Dept of Bantu Administration.

Wassmann, J. (1994). The Yupno as post-Newtonian scientists: The question of what is natural in spatial description. *Man*, 29, 1–24.

Yates, F. (1966). *The Art of Memory*. Chicago: University of Chicago Press.

Index

Printed in the United States of America